A Text Book on ENGINEERING MECHANICS

for

GATE

PSUs & Other Competitive Exams

Dr. U.C. Jindal

M.Tech, Ph.D.

Former Professor & Head of the Department

Department of Mechanical Engineering

Delhi College of Engineering, Delhi

MADE EASY Publications

MADE EASY Publications

Corporate Office: 44-A/4, Kalu Sarai (Near Hauz Khas Metro Station), New Delhi-110016
E-mail: infomep@madeeasy.in
Contact: 011-45124660, 8860378007
Visit us at: www.madeeasypublications.org

Engineering Mechanics for GATE, PSUs and Other Competitive Exams

First Edition : 2015
Second Edition : 2016
Reprint : 2017
Reprint : 2019

Preface

At the outset I must thank Mr. B. Singh (Chairman and Managing Director of MADE EASY Group) who has very kindly given me an opportunity to serve the student community through the publication of this book on Engineering Mechanics.

The present book has been designed as a self study book taking into account the severe shortage of technical teachers in engineering colleges and technical institutions.

The text in the book is well explained through examples supplemented by self explanatory illustrations, exercises supplemented by hints, key points to remember, thought provoking multiple choice questions, special problems so that a student can learn this basic subject in the shortest possible time.

The book covers all the syllabi in Engineering Mechanics of GATE, PSUs, all the universities, IITs, NITs, deemed universities. Students appearing in competitive examinations and other competitive examinations will find the book as an asset to them. The book also serves the purpose of AMIE students.

The book will greatly help the students who could not grasp the subject in the class room.

Any suggestion for the improvement in the text of the book will be thankfully acknowledged.

Dr. U. C. Jindal
Author

Contents

Engineering Mechanics

Chapter 1

Vector Quantities in Mechanics 1

1.1 Introduction 1
1.2 Position Vector 1
1.3 Displacement Vector 2
1.4 Force Vector 3
1.5 Moment of a Force about a Point 4
1.6 Varignon's Theorem 5
1.7 Moment of a Force about an Axis 7
1.8 Couple and its Moment 8
1.9 Component of a Couple about a Line 10
Problems 11
Practice Problems 12
Multiple Choice Questions 12
Explanations 13

Chapter 2

Equivalent Coplanar Force Systems 15

2.1 Introduction 15
2.2 Resultant of a Force by an Equivalent Couple and a Force at any other Point 17
2.3 Resultant of a Coplanar Force System 17
2.4 Resultant of a Parallel Forces Systems 19
2.5 System of Distributed Forces 20
Problems 21
Practice Problems 23
Multiple Choice Questions 24
Explanations 26

Chapter 3

Equilibrium of Rigid Bodies 28

3.1 Introduction 28
3.2 Support Reactions 29
3.3 Free body Diagrams 30
3.4 Conditions of Equilibrium 31
3.5 System of Concurrent Coplanar Forces 33
3.6 Coplanar Force Systems 34
3.7 Equilibrium of a Two Force Body 35
3.8 Equilibrium of a Three Force Body 36
3.9 Lami's Theorem 37
3.10 Two Parallel Forces 38
3.11 Several Parallel Forces in a Plane 40
Problems 41
Practice Problems 45
Special Problems 46
Multiple Choice Questions 47
Explanations 50

Chapter 4

Plane Trusses 53

4.1 Introduction 53
4.2 Types of Plane Trusses 54
4.3 Determination of Support Reactions 54
4.4 Determination of Forces in Members of the Truss 56
4.5 Method of Joints 56
4.6 Method of Sections 58
Practice Problems 63
Multiple Choice Questions 64
Explanations 69

Chapter 5

Centroid and Center of Gravity.......................... 73

5.1 Introduction 73

5.2 Location of Centroid 73

5.3 Centroid of Simple Sections 75

5.4 Centre of Gravity of 3 Dimensional Bodies 78

5.5 Theorems of Pappus-Guldinus 81

Multiple Choice Questions 83

Explanations 85

Chapter 6

Friction 86

6.1 Introduction 86

6.2 Laws of Dry Friction or Coulomb's Friction 87

6.3 Angle of Friction 87

6.4 Block on a Horizontal Plane 88

6.5 Two Mating Blocks 89

6.6 Angle of Repose 90

6.7 A Force Applied on a Body Lying on an Inclined Plane 90

6.8 Bodies on Inclined Plane Connected Through a Link ..91

6.9 Ladder 93

6.10 Friction in Wheels 94

Problems 95

Practice Problems 97

Multiple Choice Questions 98

Explanations 101

Chapter 7

Application of Friction in Machines 104

7.1 Introduction 104

7.2 Screw Threads 104

7.3 Flat Belt Drive 106

7.4 V-Belt and Rope Drive 107

7.5 Rolling Resistance 108

Practice Problems 110

Multiple Choice Questions 110

Explanations 111

Chapter 8

Second Moment of Area (Moment of Inertia).........112

8.1 Introduction 112

8.2 Rectangular Section 112

8.3 Triangular Section 113

8.4 Circular Section 113

8.5 Polar Moment of Inertia 114

8.6 Parallel Axes Theorem 115

8.7 Principal Axes and Product of Inertia 116

8.8 Parallel Axes Theorem for Product of Inertia 117

Multiple Choice Questions 119

Explanations 121

Chapter 9

Virtual Work 122

9.1 Introduction 122

9.2 Virtual Work 122

9.3 Work Done by a Couple 123

9.4 Principle of Virtual Work Applied to Beams 124

9.5 Potential Energy and Stability 125

9.6 Multiple Degrees of Freedom 128

Problems 130

Practice Problems 132

Multiple Choice Questions 134

Explanations 135

Chapter 10

Rectilinear Motion 137

10.1 Introduction 137

10.2 Rectilinear Motion with Constant Acceleration 138

10.3 Motion Under Gravity 139

10.4 Acceleration as a Function of Time 140

Problems 141

Practice Problems 142

Multiple Choice Questions 143

Explanations 144

Chapter 11

Curvilinear Motion 146

11.1 Introduction 146

11.2 Plane Curvilinear Motion 146

11.3 Rectangular Coordinates 147

11.4 Tangential and Normal Co-ordinates 148

Problems 149

Practice Problems 150

Multiple Choice Questions 152

Explanations 153

Chapter 12

Relative Velocity 154

12.1 Introduction 154

12.2 Motion Relative to a Frame in Translation 154

12.3 Absolute and Relative Velocity in Plane Motion 156

Problems 157

Practice Problems 160

Multiple Choice Questions 160

Explanations 162

Chapter 13

Motion of Rotation 163

13.1 Introduction 163

13.2 Rotation about a Fixed Axis 163

13.3 Various Types of Motion of Rotation 165

13.4 Instantaneous Centre of Rotation 166

Problems 169

Practice Problems 170

Multiple Choice Questions 171

Explanations 172

Chapter 14

Kinetics of a Particle 173

14.1 Introduction 173

14.2 Newton's Second Law of Motion 173

14.3 Types of Problems 174

14.4 Motion of Two Bodies in Contact 177

14.5 Tangential and Normal Forces in Curvilinear Motion 178

14.6 Motion of a Bicycle Rider Along a Circular Track 179

Problems 180

Practice Problems 182

Multiple Choice Questions 183

Explanations 184

Chapter 15

Dynamics of Rigid Bodies in Translation 186

15.1 Introduction 186

15.2 D' Alembert's Principle 187

Practice Problems 192

Multiple Choice Questions 193

Explanations 194

Chapter 16

Mass Moment of Inertia 195

16.1 Introduction 195

16.2 Parallel-Axes Theorem 196

Multiple Choice Questions 199

Explanations 200

Chapter 17

Dynamics of Rigid Bodies in Plane Motion 201

17.1 Introduction 201

Problems 203

Practice Problems 206

Multiple Choice Questions 207

Explanations 208

Chapter 18

Rotation about Fixed Axis 210

18.1 Introduction 210

18.2 Rotation about Centroidal Axis 212

18.3 Inertia Force Method 213

18.4 Centre of Percussion 214

Practice Problems 216

Multiple Choice Questions 217

Explanations 218

Chapter 19

Work and Energy 220

19.1 Introduction 220

19.2 Work done by a Force during Sliding Motion 222

19.3 Work done by a Moment during Motion of Rotation 223

19.4 Change of Potential Energy in a Spring 224

19.5 Rolling Motion under Gravity 225

Practice Problems 227

Multiple Choice Questions 228

Explanations 230

Chapter 20

Impulse and Momentum 232

20.1 Introduction 232

20.2 Angular Impulse and Angular Momentum Principle 234

20.3 Pile and Pile Hammer 235

Problems 237

Practice Problems 238

Multiple Choice Questions 239

Explanations 240

Chapter 21

Impact : Collision of Elastic Bodies 241

21.1 Introduction 241

21.2 Coefficient of Restitution 242

21.3 Impact of a Body on another Stationery Body 244

Problems 245

Practice Problems 247

Multiple Choice Questions 248

Explanations 250

■■■■

01

CHAPTER

Vector Quantities in Mechanics

1.1 Introduction

Study of engineering mechanics is incomplete if the vector quantities are not understood thoroughly for the correct solution of any problem in engineering mechanics *specially three-dimensional problems.*

Vector quantities as position vector, displacement vector, force vector, moment of a force about a point or about an axis, couple vector, addition and subtraction of couple vectors form the text of this chapter.

1.2 Position Vector

A position vector of a point is defined by the *position of a point in any coordinate system.* A vector starting from the origin of a coordinate system to the point in space is termed as *position vector.*

Fig.1.1 **(*a*)** shows a point P in x-y coordinate system with coordinates $P(x_1, y_1)$. Position vector of P is OP starting from the origin of coordinate system. Position vector R_1 from O to P can be expressed as

$$R_1 = x_1 i + y_1 j$$

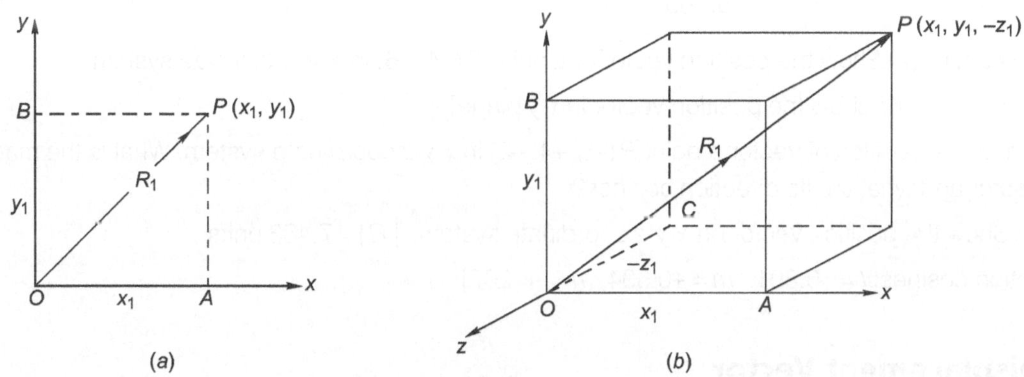

Fig. 1.1

Similarly **Fig. 1.1 (*b*)** shows a point P in x-y-z coordinate system with coordinates $P(x_1, y_1, -z_1)$, $\overrightarrow{OP}$ is the position vector of point P and can be expressed as

$$R_1 = x_1 i + y_1 j - z_1 k.$$

Example 1.1 (*a*) Mark the position vector of a point $P(4, -3)$, (*b*) Show the position vector of a point $P(4, -3, +2)$. What are its direction cosines?

Solution (*a*) Take *x-y* coordinates system as shown in **Fig. 1.2**. Take $OA = +4$ units on some scale, and $AP = -3$ units (in the negative direction of *y*-axis). Vector, *R* is position vector of point *P*, *i.e.*,

$$R = 4i - 3j$$

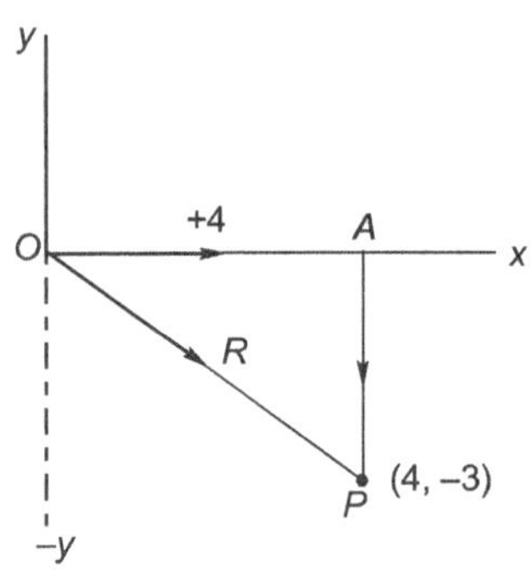

Fig. 1.2

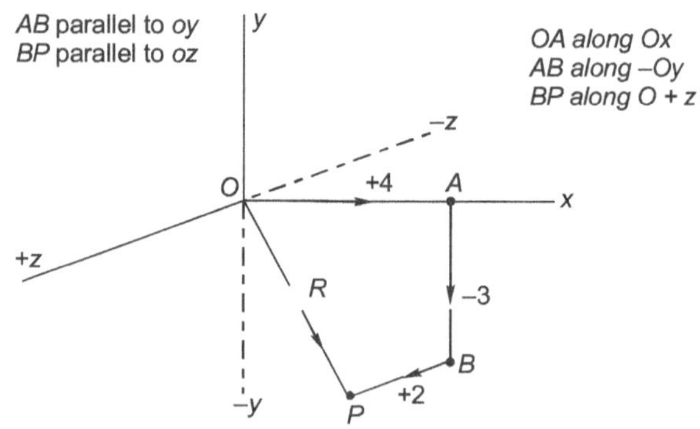

Fig. 1.3

(*b*) Consider *x-y-z* coordinate axes as shown in **Fig. 1.3**. To reach the point $P(4, -3, 2)$, take $OA = +4$ units along positive *x*-axis, from *A* to *B*, take –3 units along direction of the negative *y*-axis as shown in Fig. 1.3. From *B* draw a line parallel to *z*-axis and take $BP = +2$ units on same scale. Vector from *O* to *P*, is a position vector of point *P*,

i.e., $R = 4i - 3j + 2k$

Magnitude of position vector

$$|R| = \sqrt{4^2 + (-3)^2 + 2^2} = \sqrt{29} = 5.385 \text{ units}$$

Direction cosines $l = \frac{4}{5.385} = +0.743, \quad m = \frac{-3}{5.385} = -0.557$

$$n = \frac{2}{5.385} = +0.371$$

Exercise 1.1 (*a*) Show the position vector of a point $P(-4, +6)$ in *x-y* coordinate system.

[**Ans:** $R = -4i + 6j$, show the position vector in *x-y* plane].

(*b*) Show the position of vector a point $P(-6, +4, -2)$ in *x-y-z* coordinate system. What is the magnitude of position vector and what are its direction cosines?

[**Ans:** Show the position vector on *x-y-z* coordinate system, $|R| = 7.483$ units

direction cosines; $l = -0.801$, $m = +0.534$, $n = -0.267$].

1.3 Displacement Vector

Displacement vector between two points *P* and *Q* in space is defined as $\vec{S}_{PQ}$, displacement from point *P* to point *Q* (**Fig. 1.4**).

Point *P* and *Q* in space have coordinates $P(x_1, y_1, z_1)$ and $Q(x_2, y_2, z_2)$.

Then displacement vector $S_{PQ} = (x_2 - x_1)\,i + (y_2 - y_1)\,j + (z_2 - z_1)k$

$$S_{PQ} = R_2 - R_1$$

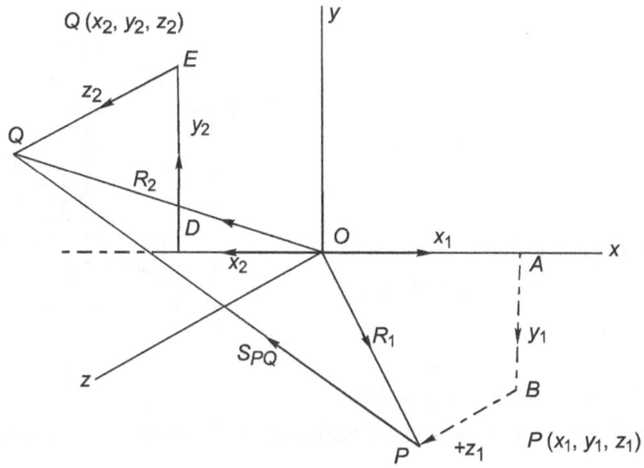

Fig. 1.4

$$= (x_2\,i + y_2\,j + z_2\,k) - (x_1\,i + y_1\,j + z_1\,k)$$

$$= (x_2 - x_1)\,i + (y_2 - y_1)\,j + (z_2 - z_1)\,k$$

Please note that arrow of displacement vector S_{PQ}, is pointing towards Q from point P.

Example 1.2 (*a*) What is the displacement vector from P (6, –4) m to Q (–3, 6) m? What are direction cosines of this vector? What is unit vector along displacement vector S_{PQ}?

Solution (*a*) Points P (6, –4) m and Q (–3, 6) m have position vectors as follows

$$R_1 = 6i - 4j \text{ m}, \quad R_2 = -3i + 6j \text{ m}$$

Displacement vector, $S_{PQ} = R_2 - R_1 = (-3i + 6j) - (6i - 4j)$ m

$$= -9i + 10j \text{ m}$$

$$|S_{PQ}| = \sqrt{(-9)^2 + (10)^2} = 13.453 \text{ m}$$

Direction cosines, $l = \dfrac{-9}{13.453} = -0.668$, $m = \dfrac{+10}{13.453} = +0.743$

Unit vector along S_{PQ}, $\bar{r} = -0.668i + 0.743j$

Exercise 1.2 (*a*) What is the displacement vector between points A (4, –6) and B (5, 3) m? What is unit vector along displacement vector S_{AB}?

[**Ans:** $1i + 9j$; $\bar{r} = 0.11i + 0.994j$].

1.4 Force Vector

A force vector is represented by its rectangular components F_x, F_y, F_z along x, y, z cartesian coordinates as shown in Fig. 1.5.

$$F = F_x\,i + F_y\,j + F_z\,k$$

Magnitude of force vector

$$|F| = \sqrt{F_x^2 + F_y^2 + F_z^2}$$

Direction cosines

$$\cos\alpha = \frac{F_x}{|F|} = l$$

$$\cos\beta = \frac{F_y}{|F|} = m$$

$$\cos\gamma = \frac{F_z}{|F|} = n,$$

where α, β and γ are the angles between the force vector *F* and coordinate axes *x*, *y* and *z* respectively.

Moreover $l^2 + m^2 + n^2 = 1$.

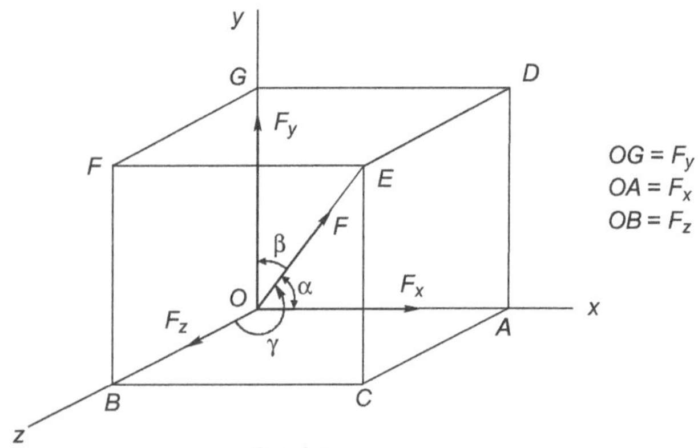

Fig. 1.5

Example 1.3 A force is, $F = 40i - 50j + 35k$ Newton. What is magnitude of *F* and what are direction cosines? Show force vector *F* in cartesian coordinate system.

Solution Fig. 1.6 shows force vector *F*

$= 40i - 50j + 35k$

to some suitable scale

$$F = \sqrt{40^2 + (-50)^2 + 35^2}$$

$$= \sqrt{1600 + 2500 + 1225}$$

$$= 72.97 \text{ N}$$

Direction cosines

$$l = \cos\alpha = \frac{40}{72.97} = +0.548$$

$$m = \cos\beta = -\frac{50}{72.97} = -0.685$$

$$n = \cos\gamma = \frac{35}{72.97} = +0.48$$

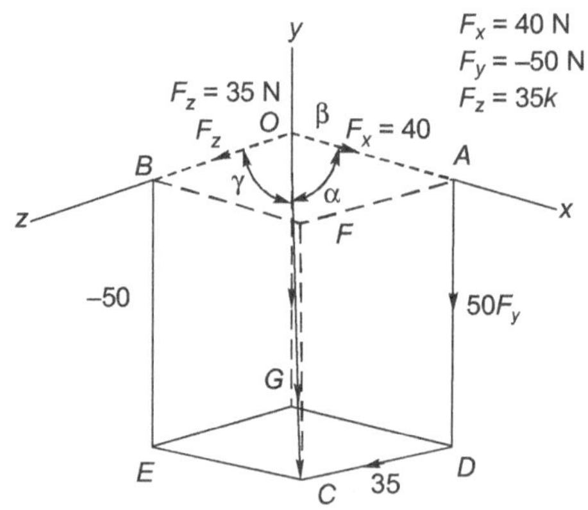

Fig. 1.6

Exercise 1.3: Show a force vector $-40i + 60j - 30k$ along cartesian coordinates. What are its magnitude and direction cosines?

[**Ans:** 78.10 N, –0.512, +0.768, –0.384].

1.5 Moment of a Force about a Point

Moment of a force vector *F* about point *O*, is a moment vector *M* as shown in the Fig. 1.7.

Moment = force × perpendicular distance from point *O* on force *F*

$= d \times F$

d = perpendicular distance from *O* on the line of force vector *F* = (OA)

Physically, moment *M* represents the tendency of the force *F* to rotate the body (on which force acts) about an axis which is passing through *O*, and this axis is perpendicular to the plane

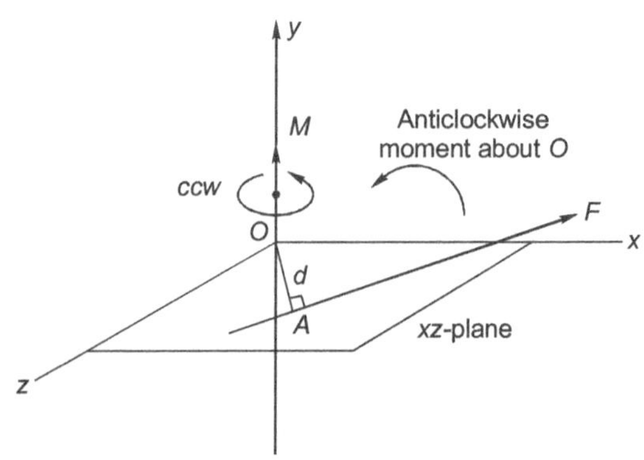

Fig. 1.7

of force F. In the **Fig. 1.7** xz is the plane of force F and M vector is perpendicular to xz plane *i.e.*, in the direction of y-axis. *If moment is anticlockwise, then moment vector is positive*, *i.e.*, it is represented along positive direction of y-axis as shown in **Fig. 1.7.**

(Using the right hand rule, the direction of vector M is determined).

Another approach to determine moment vector M, is to use a position vector r of any point P on the line of action of force F. Then moment vector M is cross product of r and F *i.e.*, $r \times F$.

$$M = r \times F = \text{position vector from } O \times \text{force vector } F$$

$$\text{Magnitude of } |M| = |r||F| \sin\beta = |F||r| \sin\alpha$$

$$= F \cdot d$$

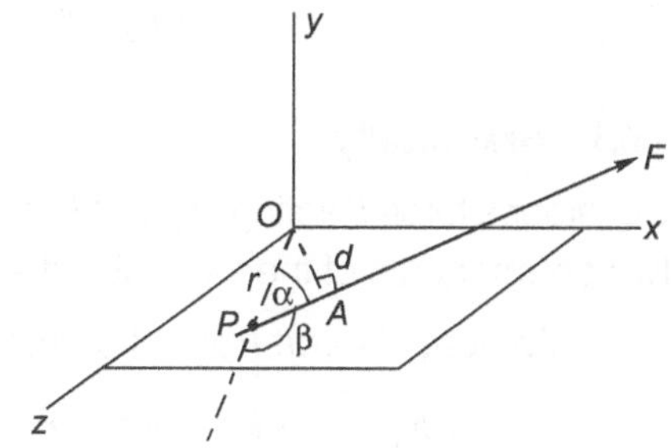

Fig. 1.8

as shown in **Fig. 1.8**. In the **Fig. 1.8**, OP is position vector of any point P from the origin, on the force vector F.

In this case $|r| \sin\alpha = d$, perpendicular distance from origin O to the line of action of F.

Example 1.4 A cantilever beam OAC, is fixed in wall at left end O. A force of 80 N inclined at an angle of 45° as shown is applied on the beam. If $OA = 2$ m, what is the moment of the force about point O. Verify this value of moment by taking position vector OA and force vector F in terms of unit vectors **(Fig. 1.9).**

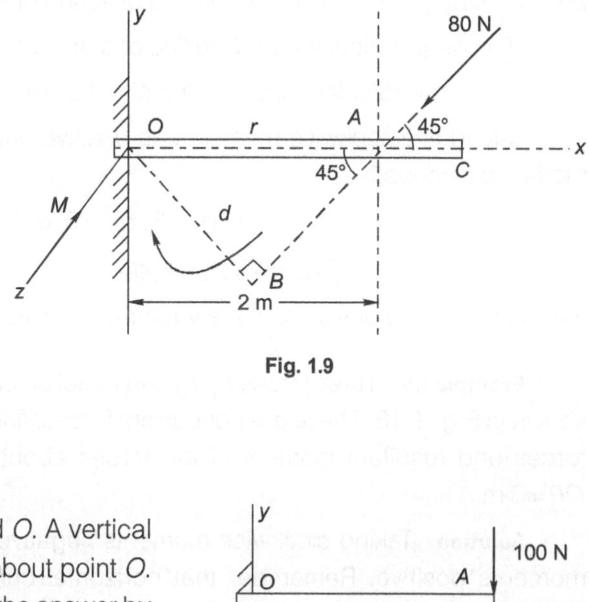

Fig. 1.9

Solution Line of action of force F is extended as shown. A perpendicular from O on the extended line of action of force is drawn.

Perpendicular distance

$$d = OA \sin 45°$$

$$= 2 \times 0.707 = 1.414 \text{ m}$$

Moment of the force about O,

$$M = 80 \times 1.414 \text{ Nm (clockwise)}$$

$$= 113.12 \text{ Nm (clockwise)}$$

$$= -113.12 \text{ Nm}$$

Exercise 1.4 A cantilever OA, 3 m long is fixed at end O. A vertical load of 100 N acts at end A. Determine moment of force about point O. Choose OA as position vector from O, **Fig. 1.10**, and verify the answer by vector method.

[**Ans:** 300 Nm (clockwise); $r = 3i$ m, $F = -100j$ N, $M = -300k$ Nm about z-axis].

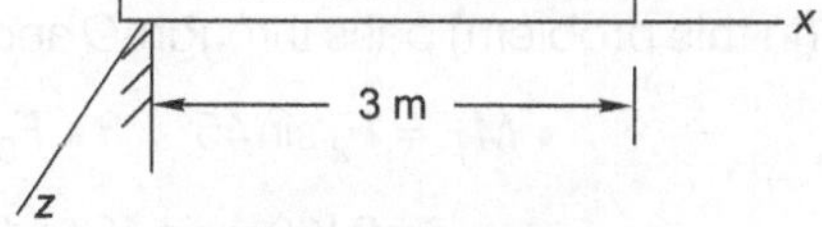

Fig. 1.10 (Ex. 1.4)

1.6 Varignon's Theorem

Consider a system of n concurrent forces F_1, F_2, F_3, F_4, ..., F_n acting at a point P in x-y-z co-ordinate system. From O to P is the position vector of the point P. All the forces F_1, F_2, ..., F_n are passing through the point P **(Fig. 1.11).**

Moment of the n forces about the origin O,

$$M_0 = r \times F_1 + r \times F_2 + r \times F_3 + r \times F_4, ..., r \times F_n$$

$$= r \times [F_1 + F_2 + F_3 + F_4, ..., F_n]$$
$$= r \times F_R$$

where F_R is the resultant of n forces F_1, F_2, ..., F_n.

It can be concluded that the *sum of the moments about a point* of a system of concurrent forces is the same as the moment of resultant of all these forces about the same point as

$$F_R = F_1 + F_2 + F_3 + F_4, ..., F_n.$$

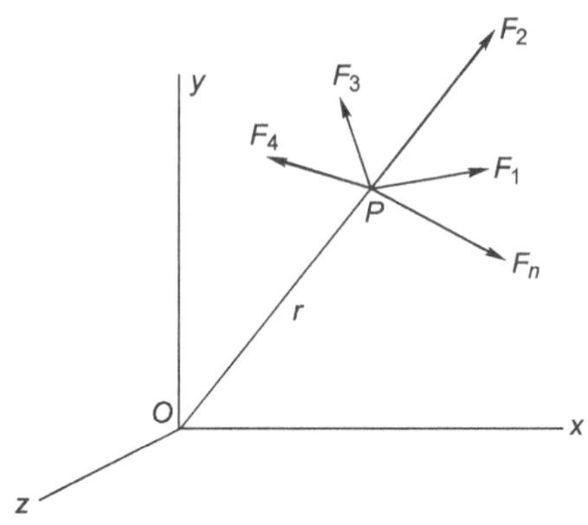

Fig. 1.11

1.6.1 Graphically

Consider a force F acting at point P of a body. Forces F_1 and F_2 represent two components of the force about any two directions **(Fig. 1.12)**.

Moment of force F about a point O,

$$M_0 = r \times F, \text{ but } F = F_1 + F_2 \text{ (vectorial sum)}$$

$$M_0 = r \times (F_1 + F_2) = r \times F_1 + r \times F_2$$

Lines of forces F_1, F_2 and F are extended and perpendiculars drawn on these lines from the point O,

i.e., d_1 is perpendicular from O to line of action of F_1

d_2 is perpendicular from O to line of action of F_2

d is perpendicular from O to line of action of F

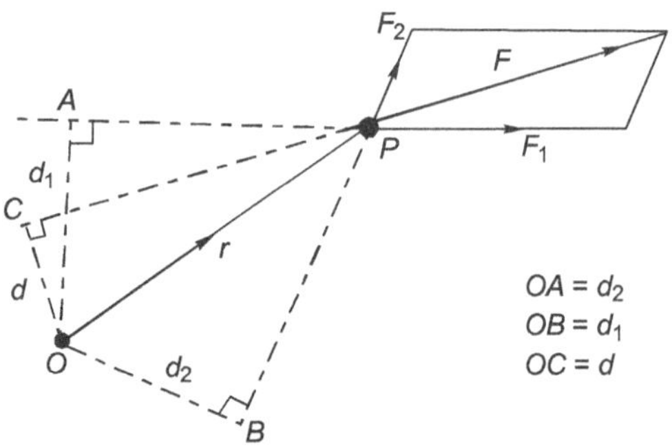

Fig. 1.12

Taking anticlockwise moments as positive and clockwise moments as negative

$$M_0 = -F_1 d_1 + F_2 d_2 = F.d$$
$$F.d = F_2 d_2 - F_1 d_1$$

this is the *scalar equivalent* of the vector expression of M_0.

Example 1.5 Three forces F_1, F_2 and F_3 act on point P in x-y plane, as shown in **Fig. 1.13**. These are concurrent forces, find the resultant of the forces and resultant moment of the forces about origin O. Distance $OP = 3$ m.

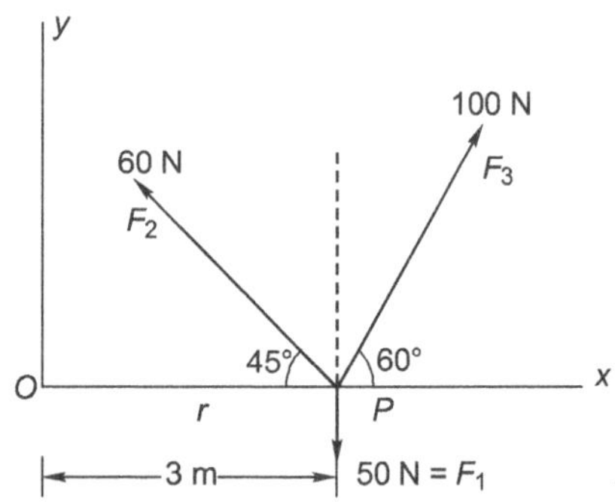

Fig. 1.13

Solution Taking clockwise moments negative and anticlock-wise moments positive. Remember that horizontal components of forces (in this problem) pass through O and will not produce any moment. So

$$M_0 = F_2 \sin 45° \times 3 + F_3 \sin 60° \times 3 - F_1 \times 3$$
$$= 3\,[60 \times \sin 45° + 100 \sin 60° - 50]$$
$$= 3\,[42.42 + 86.6 - 50] = 237.06 \text{ Nm}$$

(a positive anticlockwise moment)

Exercise 1.5 Three forces of 120 N, 80 N and 150 N act at point P, in x-y plane as shown in **Fig. 1.14**. What is the moment of these forces about origin O if $OP = 5$ m?

[**Ans:** –249.5 Nm, or –249.5k Nm along z-axis].

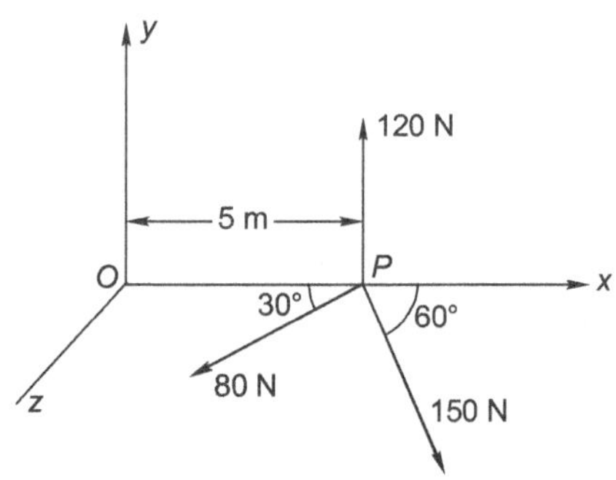

Fig. 1.14

1.7 Moment of a Force about an Axis

Let us consider a force F acting in a plane D, as shown in **Fig. 1.15**. We have to find moment of this force about an axis XX as shown. Draw a line ab in a direction perpendicular to axis XX passing through point P, *i.e.*, point of application of force P. Draw another line Pc parallel to axis XX.

Using parallelogram law of forces to find components of force F along directions ab and Pc. Component along ab is F_n, normal component and component along PC is F_p, parallel component to axis XX.

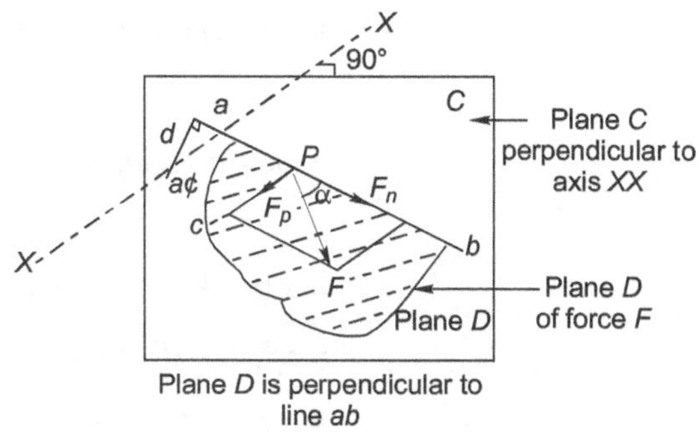

Fig.1.15

Since the component F_p is *parallel to axis XX, it will not produce any moment about the axis XX.*

Moment of force component F_n about axis XX,

$$M_{XX} = F_n \times d = |F| \cos\alpha . d$$

where $d = aa'$,

perpendicular on XX from the line ab

α = Angle of force F with line ab

Moment about an axis is a scalar quantity even though moment is associated with a particular axis. Moreover the force F_n can be decomposed into a pair of components in plane C (plane perpendicular to axis XX) and these components can be used to determine moment about axis XX.

Let us consider a force $F = F_x i + F_y j + F_z k$

in Cartesian co-ordinates as shown in **Fig. 1.16.** Force is applied at point P, where OP is position vector r. An axis AA is along the z-axis of the co-ordinate system. Co-ordinates of point P are (x, y, z).

Plane x, y is perpendicular to z-axis, containing forces.

Plane xy, plane of forces F_x, F_y.

For point P, co-ordinates are $AM = x$, $MP = y$, $ML = z$. Taking OL equal to x, LM equal to z and MP equal to y gives the position vector r.

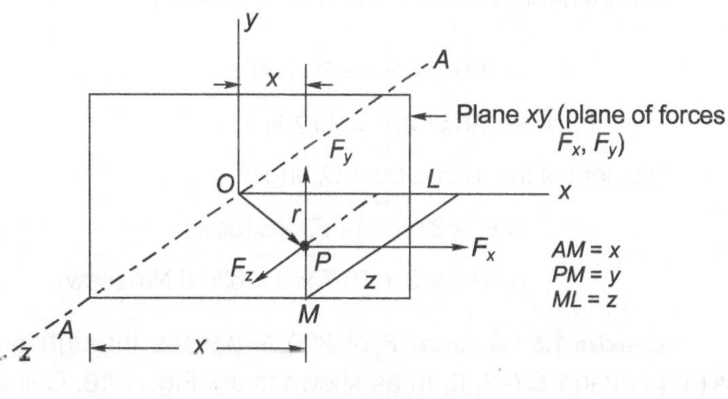

Fig. 1.16

Moment about z-axis, force F_z being parallel to z-axis, will not produce any moment. Force F_y produces xF_y (anticlockwise moment) and force F_x produces $y . F_x$ (clockwise moment), please note that force F_x acts above the axis z, by a vertical distance $+y$ (given by MP).

So moment about z-axis, $M_z = x F_y - y F_x$

Similarly moment about x-axis,

$$M_x = y F_z - z F_y$$

Please note F_x will not produce any moment about x-axis, $y F_z$ is anticlockwise moment and $z F_y$ is clockwise moment about x-axis.

Moment about y-axis, $M_y = z F_x - x F_z$

M_x, M_y, M_z are scalar components of moment M_0 of the force about origin O.

Moment vector,

$$M_0 = M_x i + M_y j + M_z k$$

$$= (yF_z - zF_y)\,i + (zF_x - xF_z)\,j + (xF_y - yF_x)\,k$$

Moment about origin O, can also be obtained by cross product of $r \times F$

$$r \times F = (xi + yj + zk) \times (F_x i + F_y j + F_z k)$$

$$= \text{Determinant of} \begin{vmatrix} x & y & z \\ F_x & F_y & F_z \\ i & j & k \end{vmatrix}$$

$$= x\,(F_y.k - jF_z) + y\,(F_z.i - F_x k) + z\,(F_x.j - iF_y)$$

$$= xF_y k - xF_z j + yF_z i - F_x y k + zF_x j - zF_y.i$$

$$= (yF_z - zF_y)i + (zF_x - xF_z)\,j + (xF_y - yF_x)k$$

The reader may realise that moment vector can be easily obtained by cross product of position vector and force vector.

Example 1.6 A force vector of 100 N is passing through two points A (6, 3) m and B (3, 6) m. Determine moment of the force vector about the origin O of co-ordinate system, as shown in **Fig. 1.17**.

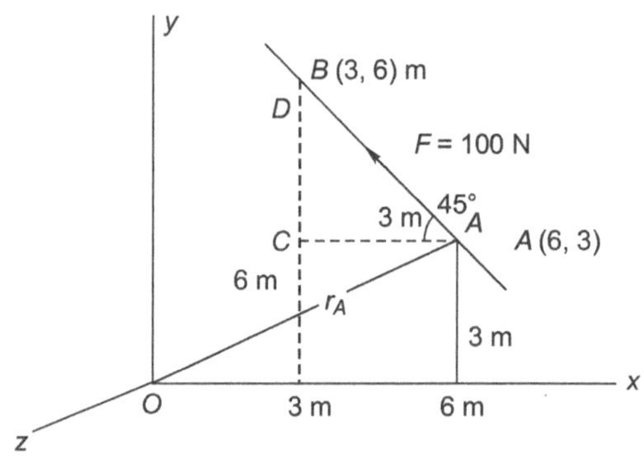

Fig. 1.17

Solution **Scalar method:** Force F is inclined with x-axis at 45°, because AC = 3 m, CD =3 m.

Components of force F along x- and y-axes.

$$\overleftarrow{F_x} = 100\cos 45° = \overleftarrow{70.7}\text{ N}$$

$$F_y \uparrow = 100\sin 45° = 70.7\text{ N}\uparrow$$

Moment of the force about O, origin

$$= F_x \times 3\,(\text{ccw}) + F_y \times 6\,(\text{ccw})$$

$$= 70.7 \times 3 + 70.7 \times 6 = 636.3\text{ Nm (ccw)}$$

Exercise 1.6 A force F of 200 N passes through points A (4, 4) m and B (–2, 6) m as shown in the **Fig. 1.18**. Calculate moment of the force about origin O. Verify by vector method.

[**Ans:** +1012 Nm or +1012k Nm].

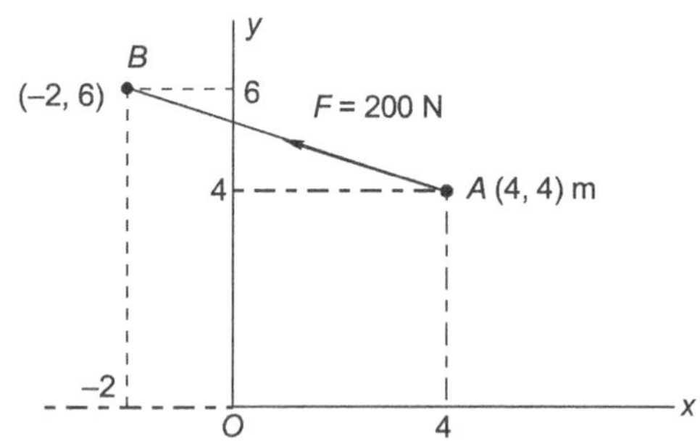

Fig. 1.18

1.8 Couple and its Moment

A couple is formed by any *two equal and opposite forces* as shown in **Fig. 1.19**. On a rigid body, a couple has an effect of turning. The turning action is given by moment of forces P and $-P$ about a point or about an axis is called the *Couple Moment.*

Scalar Method: Moment of the couple

$= P \times d$ = magnitude of the couple

= Force × Perpendicular distance between the forces as shown in **Fig. 1.19**

$= P\,.\,d$ in Nm if P is in Newton and d is in metre.

If the turning action is clockwise, then it is a negative couple and if turning action is anticlockwise, then it is a positive couple. In the figure $P \cdot d$ produces clockwise moment, so it is a negative couple moment.

Vector Method: Position vectors r_1 and r_2 are drawn, $OA = r_1$, for a point on force $+P$ and r_2 is OB, for a point B on force $-P$.

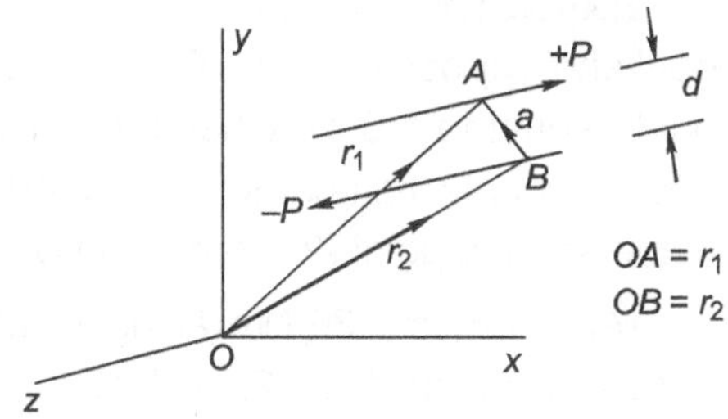

Fig. 1.19

Vector $\qquad a = r_1 - r_2 = \text{vector } \overrightarrow{BA}$

Moment of forces about the origin O

$$M_0 = r_1 \times P + r_2 \times (-P) = (r_1 - r_2) \times P$$

or couple moment, $\qquad M_0 = a \times P$

Displacement vector *'a'* lies in the plane of the forces P and $-P$. Cross product of $a \times P$ is a vector, direction of this vector is perpendicular to the plane of the couple. In the figure *x-y* is the plane of the couple, and the couple vector is represented along *z*-direction. Moreover rotation of displacement vector a to the force vector $+P$ is in the same direction as the turning action of the couple.

1.8.1 Couple-a Free Vector

Couple is a free vector and its moment about any point in space is the same as about the origin of co-ordinate system.

Moment of the couple (P and $-P$ force) about any point O' in space

$$M_0' = r_1' \times P - r_2' \times P, \textbf{ (Fig. 1.20)}$$

where $\qquad r_1' = O'A$, position vector

$\qquad r_2' = O'B$, position vector

but $\qquad r_1' - r_2' =$ *'a'*, displacement vector from B to A

where point A lies on line of action of force $+P$ and point B lies on line of action of force $-P$.

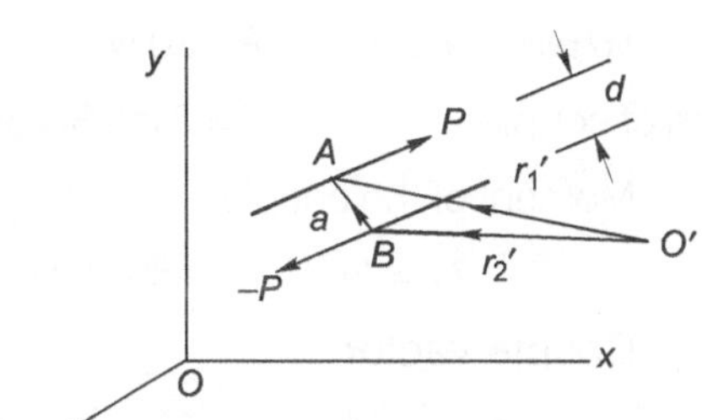

Fig. 1.20

$$M_0' = a \times P \text{ same as } M_0$$

$= P \times d =$ force $P \times$ perpendicular distance d between forces forming the couple.

Example 1.7 A couple is shown in *x-y* plane, what is the moment of this couple about the origin? What is the moment of the couple about a point (2, 3, –3) m **(Fig. 1.21)?**

Solution Moment of the couple,

$$C = 20 \times 0.4 = 8 \text{ Nm}$$

(anticlockwise) a positive couple, as d = 0.4 m

Couple vector C lies in direction, perpendicular to the plane of couple (*xy* plane) *i.e.*, *z*-direction.

So couple, $\qquad C = 8$ kNm

Couple vector +8 Nm along *z*-direction is shown in **Fig. 1.21.** Since couple is a free vector, *its moment will be the same about any point in space*, so about the point (2, 3, –3) couple moment is the same *i.e.*, +8 Nm.

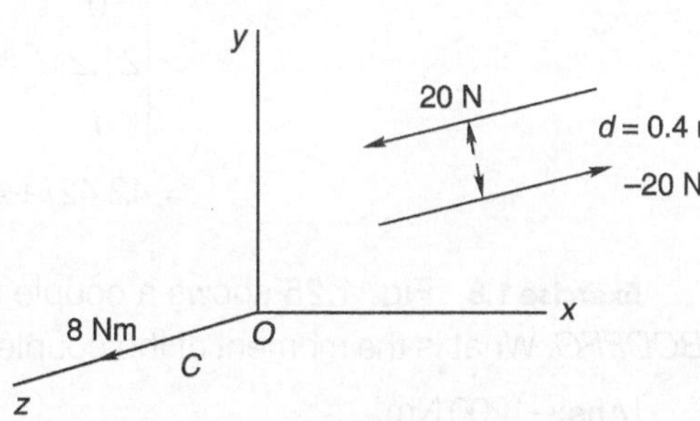

Fig. 1.21

Exercise 1.7 A couple is shown in Fig. 1.22 in z-x plane. Two equal and opposite forces ±60 N make this couple with arm of the couple equal to 0.6 m. What is moment of the couple? In which direction couple moment is represented? What is the moment of this couple about a point (2, –2) m point lying in xy plane?

[**Ans:** –36 Nm, –36j Nm (along negative direction of y-axis, same moment about point (2, –2) m also].

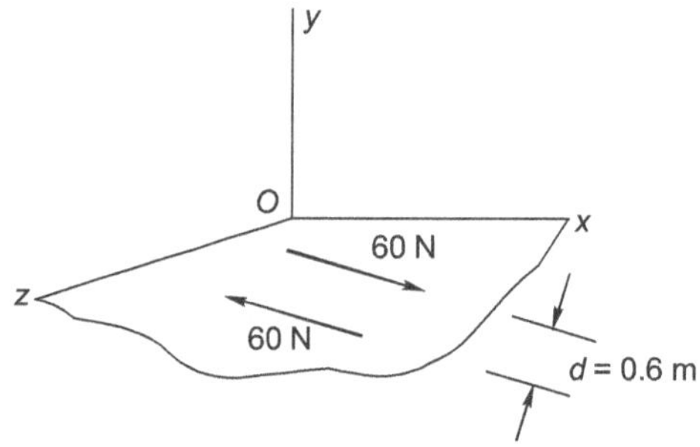

Fig. 1.22

1.9 Component of a Couple about a Line

Let us determine the component of a couple about any arbitrary line. Fig. 1.23 shows a couple vector C and a line AA. Take some point P on arbitrary line AA. Make any position vector r upto the couple vector C.

If $\bar{r}$ is the unit vector along the line AA, then component of the couple along line AA

$$M_{AA} = C.\bar{r}$$

(a dot product of couple vector C and unit vector $\bar{r}$)

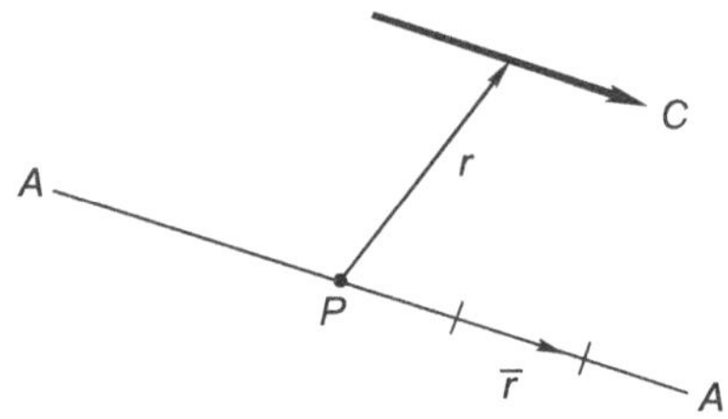

Fig. 1.23

C, couple is a free vector, the component of C, *couple along all lines parallel to line AA* will remain the same.

Example 1.8 Two equal and opposite forces are directed along the diagonals on the faces of a cube as shown in Fig. 1.24. If side of cube is 2 m and $P = 30$ N.

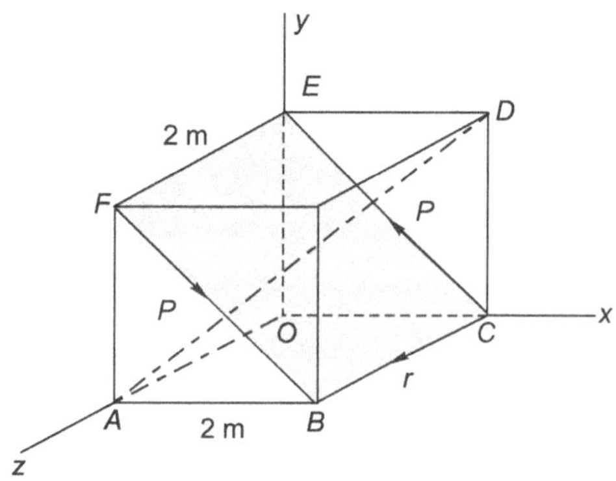

Fig. 1.24

Solution Force $P = 30$ N

arm of couple $= 2$ m (BC or FE)

Moment of the couple,

$$C = 30 \times 2 = +60 \text{ Nm (ccw)}$$

Couple vector

Force along FB,
$$P = \left(\frac{2i - 2j}{\sqrt{4+4}}\right) \times 30$$
$$= 21.21i - 21.21j \text{ N}$$

Position vector CB, $r = +2k$ m

Couple vector, $C = r \times P$

$$= (2k) \times (21.21i - 21.21j)$$

$$= \begin{vmatrix} 0 & 0 & 2 \\ 21.21 & -21.21 & 0 \\ i & j & k \end{vmatrix}$$

$$= 42.42j + 42.42i \text{ Nm}$$

Exercise 1.8 Fig. 1.25 shows a couple acting on a parallelepiped $ABCDEFG$. What is the moment of the couple?

[**Ans:** –1200 Nm].

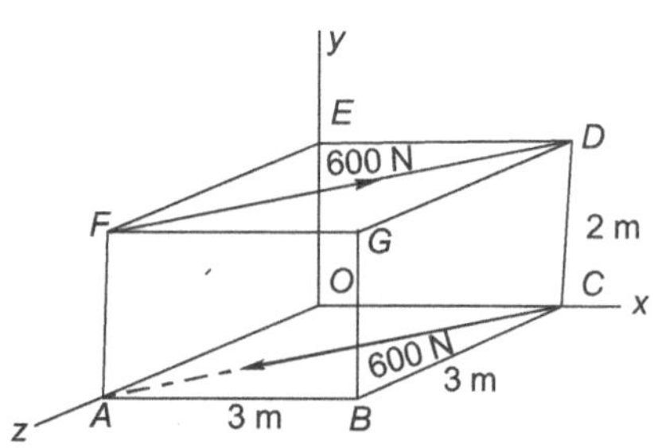

Fig. 1.25

PROBLEMS

1.1 A particle moves along a circular path of radius R in x-y plane as shown in **Fig. 1.26**. What is the position vector R of this particle in terms of y-co-ordinate?

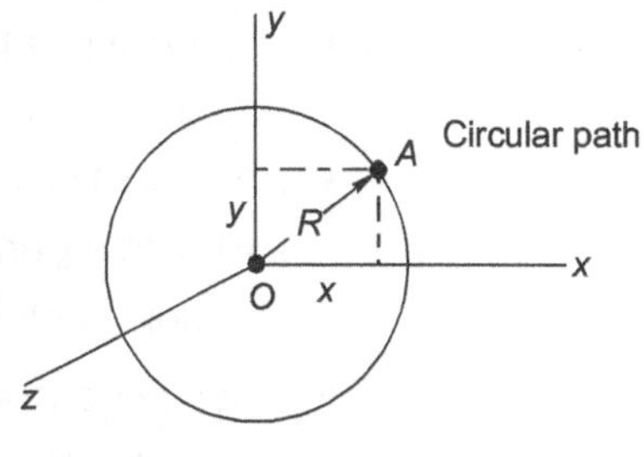

Fig. 1.26

Solution:
$$R^2 = x^2 + y^2$$
$$x^2 = R^2 - y^2$$
$$x = \sqrt{R^2 - y^2}$$

Position vector $R = xi + yj = \sqrt{R^2 - y^2}\,.i + yj.$

1.2 A particle moves along a parabolic path in xy plane. If the particle has at one point a position vector $R = 4i + 3j$, give the position vector at any point on the path as a function of x-co-ordinate.

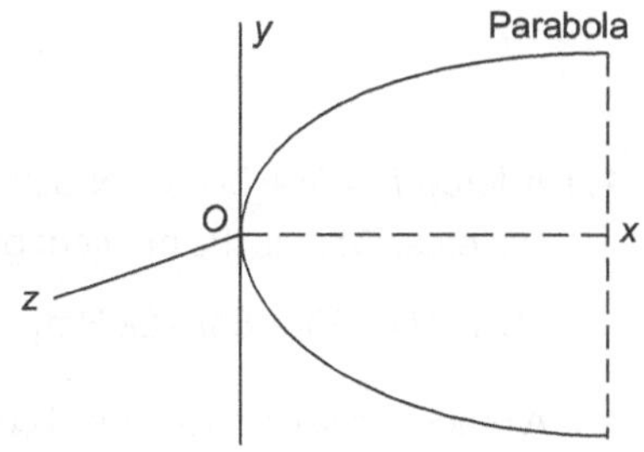

Fig. 1.27

Solution: Equation of a parabola is

$$y^2 = cx$$

Fig.1.27 shows a parabola $y^2 = cx$

Putting the value of y and x $3^2 = c \times x = 4c$, as $x = 4$

$$c, \text{ constant } = \frac{9}{4} = 2.25$$

$$y = \sqrt{cx} = 1.5\sqrt{x}$$

Therefore, position vector of path, $R = xi + 1.5\sqrt{x}\,j.$

1.3 The boom of a crane is 12 m long, it is hinged at end A. A load of 20 kN is being lifted by the crane with the help of a wire rope. What is the moment of the force about end A. Express moment as vector **(Fig. 1.28)**?

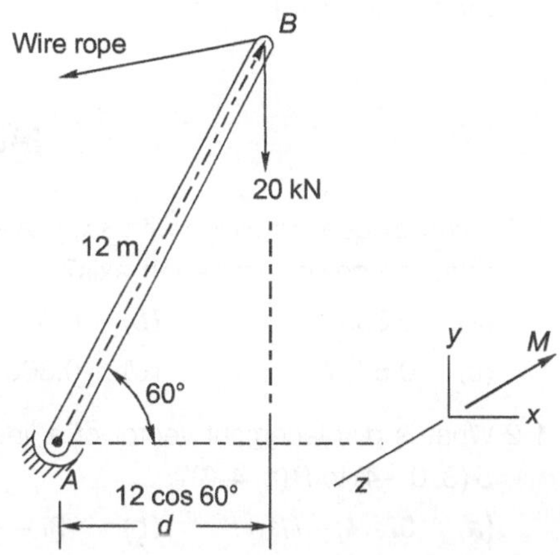

Fig. 1.28

Solution: Force = 20 kN ↓

Perpendicular distance from A to line of action of force,

$$d = 12 \cos 60° = 6 \text{ m}$$

Moment of the force about A,

$$M = 6 \times 20 = 120 \text{ kNm (cw) (a negative moment).}$$
$$= -120 \text{ kNm}$$

Remember

- Vector from origin of a co-ordinate system to the point (x, y, z) is termed as position vector, $R = xi + yj + zk.$
- Displacement vector between points $P(x_1, y_1, z_1)$ and $Q(x_2, y_2, z_2)$ is $S_{PQ} = (x_2 - x_1)i + (y_2 - y_1)j + (z_2 - z_1)k.$
- Force vector, $F = F_x.i + F_y.j + F_z.k$
- Moment of a force about a point A,
  $M_A = r \times F$
  = cross product of position vector and force vector
  = Magnitude of force × perpendicular distance from the point A to the line of action of the force.

- Varignon's theorem—algebraic sum of the moments of any number of concurrent forces about a point in their plane is equal to the moment of their resultant about the same point.
- Moment of a force about an axis is a scalar component, is equal to the component of the moment along that axis.
- A couple is formed by any two equal and opposite forces
  = Force × perpendicular distance between the forces in the plane of the couple.
- Couple Vector lies in a direction perpendicular to the plane of the couple.
- Component of a couple about any line $= C.\bar{r}$ dot product of couple moment and unit vector along the line.
- Couples can be added vectorially.

PRACTICE PROBLEMS

1.1 A force $F = 6i + 3j - 6k$ N acts at position (5, 3, 4) m relative to a co-ordinate systems. What is the moment of the force about the origin?

[**Ans:** $M = -30i + 54j - 3k$ Nm].

1.2 A bracket is fixed on a vertical column, 6 m high from the ground. Bracket supports a load of 15 kN at its end, 2 m from axis of column as shown in **Fig. 1.29**. What is moment of the force at the fixed end *B* of column? Express the moment as vector.

[**Ans:** $-30k$ kNm].

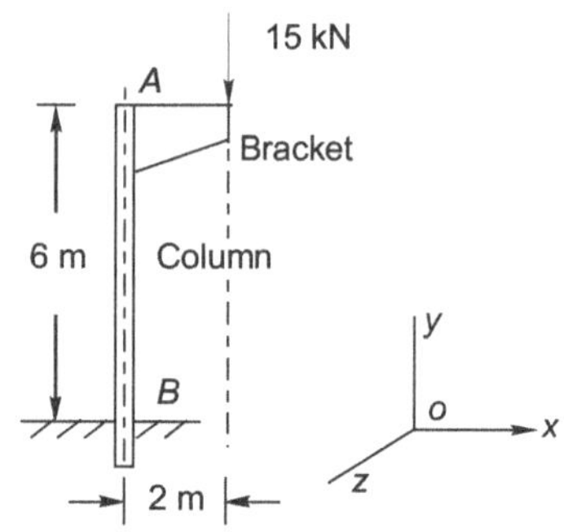

Fig. 1.29

MULTIPLE CHOICE QUESTIONS

1.1 A line is equally inclined with 3 *x*, *y*, *z*-axes. What is its direction cosine with each axis?

(*a*) 0.333 (*b*) 0.52
(*c*) 0.5777 (*d*) 0.866.

1.2 What is displacement vector of a line passing from *S* (3, 0, –4) to *P* (0, 4, 3)?

(*a*) $3i - 4j - 7k$ (*b*) $-3i + 4j + 7k$
(*c*) $-i - 3j + 4k$ (*d*) None of these.

1.3 A force $F = +5j$ kN is passing through a point (4, 0, 0), what is the moment of the force?

(*a*) $+20k$ (*b*) $-20k$
(*c*) $20j$ (*d*) None of these.

1.4 A force $P = 4i - 3j$ kN is passing through a point (4, 0, 0) m in *xyz* coordinate. What is moment of the force about the origin?

(*a*) $+12k$ kNm (*b*) $+16k$ kNm
(*c*) $-12k$ kNm (*d*) None of these.

1.5 Three concurrent forces are passing through a point *P* on *x*-axis as shown in **Fig. 1.30**, what is moment of these forces about the origin *O*?

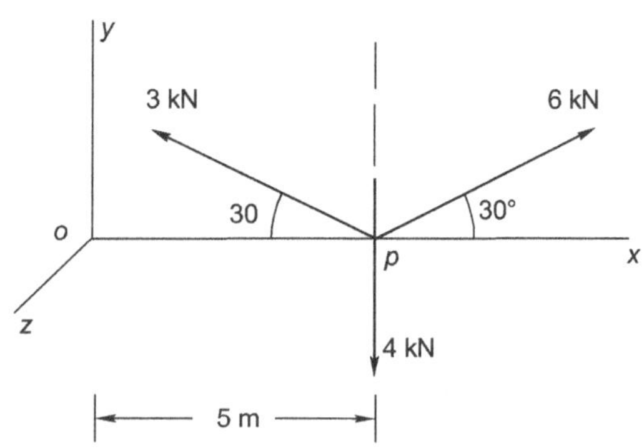

Fig. 1.30

(*a*) +22.5 kNm (*b*) +2.5 kNm
(*c*) –20 kNm (*d*) None of these

1.6 Which is a correct statement?

(*a*) Couple is a fixed vector
(*b*) Moment of couple changes with point in space
(*c*) Couple is a free vector
(*d*) All of the above

1.7 Two forces +*P* and –*P* act along two edges of a cube of side *a*. What is moment of the couple **(Fig. 1.31)**?

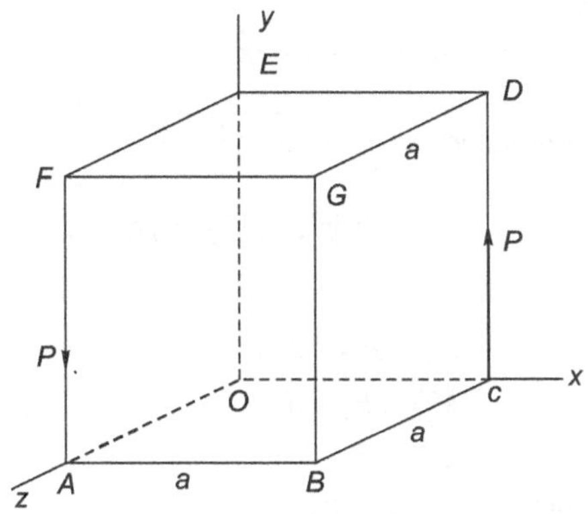

Fig. 1.31

(a) $+Pa$ (b) $+\sqrt{2}Pa$

(c) $-\sqrt{2}Pa$ (d) $+2Pa$.

1.8 What is the unit vector of the resultant of the following forces

$\bar{F}_1 = 2\hat{i}+3\hat{j}+4\hat{k}$, $\bar{F}_2 = 4\hat{i}+3\hat{j}+2\hat{k}$

(a) $6\hat{i}+6\hat{j}+6\hat{k}$ (b) $\frac{\hat{i}}{\sqrt{3}}+\frac{\hat{j}}{\sqrt{3}}+\frac{\hat{k}}{\sqrt{3}}$

(c) $-2\hat{i}+2\hat{k}$ (d) $2i-2\hat{k}$

[CSE, Prelims, CE, 2009]

1.9 What is the magnitude of the resultant of forces

$\vec{A} = 2\hat{i}+5\hat{j}$, $\vec{B} = 6\hat{i}-7\hat{k}$, $\vec{C} = 2\hat{i}-6\hat{j}+10k$

(a) 10.5 (b) 10

(c) 11.5 (d) 12.5

[CSE, Prelims, CE, 2008]

1.10 A force $\vec{F} = 2i+5j-k$ is passing through the origin, what is the moment about point (1, 1, 0)

(a) $i-j-k$ (b) $-i+j+k$

(c) $i+2k$ (d) $i++2j-3k$

[CSE, Prelims, CE, 2010]

1.11 A and B are the end points of a diameter of a disc rolling along a straight line with a counter clockwise angular velocity as shown in **Fig. 1.32**. Referring to the velocity vectors $\bar{V}_A$ and $\bar{V}_B$ shown in figure.

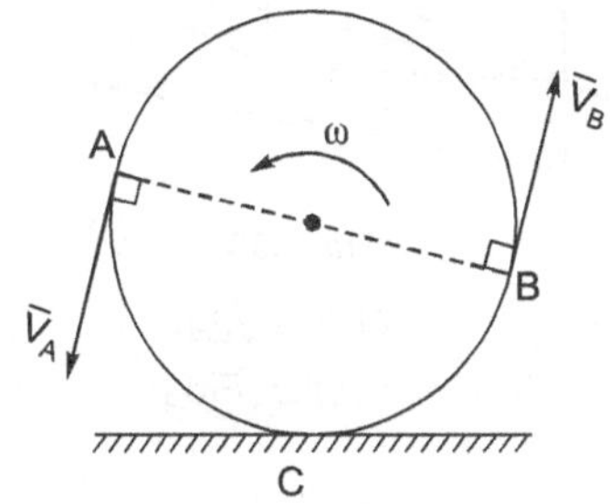

Fig. 1.32

(a) $\bar{V}_A$ and $\bar{V}_B$ both are correct

(b) $\bar{V}_A$ is correct but $\bar{V}_B$ is incorrect

(c) $\bar{V}_A$ and $\bar{V}_B$ both are incorrect

(d) $\bar{V}_A$ is correct but $\bar{V}_B$ is incorrect

[GATE : 1990, 1 Marks]

Answers

1.1 (c)	1.2 (b)	1.3 (a)	1.4 (c)	1.5 (b)
1.6 (c)	1.7 (b)	1.8 (b)	1.9 (a)	1.10 (b)
1.11 (c)				

EXPLANATIONS

1.1 (c)

$\sqrt{0.33} = 0.577$.

1.2 (b)

$(0-3)i + (4-0)j + (3+4)k = -3i + 4j + 7k$.

1.3 (a)

$F = +4j$ and point (4, 0, 0) in x, y plane moment $= +20k$ (anticlockwise) **(Fig. 1.33)**.

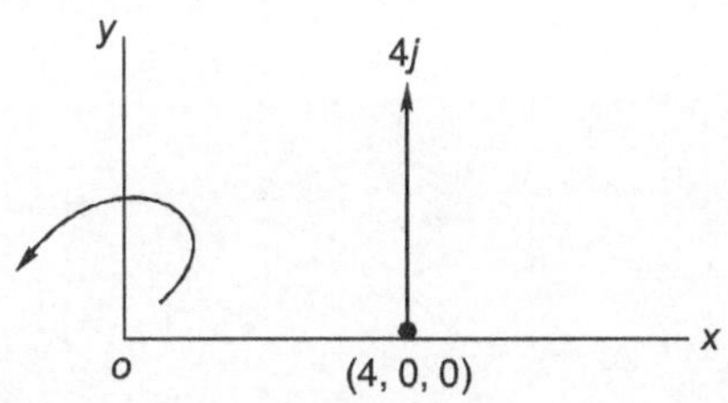

Fig. 1.33

1.4 (c)

$M_0 = -12k$ kNm **(Fig. 1.34)**.

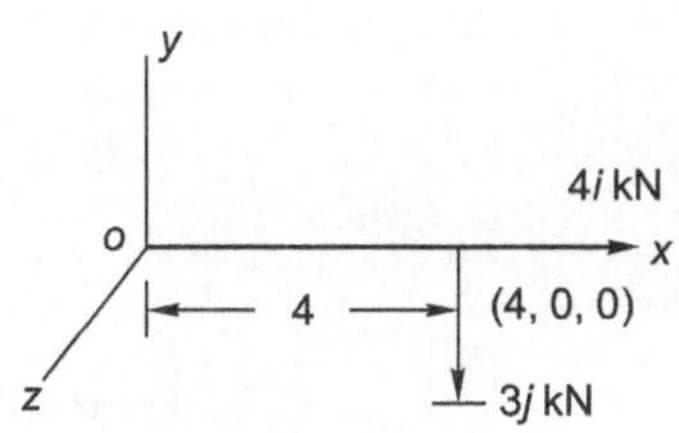

Fig. 1.34

1.5 (b)

Net force in y-direction $= 3 \sin 30^\circ + 6 \sin 30^\circ - 4$
$= 0.5$ kN

$M_0 = +5 \times 0.5$ kNm (ccw) $= +2.5$ kNm (ccw).

1.6 (c)

Couple is a free vector **(Fig. 1.35)**

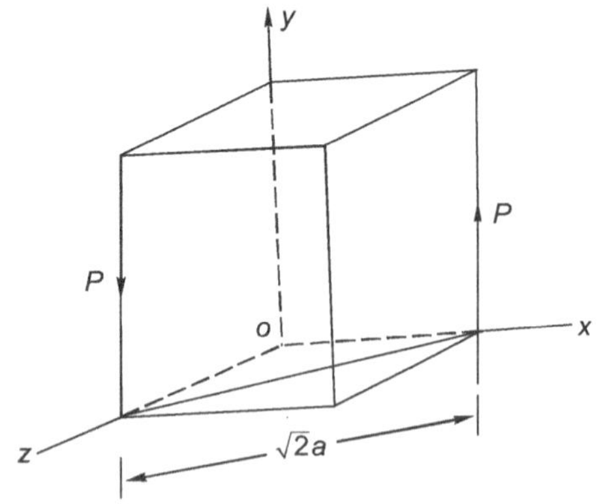

Fig. 1.35

$$\text{arm} = \sqrt{2}a$$
$$M = \sqrt{2}Pa.$$

1.7 (b)

$$\text{arm} = \sqrt{a^2 + a^2} = \sqrt{2}.a$$
$$\text{couple} = \sqrt{2}\,Pa \text{ (anticlockwise)}$$

1.8 (b)

$$\bar{F}_R = 6\hat{i} + 6\hat{j} + 6\hat{k}$$
$$\text{Unit vector} = \frac{6\hat{i}}{\sqrt{108}} + \frac{6\hat{j}}{\sqrt{108}} + \frac{6\hat{k}}{\sqrt{108}} = \frac{\hat{i}}{\sqrt{3}} + \frac{\hat{j}}{\sqrt{3}} + \frac{\hat{k}}{\sqrt{3}}$$

1.9 (a)

$$\bar{R} = 10\hat{i} - \hat{j} + 3\hat{k}$$
$$\text{Magnitude} = \sqrt{10^2 + (-1)^2 + 3^2}$$
$$= \sqrt{100 + 1 + 9} \simeq 10.5$$

1.10 (b)

$$M_x = yF_z - zF_y$$
$$M_y = zF_x - xF_z$$
$$M_z = xF_y - yF_x$$

where, $x = 1$, $y = 1$, $z = 0$

$F_x = 2$, $F_y = 3$, $F_z = -1$

$M_x = -1$, $M_y = +1$, $M_z = 3 - 2 = 1$

$M = -i + j + k$

1.11 (c)

$\bar{V}_A$ and $\bar{V}_B$, both are incorrect

Disc rotating about point of contact C

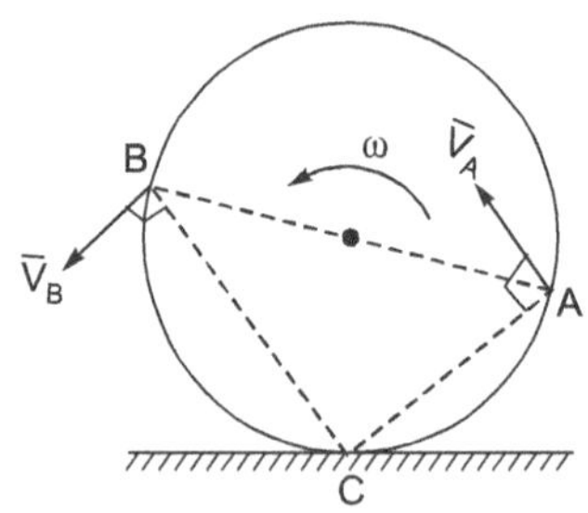

Fig. 1.36

■■■■

02

CHAPTER

Equivalent Coplanar Force Systems

2.1 Introduction

Several forces in a body can be represented by *an equivalent force* on the same body. Similarly an *equivalent couple* of several couple moments can be determined. In this chapter we will study about following cases:

1. Equivalence of a force at one point represented by a force and a couple at some other point.
2. Resultant of a system of concurrent and coplanar forces.
3. Resultant of a system of coplanar forces and couples.
4. Resultant of parallel forces.
5. Resultant of coplanar distributed forces.

For determining the equivalence of any force system, following basic principles are followed:

(*a*) Sum of a set of concurrent forces in a plane is *a single force* (equivalent to original system of forces)

Three forces F_1, F_2 and F_3 act a point O in x-y coordinates system as shown in **Fig. 2.1 (*a*)**. To find equivalent of these forces, we can use 3 methods:

(*i*) Making a Force Polygon

(*ii*) Taking components of forces in xyz-directions

(*iii*) By addition of force vectors.

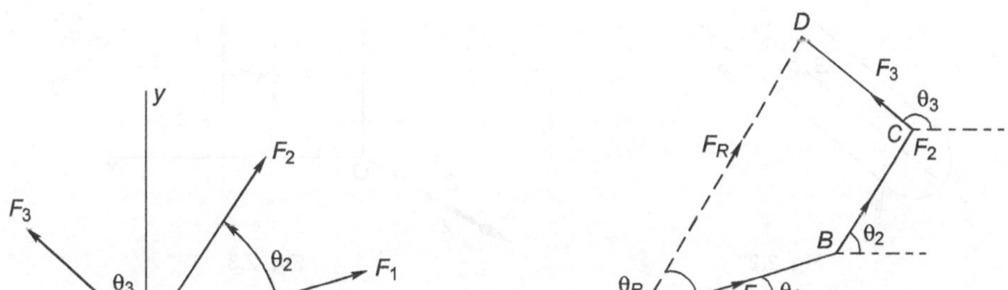

Fig. 2.1

Figure 2.1 (*b*) shows the force polygon. At point O, draw a vector OB equal in magnitude and direction of force F_1, then from point B, draw vector BC parallel to force F_2 and equal in magnitude of F_2. From C draw a line CD parallel to force F_3 and equal in magnitude of F_3. Join OD, from O to D, it is a *resultant force*, F_R = resultant of forces F_1, F_2 and F_3.

(*i*) This force polygon or graphical method takes more time and involves graphical error, therefore other two methods are preferred.

(*ii*) Components of forces in *x*- and *y*-directions.

Let us resolve the forces along *x* and *y* coordinates, *i.e.*, $F_1\cos\theta$, $F_1\sin\theta_1$, $F_2\cos\theta_2$, $F_2\sin\theta_2$, $F_3\cos\theta_3$, $F_3\sin\theta_3$. If F_R is resultant force then

$$F_{Rx} = \text{component of } F_R \text{ in } x\text{-direction}$$
$$= F_1\cos\theta_1 + F_2\cos\theta_2 + F_3\cos\theta_3$$
$$R_{Ry} = \text{component of } F_R \text{ in } y\text{-direction}$$
$$= F_1\sin\theta_1 + F_2\sin\theta_2 + F_3\sin\theta_3$$

Forces and directions are given, therefore resultant force can be determined

$$F_R = \sqrt{F_{Rx}^2 + F_{Ry}^2}$$
$$\tan\theta_R = \frac{F_{Ry}}{F_{Rx}}$$

(*iii*) Forces are written in vector form as follows:

$$F_1 = F_1\cos\theta_1 i + F_2\sin\theta_1 j$$
$$F_2 = F_2\cos\theta_2 i + F_2\sin\theta_2 j$$
$$F_3 = F_3\cos\theta_3 i + F_3\sin\theta_3 j$$

Resultant force $F_R = (F_1\cos\theta_1 + F_2\cos\theta_2 + F_3\cos\theta_3)i + (F_1\sin\theta_1 + F_2\sin\theta_2 + F_3\sin\theta_3)j$

(*b*) A force can be moved along its line of action, so *long as the point of application remains on the body.*
A force *F* is applied on a body along line of action *ABC*, as shown in **Fig. 2.2**. The point of application can be moved to *B* or to *C*, along the line of action and the effect of the force on body remains unchanged. This is also called the *law of transmissibility of a force.*

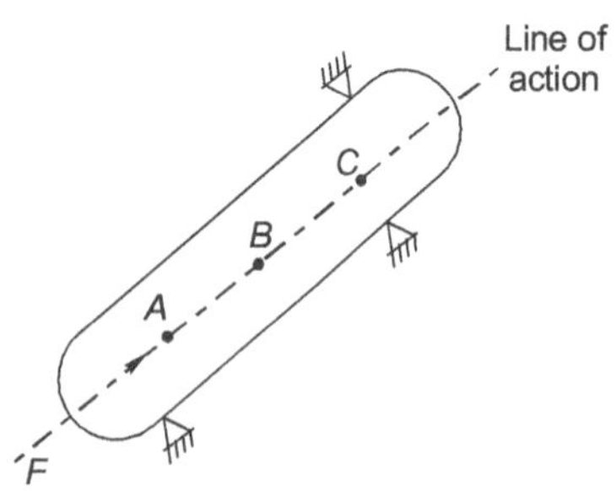

Fig. 2.2

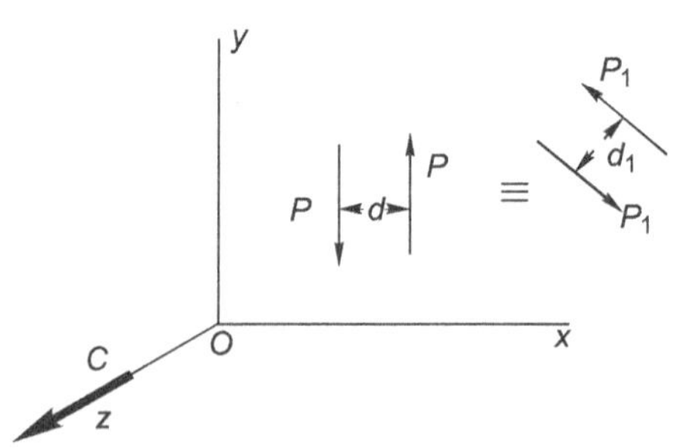

Fig. 2.3

(*c*) *Effect of a couple is only a couple moment.* Couple is a free vector, therefore a couple can be changed in any manner so long as its moment remains the same *i.e.*, (*i*) by changing the magnitude of force, (*ii*) by changing the arm of the couple, (*iii*) by changing the direction of forces **(Fig. 2.3).**
A couple $P \times d$ (*ccw*) is shown in *x*-*y* plane. Its vector *C* is in a direction perpendicular to *xy* plane *i.e.*, *z*-direction since it is counterclockwise, it is a positive couple *C*. This couple is equivalent to a couple $P_1 \times d_1$ (*ccw*) in same *xy* plane, with direction of forces changed, arm of the couple changed and magnitude of the forces *P* and $-P$ changed to P_1 and $-P_1$. Couple vector *C* remains the same.

2.2 Resultant of a Force by an Equivalent Couple and a Force at any other Point

Let us consider a force *P* acting at a point *A*. It is to replaced to a point *B* in *x*-*y* plane as shown in **Fig. 2.4.** At *B*, two equal and opposite forces (in magnitude) *P* and $-P$ are applied. (one cancels the other because of line of action is the same) Force $+P$ at *A* and $-P$ at *B* constitute a couple $P \times d$ (*cw*), where *d* is the perpendicular distance between *P* at *A* and $-P$ at *B*.

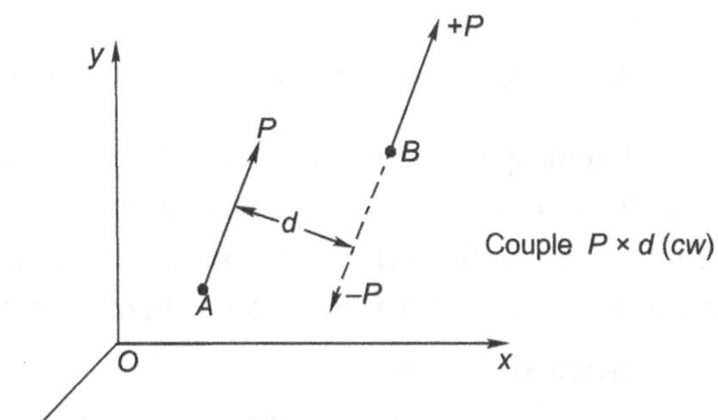

Fig. 2.4

So force *P* at *A*

$$\equiv \text{Force } P \text{ at } B + \text{a couple } P \times d \text{ at } B$$

$$= +P \text{ at } B + C \text{ (couple)}$$

where $C = P \times d\ (cw)$

Similarly a force and a couple in one plane can also be reduced to an equivalent single force on the body in the same plane.

Example 2.1 A vertical force $P = 80$ N ↑ acts at a point *A* (6, 4) *m*. Determine its equivalent at a point *B* (8, 10) m as shown in **Fig. 2.5.**

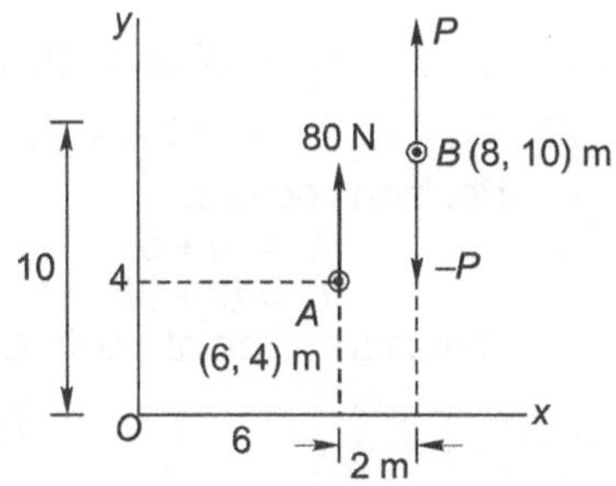

Fig. 2.5

Solution Apply forces $+P$ (80 N ↑)and $-P$ (80 N ↓) at point *B*. *P* at *A* and $-P$ at *B* constitute a couple with arm $8 - 6 = 2$ m.

Equivalent system at $B \equiv$ A force *P* (80 N ↑)

and a couple $80 \times 2 = 160$ Nm (*cw*), a negative couple.

Exercise 2.1 A force of 100 N (horizontal) is applied at *A* (2, 4) m on a body. Find its equivalent system at point *B* (4, –2) m **(Fig. 2.6).**

[**Ans:** A force $+\overrightarrow{100}$ and -600 Nm (*cw*) couple at *B*)].

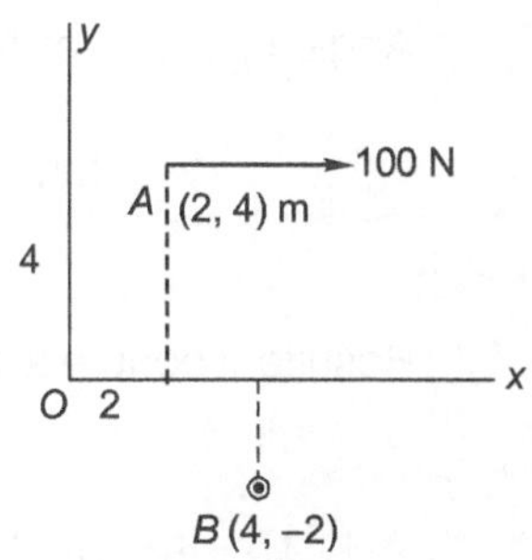

Fig. 2.6

2.3 Resultant of a Coplanar Force System

In the previous article we have learnt that any force system can be replaced by a single force and a couple. Any force or any couple can be moved from one point to another point so as to have a simple equivalent force system.

Fig. 2.7 shows forces F_1, F_2, F_3, ..., F_n and couples C_1, C_2, C_3, ..., C_n acting in one plane *i.e.*, *xy* plane of a body.

Force F_1 acts at point *A*, with position vectors r_1.

Force F_2 acts at point *B*, with position vector r_2.

Force F_3 acts at point *C*, with position vector r_3.

Force F_n acts at point *N*, with position vector r_n.

Force F_1 at *A* can be replaced by a force F_1 at *O* (origin) and a couple $r_1 \times F_1$, similarly other forces can be replaced at the origin alongwith a couple vector.

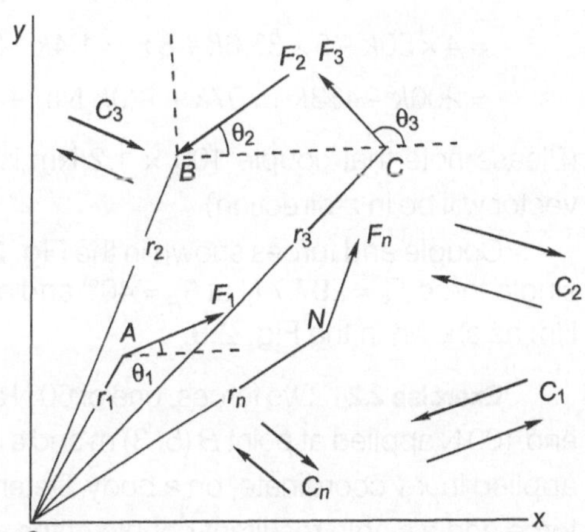

Fig. 2.7

Resultant force at origin

$$F_R = F_1 + F_2 + F_3, \dots F_n \text{ (vectorially)}$$

Resultant couple $C_R = [r_1 \times F_1 + r_2 \times F_2 + r_3 \times F_3, ..., r_n \times F_n] + [C_1 + C_2 + C_3, ..., C_n]$ vectorially

Therefore *any force system however complex it may be can be represented by a single resultant force and a single resultant couple.*

If a body subjected to a system of forces and couples is in equilibrium then $F_R = 0$ and $C_R = 0$.

Example 2.2 Force 100 N at $A(4, 5)$ m, 200 N at $B(8, 3)$ and a couple of forces ±100 N at an arm length of 1.2 m act in xy plane on a body, **Fig. 2.8**. Replace these forces and a couple by a single equivalent force and a single equivalent couple.

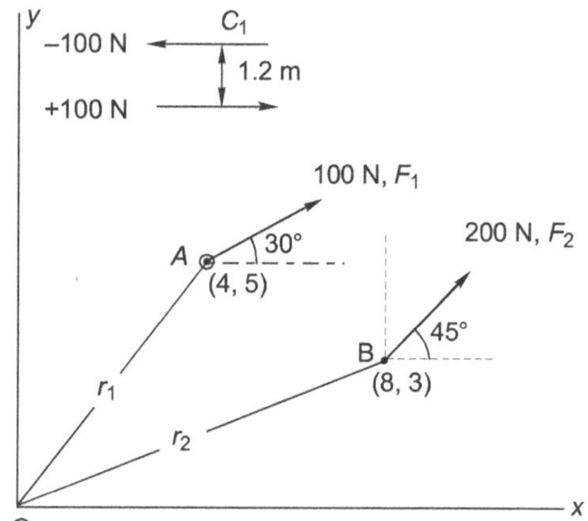

Fig. 2.8

Solution Force

$$F_1 = 100\cos 30° i + 100 \sin 30° j$$
$$= 86.6i + 50j \text{ N}$$
$$F_2 = 200\cos 45° i + 200 \times \sin 45° j$$
$$= 141.4i + 141.4j \text{ N}$$

Position vectors,
$$r_1 = 4i + 5j$$
$$r_2 = 8i + 3j$$

Resultant force at origin O,

$$F_R = F_1 + F_2 = 86.6i + 50j + 141.4i + 141.4j$$
$$= 228i + 191.4j \text{ N}$$
$$|F_R| = \sqrt{228^2 + 191.4^2} = \sqrt{51984 + 36633.96} = 297.7 \text{ N}$$

Angle θ_R with x-axis,

$$\tan^{-1}\frac{191.4}{228.0} = \tan^{-1} 0.8395$$
$$\theta_x = 40°$$

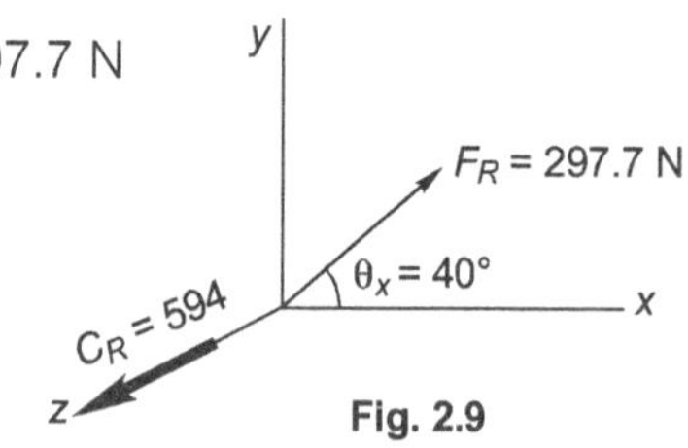

Fig. 2.9

Resultant couple at origin, $C_R = r_1 \times F_1 + r_2 \times F_2 + C_1$

$$= (4i + 5j) \times (86.6i + 50j) + (8i + 3j) \times (141.4i + 141.4j) + 100 \times 1.2k\ (ccw) \text{ Nm}$$

$$= \begin{vmatrix} 4 & 5 & 0 \\ 86.6 & 50 & 0 \\ i & j & k \end{vmatrix} + \begin{vmatrix} 8 & 3 & 0 \\ 141.4 & 141.4 & 0 \\ i & j & k \end{vmatrix} + 120k \text{ Nm}$$

$$= 4 \times 50k - 5 \times 86.6k + 8 \times 141.4k - 3 \times 141.4k + 120k \text{ Nm}$$
$$= 200k - 433k + 707k + 120k \text{ Nm} = +594k \text{ Nm}$$

(Please note that couple 100 × 1.2 Nm is in xy plane and couple vector will be in z-direction)

Couple and forces shown in the **Fig. 2.8** can be replaced by a single force $F_R = 297.7$ N at $\theta_R = 40°$ and a single couple $C_R = 594k$ Nm as shown in the **Fig. 2.9.**

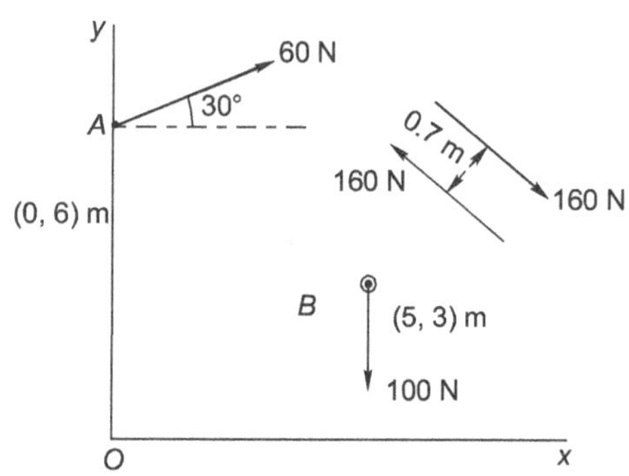

Fig. 2.10

Exercise 2.2 Two forces, one of 60 N applied at point $A(0, 6)$ m and 100 N applied at point $B(5, 3)$ m and a coupe 160 N × 0.7 m (cw) applied in x-y coordinate, on a body. Determine the single resultant force and a single resultant couple of the system **(Fig. 2.10).**

[**Ans:** $F_R = 87.18$ N, $\theta_x = -53.4°$, $C_R = -663.96k$ Nm].

2.4 Resultant of a Parallel Forces Systems

Let us consider a system of forces $F_1, F_2, F_3, ..., F_n$ all parallel in y-direction and couples $C_1, C_2, ..., C_m$ whose couple forces are lying in zx plane or the *forces forming the couples are all along y-direction*, as shown in Fig. 2.11. Resultant of the parallel forces

$$F_R = \left[\sum_{i=1}^{n} F_i\right] = (F_1 + F_2 + F_3 ... F_n)\, j$$

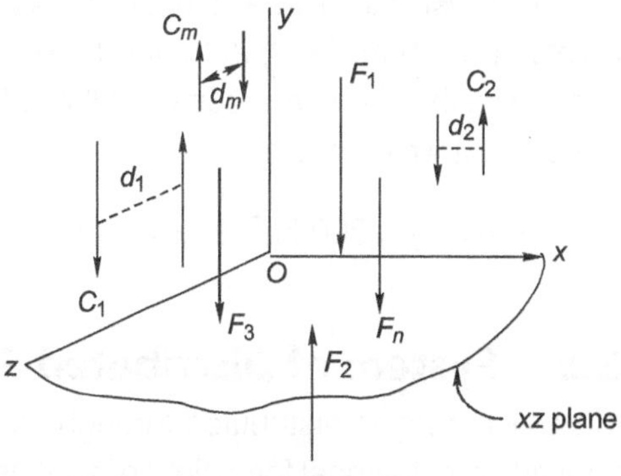

Fig. 2.11

Resultant couple of the parallel force systems

$$C_R = \left[\sum_{i=1}^{n}(x_i i + z_i k) \times F_i j\right] + \sum_{i=1}^{m}[(C_i)_x i + (C_i)_z k]$$

$$C_R = (x_i i + z_i k) \times F_1 j + (x_2 i + z_2 k) \times F_2 j ... (x_n i + z_n k) \times F_n j + (C_{1x} i + C_{1z} k) + (C_{2x} i + C_{2z} k) + ...(C_{mx} i + C_{mz} k)$$

where x_i, z_i are coordinates of the points of application of the forces F_i. Moreover $C_x . i$, $C_z . k$ are components of couple vectors C_i.

If $F_R \neq 0$, then F_R can be moved to the origin O of the system such that

$$|C_R| = |F_R|\, d$$

where d is the perpendicular distance from O to resultant F_R. Moreover if resultant couple, $C_R = 0$, then there is only resultant force F_R at the origin.

If $F_R = 0$, then there will be only resultant couple C_R. Therefore *the simplest resultant of a parallel force systems is either a force or a couple moment.*

Example 2.3 Determine the simplest resultant of 3 forces acting in y-direction as shown in Fig. 2.12.

$F_1 = 400$ N acts downwards at A (2, 0, 3) m

$F_2 = 150$ N acts upwards at B (4, 0, 0)

$F_3 = 200$ N acts upwards at C (4, 0, 4) m

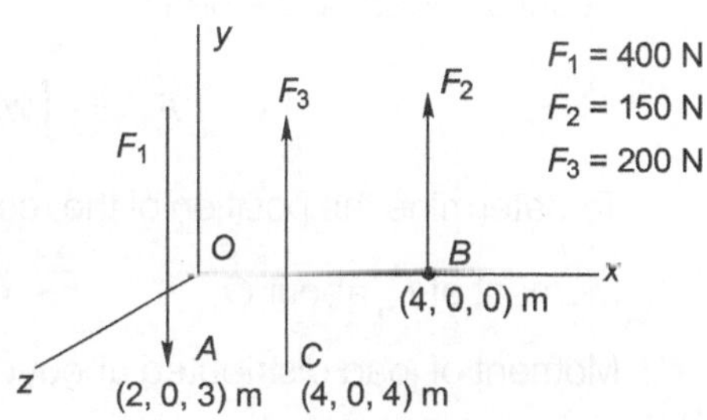

Fig. 2.12

Replacing each force by an equal force at origin and a couple.

Resultant force,

$$F_R = -F_1 + F_2 + F_3$$
$$= -400 + 150 + 200 = -50 \text{ N} \downarrow$$
$$= -50j \text{ N (at the origin)}$$

Position vectors, $r_1 = 2i + 3k$

$r_2 = 4i, \quad r_3 = 4i + 4k$

Say the simplest resultant passes through coordinates $\bar{x}, \bar{z}$, then taking moments about z-axis

$$-400 \times 2 + 150 \times 4 + 200 \times 4 = -50\bar{x} = F_R . \bar{x}$$
$$-800 + 600 + 800 = -50\bar{x}$$
$$600 = -50\bar{x}$$
$$\bar{x} = -12 \qquad ...(1)$$

Taking moments about x-axis

$+3 \times 400 + 150 \times 0 - 200 \times 4 = +50\bar{z}$, taking ccw moments as positive

$$400 = +50\bar{z}$$
$$\bar{z} = +8 \text{ m}$$

But the simplest resultant of the force system is 50 N ↓ acting at point (–12, 0, 8) m.

If this resultant is shifted to origin then

$F_R = -50j$ N + couple $C_R = +400i + 600k$ Nm, $|C_R| = 721$ Nm.

Exercise 2.3 Three vertical forces 200, 400, –300 N are acting at points (1, 0, 3) m, (3, 0, 2) m and (2, 0, –2) m as shown in **Fig. 2.13**. Determine the simplest resultant of these forces and the coordinates of the point at which it acts.

[**Ans:** $F_R = 300$ N ↑, $\bar{z}$ $\bar{x} = 6.66$ m, $\bar{z} = 2.66$ m].

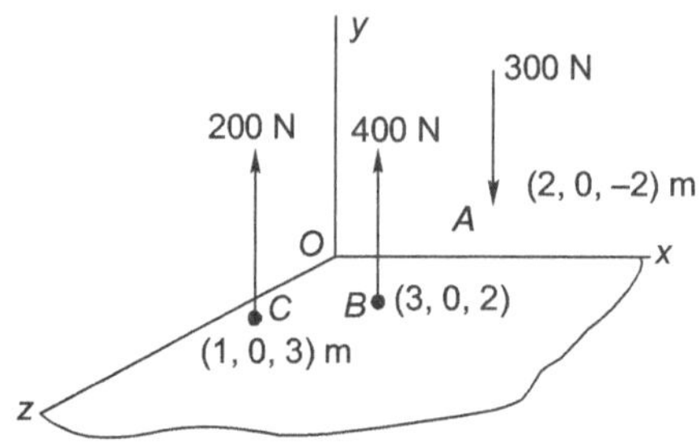

Fig. 2.13

2.5 System of Distributed Forces

Mass of a body is distributed throughout the volume of a body and mass per unit volume is known as *density*. Similarly gravitational force field of earth is distributed as a force vector. This gravitational force exerts influence directly on the elements of mass distributed throughout the body and are known *as body force distributions*. Force distribution over a surface are called *surface force distributions*, given by force per unit area of the surface directly influenced by the forces. Force due to hydrostatic pressure is distributed on the surface of an immersed body, and is known as *pressure*. However pressure is a scalar quantity.

The weight of the roof in a building is distributed over the area of the roof. Moreover beams and cantilevers are subjected to transverse loads which may be considered as uniformly distributed along the length of the beam. Let us take a general case, if the load is represented as a function of *x*, along the axis of the beam, as shown in **Fig. 2.14**.

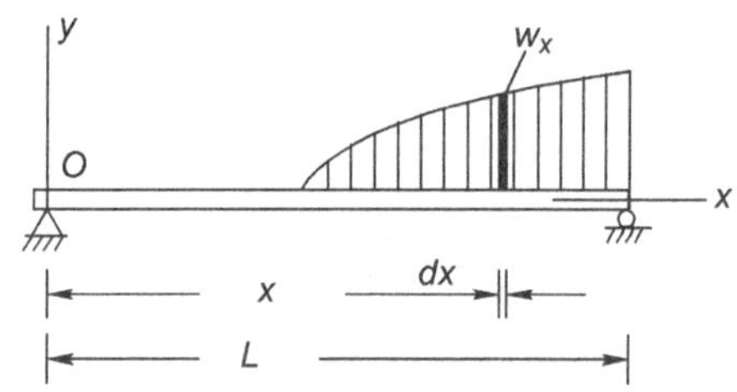

Fig. 2.14

Intensity of transverse loading (perpendicular to the axis of the beam) = $w(x)$, as a function of *x*.

Load over length, dx, $dF = -w(x)\,dx\,j$ (in the negative direction of *y*-axis)

Resultant load on the beam,

$$F_R = -\int w(x)\,dx$$

To determine the position of the resultant force, F_R without a couple moment,

Moment of F_R about O $= F_R.\bar{x} = -\bar{x}\int w(x)\,dx$

Moment of load distributed about point O

$$= -\int w(x)\,x\,dx$$

Distance, $$\bar{x} = \frac{-\int w(x).x.dx}{-\int w(x)\,dx} = \frac{\text{Moment of load about origin}}{\text{Total distributed load}}$$, distance of resultant from 0

Example 2.4 A beam *ABCD*, 7 m long is subjected to distributed load and a concentrated load as shown in **Fig. 2.15**. Determine the simplest resultant of forces on beam and the distance of its line of action from end *A*. If rate of loading *w* = 4 kN/m at *B*, at a distance of 4 m from *A*.

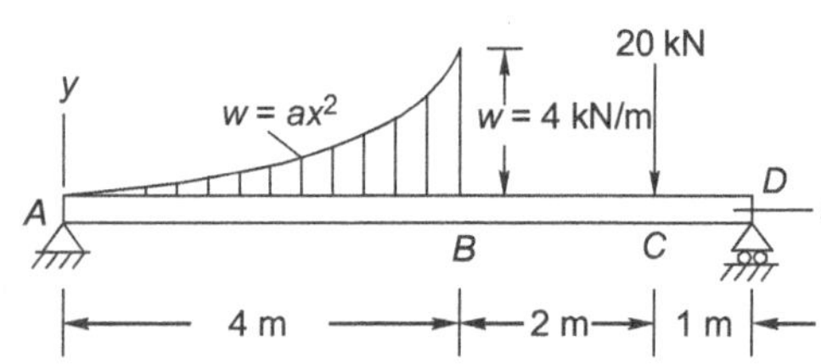

Fig. 2.15

Solution Intensity of loading,

$$w = ax^2$$
$$= 0 \text{ at } x = 0$$
$$= 4 \text{ kN/m at } x = 4 \text{ m}$$

Therefore $4 = a \times 4^2$

$$a = 0.25$$

or $w = 0.25x^2$ kN/m

Total distributed load, $= \int_0^4 w\,dx = \int_0^4 0.25x^2\,dx = \left|\frac{x^3}{12}\right|_0^4 = \frac{64}{12} = 5.33$ kN

Moment of distributed load, about end A

$$M_w = \int_0^4 xw\,dx = \int_0^4 0.25x^3\,dx = \left|\frac{x^4}{16}\right|_0^4 = 16 \text{ kNm}$$

F_R, simplest resultant of forces on beam,

$$F_R = 5.33 + 20 = 25.33 \text{ kN}$$

Taking moments about end A,

$$M_A = M_w + 20 \times 6 = 16 + 120 = 136 \text{ kNm } (cw)$$
$$\bar{x} F_R = 136$$
$$\bar{x} = \frac{136}{25.33} = 5.37 \text{ m from end } A$$

Simplest resultant is 25.33 kN acting at a distance of 5.37 m from A.

Exercise 2.4 A beam 6 m long carries a linearly increasing distributed load, with $w = 0$ at C to $w = 6$ kN/m at D and a concentrated load of 6 kN at B. Determine the simplest resultant and its line of action **(Fig. 2.16)**.

[**Ans:** $F_R = 15$ kN, $\bar{x} = 3.6$ m from A].

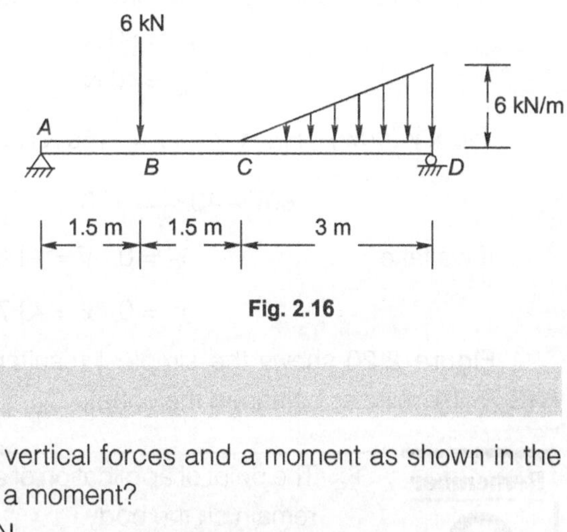

Fig. 2.16

PROBLEMS

Problem 2.1 A beam $ABCDEF$, 7 m long is subjected to vertical forces and a moment as shown in the Fig. 2.17. What is the simplest resultant of all these forces and a moment?

Solution: Total forces on the beam = +10 – 30 – 80 = –100 N

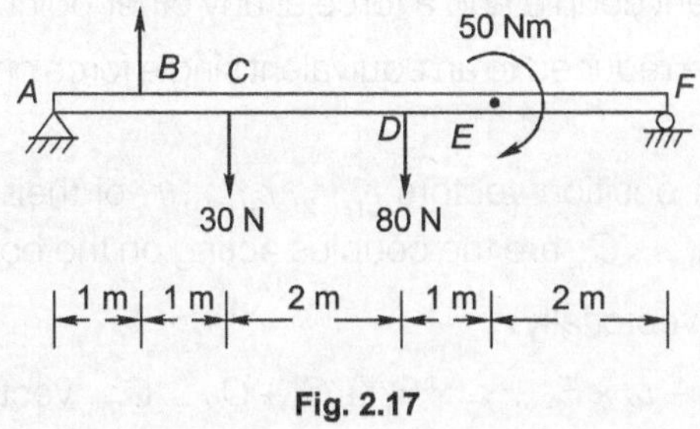

Fig. 2.17

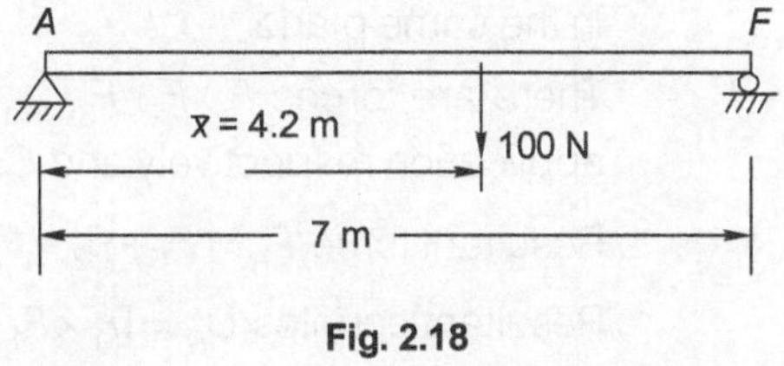

Fig. 2.18

Taking moments about A, anticlockwise moments as positive

$$M_A = +10\times1-30\times2-80\times4-50 = -420 \text{ Nm } (cw)$$

Simplest resultant $= -100$ N, say it acts at a distance of $\bar{x}$ from end A, then

$$-100\bar{x} = -420$$

$$\bar{x} = \frac{400}{100} = 4.2 \text{ m from end } A\text{, as shown in **Fig. 2.18.**}$$

Problem 2.2 Determine the resultant of 3 forces at E, A and B and a couple 150 Nm (ccw) at D. Determine the resultant force and the point of application of resultant force **(Fig. 2.19).**

Solution Let us take origin at A. All the forces can be shifted to the origin, (equivalent to a force and a couple at origin). Resultant force,

$$F_R = 80i + 100j - 60j \text{ N} = 80i + 40j \text{ N}$$

$$|F_R| = \sqrt{80^2 + 40^2} = 89.4 \text{ N}$$

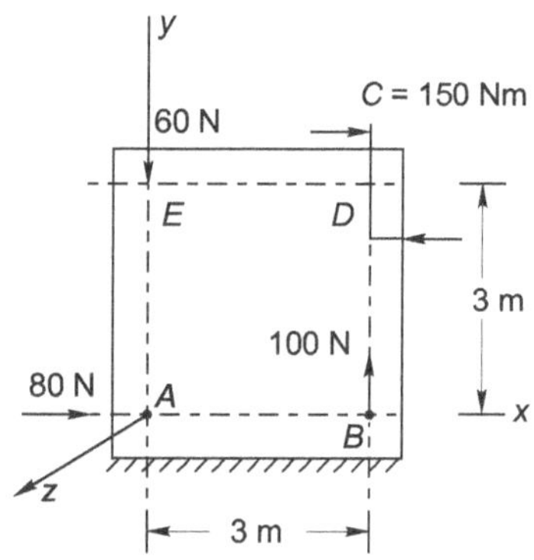

Fig. 2.19

Moments about origin,

$$M_A = 100 \text{ N}\times3 \text{ m } (ccw) - 150 \text{ Nm } (cw)$$

or $C_R = 150$ Nm (ccw) resultant couple

Note that forces 60 N vertical and 80 N horizontal are passing through A, these forces will not produce any moment.

$$C_R = 150 \text{ Nm } (ccw) \text{ in plane } xy$$

(Direction of couple vector along z-axis)

Say the resultant F_R passes through $\bar{x}$ and $\bar{y}$, then

$$F_{Rx} = 80 \text{ N}$$

$$F_{Ry} = 40 \text{ N}$$

$$F_{Rx}\times\bar{y}\ (ccw) - F_{Ry}.\bar{x}\ (ccw) = -150 \text{ N } (cw) \text{ for balancing}$$

$$80\bar{y} - 40\bar{x} = -150$$

If we take $\bar{x} = 0,\ \bar{y} = -1.875$ m

$$\bar{y} = 0,\ \bar{x} = +3.75 \text{ m}$$

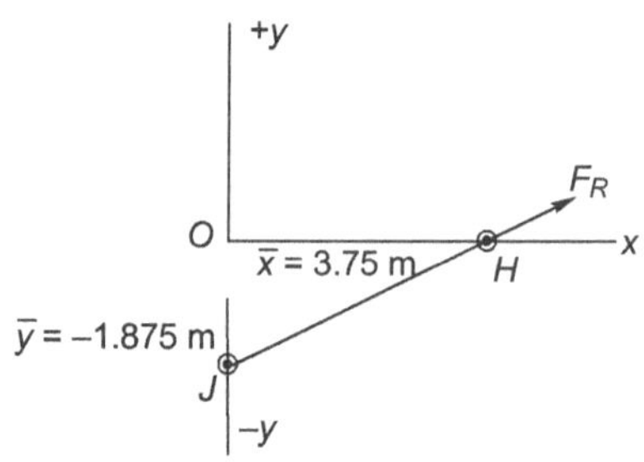

Fig. 2.20

Figure 2.20 shows the simplest resultant, $|F_R| = 89.4$ N passing through points $J(0, -1.875)$ m and $H(3.75, 0)$ m, it has balanced the couple C_R.

Remember

- The point of application of a force can be shifted along its line of action so long as the point remain on the body.
- Couple is a free vector. Effect of a couple is only a couple moment.
- A force can be replaced by an equivalent couple and a force at any other point.
- A force and a couple in one plane can be reduced to an equivalent single force on the body in the same plane.
- There are forces $F_1, F_2, F_3, \ldots, F_n$, with position vectors $r_1, r_2, r_3, \ldots, r_n$ of their points of application respectively and $C_1, C_2, C_3, \ldots, C_m$ are the couples acting on the body then

  Resultant force, $F_R = F_1 + F_2 + F_3 \ldots F_n$ (vectorially)

  Resultant couples, $C_R = [r_1\times F_1 + r_2\times F_2 + r_3\times F_3 \ldots r_n\times F_n] + C_1 + C_2 \ldots C_m$, vectorially.

- The simplest resultant of a parallel force system is either a force or a couple.
- Load distributed along the length of a beam, such that rate of loading $w_x = w(x)$, a function of x.

  Resultant force, $F_R = -\int w(x)\,dx$

  Point of application of F_R, $\bar{x} = \dfrac{-\int w(x)\,x\,dx}{-\int w(x)\,dx}$.

PRACTICE PROBLEMS

2.1 A force of 200 N is applied at A on rigid body ABC as shown in **Fig. 2.21**. Replace this force by equivalent couple and force at C.

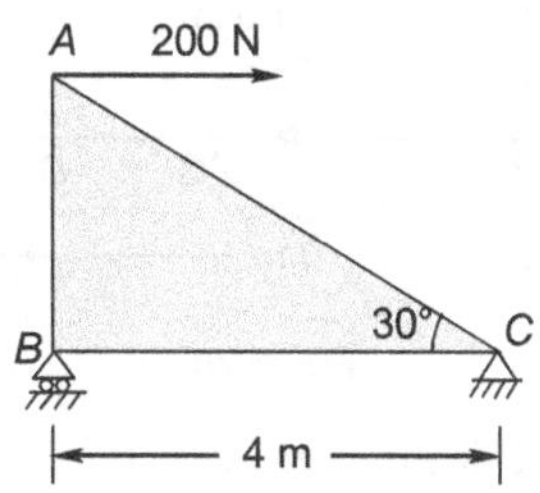

Fig. 2.21

[**Ans:** At C, 200 N force in the same direction –461.8 Nm (cw)].

2.2 A force of 600 N is applied at A (0, 6) m of a rigid body. What is the equivalent force and couple at point B (7, 0) m as shown in **Fig. 2.22**?

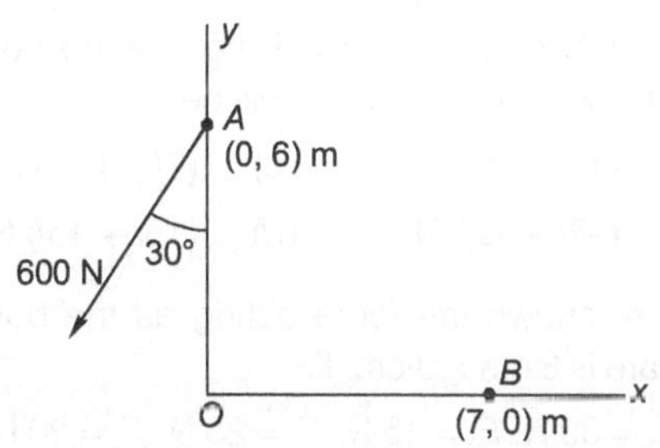

Fig. 2.22

[**Ans:** Force P=–300i – 519.6j N; Couple=+5437.2k Nm].

2.3 A beam $ABCDEF$, 6 m long with vertical projection DE = 1 m, is subjected to forces and couples as shown in the **Fig. 2.23**. Determine the simplest resultant of this force and couple system.

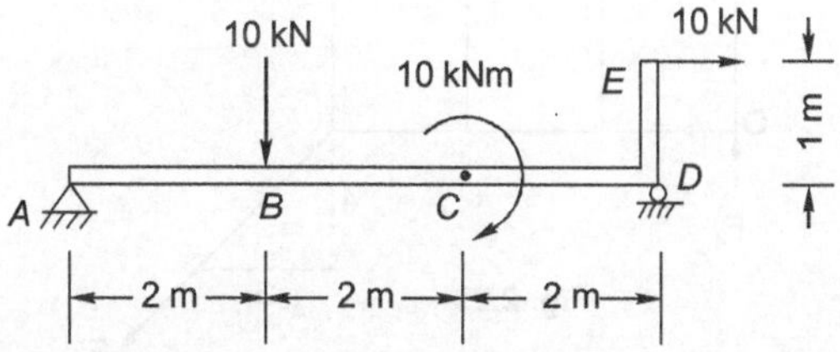

Fig. 2.23

[**Hint:** Moment at D is 10 kNm (cw)]

[**Ans:** 10 kN at point C].

2.4 Determine the resultant of 3 forces acting at B, F and D and a couple at E as shown in **Fig. 2.24**. Determine the resultant and the point of application of resultant.

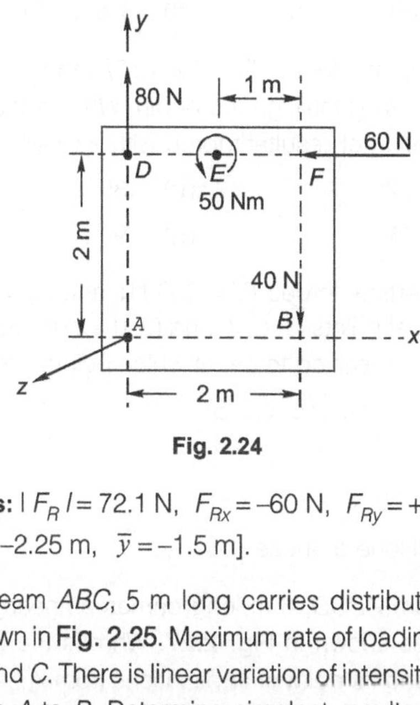

Fig. 2.24

[**Ans:** $|F_R| = 72.1$ N, $F_{Rx} = -60$ N, $F_{Ry} = +40$N, $\bar{x} = -2.25$ m, $\bar{y} = -1.5$ m].

2.5 A beam ABC, 5 m long carries distributed load as shown in **Fig. 2.25**. Maximum rate of loading is 4 kN/m at end C. There is linear variation of intensity of loading from A to B. Determine simplest resultant of forces and line of action of simplest resultant from end A.

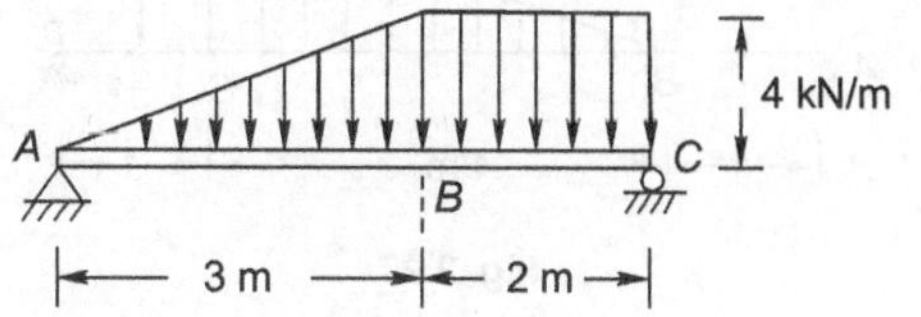

Fig. 2.25

[**Ans:** $F_R = 14$ kN, $\bar{x} = 3.143$ m].

2.6 A cantilever 7 m long carries distributed loads as shown in **Fig. 2.26**, constant rate from A to B, equates to 8 kN/m and linearly varying load from B to C with w = 6 kN/m at B and w = 0 at C in the upward direction.

Determine simplest resultant of forces and its line of action.

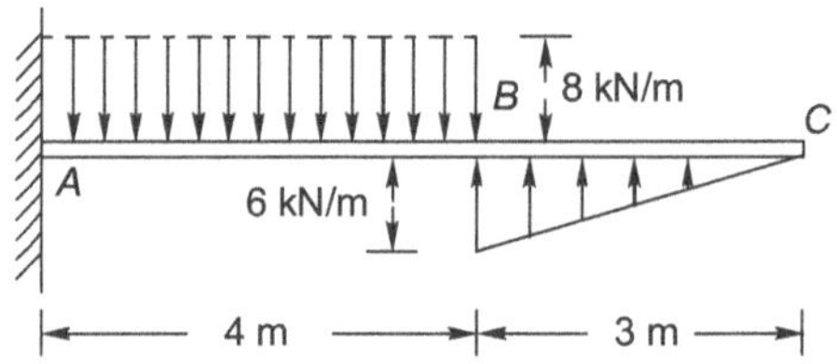

Fig. 2.26

[**Ans:** 23 kN, $\bar{x} = 0.826$ m from *A*].

MULTIPLE CHOICE QUESTIONS

2.1 In *x-y* plane two forces $F_1 = 40i - 30j$ N and $F_2 = -30i + 50j$ N are acting. What is the resultant force (forces are passing through the origin)?

(*a*) $10i + 20j$ N (*b*) $70i + 80j$ N
(*c*) $40i + 20j$ N (d) None of these.

2.2 In *x-y* plane, forces $F_1 = 40i + 50j$ and $F_2 = -30i - 10j$ are passing through the origin. What is the angle of inclination of resultant force with *x*-axis?

(*a*) 14° (*b*) 35°
(*c*) 76° (*d*) 90°.

2.3 Two vertical forces $F_1 = 300$ N↓ and $F_2 = 100$↑ are acting at points (1, 0, 3) and (2, 0, –2) m respectively. What is resulting force, at which points it is acting?

(*a*) 200 N, (2.5, 0, 3.5)
(*b*) 200 N (0, 2.5, –3.5)
(*c*) 300 N (2.5, 0, 3.5)
(*d*) None of these.

2.4 A beam *ABCD*, 6 m long carries a linearly variable load as shown in **Fig. 2.27**. What is the equivalent load and its point of applications from *A*?

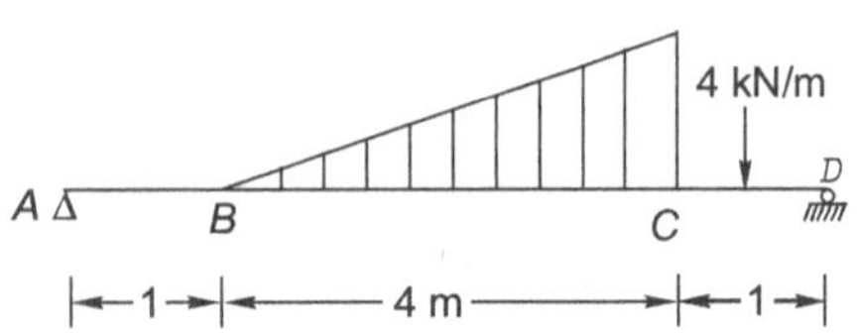

Fig. 2.27

(*a*) 8 kN, 3 m from *A*
(*b*) 8 kN, 3.67 m from *A*
(*c*) 8 kN, 4 m from *A*
(*d*) 8 kN, 5 m from *B*

2.5 A beam 10 m long carries loads and a moment as shown in **Fig. 2.28**. What is equivalent load and point of its application from *A*?

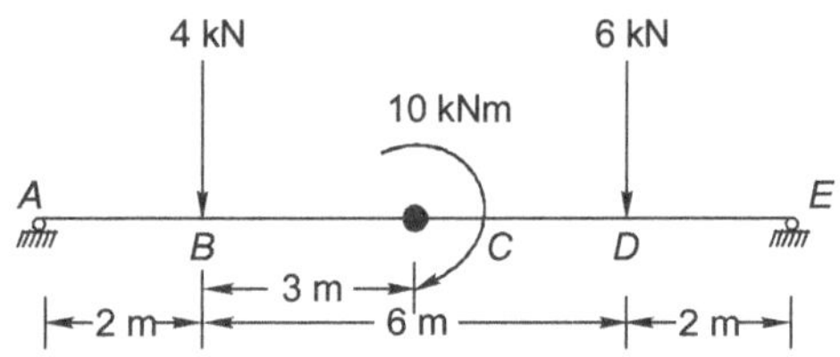

Fig. 2.28

(*a*) 20 kN, 6.6 m from *A*
(*b*) 10 kN, 6.6 m from *A*
(*c*) 0 kN, 5 m from *A*
(*d*) None of these.

2.6 Three forces activity at a point *O* are

$P_1 = (3i + 6j)$ N, $P_2 = (-1.5i + 4.5j)$ N and

$P_3 = (-10.5i + 1.5j)$ N

If a fourth force P_4 is added such that the point *O* is in equilibrium, then force P_4 will be

(*a*) $(15i - 15j)$ N (*b*) $(9i - 12j)$ N
(*c*) $(-9i + 12j)$ N (*d*) $(15i + 15j)$ N

2.7 Figure shown the force acting as the body. Each square is 3 cm × 3 cm. If

$F_1 = 30$ N, $F_2 = 15$ N, $F_3 = 25$ N, $F_4 = 20$ N

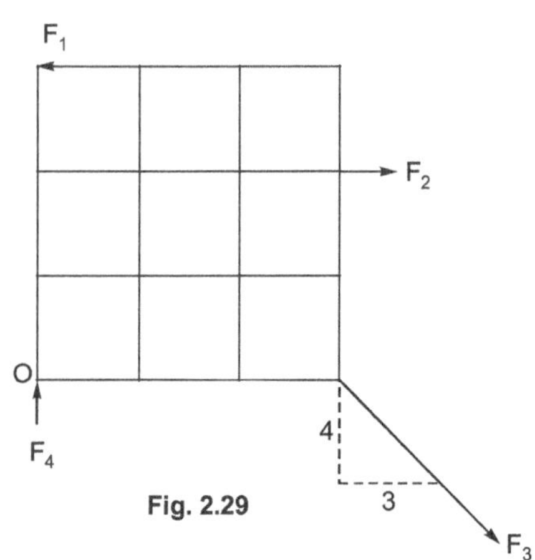

Fig. 2.29

The resultant force and the moment about point O respectively

(a) 0 N, 0 N cm
(b) 90 N, 80 N cm
(c) 60 N, 60 N cm
(d) 30 N, 30 N cm

[CSE, Prelim, CE : 2003]

2.8 Two men, one stronger than the other have to lift a load of 1200 N which is suspended from a light rod of length 3 m. The load is suspended between the two persons positioned at the two ends of the rod. The weaker of the two persons can carry a load upto 400 N only. The distance of the load to be suspended from the stronger person such that the weaker person can has the full share of 400 N

(a) 0.5 m (b) 1.0 m
(c) 1.5 m (d) 2.0 m

[CSE, Prelim, CE : 2002]

2.9 A horizontal rod AB carries three loads of 3.0 kg, 7.0 kg and 10 kg at distance of 2.0 cm, 9 cm and 15 cm respectively from A, where A is hinged. Neglected the weight of the rod which is the point at which the rod will balance?

(a) 10.95 cm from A
(b) 11.95 cm from A
(c) 12.55 cm from A
(d) 13.25 cm from A

[CSE, Prelim, CE : 2007]

2.10 When two forces A and B are mutually at right angles, their resultant is 10 kN. When they are inclined at 60°, the resultant is $5\sqrt{6}$ kN. Individual magnitude of forces are

(a) 4 kN, 6 kN (b) $\sqrt{50}$ kN, $\sqrt{50}$ kN
(c) 5 kN, $\sqrt{75}$ kN (d) 6 kN, 8 kN

[CSE, Prelim, CE : 2004]

2.11 If the magnitude of the resultant of two forces F and $2F$ acting at a point O is 50 N, and the included angle between F and $2F$ is 90°, then what is magnitude of F

(a) 10 N (b) $50\sqrt{2}$ N
(c) $25\sqrt{2}$ N (d) $10\sqrt{5}$ N

[CSE, Prelim, CE : 2006]

2.12 What is the moment (approximate) of the force acting on a railway sign post shown in the **Fig. 2.30.**

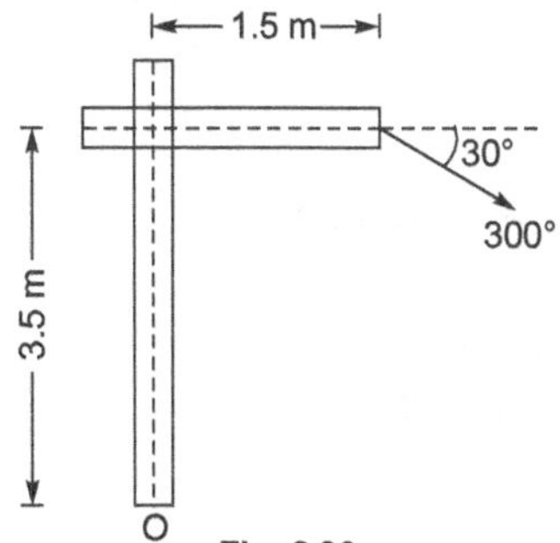

Fig. 2.30

(a) 1135 Nm (b) 1205 Nm
(c) 915 Nm (d) 1300 Nm

[CSE, Prelim, CE : 2007]

2.13 What is the magnitude of the parallel force systems shown in the **Fig. 2.31.**

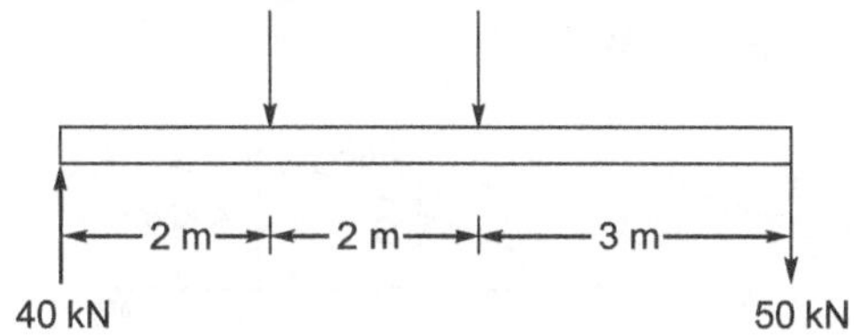

Fig. 2.31

(a) 70 kNm (clockwise)
(b) zero
(c) 70 kNm couple (anticlockwise)
(d) 90 kN vertical force and a moments of 70 kNm (clockwise)

[CSE, Prelim, CE : 2007]

2.14 A string 2 m long is tied to the ends of a uniform rod that weighs 60 N and 1.6 m long. The string passes over a nail so that the rod hangs horizontally. What is the tension in the string?

(a) 24 N (b) 30 N
(c) 42 N (d) 50 N

[CSE, Prelim, CE : 2008]

2.15 If two forces each of 10 N act at an angle θ, then what is their resultant

(a) $20\cos\theta$ (b) $20\cos\frac{\theta}{2}$
(c) $20\cos2\theta$ (d) $20\sin\frac{\theta}{2}$

[CSE, Prelim, CE : 2010]

Answers

2.1 (a)	2.2 (c)	2.3 (a)	2.4 (b)	2.5 (b)
2.6 (b)	2.7 (a)	2.8 (b)	2.9 (a)	2.10 (b)
2.11 (d)	2.12 (c)	2.13 (c)	2.14 (d)	2.15 (b)

EXPLANATIONS

2.1 (a)

$F_A = F_1 + F_2 = 10i + 20j$ N.

2.2 (c)

$F_R = 10i + 40j$

$\tan \alpha = 4,\ \alpha = 76^\circ$.

2.3 (a)

$$F_R = 200 \text{ N}$$
$$200\bar{x} = 300 + 200 = 500$$
$$\bar{x} = 2.5 \text{ m}$$
$$200\bar{y} = 0$$
$$200\bar{z} = 900 - 200 = 700$$
$$\bar{z} = 3.5.$$

2.4 (b)

See **Fig. 2.32**.

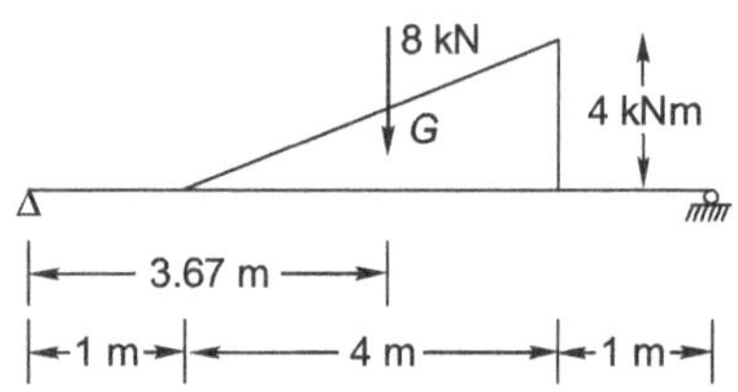

Fig. 2.32

2.5 (b)

See **Fig. 2.33**.

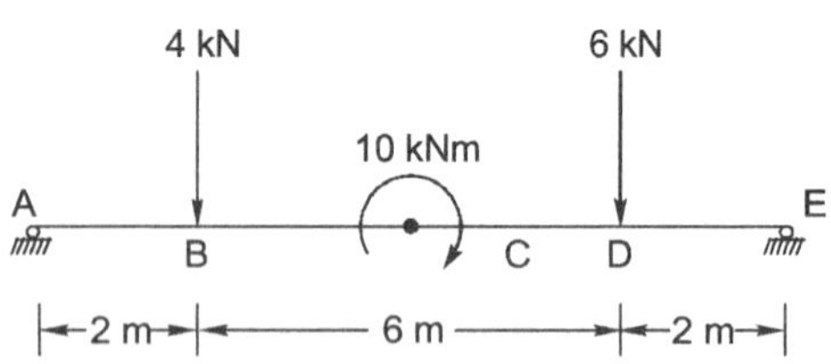

Fig. 2.33

$$M_A = 8 + 48 + 10$$
$$= 66k \text{ Nm}$$
$$F_R = 10 \text{ kN}$$
$$\bar{x} = 6.6 \text{ m from } A.$$

2.6 (b)

Resultant $= (3 - 1.5 - 10.5)\, i + (6 + 4.5 + 1.5) j$

$= -9i + 12j$

$P_4 = -R = 9i - 12j$ (to balance)

2.7 (a)

$$F_{3x} = 15 \text{ N}$$
$$F_{3y} = 20 \text{ N}$$
$$F_{Rx} = 30 - 15 - 15 = 0$$
$$F_{Ry} = 20 - 20 = 0$$
$$F_R = 0$$

Moments, $M_0 = 30 \times 9\,(ccw) + 15 \times 6\,(cw)$

$+20\,(9)\,(cw)$

$= 270 - 90 - 180 = 0$

2.8 (b)

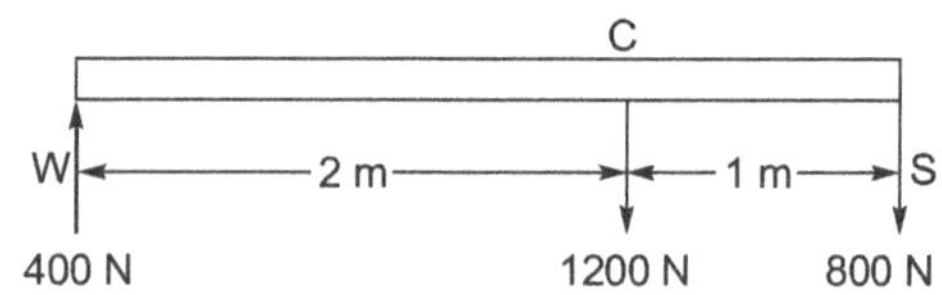

Fig. 2.34

Load shared by stronger person

$= 1200 - 400 = 800$ N

$800 \times 1 \text{ m} = 400 \times 2 \text{ m}$

Load at C, 1 m from stronger person

2.9 (a)

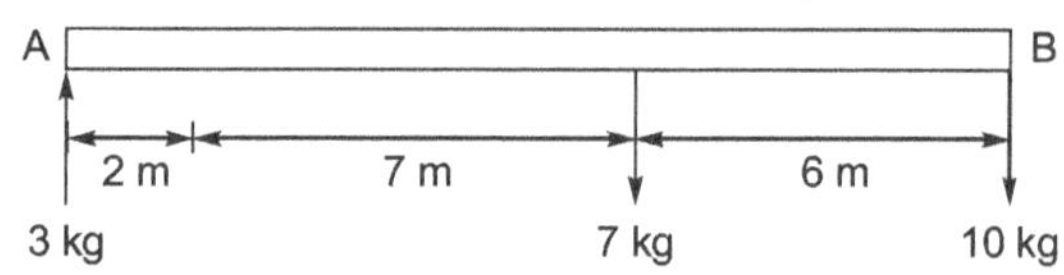

Fig. 2.35

Total load = 20 kg

Moment about $A = 2 \times 3 + 7 \times 9 + 15 \times 10$

$= 6 + 63 + 150 = 219$ kg cm

$$\bar{x} = \frac{219}{20} = 10.95 \text{ cm}$$

to balance from A

2.10 (b)

$$A^2 + B^2 = 100 \quad \text{...(i)}$$
$$\theta = 60^\circ$$
$$A^2 + B^2 + 2AB\cos\theta = (5\sqrt{6})^2 = 150$$
$$2AB\cos\theta = 50$$
$$\cos\theta = \cos 60^\circ = 0.5$$
$$AB = 50 \quad \text{...(ii)}$$

If $A = B = \sqrt{50}$

$AB = 50$

2.11 (d)

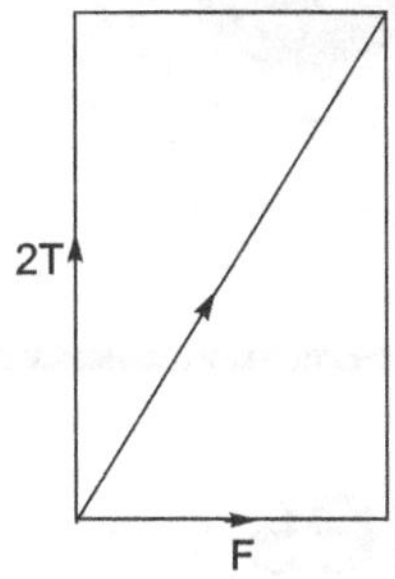

Fig. 2.36

$$R = \sqrt{F^2 + 4F^2} = F\sqrt{5} = 50 \text{ N}$$

$$F = \frac{50}{\sqrt{5}} = \frac{50}{\sqrt{5}} \times \frac{\sqrt{5}}{\sqrt{5}} = 10\sqrt{5} \text{ N}$$

2.12 (c)

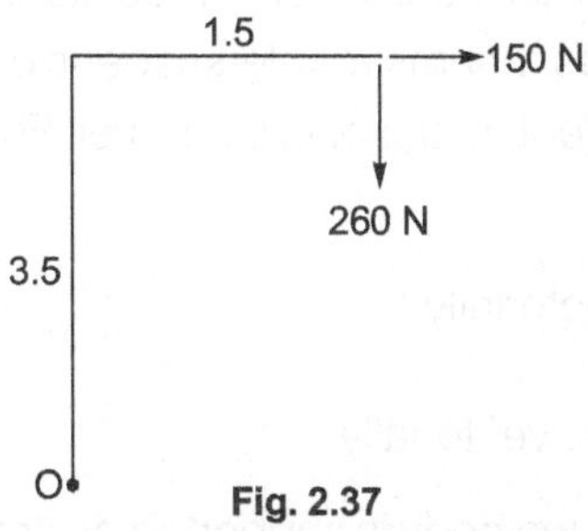

Fig. 2.37

$$M_O = 150 \times 3.5 \text{ (cw)} + 260 \times 1.5 \text{ (}cw\text{)}$$
$$= 525 + 390 = 915 \text{ Nm}$$

2.13 (c)

Two couple acting on the body

$$= 350 \text{ kNm } (ccw) + 280 \text{ kNm } (cw)$$
$$= 70 \text{ kNm } (ccw)$$

Net force = zero

2.14 (d)

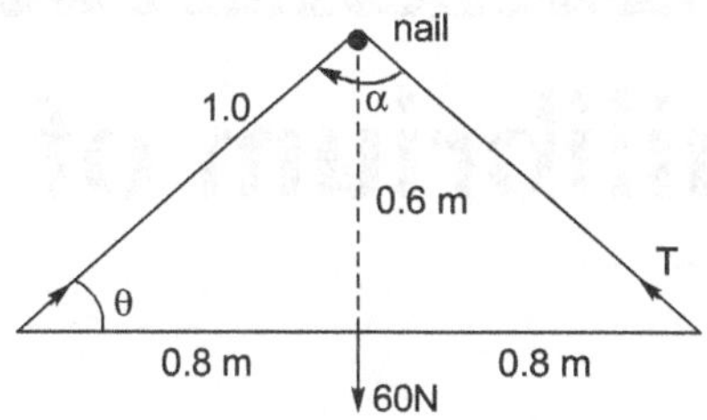

Fig. 2.38

$$\sin\theta = 0.6$$
$$\cos\theta = 0.8$$
$$T\cos\theta = \text{Balanced}$$
$$T\sin\theta = 30 \text{ N}$$

Tension, $T \times 0.6 = 30$ N, $T = 50$ N

2.15 (b)

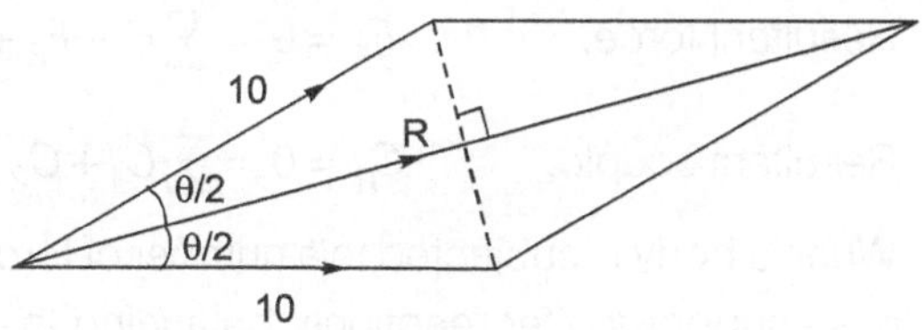

Fig. 2.39

$$R = \left(10\cos\frac{\theta}{2}\right)2 = 20\cos\frac{\theta}{2}$$

■■■■

03

CHAPTER

Equilibrium of Rigid Bodies

3.1 Introduction

The study of equilibrium conditions of rigid bodies in static condition, when there is neither resultant moment nor any resultant force on a rigid body, such as trusses, bridges, structures, beams, cable-suspensions all remain stationery when forces and or moments are applied on them, is the subject matter of this chapter. For equilibrium of any rigid body.

Resultant force, $F_R = 0 = \sum F_1 + F_2 + F_3 \ldots F_n,$ forces, vectorially

Resultant couple, $C_R = 0 = \sum C_1 + C_2 + C_3 \ldots C_m,$ couples, vectorially

When a body is subjected to a number of forces and or moments, the body is supported on some supports and these supports offer reactions, balancing the applied forces. These support reactions must be analysed completely for the proper and efficient design of foundations or supports, because if the support is not designed properly it may fail or it may sink.

To determine the unknown support reactions or unknown external forces, equations of equilibrium are established along *x-y-z* coordinate axes. There are following six scalar equations:

$$\sum F_x = 0, \quad \sum F_y = 0, \quad \sum F_z = 0$$

$$\sum M_x = 0, \quad \sum M_y = 0, \quad \sum M_z = 0$$

These equations state the fact that components of external forces along *x*, *y*, *z* directions acting on the body are balanced. Similarly components of external moments along *x*, *y*, *z* directions acting on the body are also balanced.

Following steps are followed in the analysis of a rigid boy in equilibrium:

(*a*) Identify the direction of each reaction offered by supports.

(*b*) *Isolate the rigid body from surroundings and show reactions at each point of support.*

(*c*) Write the equations of equilibrium, necessary for taking into account external forces/couples applied and support reactions/moments of each support.

(*d*) Evaluate the unknowns with the help of equations of equilibrium.

Isolation of a rigid body from surroundings (supports) and replacing the supports by support reactions shown on the body is called a free body diagram. Remember that an incorrect free body diagram will result in a wrong solution.

3.2 Support Reactions

The direction and type of reactions depend upon the nature of support provided to a rigid structure. Generally following types of supports are provided to a structural member:

(a) *Roller support*
(b) *Roller support* in a guided path
(c) Rocker at the end of a member oscillating on the *surface*
(d) Rounded end of a member against a *frictionless surface.*
(e) Rounded end of a member against a rough surface.
(f) Hinged support
(g) Fixed support.

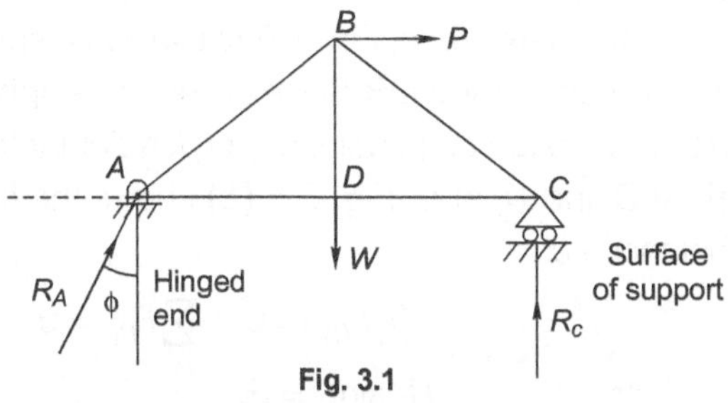

Fig. 3.1

Fig. 3.1 shows a truss *ABCD* supported on (*i*) a hinged support at *A* and (*ii*) a roller support at *C*.

At the roller support, reaction is always *perpendicular to the surface of the support*, as shown by reaction R_C at *C*. At the hinged support, (end of the structure is connected to the support with the help of a pin joint), the reaction can be horizontal, vertical or inclined as shown by reaction R_A, depending upon the type of the load. In the case shown, the reaction is inclined to vertical plane, because there are horizontal and vertical loads on the structure (as shown by *P* and *W*).

Fig. 3.2 (a) shows a roller at the end of a member sliding in *a guided path* (a frictionless slot) reaction will be perpendicular to the surface of the path at the point of contact. **Fig. 3.2 (b)** shows a rocker at the end of a member oscillating on a surface. Reaction will be perpendicular to the surface at the point of contact of rocker on surface.

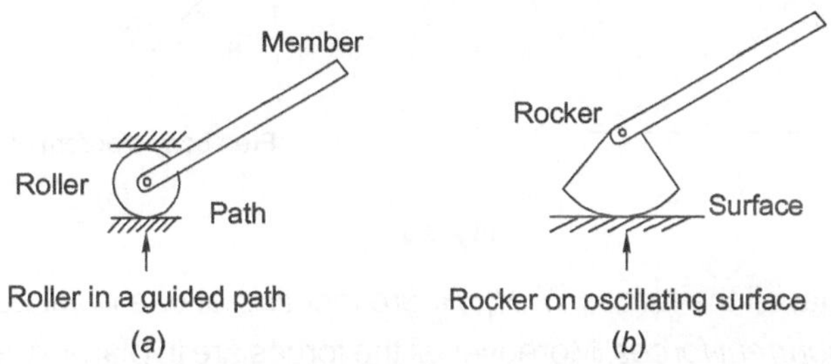

Fig. 3.2

Fig. 3.3 (a) shows a rounded end member supported on a **frictionless surface**, and the reaction will be perpendicular to the surface as shown. **Fig. 3.3 (b)** shows a rounded end member supported on *a* ***rough surface***, in this case *reaction will be inclined to the vertical axis* as shown because of frictional force that comes into play when there is sliding motion of member on rough surface.

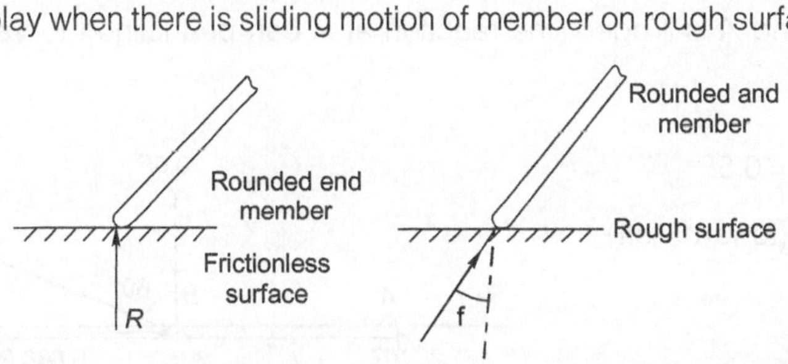

Fig. 3.3

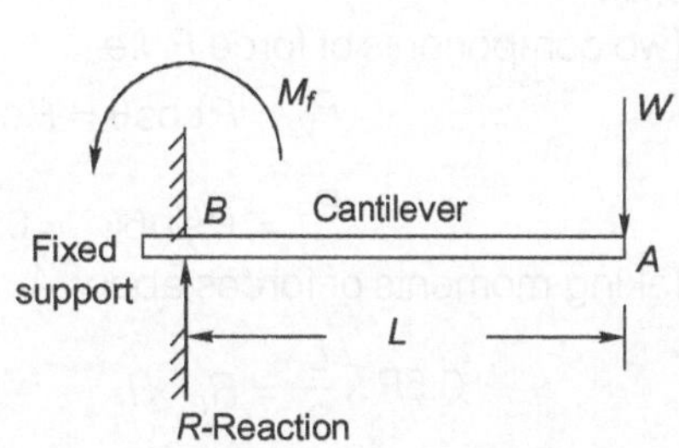

Fig. 3.4

A fixed support, supporting the end *B* of a cantilever beam is shown in Fig. 3.4. This type of support offers a reaction, *R* and a fixing couple, M_f to maintain the equilibrium of the cantilever beam.

3.3 Free body Diagrams

While solving the problems in Mechanics involving equilibrium, the first step towards solution is to draw the *free body diagram i.e.*, body is detached from the supports and reactions from the supports are shown on the body, the direction of reactions depending upon the type of support as already discussed in Article 3.2.

As an example, **Fig. 3.5 (*a*)** shows a sphere of weight W acting at its CG *i.e.*, point G, supported between two planes AO and OB. Since the object is spherical, there will be a reaction at point B, perpendicular to the plane AO and a reaction at point C, perpendicular to the plane OB. Planes AO and OB are now replaced by reactions R_1 at B and R_2 at C. **Fig. 3.5 (*b*)** shows the free body diagram of the sphere. For equilibrium, the forces must balance *i.e.*,

$$\sum F_{Rx} = 0, \quad \sum F_{Ry} = 0$$

or
$$R_2 \sin\theta = R_1$$
$$R_2 \cos\theta = W$$

or Reaction, $$R_2 = \frac{W}{\cos\theta} = W\sec\theta$$

Reaction $$R_1 = R_2 \sin\theta = \frac{W}{\cos\theta} \times \sin\theta = W\tan\theta$$

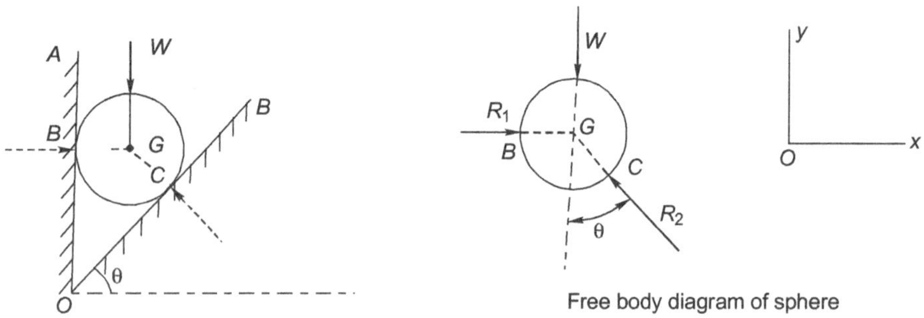

Free body diagram of sphere

(*a*) (*b*)

Fig. 3.5

The reader may note that all the forces W, R_1, R_2 are meeting at one point *i.e.*, G, *i.e.*, centre of gravity of sphere, these are called *concurrent forces*. Moreover all the forces are in plane xy as shown, these are coplanar forces also. The above example shows the reactions from roller support, *i.e.*, perpendicular to the plane of the support.

Taking another example of a beam ABC of length L, hinged at end A and roller supported at end C, with an inclined load P at B. **Since C is roller supported there will be only reaction perpendicular to horizontal surface, on which the roller is placed**; but end A is hinged, the reaction at A can be inclined to **vertical (Fig. 3.6).**

Two components of force P, *i.e.*,

$$P_V = P\cos\theta = P\cos 60^\circ = 0.5P$$

$$P_H = P\sin 60^\circ = \overleftarrow{0.866P} \text{ (towards left)}$$

Taking moments of forces about A

$$0.5P \times \frac{L}{2} = R_C \times L$$

Reaction, $$R_C = 0.25P \uparrow$$

Reaction, $$R_{AV} = 0.5P - 0.25P = 0.25P \uparrow$$

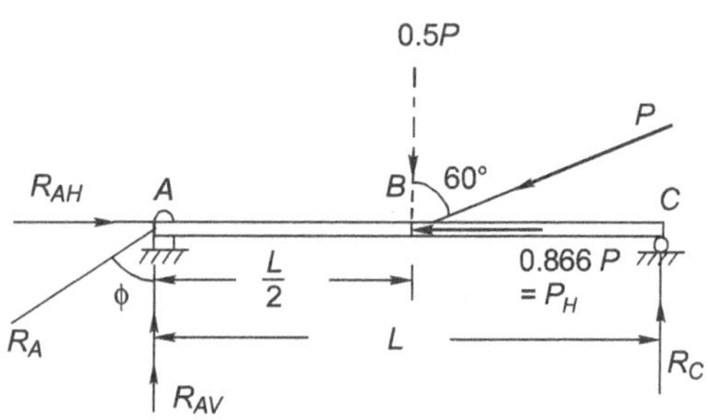

Fig. 3.6

Horizontal reaction at A to balance $0.866P$, *i.e.*, horizontal component of P.

So $$R_{AH} = \overrightarrow{0.866P}$$

Resultant reaction $$R_A = \sqrt{R_{AH}^2 + R_{AV}^2} = \sqrt{(0.866^2 + 0.25^2)}\,P = 0.90P$$

Angle of inclination of R_A,

$$\tan\varphi = \frac{R_{AH}}{R_{AV}} = \frac{0.866P}{0.25P} = 3.464, \text{ angle, } \varphi = 73.9°$$

Example 3.1 A cylinder is supported by a bar and a string as shown in **Fig. 3.7 (*a*)**. Draw the free body diagram of the cylinder and bar. Only mark the direction of reactions and applied forces on bar and cylinder.

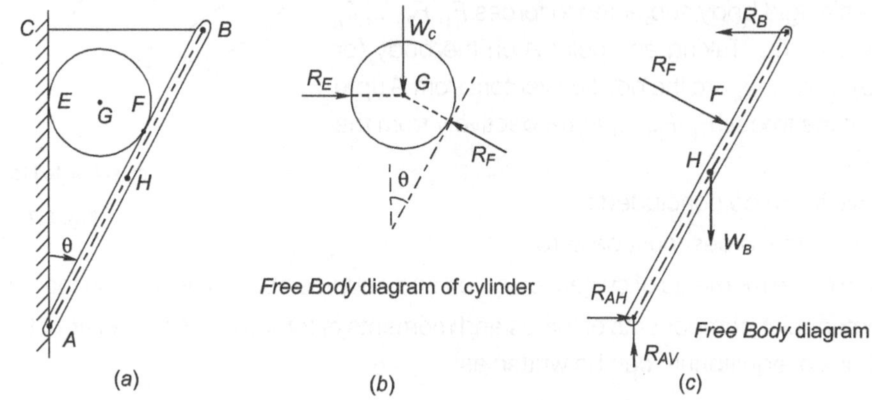

Fig. 3.7

Solution The bar is hinged at A, supports the cylinder and cylinder touches the vertical surface at E and bar at F, with the help of string BC (horizontal). There will be tension in string BC.

Cylinder: Weight of cylinder, W_C acts at its CG *i.e.*, G.

Reaction at E is horizontal as surface is vertical.

Reaction at F, is perpendicular to surface of bar.

Bar: End A is hinged, there are two components of reaction R_A *i.e.*, R_{AV} and R_{AH} as shown. Weight of bar W_B acts at its CG *i.e.*, H, in vertical direction as shown in **Fig. 3.7 (*c*)**. Pull in the string at B *i.e.*, R_B. Reaction of the cylinder on bar is equal and opposite to R_F (on cylinder), perpendicular to bar surface at F.

Exercise 3.1 A uniform heavy plate of rectangular shape of mass M, is supported in vertical plane by a cable at B and it is hinged at A. **Fig. 3.8** shows, CG of the plate at G. Draw free body diagram of rectangular plate.

[**Ans:** R_{AV}, R_{AH}; $M_g \downarrow$; Tension R_B along cable].

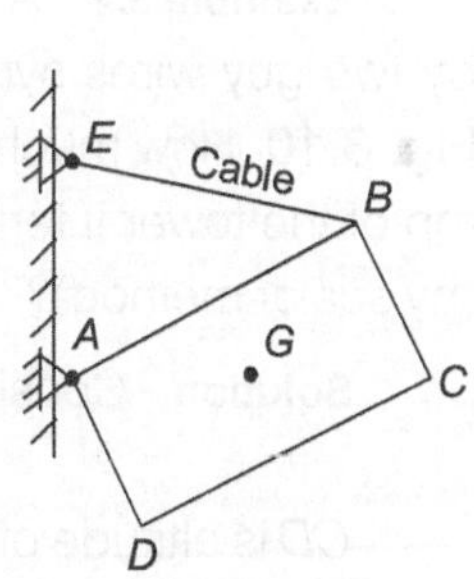

Fig. 3.8

3.4 Conditions of Equilibrium

A rigid body subjected to a number of forces $F_1, F_2, F_3, \ldots, F_n$ and couples $C_1, C_2, C_3, \ldots, C_m$ is in equilibrium if following conditions are satisfied:

(*i*) Resultant of all forces is zero *i.e.*, $F_R = 0$;

$$\sum_{i=1}^{n} F_i = 0 \qquad \ldots(1)$$

(*ii*) Resultant of all the moments about any point is zero

$$\sum_{i=1}^{n} r_i \times F_i + \sum_{i=1}^{m} C_i = 0$$

where r_i is the **position vector from the point upto the line of action of the respective force F_i**

or $\quad F_1 + F_2 + F_3 \ldots F_n = 0$

$$(r_1 \times F_1 + r_2 \times F_2 + r_3 \times F_3 + \ldots r_n \times F_n) + \ldots (C_1 + C_2 \ldots C_m) = 0$$

Above two conditions are sufficient and necessary for a body to be in equilibrium. So there are only *two independent vector equations of equilibrium for any single free body.*

Fig. 3.9 shows a rigid body subjected to forces $F_1, F_2, \ldots, F_n$ and couples $C_1, C_2, \ldots, C_m$. Taking any point A on the body (or anywhere in space) $r_1, r_2, \ldots, r_n$ are the position vectors from A upto the lines of action of the forces $F_1, F_2, \ldots, F_n$ respectively from the point A.

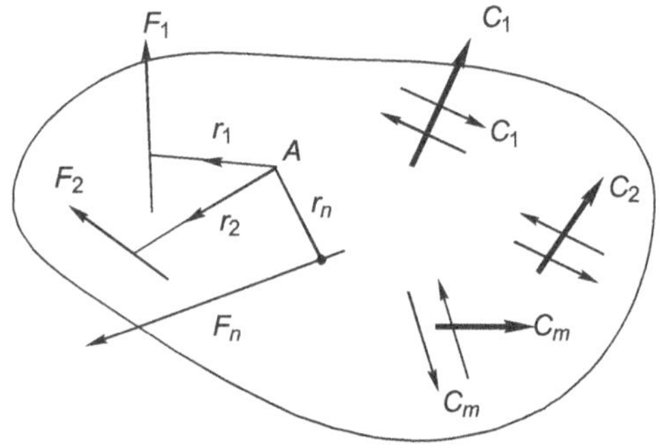

Body subjected to forces and couples

Fig. 3.9

From the above it can be concluded that:

(*i*) Vector sum of the forces must be zero.

(*ii*) Vector sum of the moments of system of forces and couples about any point on the body must be zero.

Taking the rectangular components of forces and moments of forces and couples about any point on body or in space, equations of equilibrium can be written as:

$$\sum_{i=1}^{n} (F_x)_i = 0, \quad \sum_{i=1}^{n} (F_y)_i = 0, \quad \sum_{i=1}^{n} (F_z)_i = 0$$

$$\sum_{i=1} (M_x)_i = 0, \quad \sum_{i=1} (M_y)_i = 0, \quad \sum_{i=1} (M_z)_i = 0$$

These are the equations of equilibrium in a general case and not more than six unknown quantities (*i.e.*, forces and couples) can be solved by the method of statics for a single free body.

Example 3.2 A vertical tower of height 20 m is supported by two guy wires symmetrically (of same length) as shown in Fig. 3.10. How much horizontal force, F can be applied on the top of the tower if tensions in guy wires is not to exceed 30 kN (by scalar method)?

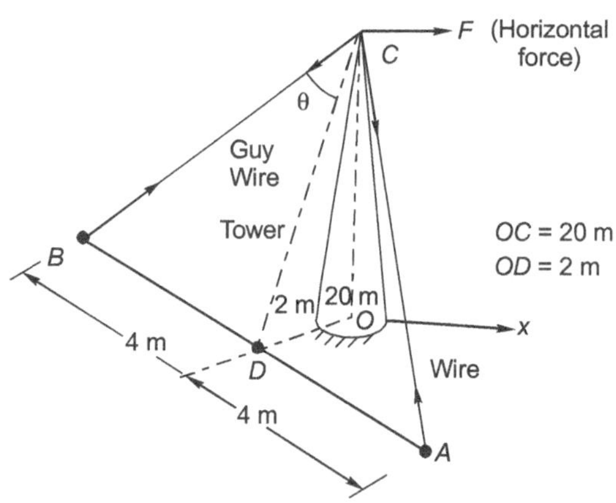

Fig. 3.10

Solution Consider plane *CBDA*,

$$CB = CA = \text{length of guy wires.}$$

CD is altitude of triangle *CBDA*, where

$$CD = \sqrt{CO^2 + OD^2} = \sqrt{20^2 + 3^2}$$

$$\sqrt{409}\text{ m} = 20.22\text{ m}$$

Fig. 3.11 shows plane *CBDA*, in which

$$BD = DA = 4\text{ m}$$

$$CD = 20.22\text{ m}$$

$$\tan\theta = \frac{4}{20.22} = 0.1978$$

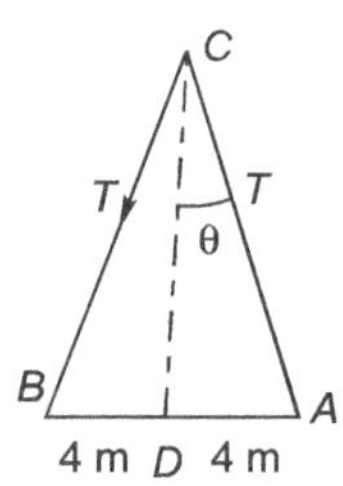

Plane *CBDA*

Fig. 3.11

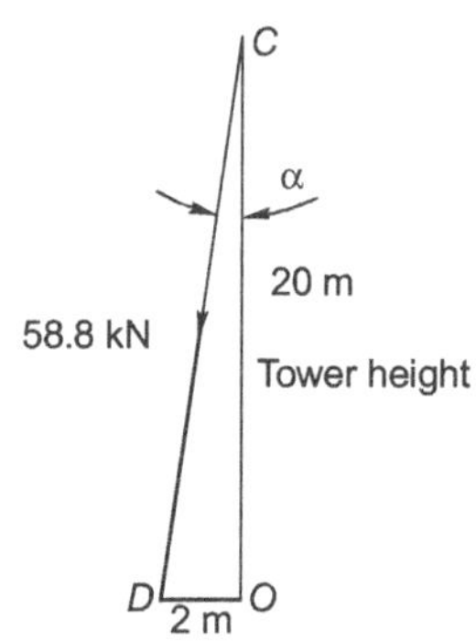

Fig. 3.12

$$\theta = 11.19^\circ$$

$$\cos\theta = 0.980$$

Component of tensions along *CD* is added,

$$T_{CD} = 2 \times T \times \cos\theta = 2 \times 30 \times 0.980 = 58.8 \text{ kN}$$

Now let us consider triangle *CDO*, in which

$$\tan\alpha = \frac{2}{20} = 0.1$$

$$\alpha = 5.71^\circ$$

$$\sin\alpha = 0.0995$$

Component of 58.8 kN along horizontal direction

$$= 58.5 \times \sin\alpha = 5.85 \text{ kN}$$

A horizontal force of 5.85 kN can be applied at point *C* of tower.

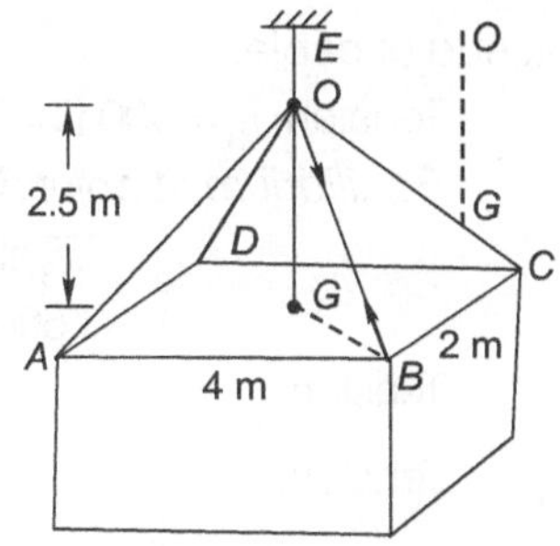

Fig. 3.13

Exercise 3.2 A rectangular block weighing 10 kN is held by 4 cables as shown in **Fig. 3.13**. All 4 cables are identical in size and identically connected at ends of the block. Find the tension in cables.

[**Ans:** $T_{OA} = T_{OB} = T_{OC} = T_{OD} = 3.35$ kN]

[**Hint:** Consider the triangle *OGB*, vertical component of tension in each cable is $\frac{10}{4} = 2.5$ kN].

3.5 System of Concurrent Coplanar Forces

In a coplanar force system, when forces are concurrent (lines of action of forces meeting at one point), there are only two independent equations. If *xy* is the plane of concurrent forces, then

$$\sum_{i=1}^{n} (F_i)_x = 0, \quad \sum_{i=1}^{n} (F_i)_y = 0$$

or $$F_{1x} + F_{2x} + F_{3x} \ldots F_{nx} = 0 \qquad \ldots(1)$$

$$F_{1y} + F_{2y} + F_{3y} \ldots F_{ny} = 0 \qquad \ldots(1)$$

Summation of components of forces in *x*-direction and summation of components of forces in *y*-direction are independently zero.

For example, a ball is suspended from a string, and ball is pulled in one direction by a force *P*, then three forces *i.e.*, *mg*, weight of the ball, pull applied *P* and tension *T* in the string meet at point *G*, centre of the ball as shown in the **Fig. 3.14 (*a*)**.

For equilibrium

$$T\sin\theta = P \qquad \ldots(1)$$

$$T\cos\theta = mg \qquad \ldots(2)$$

If θ and *mg* are known, then

$$\frac{P}{mg} = \tan\theta$$

Pull, $$P = mg\tan\theta \qquad \ldots(3)$$

$$T = \frac{P}{\sin\theta} = \frac{mg\tan\theta}{\sin\theta}$$

$$= mg\sec\theta \qquad \ldots(4)$$

Unknown forces *P* and *T* are determined).

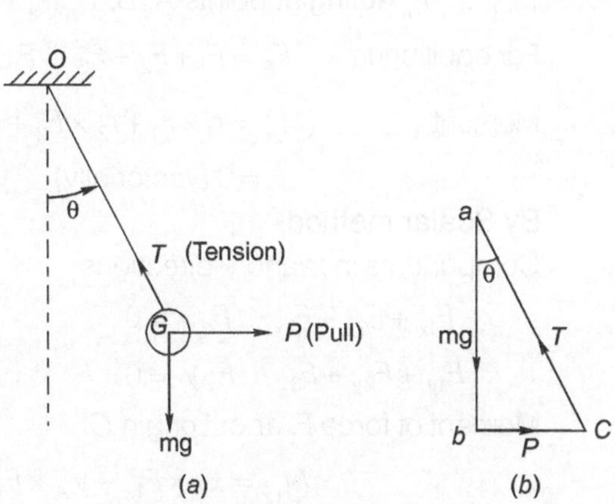

Fig. 3.14

Alternative method to determine unknown forces is through force polygon at point *G*, force polygon must close because ball is at rest. To some suitable scale, take $ab = mg$ in a vertical direction *i.e.*, parallel to force *mg*. From *a* draw a line parallel to *OG* (representing tension *T*) and from *b* draw a line *bc* parallel to force *P*. Triangle *abc* is known as *force polygon* at point *G* of ball as shown in **Fig. 3.14 (*b*).**

From the scale taken for *mg*, side *bc* represents force *P* and side *ca*, taken in order (*c* to *a*) as shown, represents tension *T* in the string.

Example 3.3 A ring *B* is held by cables *AB*, *BC* and *BD* (passing over a pulley) weight applied at the end of cable *BD* is 500 N. Determine tensions T_1, T_2 and T_3 **(Fig. 3.15).**

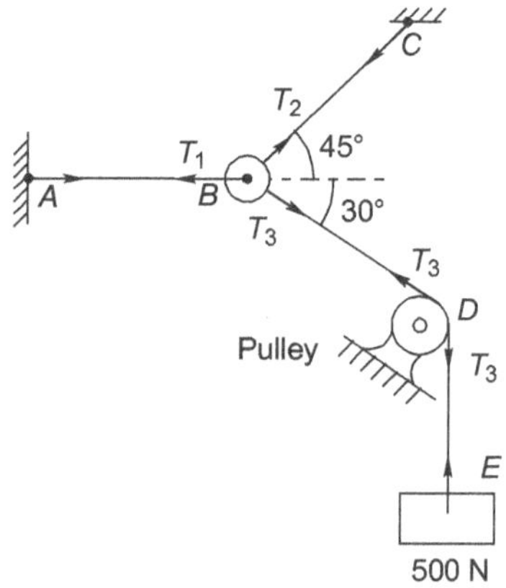

Fig. 3.15

Solution Same cable *BD* passes over pulley and a load 500 N is applied at end of cable.

Tension $T_3 = 500$ N = Tension in part *DE* of same cable.

Equilibrium at point B

$$T_3 \sin 30^\circ = T_2 \sin 45^\circ$$

$$500 \times 0.5 = T_2 \times 0.707$$

Tension $\quad T_2 = 353.6$ N

Tension, $\quad T_1 = T_3 \cos 30^\circ + T_2 \cos 45^\circ$

$$= 500 \times \cos 30^\circ + 353.6 \cos 45^\circ$$

$$= 500 \times 0.866 + 353.6 \times 0.707$$

$$= 433 + 250 = 683 \text{ N}.$$

Exercise 3.3 A ball of weight 500 N rests upon a smooth horizontal plane. Two strings *AB* and *AC* are attached to the centre of the ball as shown in **Fig. 3.16**. If the string *AB* is horizontal, determine angle α of string *AC* with the horizontal. What is the reaction of the smooth plane on ball?

[**Hint:** Vertical force at *A* is weight of the ball minus reaction at *D*]

[**Ans:** $\alpha = 70.55^\circ$, $R_D = 217.12$ N].

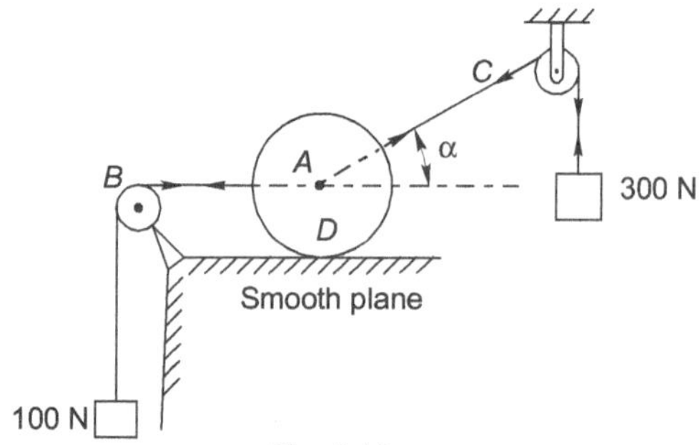

Fig. 3.16

3.6 Coplanar Force Systems

In a coplanar force system, all the forces acting on a body are in one plane, as shown in **Fig. 3.17**. Say forces are $F_1, F_2, F_3, \ldots, F_n$ acting at points *A*, *B*, *C*, ..., *N* as shown.

For equilibrium $\quad F_R = F_1 + F_2 + F_3 \ldots F_n = 0$

Moment $\quad (M_z) = r_1 \times F_1 + r_2 \times F_2 + r_3 \times F_3 \ldots r_n \times F_n$, about z-axis

$= 0$ (vectorially)

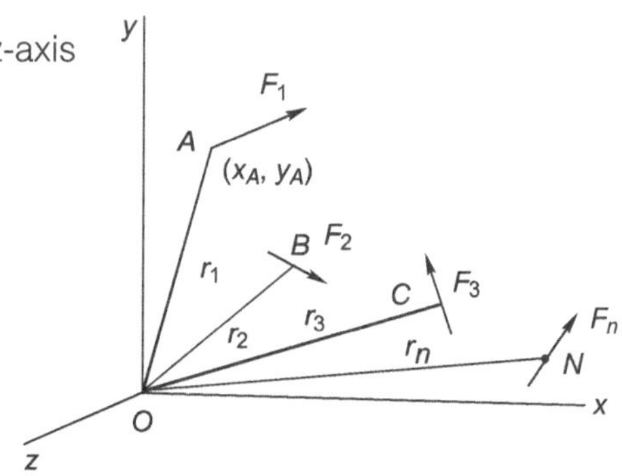

Fig. 3.17

By Scalar method

Components in *x*- and *y*-directions

$$F_{1x} + F_{2x} + F_{3x} \ldots F_n x = 0$$

$$F_{1y} + F_{2y} + F_{3y} \ldots F_x y = 0$$

Moment of force F_1 about origin *O*

$$M_{1z} = x_A \times F_{1y} - y_A \times F_{1x}$$

(anticlockwise moments are positive)

Similarly $M_{2z} = x_B \times F_{2y} - y_B \times F_{2x}$

$M_{3z} = x_c \times F_{3y} - y_c \times F_{3x}$

$M_{nz} = x_n \times F_{ny} - y_n \times F_{nx}$

For Equilibrium, moments $M_{1z} + M_{2z} + M_{3z} \ldots M_{nz} = 0$

Please note that plane of moments is, *xy* and *z* is the axis of moment vector, along which moment vector can be represented.

Example 3.4 A corner plate *ABDC* is hinged at end *A*, and roller supported at *C*. A force of 500 N acts at point *B* as shown in **Fig. 3.18**. Find reactions at supports.

Solution Corner plate is subjected to a force of 500 N at point *B*, reactions at supports *C* and *A* will be developed to maintain equilibrium, *C* is a roller support, therefore reaction at *C* will be perpendicular to vertical surface of support *i.e.*, reaction R_C will be in horizontal direction. Line of application of reaction R_C and force 500 N meet at point *O*, 0.2 m away from *C*. *These are concurrent forces, therefore line of action of reaction R_A will pass through O. Therefore*

$$\tan\alpha = \frac{0.4}{0.6} = 0.667; \quad \alpha = 33.70°$$

Now $R_A \times \sin\alpha = R_C$

$R_A \cos\alpha = 500\text{ N}$

$$R_A = \frac{500}{\cos 33.7} = \frac{500}{0.832} = 601\text{ N}$$

(as shown in **Fig. 3.18**)

Reaction, $R_C = 601 \times \sin\alpha = 601 \times \sin 33.7°$

$= 601 \times 0.5548 = 333.5\text{ N}$

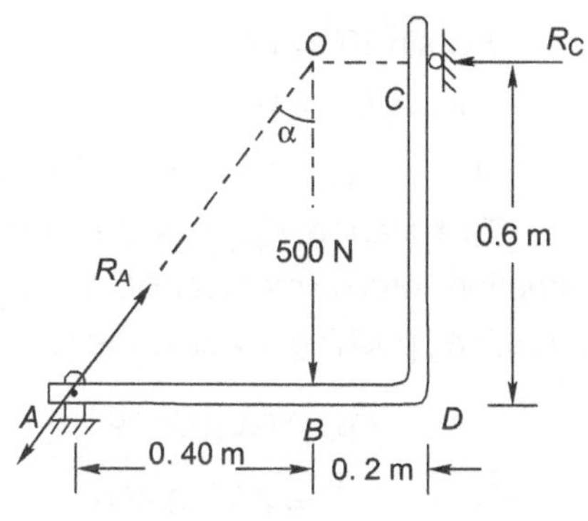

Fig. 3.18

Exercise 3.4 A beam *ABC*, 4 m long is hinged at end *A* and roller supported at *C*, is subjected to an inclined force of 1 kN at point *B* as shown in **Fig. 3.19**. Determine reactions at supports.

[**Hint:** Reaction at *C* is vertical, line of action of 1 kN force and reaction R_C meet at some point *O*. Reaction at *O* will be in line *AO*].

[**Ans:** R_C = 353.5 N↑, R_A = 790.8 N; inclination of R_A with horizontal = 26.56°].

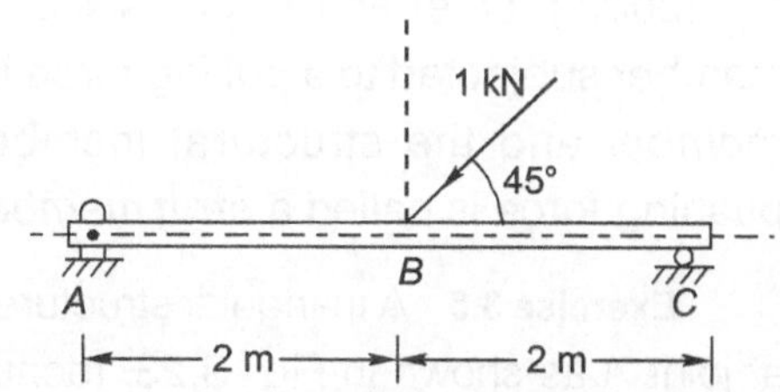

Fig. 3.19

3.7 Equilibrium of a Two Force Body

When a body is subjected to only two forces and it is in equilibrium, then it is termed as a *two force body*. **Fig. 3.20 (*a*)** shows a body subjected to two forces F_A and F_B at points *A* and *B* respectively. The lines of action of the forces intersect at point *C*, therefore, sum of the moments of forces at the point *C* is zero. But the two forces are not in one line, **therefore there is resultant of these two forces which is not zero. Therefore the body is not in equilibrium.**

At the point *A*, moment due to force F_A is zero (because force F_A is passing through *A* and for equilibrium moment of force F_B about point *A* should also be zero, **this means that force F_B must pass through point *A*.** Similarly taking moments about point *B*, we can prove that force F_A must also pass through point *B* for equilibrium. **In other words,** *both forces F_A and F_B must have the same line of action i.e., AB*, as shown in **Fig. 3.20 (*b*)**.

Moreover for equilibrium

Forces, $F_{Ax} + F_{Bx} = 0$

Forces, $F_{AY} + F_{By} = 0$

which puts the condition that both forces must have same magnitude but opposite in sense (direction).

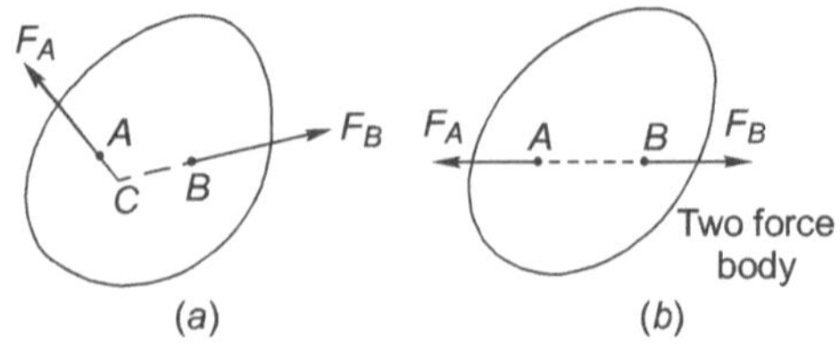

Fig. 3.20

Therefore for equilibrium of a two-force body, the two forces must be collinear (*i.e.*, same line of action), same magnitude but opposite in sense so that resultant, $F_R = 0$.

Example 3.5 Fig. 3.21 **(*a*)** shows a simple structure *ACB*, supporting a load *W* at joint *C*. There are two members *AC* and *BC* having two forces each. Identify these forces.

Solution Fig. 3.21 **(*b*)** shows forces at the joint *C* of structure and joint is in equilibrium. Force in member *AC* must be a *pulling force* at *C*.

$$F_{AC} \sin 30^\circ = W$$

$$F_{AC} \times 0.5 = W$$

or $\quad F_{AC} = 2W$ (internal force)

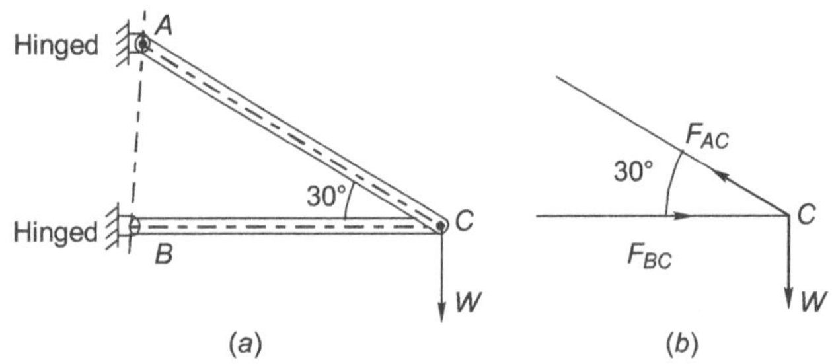

Fig. 3.21

To balance F_{AC} cos 30° force in horizontal direction, force in member *BC*, *i.e.*, F_{BC} must be a *pushing force*, *i.e.*, pushing towards the joint *C*.

$$F_{BC} = F_{AC} \cos 30^\circ$$

$$= 2W \times 0.866$$

$$= 1.732\ W \text{ (internal force)}$$

Fig. 3.22 shows members *AC* and *BC* with forces F_{AC} (pulling force) and F_{BC} (pushing force). **Structural member subjected to a pulling force is termed as a *tie member* and the structural member subjected to pushing force is called a *strut member*.**

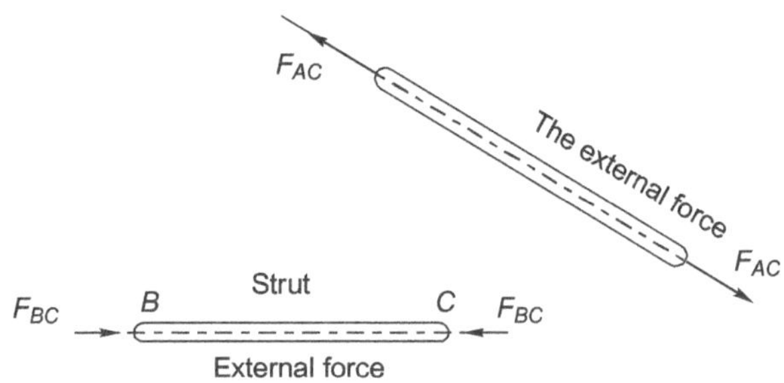

Fig. 3.22

Exercise 3.5 A triangular structure *ABC* supports a load *W* at joint *A* as shown in Fig. 3.23. Identify the type of forces in members *AB* and *AC*. Determine the magnitude of these forces.

[**Ans:** $F_{AB} = W$ (Pulling force),

$F_{AC} = 1.732$ W (pushing force)

[**Hint:** (Horizontal component of forces in *AB* and *BC* must balance or draw a force polygon at joint *A*)].

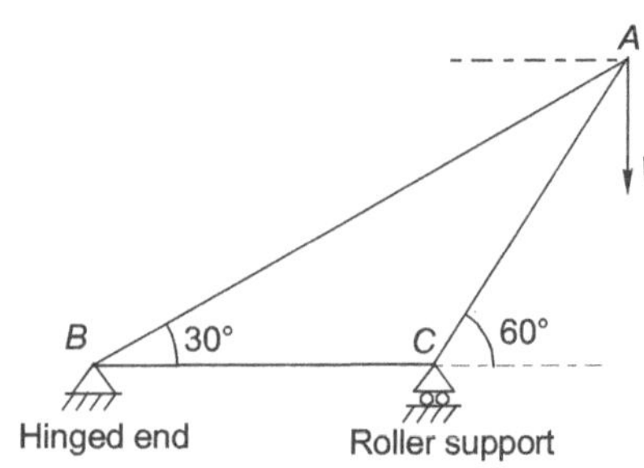

Fig. 3.23

3.8 Equilibrium of a Three Force Body

A three force body is of utmost importance, because in many engineering applications, only 3 forces are applied and the body remains in equilibrium.

A rigid body subjected to **3 forces at three points is known as three force body and a number of engineering applications fall into this category.**

Consider a body subjected to 3 forces F_1, F_2 and F_3 at points *A*, *B* and *C* respectively as shown in **Fig. 3.24**. Say lines of action of F_1 and F_2 intersect at point *D*, for equilibrium, the moment of force F_3 should be zero about *D*. **Therefore line of action of force F_3 also must pass through the point *D***, or *for equilibrium, the lines of action of the three forces acting on a body must be concurrent.* However this statement refers to only *nonparallel set of forces.*

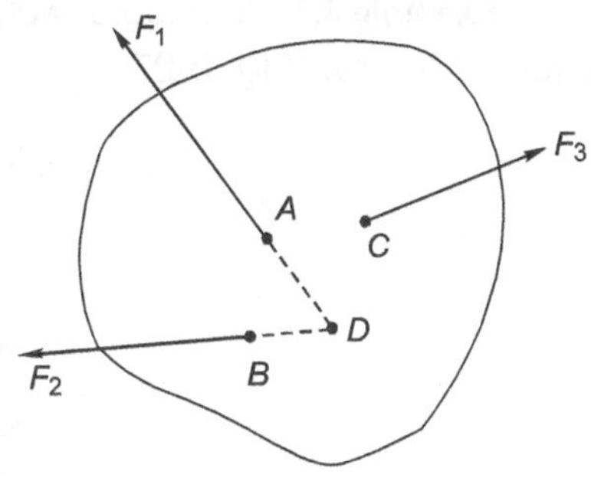

Fig. 3.24

Exercise 3.6 A force $P = 150$ N is applied at end *A* of a rod *ABC*, which is supported by a pin and bracket at *C* and rests against a frictionless peg at *B* as shown in **Fig. 3.25**. Force *P* is vertical. Determine reactions R_B and R_C.

[**Hint:** Reaction at *B* will be perpendicular to the surface on which peg is fitted. Lines of action of R_C, R_B and force *P* meet at one point].

[**Ans:** $R_C = 150$ N, inclination, with horizontal = –26°

$R_B = 159$ N inclined to horizontal at 32°].

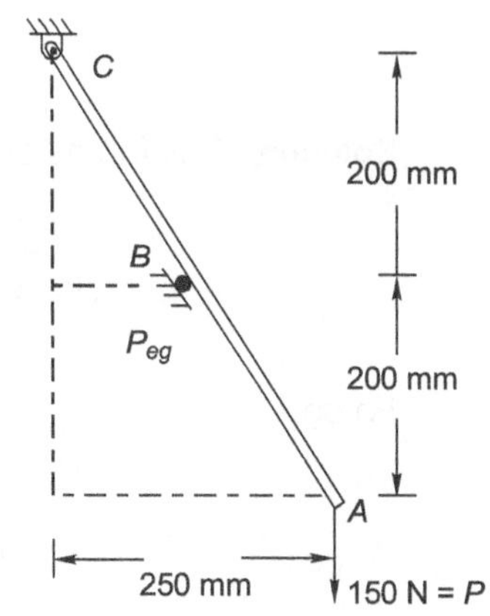

Fig. 3.25

3.9 Lami's Theorem

Lami's theorem is important in the solution of unknown forces and their directions in the case of a 3 force body. Lami's theorem states that if the forces acting at a point of a body are in equilibrium, then *each force is proportional to the sine of the angle between other two forces* or sine of angle opposite to the force under consideration. Figure 3.26 shows three forces F_1, F_2 and F_3 acting at point *O* of a body. Angles between F_2 and F_3, between F_3 and F_1 and between F_1 and F_2 are respectively α, β and γ as shown in the figure, then as per Lami's theorem.

$$\frac{F_1}{\sin\alpha} = \frac{F_2}{\sin\beta} = \frac{F_3}{\sin\gamma}$$

The body is in equilibrium, the resultant of these forces will be zero, or the resultant of forces F_2 and F_3 will be in line with the force F_1 and equal in magnitude of F_1 and opposite in the direction of F_1, as shown in **Fig. 3.26.**

In triangle *OAC*, $$\frac{OA}{\sin\angle OCA} = \frac{OC}{\sin\angle OAC} = \frac{AC}{\sin\angle AOC}$$

$$\frac{F_2}{\sin(180-\angle OCA)} = \frac{F_1}{\sin(180-\angle OAC)} = \frac{BC}{\sin(180-\angle AOC)}$$

$$\angle OAC + \angle AOB = 180^\circ$$

$$\angle AOB = 180^\circ - \angle OAC = \text{angle } \alpha$$

Similarly angle $$\beta = 180^\circ - \angle OCA$$

$$\gamma = 180^\circ - \angle AOC$$

or $$\frac{F_1}{\sin\alpha} = \frac{F_2}{\sin\beta} = \frac{F_3}{\sin\gamma}.$$

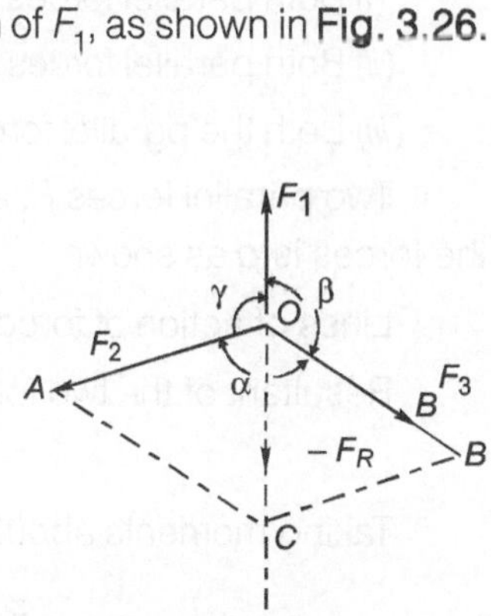

Fig. 3.26

Example 3.6 In a tug of war, team *A* pulls with a force of 1800 N, how much forces must teams *B* and *C* exert for a draw **(Fig. 3.27)?**

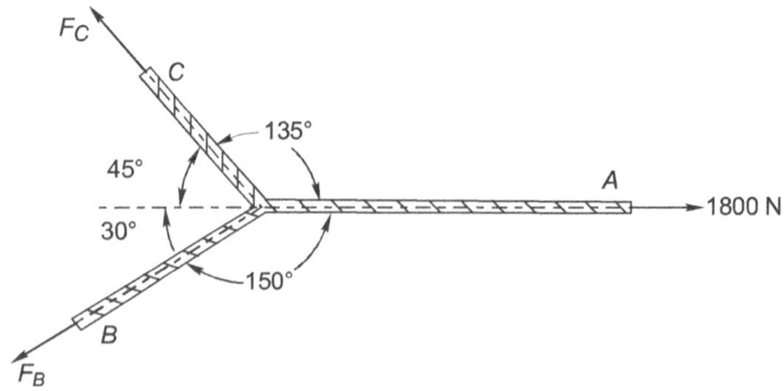

Fig. 3.27

Solution Say forces applied by teams *B* and *C* are F_B, F_C, then using Lami's theorem

$$\frac{F_B}{\sin 135°} = \frac{F_C}{\sin 150°} = \frac{1800\text{ N}}{\sin 75°}$$

Force, $$F_B = 1800 \times \frac{\sin 135°}{\sin 75°} = \frac{1800 \times 0.707}{0.966} = 1317.5\text{ N}$$

Force, $$F_C = 1800 \times \frac{\sin 150°}{\sin 75°} = \frac{1800 \times 0.5}{0.966} = 931.7\text{ N.}$$

Exercise 3.7 A lamp shade weighing 200 N is hung from two strings *AB* and *BC* connected to the wall and ceiling of the room respectively as shown in **Fig. 3.28**. Using Lami's theorem determine tension in strings *AB* and *BC*.

[**Ans:** $T_{AB} = 10$ N, $T_{BC} = 17.32$ N].

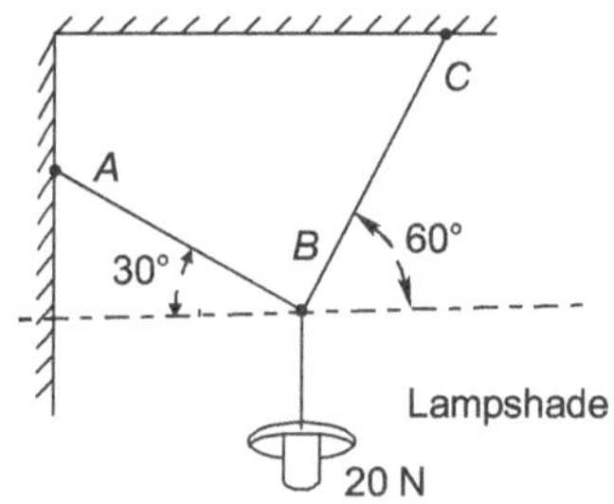

Fig. 3.28

3.10 Two Parallel Forces

Let us consider the effect of two parallel forces on a rigid body. Three cases may occur in such a situation:

(*i*) Both parallel forces are in the same direction.

(*ii*) Both parallel forces are unequal in magnitude and acting in opposite directions.

(*iii*) Both the parallel forces are equal in magnitude but opposite in direction (forming a couple).

Two parallel forces P_1 and P_2 acting on a body are shown in **Fig. 3.29**. Say $P_1 > P_2$, and distance between the forces is *d* as shown.

Lines of action of forces are *AC* and *BD* respectively.

Resultant of the two forces,

$$F_R = P_1 + P_2$$

Taking moments about *A*,

$$P_2 \times d = (P_1 + P_2)\, a = P_R . a$$

Distance, $$a = \frac{P_2 . d}{P_1 + P_2} \quad ...(1)$$

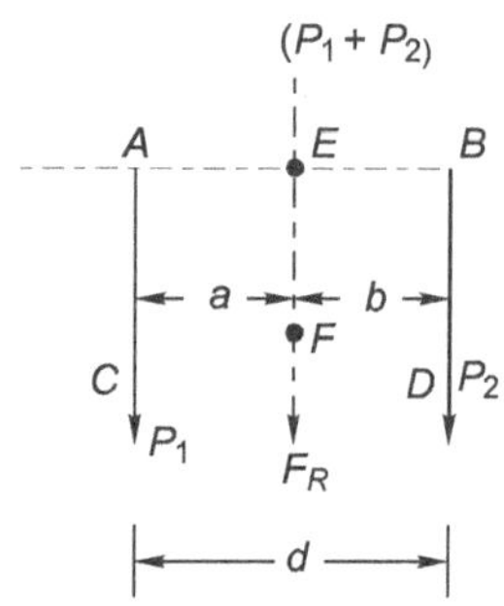

Fig. 3.29

Similarly taking moments about point *B*,

$$P_1 \times d = (P_1 + P_2)\, b$$

or distance,
$$b = \frac{P_1 . d}{P_1 + P_2} \qquad ..(2)$$

From Equations (1) and (2)

$$\frac{b}{a} = \frac{P_1}{P_2} \text{ or } \quad b > a \text{ if } P_1 > P_2$$

Line of action *EF* of F_R divides the distance between the forces P_1 and P_2 in the ratio of $a : b$ where $a < b$ and $P_1 > P_2$. So the line of action of resultant force divides the distance between the lines of application of forces in the ratio *inversely proportional to their magnitude.*

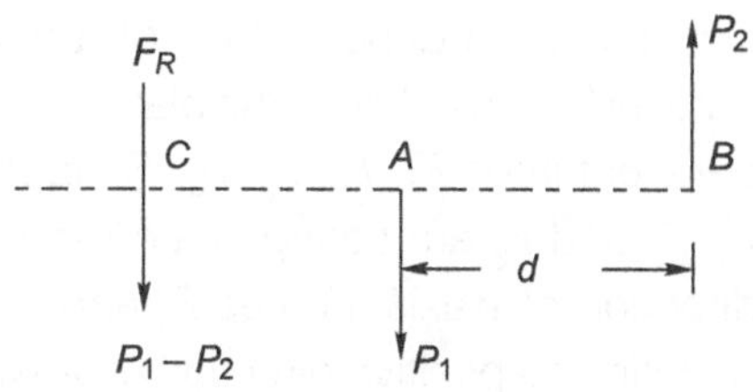

Fig. 3.30

Two parallel forces P_1 and P_2 acting at point *A* and *B* respectively are in opposite directions as shown in **Fig. 3.30**. Let us locate point *C* as moment centre such that

$$P_1 CA - P_2 CB = 0$$

$$\frac{CA}{CB} = \frac{P_2}{P_1} \qquad ...(3)$$

Again the distances of forces P_1 and P_2 from the line of action of the resultant are inversely proportional to their magnitudes but the line of action of the resultant *lies outside the space* between the forces (*i.e.*, distance *d*), distance *CB* > *AB*, or *CB* > *d*, and the *line of action of the resultant lies on the side of the larger force.*

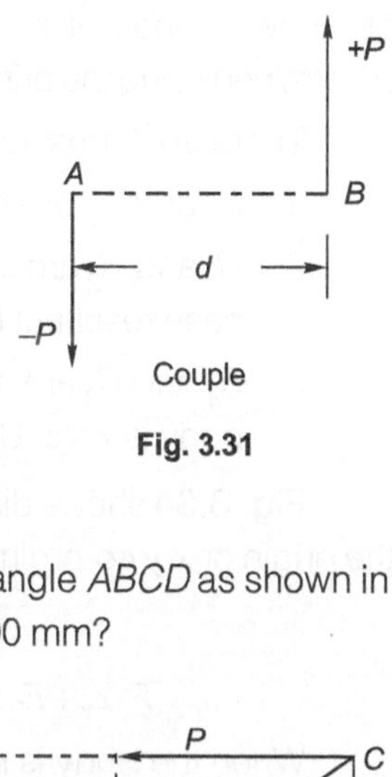

Fig. 3.31

Two equal and opposite parallel forces *P* and –*P* are shown in **Fig. 3.31**. Two such forces form a *couple* of moment *P* × *d* (arm) *ccw*, a positive couple. This system cannot be reduced to a single force. *Algebraic sum of the moments of the forces (forming a couple) is independent of the position of the moment centre.*

Couple, *C* = *P* . *d*, an anticlockwise as shown in figure is a positive couple. There has been comprehensive discussion on a couple in Chapter 1.

Example 3.7 Four forces *P* = 30 N and *Q* = 20 N are applied on the edges of rectangle *ABCD* as shown in **Fig. 3.32**. What is the arm *a* of the resultant couple if side *AB* = 500 mm, side *BC* = 300 mm?

Solution: Resultant couple on the body,

$$C_R = P \times CB - Q \times AB$$

Resultant force, $\quad F_R = \sqrt{P^2 + Q^2}$ along diagonal

Arm *a*, $\quad F_R \times a = P \times CB - Q \times AB$

Putting the values $\quad F_R = \sqrt{30^2 + 20^2} = 36.055$ N

$$C_R = 30 \times 300 - 20 \times 500$$

$$= 9000 - 10{,}000 = -1000 \text{ Nmm } (cw)$$

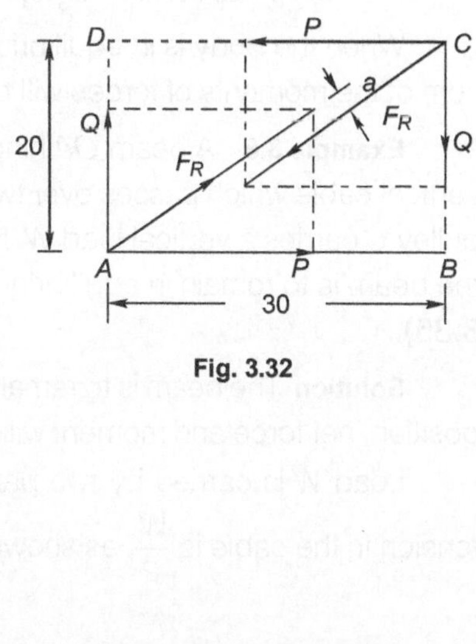

Fig. 3.32

Arm $\quad a = \dfrac{+1000}{36.055} = 27.73$ mm.

Exercise 3.8 A rigid frame *ABCD*, is supported at *B* and *C* as shown in **Fig. 3.33**. At the ends *A* and *D*, two equal and opposite forces *P* each are applied as shown. Determine reactions at the supports.

[**Hint:** Take moments about *B*, reaction at *C* (roller support) will be upwards perpendicular to horizontal surface].

[**Ans:** $R_c = \frac{P(b-a)}{L}\uparrow,\ R_B = \frac{-P(b-a)}{L}\downarrow$].

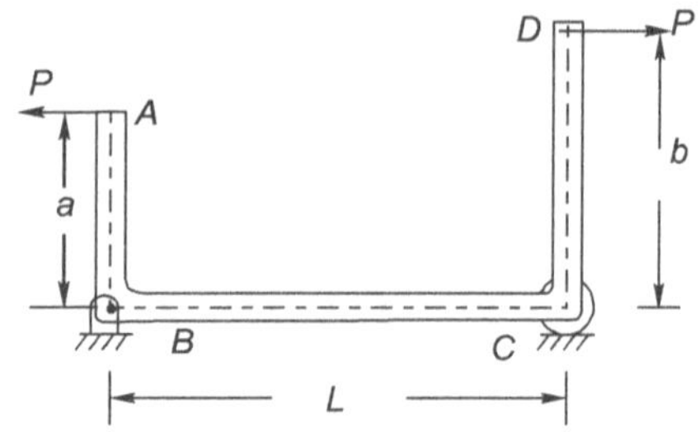

Fig. 3.33

3.11 Several Parallel Forces in a Plane

Considering a general case of a body subjected to several parallel forces acting in the plane of the body. **Fig. 3.34** shows parallel forces F_1, F_2, F_3, F_4, F_5 applied on the body. Forces F_1, F_2 and F_3 are acting in the same direction, *i.e.*, negative direction of *y*-axis. Forces F_4 and F_5 are acting in the same direction as positive direction of *y*-axis as shown.

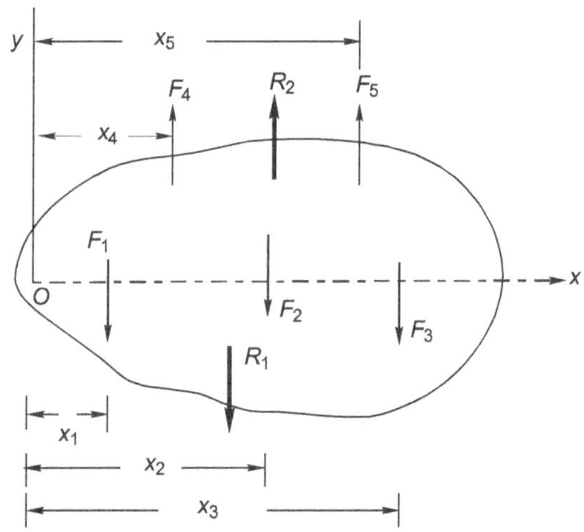

Fig. 3.34

Say R_1 is the resultant of forces F_1, F_2, F_3 *i.e.*,

$$R_1 = F_1 + F_2 + F_3$$

and R_2 is the resultant of forces F_4 and F_5 *i.e.*,

$$R_2 = F_4 + F_5$$

as shown. Lines of action of resultants R_1 and R_2 can be determined using the principle already discussed in Article 3.10.

There are 3 possibilities of these reactions:

1. R_1 and R_2 are different in magnitude and final resultant force on the body is $R_1 - R_2$.
2. R_1 and R_2 are equal but acting at some distance between them producing a couple on the body. In that case resultant force on body will be zero but a resultant couple C_R acts on the body.
3. R_1 and R_2 are equal and collinear such that final resultant on the body is zero, and resultant couple on body is zero. Under these conditions the body will be in equilibrium.

Fig. 3.34 shows distances x_1, x_2, x_3, x_4, x_5 of the lines of action of forces F_1, F_2, F_3, F_4, F_5 respectively from the origin of *x-y* co-ordinate system. Then for equilibrium

$$F_1 + F_2 + F_3 + F_4 + F_5 = 0$$

$$F_1 x_1 + F_2 x_2 + F_3 x_3 + F_4 x_4 + F_5 x_5 = 0$$

When the body is in equilibrium then any point can be selected as moment centre in space about which sum of the moments of forces will be zero.

Example 3.8 A beam *OA* hinged at *O*, is supported at *A* by a vertical cable which passes over two frictionless pulleys *B* and *C*. If pulley *C* carries a vertical load *W*, find the position *x* of the load *P* if the beam is to remain in equilibrium in the horizontal position **(Fig. 3.35)**.

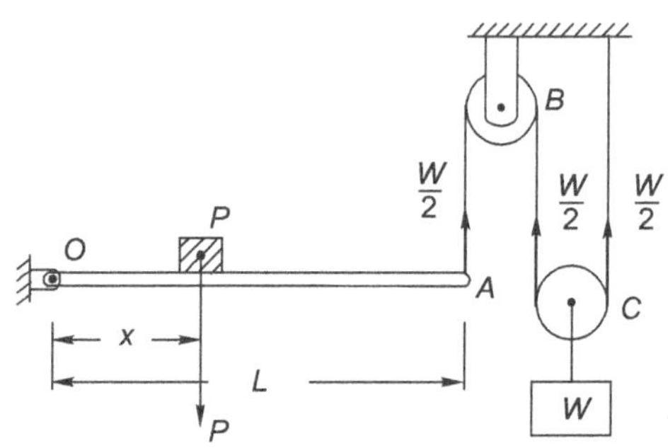

Fig. 3.35

Solution The beam is to remain in equilibrium and in horizontal position, net force and moment will be zero on the beam.

Load *W* is carried by two parts of the cable on pulley *C*, so tension in the cable is $\frac{W}{2}$ as shown.

Taking moments about O, $Px = \frac{W}{2}\cdot L$

$$x = \frac{WL}{2P} \quad ...(1)$$

Reaction at end O, $R_0 = \left(P - \frac{W}{2}\right)\uparrow$

Exercise 3.9 A beam AB, 7 m long is hinged at end A and supported at end B by a vertical cord passing over a frictionless pulley at C. Determine the force P so that bar AB remains in equilibrium. What is the reaction at A **(Fig. 3.36)**?

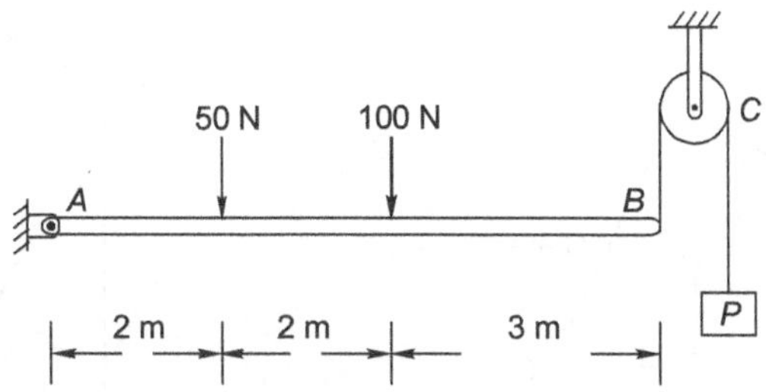

Fig. 3.36

[**Ans:** $P = 71.43$ N; $R_A = 78.57\uparrow$]

PROBLEMS

Problem 3.1 A cantilever beam AB of length L is fixed at end A as shown in **Fig. 3.37**. At the free end B, horizontal force P and vertical load W are applied.

(*a*) Draw the free body diagram of cantilever.

(*b*) Find values of reactions considering equilibrium of cantilever.

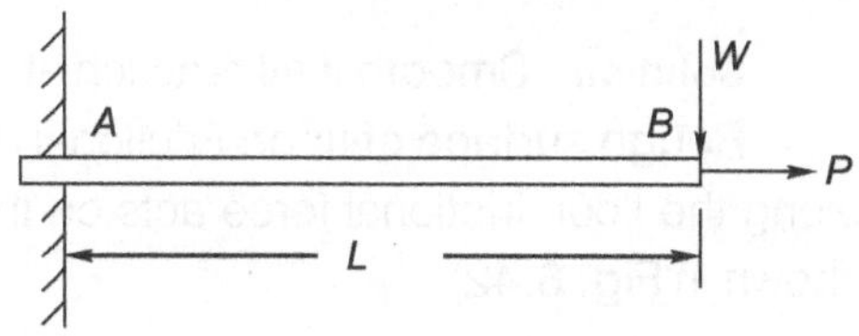

Fig. 3.37

Solution **Fig. 3.48** shows the *free body diagram* of cantilever. End A is fixed, there will be fixing couple C_f, reactions R_{AV} and R_{AH} at A as shown.

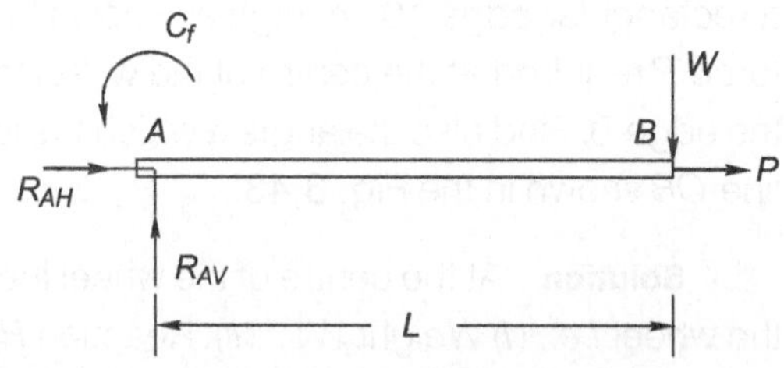

Fig. 3.38

Equilibrium: For equilibrium

Reaction $R_{AV}\uparrow = W\downarrow$

Reaction $\overleftarrow{R_{AH}} = \vec{P}$

Couple C_f(*ccw*) $= W \times L$ (*cw*).

Problem 3.2 **Fig. 3.39** shows a bar AB, of weight mg hinged at end A and tied to a string BC at B as shown.

(*a*) Draw the free body diagram of the bar

(*b*) Find direction of reaction at A, for equilibrium of the bar.

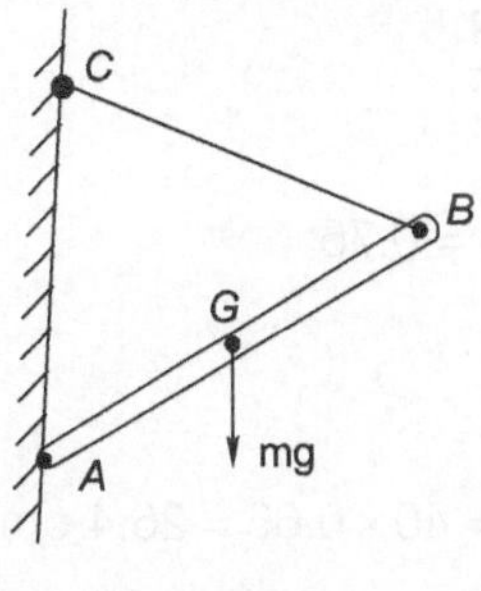

Fig. 3.39

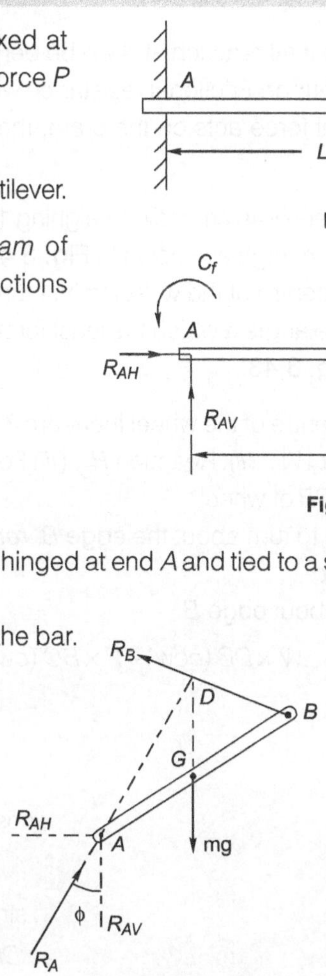

Fig. 3.40

Solution Fig. 3.40 shows the free body diagram of bar AB, with weight $mg\downarrow$ at G of the bar, reaction R_{AV} and R_{AH} at A and pull of string at B, shown by R_B.

Equilibrium: *There are 3 coplanar forces acting on the bar i.e., R_A, mg and R_B, for equilibrium these forces must be concurrent.* Therefore three forces must pass through the same point. *Two forces R_B and mg meet at point D*, reaction R_A must pass through the point D as shown. Direction of R_A or angle of inclination ϕ can be determined in an actual problem.

Problem 3.3 A rectangular crate of mass m rests against a *smooth vertical wall* and a *rough horizontal floor* as shown in Fig. 3.41. Draw the free body diagram of the crate.

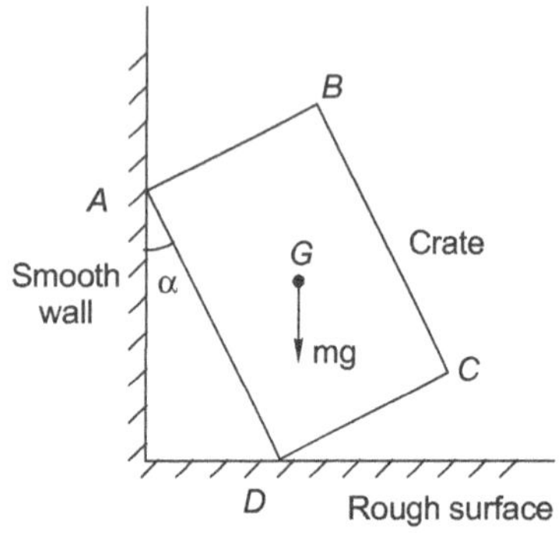

Fig. 3.41

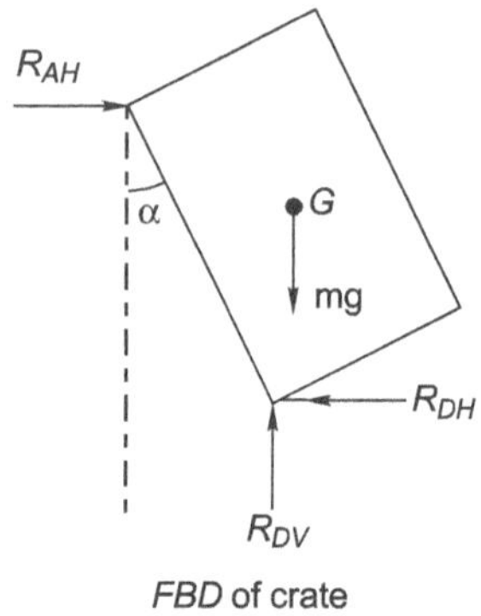

Fig. 3.42

Solution Smooth wall reaction at A will be perpendicular to wall surface.

Rough surface of floor: Frictional resistance will occur on the sliding of crate, or to prevent sliding of crate along the floor, frictional force acts on the crate, there are two reactions R_{DV} and R_{DH} (frictional resistance as shown in Fig. 3.42).

Problem 3.4 A wheel of 40 cm radius weighing 1 kN rests against a rectangular edge 10 cm high as shown in Fig. 3.43. Find the least force P required at the centre of the wheel to just turn the wheel over the edge B. Find also the angle α which the least force makes with the line OB shown in the Fig. 3.43.

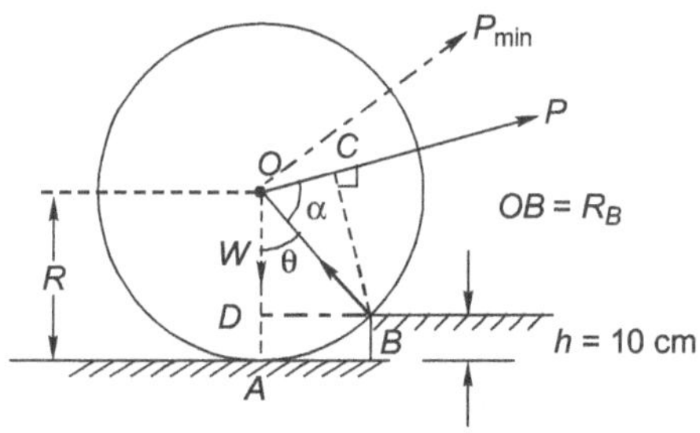

Fig. 3.43

Solution At the centre of the wheel there are 3 forces acting on the wheel *i.e.*, (*i*) Weight $W\downarrow$, (*ii*) Reaction R_B, (*iii*) Force P inclined at an angle α with radius OB of wheel.

When the wheel is to turn about the edge B, *reaction at contact point A will become zero.*

Taking moments about edge B

$$W \times DB\,(ccw) - P \times BC\,(cw) = 0$$

or

$$P = W \times \frac{DB}{BC}$$

$$\cos\theta = \frac{40-10}{40} = 0.75$$

$$\theta = 41.41^{\circ}$$

$$\sin\theta = 0.66$$

$$DB = R\sin\theta = 40 \times 0.66 = 26.4 \text{ cm}$$

$$BC = R\sin\alpha$$

Force $$P = \frac{W \times DB}{R\sin\alpha} = \frac{1000 \times 26.4}{40\sin\alpha} = \frac{660}{\sin\alpha}$$

P will be minimum when $\sin\alpha$ becomes maximum *i.e.*, when

$$\alpha = 90^\circ$$
$$P_{min} = 660 \text{ N}$$

Direction of P_{min} is shown in the figure. (direction of P_{min} is perpendicular to OB)

Problem 3.5 A beam ACB is hinged at end B with another beam BED, as shown in the **Fig. 3.44 (*a*)**. Determine reactions at support A.

Solution Consider beam BED, roller supported at E, taking moments about E,

$$W_2 \cdot L_4 = R_B \times L_3$$

or Reaction, $$R_B = \frac{W_2 L_4}{L_3} \downarrow \quad ...(1)$$

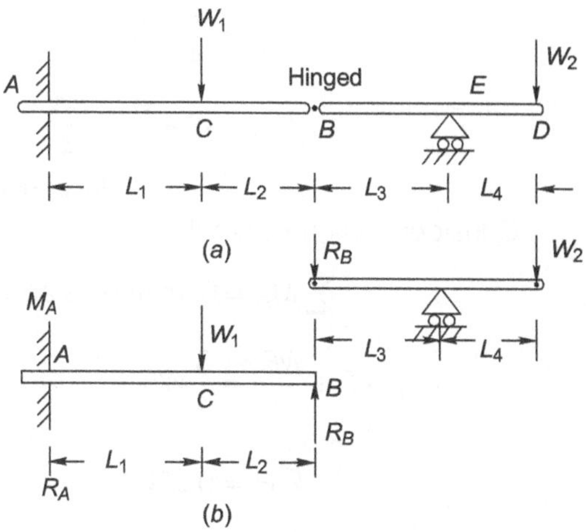

Fig. 3.44

We have not applied any force at B, reaction $R_B \downarrow$ has to be balanced by $R_B \downarrow$ for beam ACB.

Taking moments about A

$$M_A = R_B\,(L_1 + L_2) - W_1\,(L_1)$$

$$= +\frac{W_2 L_4}{L_3}(L_1 + L_2) - W_1 L_1 \quad ...(2)$$

To balance M_A, fixed support will have fixing couple

$$M_A = -\frac{W_2 L_4}{L_3}(L_1 + L_2) + W_1 L_1$$

Reaction, $$R_A = W_1 - R_B = \left(W_1 - \frac{W_2 L_4}{L_3}\right)\uparrow$$

Problem 3.6 A load W is connected to 3 strings AB, BD and BC. Determine the force P required to maintain the position when $\theta = 40^\circ$, if $W = 300$ kg. Use Lami's theorem for the solution **(Fig. 3.45)**.

Solution Force in string BC is P, Note that ABC is an isosceles triangle because $AB = AC = 2$ m each.

$$\angle ABC = 90 - \theta/2 = 90 - 20 = 70^\circ$$

because $\theta = 40^\circ$

Angle of BC with the vertical is $\theta/2 = 20^\circ$

So angle $$\angle CBD = 180 - 20 = 160^\circ$$

$$\angle ABD = 180 - \angle ABO$$

$$\angle ABO = \angle ABC - \frac{\theta}{2} = \frac{180 - 40}{2} - 20 = 50^\circ$$

$$\angle ABD = 180 - 50 = 130^\circ$$

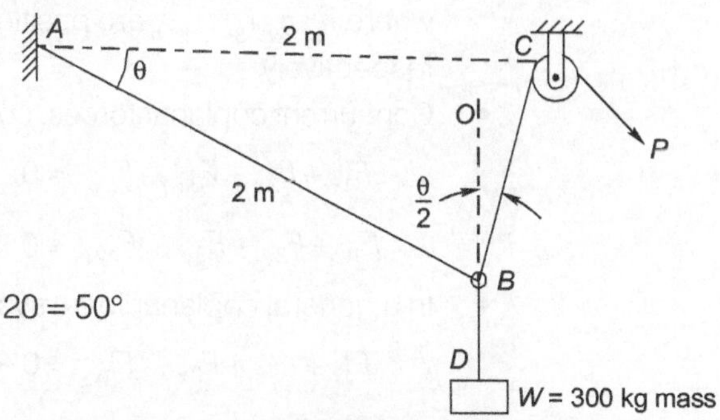

Fig. 3.45

Using Lami's theorem

$$\frac{W}{\sin 70^\circ} = \frac{P}{\sin 130^\circ}$$

$$P = \frac{W \sin 130^\circ}{\sin 70^\circ} = W \times \frac{0.766}{0.9396} = 0.8152\ W$$

$$W = 300 \times 9.81 = 2943 \text{ N}$$

So $$P = 0.8152 \times 2943 = 2399 \text{ N} = 2.399 \text{ kN.}$$

Problem 3.7 A 3-hinged arch is loaded at *C*, at a distance of *R*/2 from *A*. Find the horizontal and vertical reactions at *A* and *B* **(Fig. 3.46)**.

Solution Taking moments at end *A*, (R_{BH} passes through *A*)

$$W \times \frac{R}{2} = R_{BV} \times 2R$$

$$R_{BV} = \frac{W}{4}$$

$$R_{AV} = W - \frac{W}{4} = \frac{3W}{4}$$

(for equilibrium)

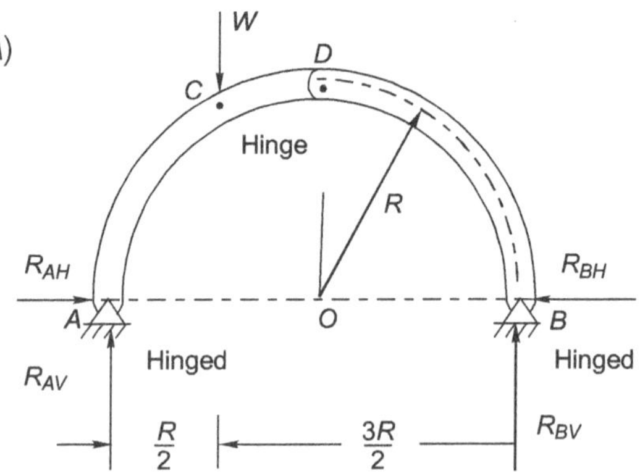

Fig. 3.46

Consider portion *ACD* only,

$$\sum M_D = 0, \text{ moments about } D$$

$$R_{AH} \times R + \frac{WR}{2} = \frac{3}{4} WR$$

$$R_{AH} = \overrightarrow{0.25W}$$

$$R_{BH} = \overleftarrow{W}/4 \text{ (for equilibrium)}$$

Remember

- Isolation of a rigid body from surroundings (supports) and replacing the support by *support reactions* shown on the body, is called a *free body diagram*.
- At the roller support, there is only one reaction, perpendicular to the surface of support.
- At the hinged end, there can be one or two reaction components.
- At the fixed end, in addition to reactions, there will be a fixing couple.
- Conditions of equilibrium of forces $F_1, F_2, F_3, ..., F_n$ and couples $C_1, C_2, ..., C_m$ acting on a body are:

$$F_1 + F_2 + F_3 \dots F_n = 0 \text{ (vector sum)}$$

$$(r_1 \times F_1 + r_2 \times F_2 + r_3 \times F_3 \dots r_n \times F_n) + (C_1 + C_2 + C_3 \dots C_m) = 0$$

where $r_1, r_2, r_3, ..., r_n$ are position vectors of points of application of forces $F_1, F_2, F_3, ..., F_n$ respectively.

- Concurrent coplanar forces, conditions of equilibrium:

$$F_{1x} + F_{2x} + F_{3x} \dots F_{nx} = 0$$

$$F_{1y} + F_{2y} + F_{3y} \dots F_{ny} = 0.$$

- In a general coplanar force system, conditions of equilibrium are:

$$F_{1x} + F_{2x} + F_{3x} \dots F_{nx} = 0$$

$$F_{1y} + F_{2y} + F_{3y} \dots F_{ny} = 0$$

$$M_{1z} + M_{2z} + M_{3z} \dots M_{nz} = 0$$

where $M_{1z} = x_A F_{1y} - y_A F_{1x}$, where x_A, y_A are co-ordinates of point of application of force F_1. Similarly moments of other forces can be determined.

- For a two force body, two forces on the body must be equal, opposite in direction and collinear.
- If 3 non-parallel forces are applied on a body, then for equilibrium these forces must be concurrent.
- Lami's theorem for 3 concurrent forces acting on a body for equilibrium:

$$\frac{F_1}{\sin\alpha} = \frac{F_2}{\sin\beta} = \frac{F_3}{\sin\gamma}$$

where α, β and γ are angles opposite in the direction of F_1, F_2 and F_3 respectively.

PRACTICE PROBLEMS

3.1 A bar *AB* of mass 4 kg is hinged at *A*, and tied to a string at *B*. Draw the free body diagram of the bar and for equilibrium find reaction R_A and pull in string at *B* (**Fig. 3.47**).

[**Ans:** Free body diagram, R_{AV}, R_{AH} at *A*, 4 g Newton at *G* of the bar. Reaction R_B at *B* in the direction of string. **Equilibrium:** *ABC* is equilateral triangle, force 4 g and R_A meet at centre of the string *BC*. R_A makes an angle of 30° with the vertical, R_B = 2 g Newton, W_B = 4 g Newton, reaction R_A = 3.464 g N].

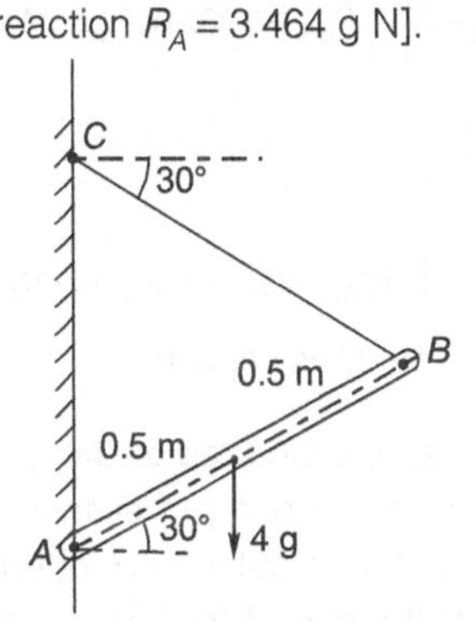

Fig. 3.47

3.2 A uniform wheel of radius *R* weighing *W* Newton rests against a rectangular edge *R*/3 high as shown in **Fig. 3.48**. Find the magnitude of the force *P* required to just turn the wheel about the edge *B*.

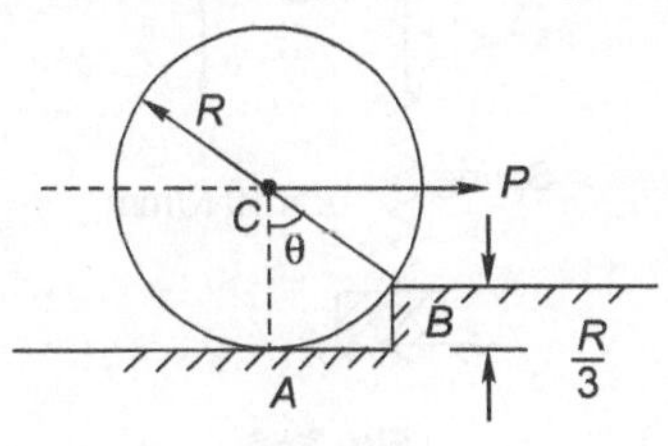

Fig. 3.48

[**Ans:** $P = 1.5\sin\theta.W$].

3.3 A pull of 200 N is applied on a pulley system shown in **Fig. 3.49**. Weights of pulleys *A*, *B* and *C* are 100 N, 30 N and 40 N respectively. Pulleys are frictionless. What weight *W* can be lifted by the system.

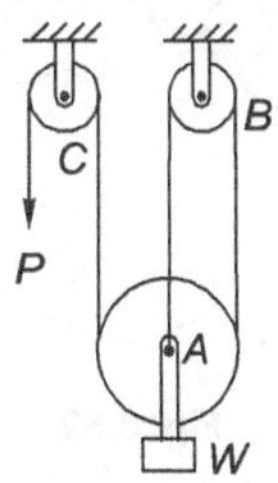

Fig. 3.49

[**Hint:** Consider equilibrium of center of pulley *A*].
[**Ans:** *W* = 500 N].

3.4 **Fig. 3.50** shows a system of levers *ABC* and *DEF* with fulcrums at *E* and *A*. How much load *W* can be lifted by an effort *P* = 80 N at *F*?

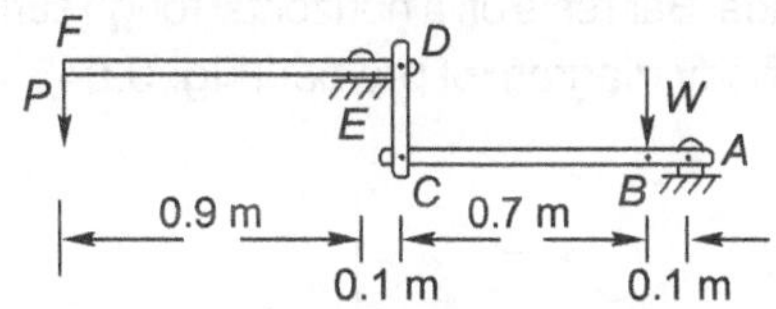

Fig. 3.50

[**Hint:** Treat *DC* at two force member].
[**Ans:** *W* = 5.76 kN].

3.5 An elastic string *AB* is just taught before a force of 500 N is applied at its centre at *C*, as shown in the **Fig. 3.51**. If the string takes 4 N/mm of elongation in string, at what angle α, equilibrium will be maintained after the application of force?

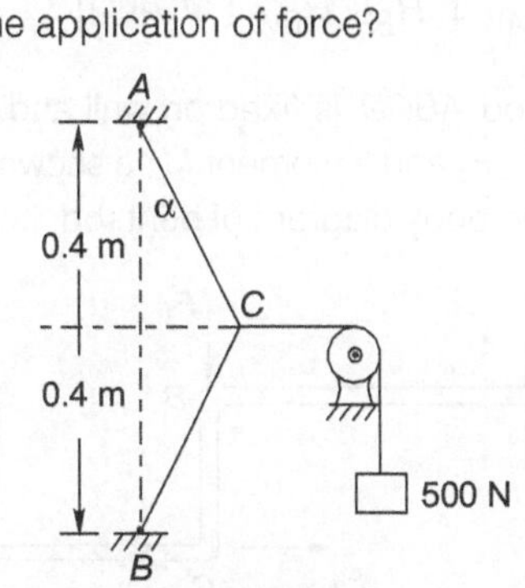

Fig. 3.51

[**Hint:** Elongation = $400(\sec\alpha - 1)$

Tension, $T = 400\,(\sec\alpha - 1) \times 4$ N]

[**Ans:** $\alpha = 37.3°$, by trial and error].

3.6 Two wheels *A* and *B* of weight W_A and W_B respectively are connected by a rod *AB* of negligible weight and are free to roll on the surfaces shown in the **Fig. 3.52**. Determine the angle α for the rod, so that rod remains in equilibrium. Take $W_B = 2W_A$.

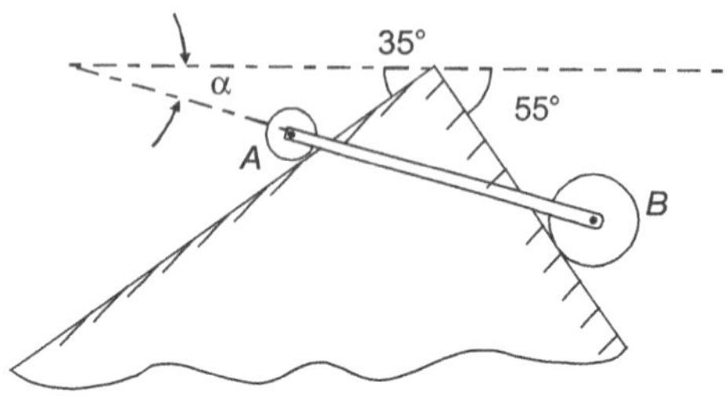

Fig. 3.52

[**Hint:** Draw free body diagrams for wheels *A* and *B*. There are 3 forces on each wheel, *i.e.*, weight of wheel, force in bar *AB* and reaction from inclined surface. In terms of W_A and angle α, find force on bar. For wheel *B*, consider 3 forces *i.e.*, force in bar, weight of wheel *B*, reaction from inclined surface, since $W_B = 2W_A$, value of angle α will be obtained, force on bar = $W_A \times \sin 145°/\cos(\alpha + 35°)$].

[**Ans:** $\alpha = 35.7°$].

3.7 A 3 hinged arch with horizontal length 4*l* and vertical height 1.5*l* is subjected to vertical load *W* at point *D* as shown in **Fig. 3.53**. Determine components of reactions at *A* and *B*.

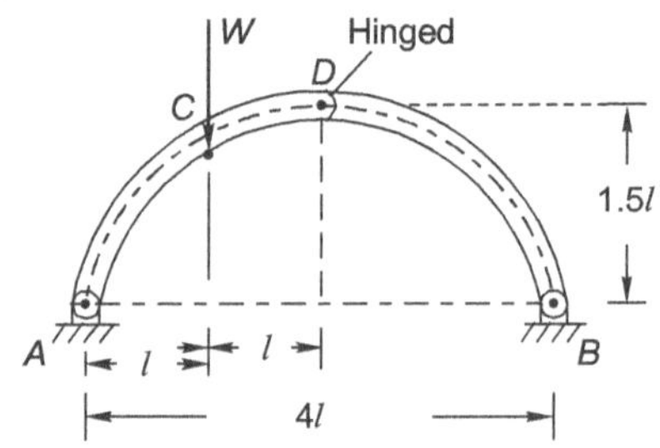

Fig. 3.53

[**Ans:** $R_{AV} = \frac{3W}{4}$, $R_{BH} = \overrightarrow{0.333}\,W$, $R_{BV} = W/4$, $R_{BH} = \overleftarrow{0.333}\,W$].

SPECIAL PROBLEMS

3.1 A pry bar is lifting a body *A* having a smooth horizontal surface. Bar rests on a horizontal rough surface. Draw free body diagram of pry bar (**Fig. 3.54**).

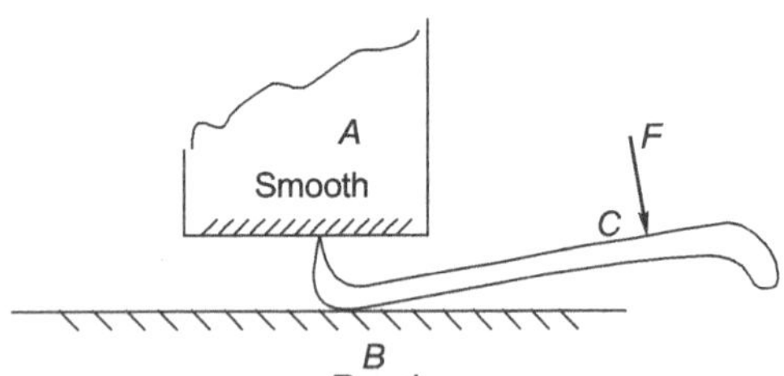

Fig. 3.54

[**Ans:** $R_{AV}\downarrow, \overleftarrow{R}_{BH}, R_{BV}\uparrow$, *F* at *C*].

3.2 A bent rod *ABCD*, is fixed on wall and subjected to forces F_1, F_2 and a moment *M* as shown in **Fig. 3.55**. Draw free body diagram of bent rod.

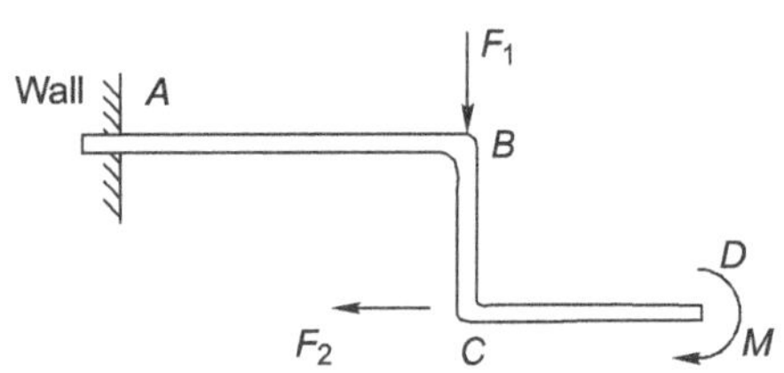

Fig. 3.55

[**Ans:** $R_{AV}\uparrow, \overrightarrow{R}_{AH}, \overleftarrow{M}_f$ (*ccw*) fixing couple at *A* and F_1, F_2 and *M* at *D* as shown].

3.3 A 50 kg mass is supported on the ground. It is tied to a rope through a spring of stiffness 10 N/mm. The rope passes over a pulley supported in a frictionless bearing. At the end of the rope a force of 200 N is applied as shown in **Fig. 3.56**.

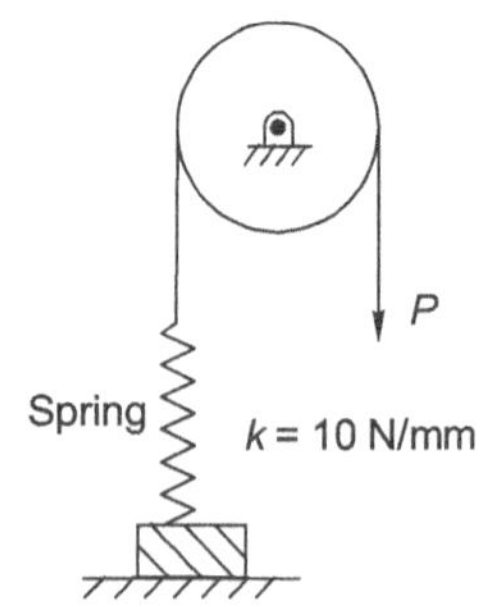

Fig. 3.56

(*a*) What is the reaction of the ground on the mass?
(*b*) What is displacement of spring?
(*c*) What is displacement of mass?

[**Hint:** Spring force is less than the weight of the mass].
[**Ans:** (*a*) 290.5 N ↑, (*b*) 20 mm, (*c*) zero].

3.4 A rope connects two masses *A* and *B* which rest on two inclined planes as shown in the **Fig. 3.57**. Planes are smooth and rope passes over a frictionless pulley. Determine tension in rope and angle θ.

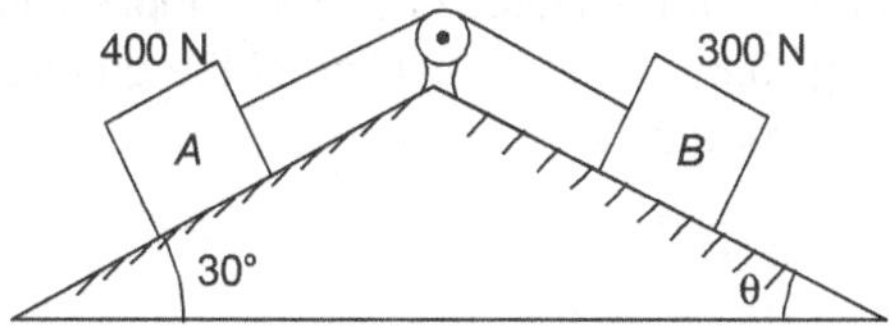

Fig. 3.57

[**Hint:** Considering equilibrium of block *A*, find *T*, tension in rope, then for block *B*, tension is given which will be equal to $W_B \sin\theta$].

[**Ans:** Tension = 200 N, θ = 41.8°].

MULTIPLE CHOICE QUESTIONS

3.1 A block of mass *m*, side *a* is shown in **Fig. 3.58**. A force *P* at an angle 45° is applied at edge *A*. What is the magnitude of force *P* so that block is just lifted from roller at *D*?

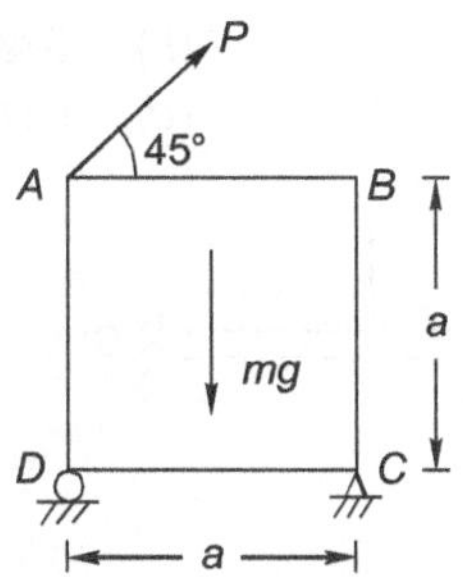

Fig. 3.58

(*a*) $\frac{mg}{2\sqrt{2}}$ (*b*) $\frac{mg}{2}$

(*c*) $\frac{mg}{\sqrt{2}}$ (*d*) None of these

3.2 A sphere of weight *W* is lying between two inclined planes at an angle of 45°. What is reaction R_1 (**Fig. 3.59**)?

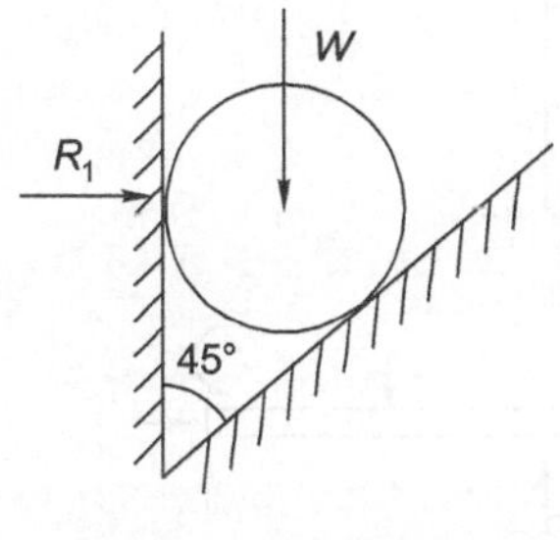

Fig. 3.59

(*a*) *W* (*b*) 0.5*W*
(*c*) 0.707*W* (*d*) 1.414*W*.

3.3 A truss *ABCD*, carries horizontal and vertical loads as shown in Fig. 3.60. What is resultant reaction at *A*?

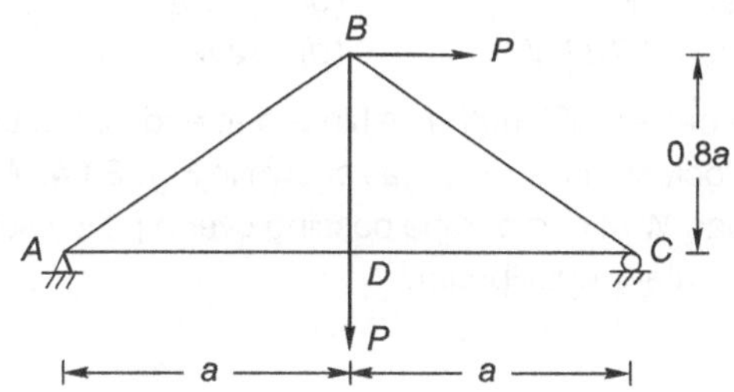

Fig. 3.60

(*a*) *P* (*b*) 1.05*P*
(*c*) 0.8*P* (*d*) 1.1*P*.

3.4 A ball of mass *m* is tied to a string of length *L* as shown in **Fig. 3.61**. It is pulled by a force *P* so that string makes an angle θ with the vertical, what is *P*?

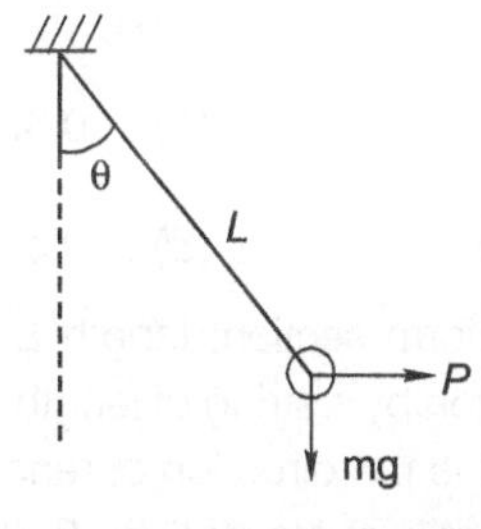

Fig. 3.61

(*a*) *mg* (*b*) *mg* sin θ
(*c*) *mg* tan θ (*d*) *mg* cot θ.

3.5 A ring is pulled by 3 strings as shown in **Fig. 3.62**. What is the ratio of T_1/T_2?

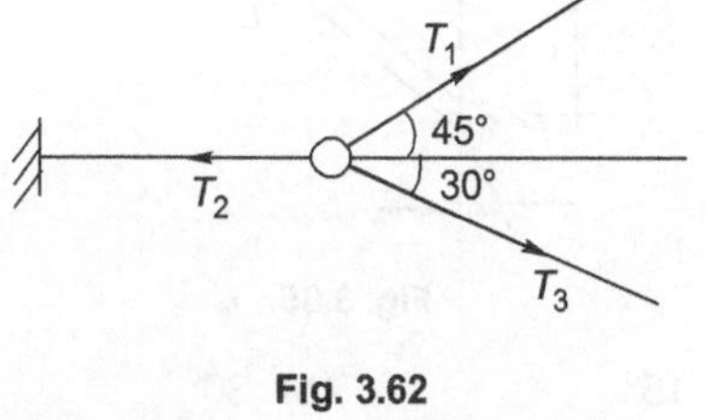

Fig. 3.62

(*a*) 0.5 (*b*) 0.517
(*c*) 1.932 (*d*) None of these

3.6 A bar *AB* hinged at *A*, carries a load *W* at its end *B*. It is held in position by a string as shown in **Fig. 3.63**. Angle *ABC* = 30°. What is tension in string?

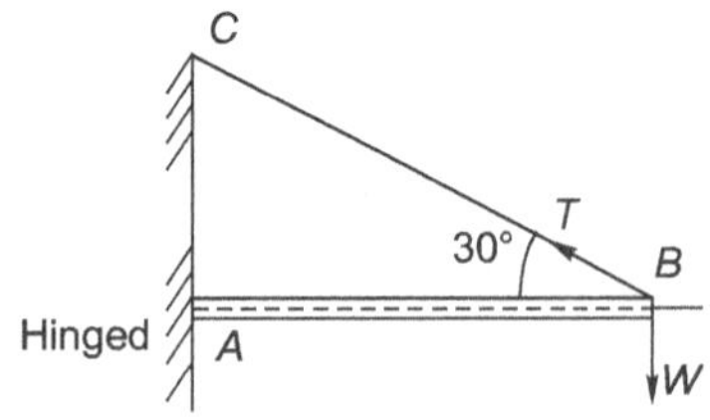

Fig. 3.63

(*a*) *W* (*b*) 1.5*W*
(*c*) 1.732*W* (*d*) 2*W*

3.7 A bar *AB* of length *L* is hinged at end *A*. It carries a block of mass *m* at *C* as shown in **Fig. 3.64**. What is load *W* at end of rope passing over a pulley so as to maintain equilibrium?

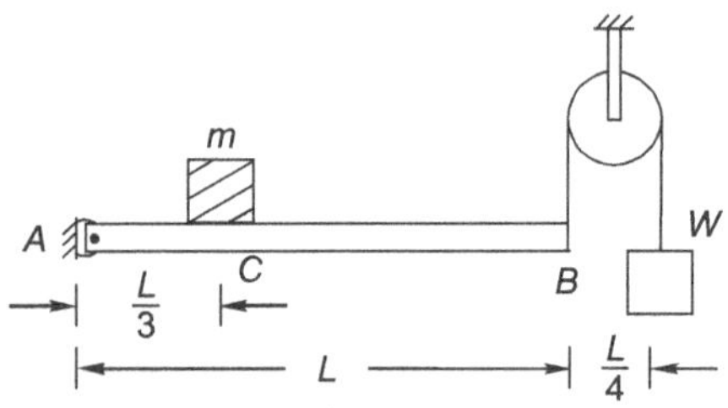

Fig. 3.64

(*a*) $\frac{mg}{3}$ (*b*) 0.48*mg*
(*c*) 0.9*mg* (*d*) *mg*

3.8 A bar of uniform section, length *L* and mass *mg* is held in position by a string of length *L*. Distance *AB* is also *L*. What is the direction of reaction at *B* with the vertical? What is angle ϕ? (**Fig. 3.65**).

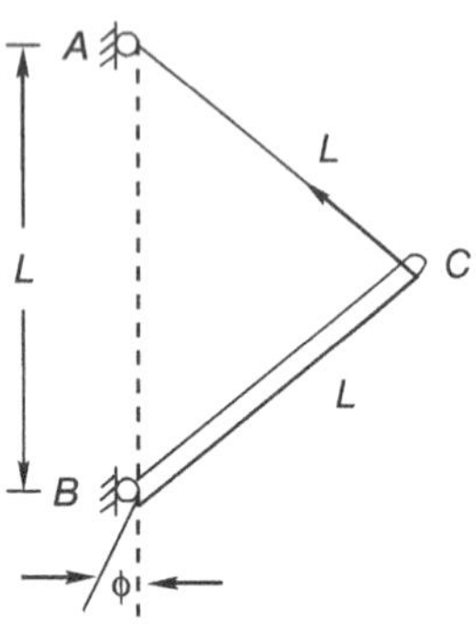

Fig. 3.65

(*a*) 15° (*b*) 30°
(*c*) 45° (*d*) 60°

3.9 A mass of 60 kg is supported on the ground. It is tied to a rope through a spring of stiffness 10 N/mm. At the end of the rope a force of 200 N is applied as shown in **Fig. 3.66**. What is reaction of the ground on the mass, $g = 9.8$ m/s^2?

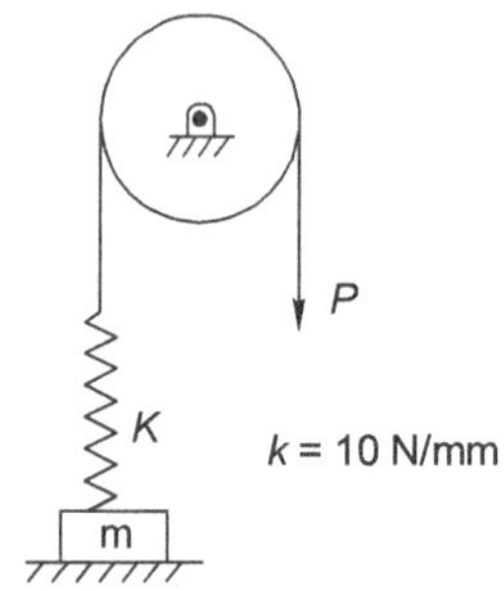

Fig. 3.66

(*a*) 788 N (*b*) 588 N
(*c*) 388 N (*d*) None of these.

3.10

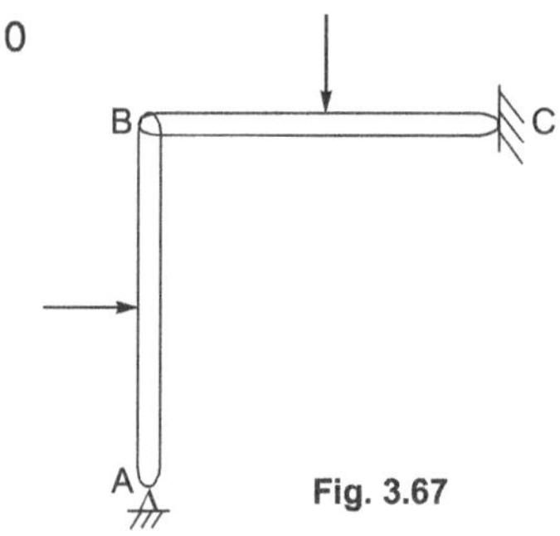

Fig. 3.67

Select the correct free body diagram for the structure shown in the figure given above **(Fig. 3.67)**. The structure is hinged at *A* and *B* while *C* is fixed.

(*a*)

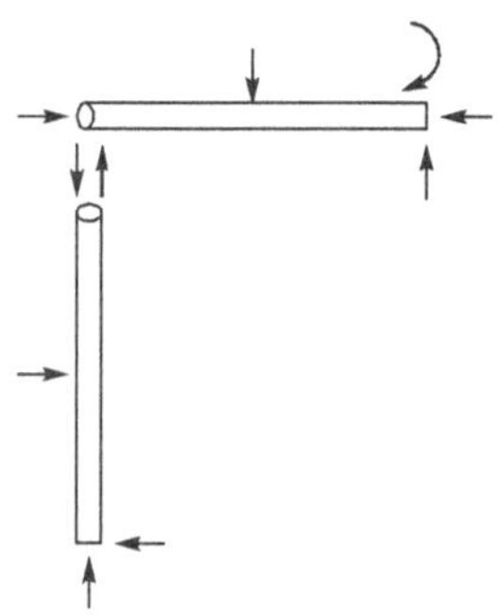

(*b*)

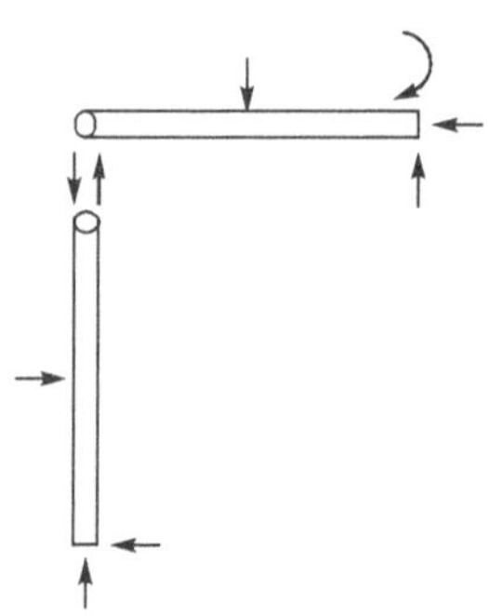

(c)

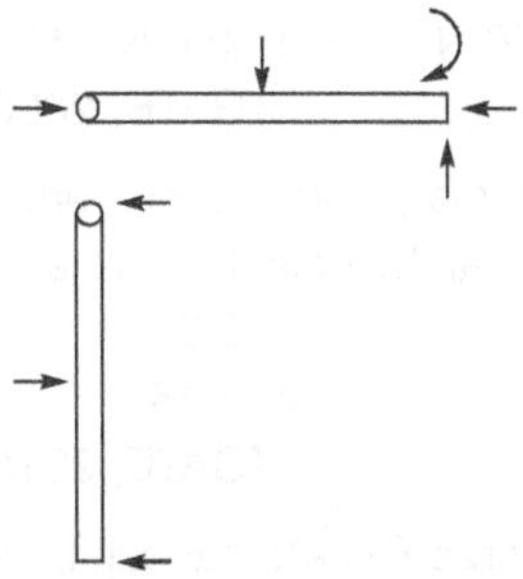

(d)

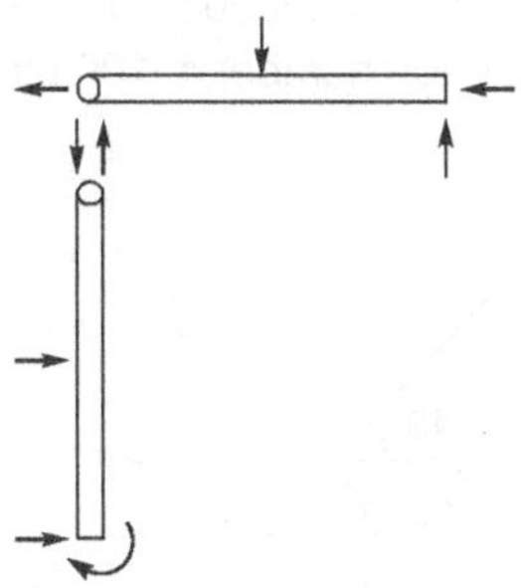

[CSE, Prelim, CE : 2002]

3.11 A solid sphere will be in stable equilibrium if its centre of gravity lies
(a) vertically above its geometric centre
(b) vertically below its geometric centre
(c) on the horizontal line through its centre
(d) at the centre

[CSE, Prelim, CE : 2003]

3.12 Consider the following statements with regard to equilibrium
1. A body is said to be in stable equilibrium when an receiving a slight displacement, it tends to go further away from its position of rest.
2. A body is said to be in unstable equilibrium when on receiving a slight displacement, it tends to go further away from its position of rest
3. A body is in neutral equilibrium when an receiving a slight displacement, it tends to come to rest in its new position

Which of the statements given above are correct?
(a) 1, 2 and 3 (b) 1 and 2
(c) 2 and 3 (d) 1 and 3

[CSE, Prelim, CE : 2004]

3.13 A ladder AB of weight W and length l, is held in equilibrium by a horizontal force P as shown in the **Fig. 3.68.** Assuming the ladder to be idealized as a homogeneous rigid bar and the surfaces to be smooth, which one of the following is correct.

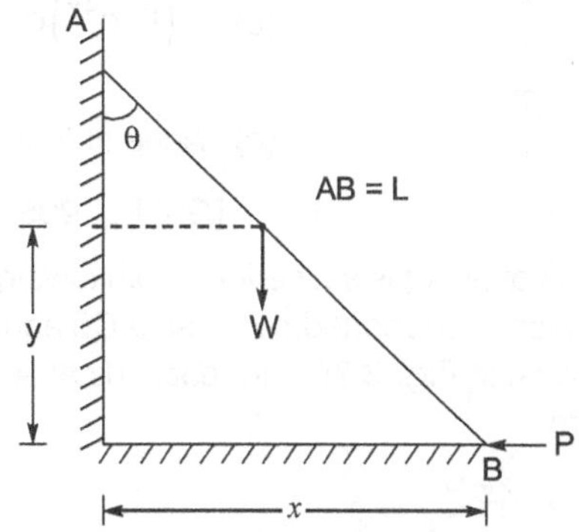

Fig. 3.68

(a) $P = \frac{1}{2} W \tan\theta$ (b) $P = \frac{1}{2} W \operatorname{cosec}\theta$
(c) $P = \frac{1}{2} W \cos\theta$ (d) $P = 2W\cos\theta$

[CSE, Prelim, CE : 2005]

3.14 Two forces P = 6 N and Q = 10 N act on a particle and their lines of action are inclined to each other at an angle of 60°. What is the magnitudes of the third force R which will keep the particle in equilibrium
(a) 13.30 N (b) 13.89 N
(c) 14.00 N (d) 14.02 N

[CSE, Prelim, CE : 2006]

3.15 Two unlike parallel forces of 10 kN and 5 kN are 45 cm apart. If the direction of 5 kN force is reversed, then the resultant shifts through what distance
(a) 90 cm (b) 60 cm
(c) 45 cm (d) 30 cm

[CSE, Prelim, CE : 2008]

3.16 A spring scale indicates a tension T in the right hand cable of the pulley system shown in **Fig. 3.69.** Neglecting the mass of the pulleys and ignoring friction between the cable and the pulley, the mass m is

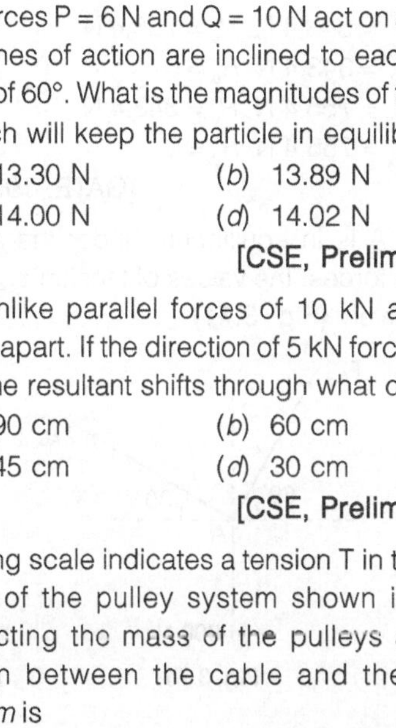

Fig. 3.79

(a) $\frac{2T}{g}$ (b) $T\left(1+e^{4\pi}\right)g$

(c) $\frac{4T}{g}$ (d) None of these

[GATE, 1995 : 2 Marks]

3.17 A mass of 35 kg is suspended from a weightless bar *AB* which is supported by a cable *CB* and a pin at *A* as shown in **Fig. 3.70.** Pin reactions at *A* on the bar *AB* are

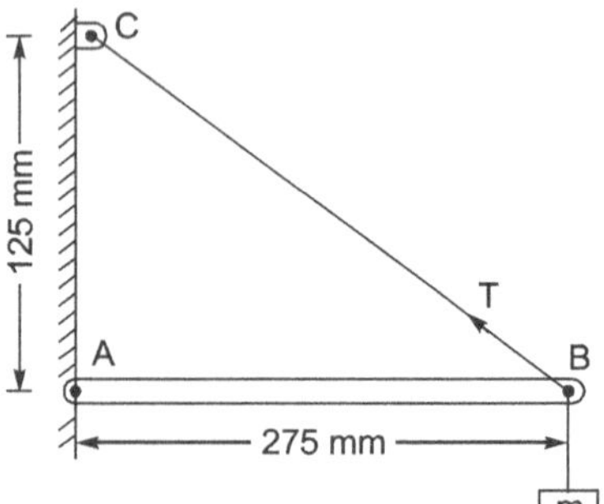

Fig. 3.70

(a) $R_x = 343.4$ N, $R_y = 755.4$ N
(b) $R_x = 343.4$ N, $R_y = 0$
(c) $R_x = 755.4$ N, $R_y = 343.4$ N
(d) $R_x = 755.4$ N, $R_y = 0$

[GATE, 1997, 2 Marks]

3.18 If point A is in equilibrium under the action of the applied forces, the values of tension τ_{AB} and τ_{AC} are respectively **(Fig. 3.71)**

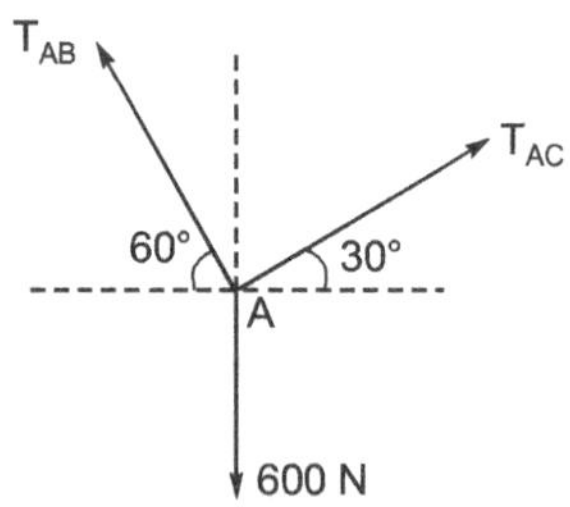

Fig. 3.71

(a) 520 N, 300 N (b) 300 N, 500 N
(c) 450 N, 150 N (d) 150 N, 450 N

[GATE, 2006 : 2 Marks]

3.19 If F = 1 kN, the magnitude of the vertical reactions force developed at the point B is kN is

(a) 0.63 (b) 0.32
(c) 1.26 (d) 1.46

[GATE, 2012 : 2 Marks]

3.20 The maximum force *F* in kN that can be applied at *C* **(Fig. 3.72)** such that the axial stress in any of the truss member does not exceed 100 MPa, if area of cross-section of each number is 100 mm^2

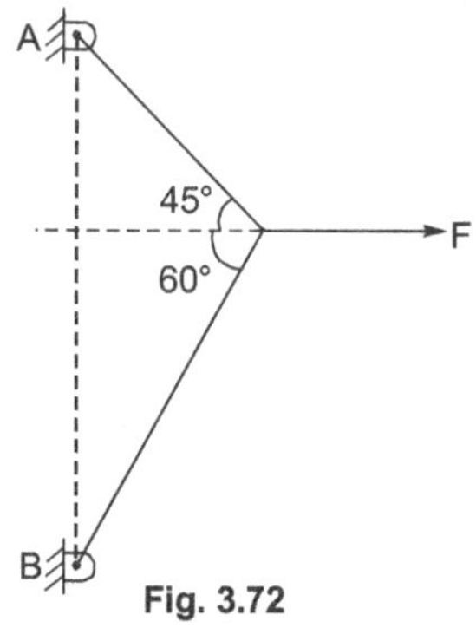

Fig. 3.72

(a) 8.17 (b) 11.15
(c) 14.14 (d) 22.30

[GATE, 2002 : 2 Marks]

Answers

3.1 (a)	3.2 (a)	3.3 (b)	3.4 (c)	3.5 (b)
3.6 (d)	3.7 (a)	3.8 (b)	3.9 (c)	3.10 (a)
3.11 (b)	3.12 (c)	3.13 (a)	3.14 (c)	3.15 (c)
3.16 (c)	3.17 (d)	3.18 (a)	3.19 (a)	3.20 (b)

EXPLANATIONS

3.1 (a) (Fig. 3.77) $P \times \sqrt{2}a = mg\ a/2$, $P = mg/2\sqrt{2}$.

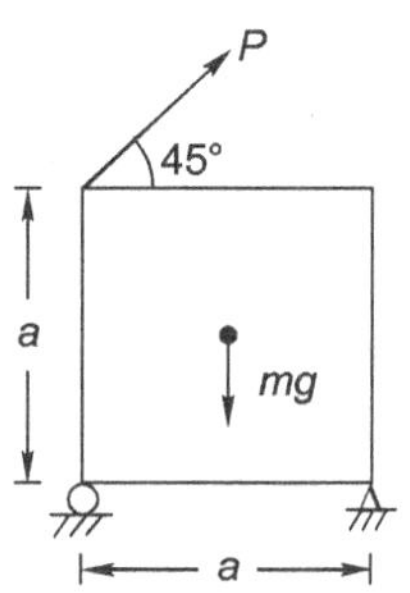

Fig. 3.73

3.2 (a)

$R_1 = W$
$R_2 = 1.414W$ **(Fig. 3.74)**

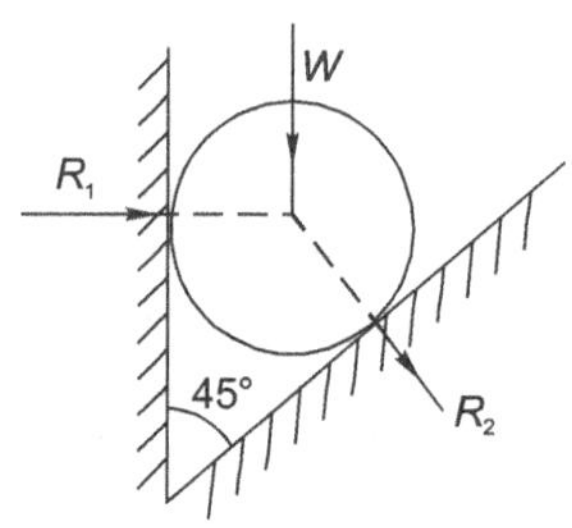

Fig. 3.74

3.3 (b) $$R_C = \frac{0.8Pa + Pa}{2a} = 0.9P$$

$$R_{AV} = 0.1P,\ R_{AH} = P$$

$$R = \sqrt{(0.1P)^2 + P^2} = 1.05P \text{ (Fig. 3.75)}$$

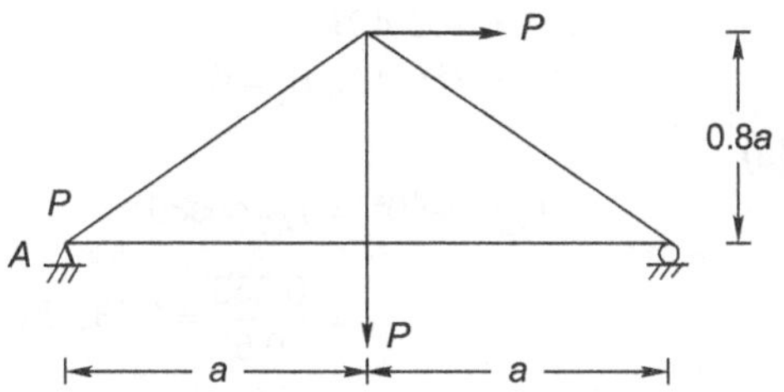

Fig. 3.75

3.4 (c) (Fig. 3.76) $P/mg = \tan\theta$

$$P = mg\tan\theta.$$

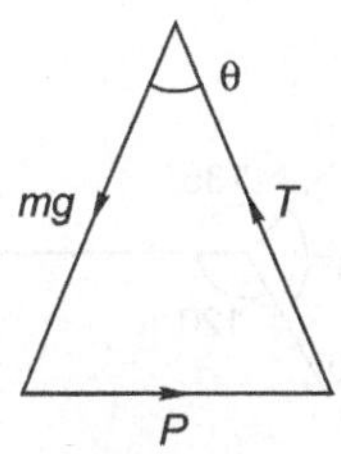

Fig. 3.76

3.5 (b) (Fig. 3.77)

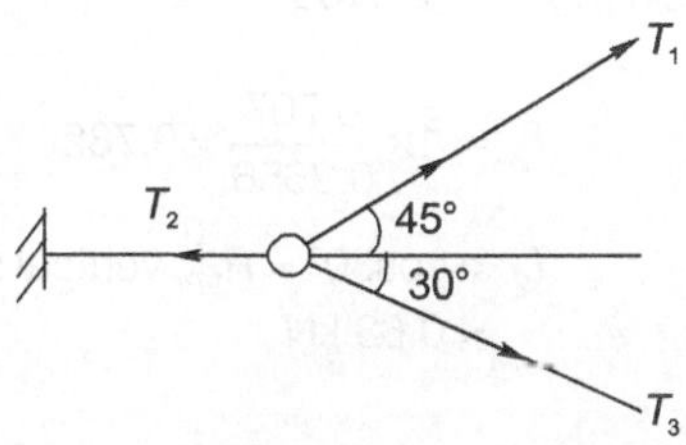

Fig. 3.77

$$T_3 \sin 30° = T_1 \sin 45°$$
$$T_3 = 1.414T_1$$
$$T_2 = T_1 \cos 45° + T_3 \times \cos 30°$$
$$= 0.707T_1 + 1.414 \times 0.866T_1$$
$$= (0.707 + 1.224)\ T_1$$
$$T_1/T_2 = 0.517.$$

3.6 (d) $T \sin 30° = W$

$$T = 2W.$$

3.7 (a) $mg \times L/3 = WL$

$$W = mg/3.$$

3.8 (b) See Fig. 3.78.

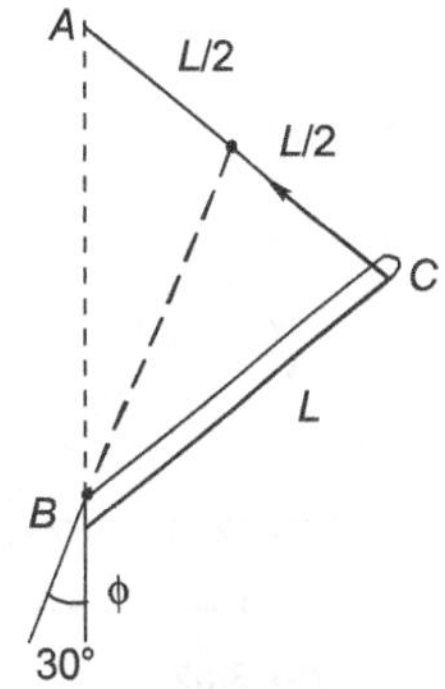

Fig. 3.78

3.9 (c) Reaction $= 60 \times 9.8 - 200$
$= 588 - 200$
$= 388$ N

3.10 (a)

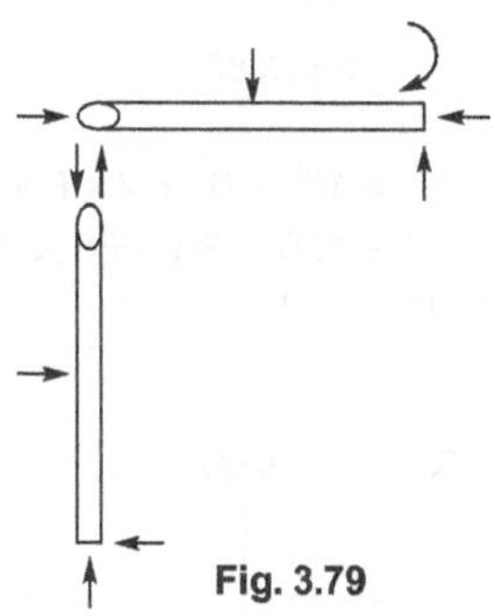

Fig. 3.79

3.11 (b)

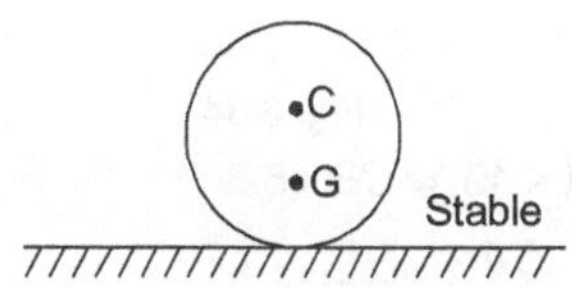

Fig. 3.80

3.12 (c)

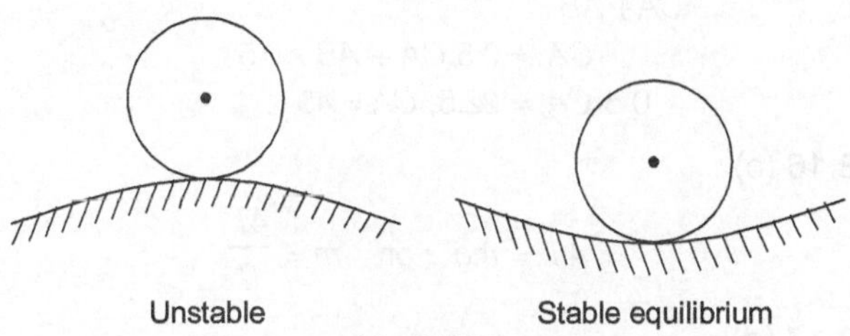

Fig. 3.81

3.13 (a) Moments about A

$$W \times \frac{l}{2}\sin\theta = P \times l\cos\theta$$

$$P = \frac{W}{2}\tan\theta$$

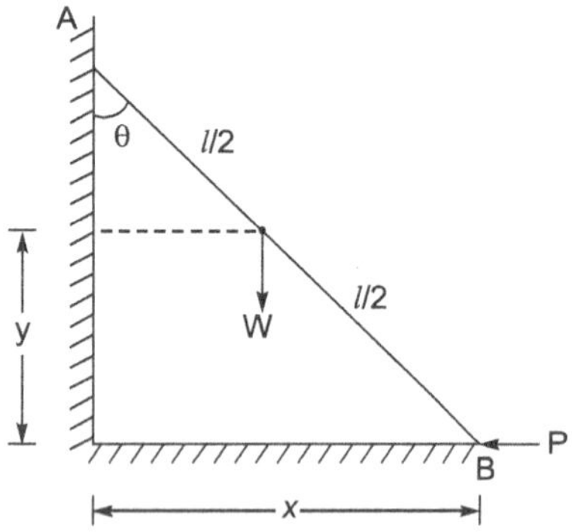

Fig. 3.82

3.14 (c)

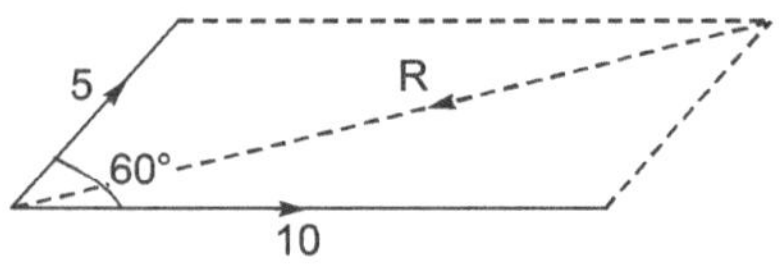

Fig. 3.83

$$R^2 = 10^2 + 6^2 + 2 \times 6 \times 10 \cos 60°$$
$$= 100 + 36 + 60 = 196$$
$$R = 14 \text{ N}$$

3.15 (c)

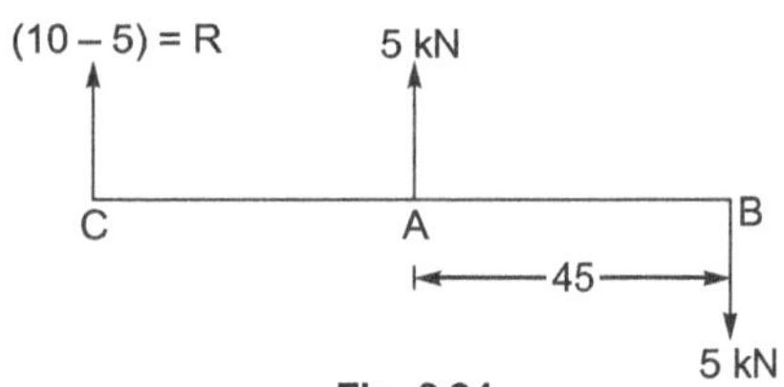

Fig. 3.84

$$CA \times 10 = CB \times 5$$
$$\frac{CA}{CB} = 0.5$$
$$\frac{CA}{CA + AB} = 0.5$$
$$CA = 0.5\, CA + AB \times 0.5$$
$$0.5\, CA = 22.5,\ CA = 45$$

3.16 (c)

$$4T = mg \quad \text{or} \quad m = \frac{4T}{g}$$

3.17 (d)

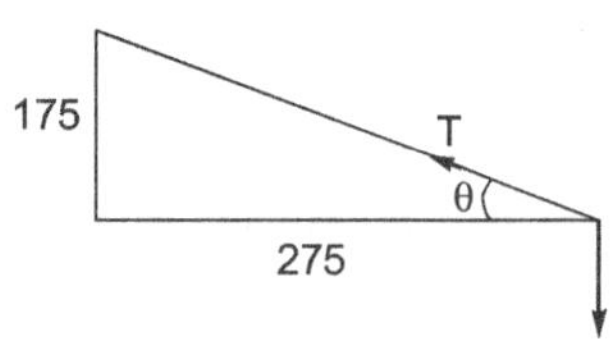

Fig. 3.85

$$\tan\theta = \frac{127}{275} = 0.4515$$
$$\theta = 24.44°$$
$$\sin\theta = 0.4137$$
$$T = \frac{343.4}{0.4137} = 830 \text{ N}$$
$$T\cos\theta = 755.4 \text{ N}$$
$$R_x = 755.4 \text{ N},\ R_y = 0$$

3.18 (a)

$$T_{AB}\cos 60° = T_{AC}\cos 30°$$
$$T_{AB} = \frac{0.866}{0.5} = 1.732\ T_{AC}$$
$$T_{AB}\sin 60° + T_{AC}\sin 30° = 600$$
$$1.5\ T_{AC} + 5 T_{AC} = 600$$
$$T_{AC} = 300 \text{ N}$$
$$T_{AB} = 520 \text{ N}$$

3.19 (a)

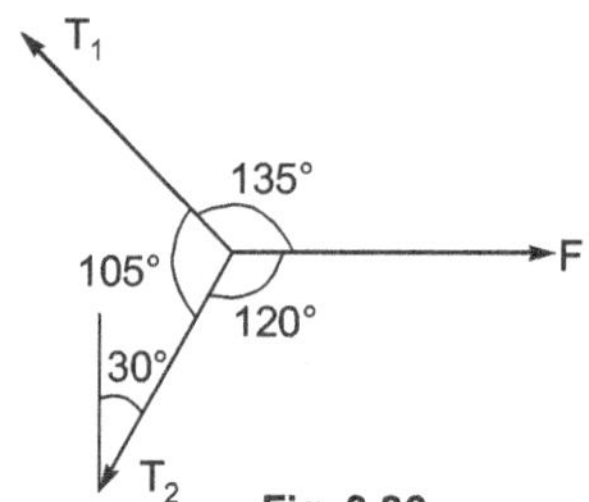

Fig. 3.86

$$\frac{T_2}{\sin 135°} = \frac{F}{\sin 105°}$$
$$T_2 = 1 \times \frac{0.707}{0.9656} = 0.732$$
$$T_2 = \cos 30° = R_{VB}, \text{ vertical reaction at } B$$
$$= 0.63 \text{ kN}$$

3.20 (b)

$$\frac{T_1}{\sin 120°} = \frac{T_2}{\sin 135°} = \frac{F}{\sin 105°}$$
$$T_1 =$$
$$F \times \frac{\sin 120°}{\sin 105°} = F \times \frac{0.866}{0.9656} = 0.8968F$$
$$\frac{0.8968}{100} = 100 \qquad T_1 > T_2$$
$$F = 11150 \text{ N} = 11.15 \text{ kN}$$

■■■■

04

CHAPTER

Plane Trusses

4.1 Introduction

Plane trusses are in common sight on workshop building and industrial sheds, so that the roof is inclined and on one side glass planes can be fitted for the provision of proper light of the workshop on ground.

A plane truss can be defined as a system of bars connected at their ends so as to form a *rigid structure*. This structure may be subjected to external loads, wind loads, gust loads due to which internal forces are developed in bars. The external forces on the structure, the internal forces in bars and the reactions at the supports all are taken to act in one and same plane, therefore *these are termed as plane trusses.*

Fig. 4.1 (*a*) shows a plane truss with five bars joined at their ends. There are 4 joints in the truss *i.e.*, 1, 2, 3 and 4. The truss carries vertical loads W_1 at joint 3 and W_2 at joint 2, while it is supported at joint 1 through a pin joint (or hinge) and at joint 4 through rollers, on a horizontal floor. Reactions developed at the supports are R_1 at joint 1 and R_4 at joint 4.

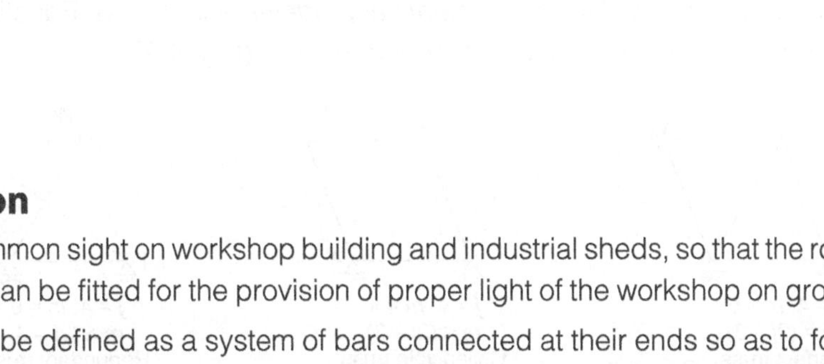

Fig. 4.1

Due to the applied loads, internal forces are developed in the members and these members (bars) offer equal and opposite reactions at joints.

In bar 2–3, tensile force F_1 is developed, showing pulling force F_1 at joints 2 and 3. This type of bar subjected to tensile force is known as a *TIE member*. In bar 3–4, there is compressive force developed say F_2, producing equal and opposite reaction at supports 3 and 4, showing *pushing at the joints*, this type of member is called a *STRUT member*. Similarly tensile force F_3 in member 1–4, tensile force F_4 in member 1–2 and compressive force F_5 in member 2–4.

Remember a tie member produces pulling force at joints and a strut member produces pushing force at joints.

4.2 Types of Plane Trusses

There are 3 types of plane trusses *i.e.*, (*a*) Perfect truss, (*b*) Collapsible truss, (*c*) Redundant truss.

In a truss or frame, if j = number of joints

For a perfect truss, number of members,

$$m = 2j - 3$$

In the truss we have analysed in Section 4.1, there are 4 joints and 5 members.

So $m = 2j - 3 = 2 \times 4 - 3 = 5$

This is a perfect truss.

But if $m < (2j - 3)$, then it is a collapsible truss, *i.e.*, under the action of external loads, the truss tends to *collapse.*

If $m > (2j - 3)$, then it is a *redundant truss or statically indeterminate truss.* Equations of *static equilibrium* are unable to determine forces in the members of redundant truss **(Fig. 4.2).**

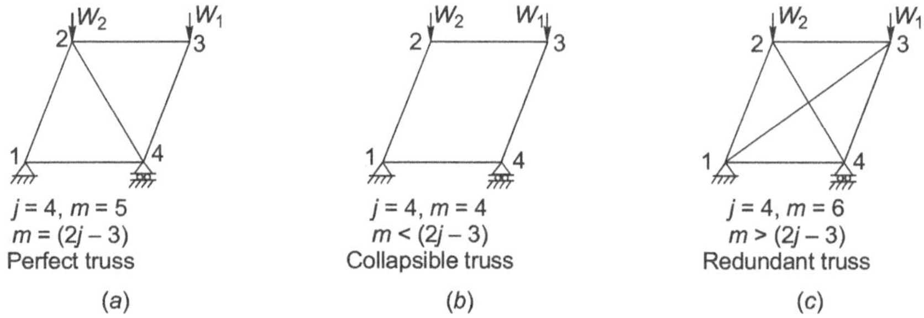

Fig. 4.2

Fig. 4.2 (*a*) shows a *perfect truss* in which $m = 2j - 3$ *i.e.*, members, $m = 5$, joints, $j = 4$. Equations of static equilibrium can be used to determine support reactions and forces in the members of the truss. **Fig. 4.2(*b*)** shows a *collapsible* truss in which $m < (2j - 3)$, *i.e.*, $m = 4$, $j = 4$, $m < (2j - 3)$, under the action of external load, the truss will collapse. **Fig. 4.2 (*c*)** shows a *redundant truss* or a statically indeterminate truss, for determining forces in members equations of static equilibrium are insufficient.

In this chapter we will study only the perfect trusses.

In figures we have shown that members are pin jointed but in actual practice these members are either riveted to a plate or are welded to a plate as shown in **Fig. 4.3**. But to simplify the analysis, the joint is assumed to be hinged, *i.e.*, all the members at the joint are *pin jointed at one location.*

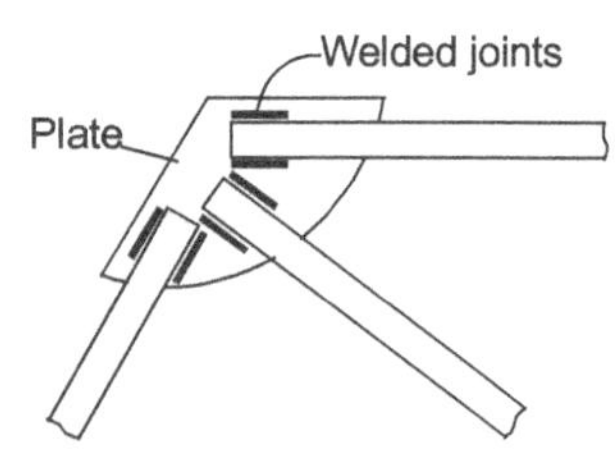

Members of a truss welded to the plate

Fig. 4.3

Therefore following assumptions are made while determining forces in members of a truss:

(*a*) Bars are connected at ends through frictionless hinges.

(*b*) All the external loads are applied at the hinges only and the reactions develop at the hinges only.

(*c*) Axes of all the bars (members) lie in one plane, *i.e.*, middle plane of truss.

(*d*) The bars are of negligible weight.

(*e*) The truss is statically determinate, *i.e.*, equations of static equilibrium can be used to determine forces in members.

4.3 Determination of Support Reactions

The truss is supported either (*a*) as hinged end or (*b*) as roller supported end. The reaction at hinged end can have two components in horizontal and vertical directions, but the reaction at roller support is *always perpendicular to the surface of the support.* Generally one end of a truss is hinged while other end is roller supported, so that when loads are applied on truss, the truss is free to undergo horizontal displacement at its ends.

Reactions at supports can be determined analytically or graphically.

Analytical method: Let us illustrate this method through an example. **Fig. 4.4** shows a truss *ABCD*, with 4 joints and 5 members, *a perfect truss.* It is hinged at end *A* and roller supported at *C*. Distance between supports is 2*a* as shown and altitude of truss *BD*= '*a*' as shown. The truss is subjected to a vertical load 2 W at *D* and a horizontal load *W* at *B* as shown. Note that reaction at *C* is only vertical *i.e.*, perpendicular to horizontal support at *C* *i.e.*, R_{CH} = 0. To determine R_{CV} (vertical component of reaction at *C*) let us take moments about end *A*

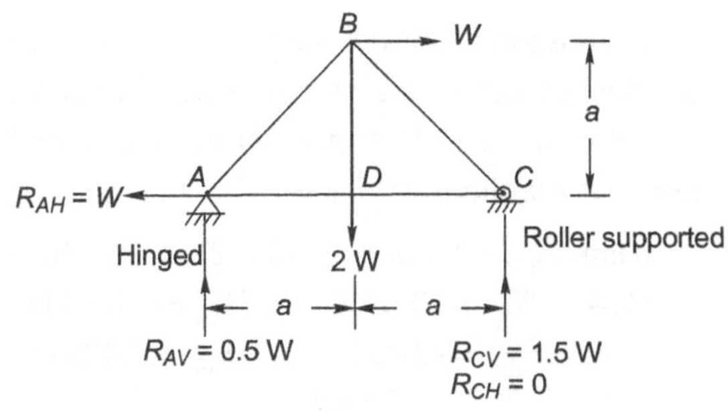

Fig. 4.4

$$W \times a\,(cw) + 2W \times a\,(cw) = R_{CV} \times 2a\,(ccw)$$

Reaction $\qquad R_{CV} = 1.5\text{ W}\uparrow$ ...(1)

Reaction at *A*

$$R_{AH} = \overleftarrow{W} \text{ (to balance horizontal load } W \text{ at } B)$$

$$R_{AV} = 2\text{ W} - 1.5\text{ W} = 0.5\text{ W}\uparrow \text{ (to balance vertical forces).}$$

Exercise 4.1 A plane truss *ABCDE*, hinged at *B* and roller supported at *A*, distance between supports 4 m, carries vertical and horizontal loads of 20 kN and 10 kN as shown in **Fig. 4.5**. Determine reactions at *A* and *B*. State whether it is a perfect truss.

[**Hint:** Take moments about *B*].

[**Ans:** $R_{AV} = 7.5\text{ kN}\uparrow$, $R_{AH} = 0$,

$R_{BV} = 12.5\text{ kN}\uparrow$, $R_{BH} = \overleftarrow{10\text{ kN}}$

A perfect truss].

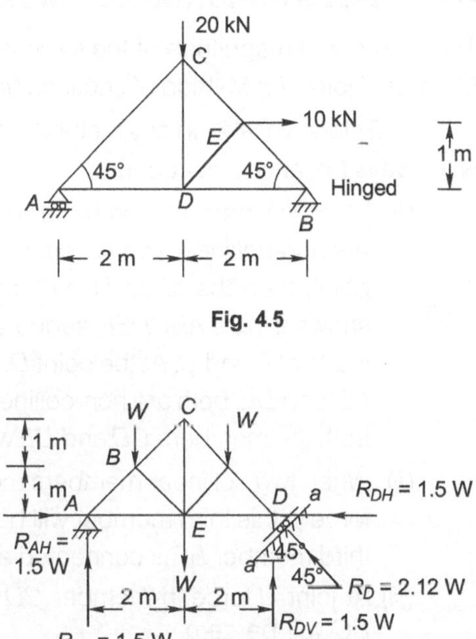

Fig. 4.5

Inclined support: In this example of plane truss, one end of truss is roller supported on inclined plane '*aa*' as shown in **Fig. 4.6**. Plane '*aa*' is inclined with horizontal at an angle 45°. Truss carries 3 loads as shown. Determine support reactions if end *A* is hinged and end *D* is roller supported on inclined plane as shown.

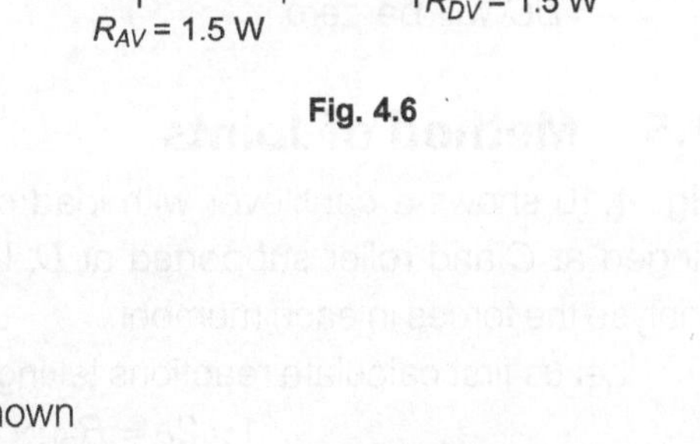

Fig. 4.6

Solution Reaction at *D* will be perpendicular to plane *aa*, so reaction R_D is inclined to horizontal at an angle 45° as shown. There are two components of reaction R_D *i.e.*, R_{DH} and R_{DV}, since R_{DH} passes through point *A*, it will not produce any moment about point *A*.

Taking moments about *A*

$$W \times 1 + 2 \times W + 3 \times W = R_{DV} \times 4$$

$$R_{DV} = 1.5\text{ W}$$

$$R_{DH} = 1.5\text{, W because angle is 45° as shown}$$

$$R_D = \sqrt{R_{DV}^2 + R_{DH}^2} = \sqrt{1.5^2 + 1.5^2} = 2.12\text{ kN.}$$

Reaction at A

$$R_{AH} = R_{DH} = \overrightarrow{1.5\text{ W}} \text{ (to balance } R_{DH})$$

$$R_{AV} = W + W + W - 1.5\text{ W} = 1.5\text{ W} \uparrow$$

Exercise 4.2 A truss *ABCDEF*, hinged at *A* and roller supported at *E* on inclined plane *'aa'*, inclined with horizontal at 30°. Truss carries 3 loads of 6 kN, 10 kN and 4 kN at *B*, *C* and *D* as shown in **Fig. 4.7.** Determine reactions at *A* and *E*.

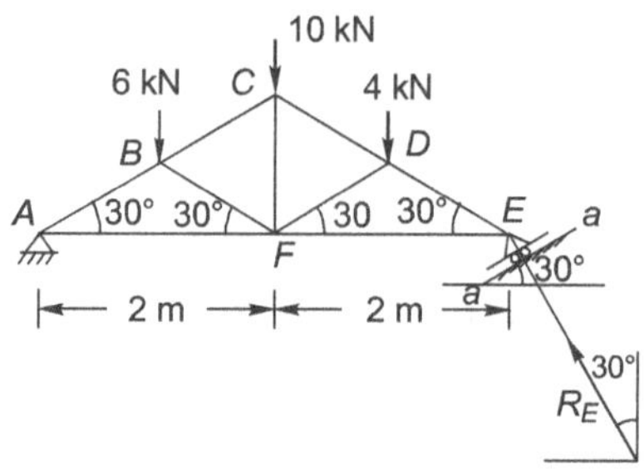

Fig. 4.7

[**Hint:** $R_{EV} \times 4 = 6 \times 1 + 10 \times 2 + 4 \times 3$ kNm].

[**Ans:** $R_{AV} = 10.5$ kN, $R_{AH} = 5.485$ kN;
$R_{EV} = 9.5$ kN, $R_{EH} = 5.485$ kN
$R_E = 10.97$ kN.

Exercise 4.3 A truss with 6 joints is shown in **Fig. 4.8**. State whether it is a perfect truss?

It carries loads of 6 kN and 2 kN at joints 3 and 4. Joint 1 is hinged while joint 5 is roller supported. Determine reactions at joints 1 and 5.

[**Hint:** Note that both reactions R_1 and R_5 are vertical].

[**Ans:** $R_1 = -1$ kN (downwards), $R_5 = +9$ kN (upwards)].

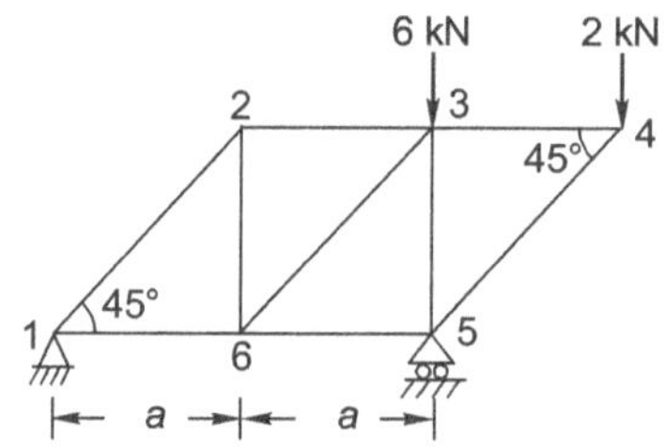

Fig. 4.8

4.4 Determination of Forces in Members of the Truss

The nature and magnitude of the forces in bars constituting the truss can be determined by 3 methods *i.e.*, (*i*) Method of joints (*ii*) Method of sections (*iii*) Graphical method.

Before we take up any method for discussion, following general rules will always be taken into account:

(*i*) When two bars (or members) are meeting at a point and the two *bars are non-collinear* and *no external force* or a reaction is acting at the point, then the *force in both the members will be zero.* **Fig. 4.9** shows a truss *ABCDEF*, supported at *E* and *A* and carrying vertical loads at *B* and *C*. At the point *D*, there is no external force. Two bars *CD* and *DE*, both are non-collinear and meet at the point *D*. Force in both the members *CD* and *DE* will be zero.

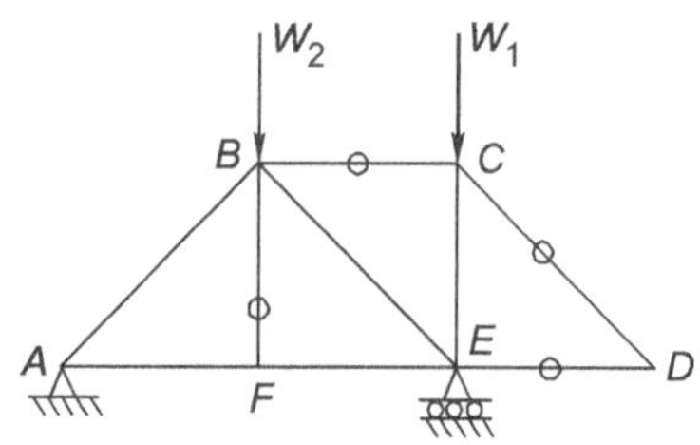

Fig. 4.9

(*ii*) When two collinear members meet at a point, and a third member is connected at this point then, the force in this third member will be zero. In the Figure 4.9, two members *AF* and *FE* are collinear and the third member *BF* is connected at the point *F*, *then force in the member BF will be zero.*

(*iii*) At joint *D* force in member *CD* is zero. Force in member *CE* will be equal to W_1. Force in member *BC* will be zero.

4.5 Method of Joints

Fig. 4.10 shows a cantilever with load of 1 kN at joint *A*, the truss is hinged at *C* and roller supported at *D*. Let us take joint by joint, and analyse the forces in each member.

Let us first calculate reactions taking moments about *C*

$$1 \times 2a = R_{DH} \times d; \quad R_{DH} = 2\text{ kN (roller support)}$$

So $$R_{CH} = \overrightarrow{2\text{ kN}} \quad R_{CV} = 1\text{ kN}\uparrow$$

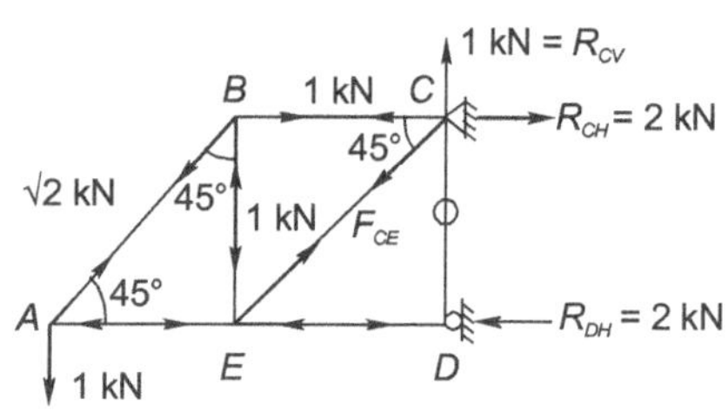

Fig. 4.10

As truss is in equlibrium, so all the joints and members will also be in equlibrium.

Joint *A*: Two members *AB* and *AE*, and a force of 1 kN. Say forces in *AB* and *AE* are F_{AB} and F_{AE} respectively.

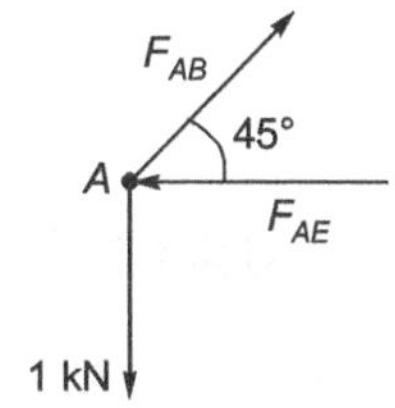

Force in member *AB*,

$$F_{AB} \sin 45° = 1 \text{ kN}$$

$$F_{AB} = \sqrt{2} \text{ kN (tension)}$$

Force in member *AE*,

$$F_{AE} = F_{AB} \cos 45°$$

$$F_{AE} = 1 \text{ kN (comp.)}$$

Joint *B*: Three forces in members *i.e.*, F_{AB}, F_{BE} and F_{BC}.

Force in member *BC*, $F_{BC} = F_{BA} \sin 45°$

$F_{BC} = 1$ kN (tension)

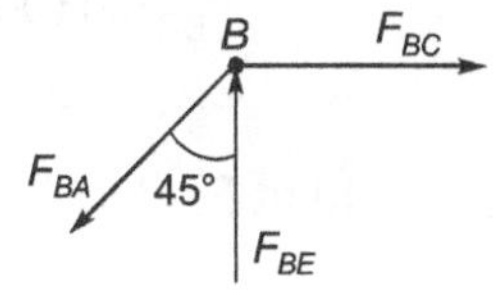

Force in member *BE*, $F_{BE} = F_{AB} \cos 45°$

$F_{BE} = 1$ kN (comp.)

Joint *D*: F_{DE} and 2 kN reaction are collinear, therefore

Force in member *CD*, $F_{CD} = 0$

Force in member *DE*, $F_{DE} = 2$ kN (comp.)

Joint *C*: $F_{BC} = 1$ kN (tension)

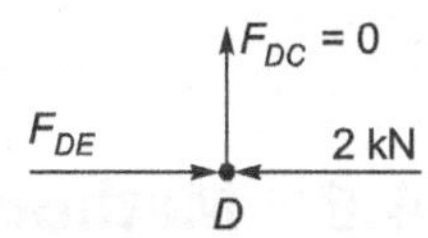

$$R_{CH} = 2 \text{ kN} \rightarrow$$

$$R_{CV} = 1 \text{ kN} \uparrow \text{(to balance 1 kN force at } A)$$

Let us take F_{CE} (tension) and force in member *CE*,

$$F_{CE} \sin 45° = 1$$

$$F_{CE} = \sqrt{2} \text{ kN (tension)}$$

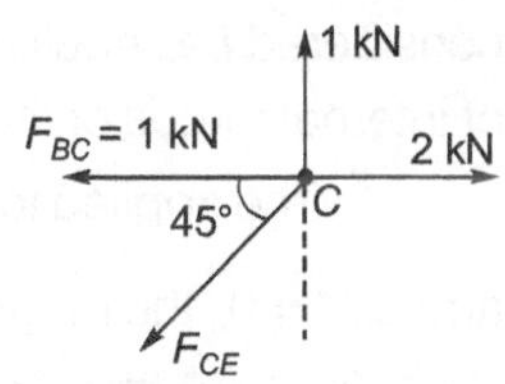

F_{CE} sin 45° balances R_{CV} of 1 kN

Forces in all the members have been determined.

+ sign – tension member (*T*)

– sign – compression members (*C*)

Member	Force	Nature
AB	$+\sqrt{2}$ kN	*T*
BC	+1 kN	*T*
BE	–1 kN	*C*
CE	$+\sqrt{2}$ kN	*T*
CD	0	
ED	–2 kN	*C*
AE	–1 kN	*C*

Exercise 4.4 Fig. .4.13 shows a cantilever truss *ABCDEF*. Determine support reactions and forces in member by method of joints.

[**Hint:** Force in members *BC* and *CD* will be zero. Moreover at joint *F*, reaction will be horizontal and force in member *FE* will be equal to horizontal reaction at *F*, so force in member *AF* will become zero].

[**Ans:** As truss is in equlibrium, so all the joints and members will also be in equlibrium.

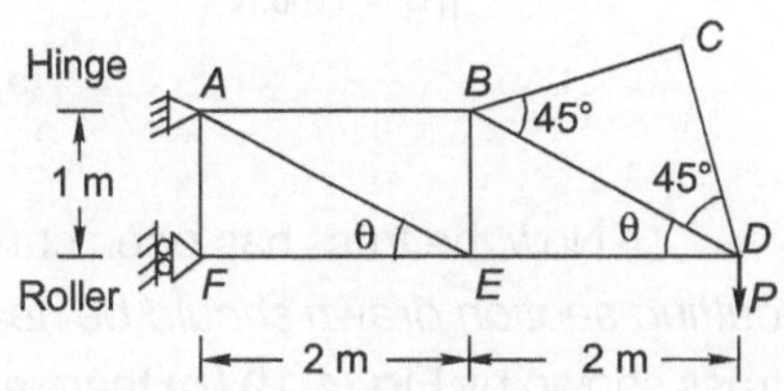

Fig. 4.13

Reactions, $R_{FH} = \overrightarrow{4P}$ (Horizontal)

$R_{AV} = P\uparrow,\ R_{AH} = \overleftarrow{4P}$ (To balance reaction at F)].

$$\tan\theta = \frac{1}{2};\ \theta = 26.56°,$$

Joint D: $F_{BD}\sin\theta = P$

$$F_{BD} = P\operatorname{cosec}\theta = 2.236\,P\text{(tension)}$$
$$F_{ED} = F_{BD}\cos\theta = 2\,P\text{(comp.)}$$

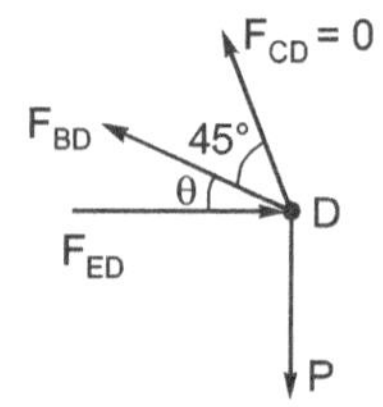

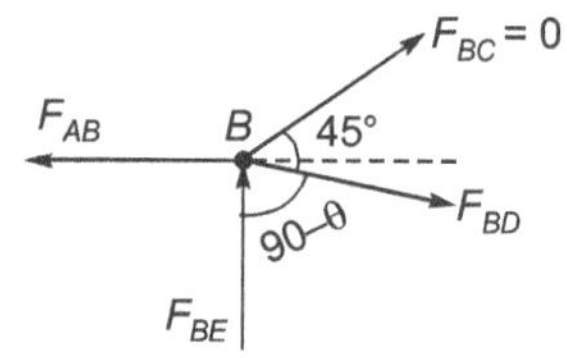

Joint B:

$$F_{BD}\cos(90° - \theta) = F_{BE}$$
$$F_{BE} = P\text{(comp.)}$$
$$F_{BD}\sin(90 - \theta) = F_{AB}$$
$$F_{AB} = 2P\text{(tension)}$$

Joint F: $F_{FE} = 4\,P$(comp.)

$$F_{AF} = 0$$

Joint E:

$$F_{BE} = F_{AE}\sin\theta$$
$$F_{AE} = F_{BE}\operatorname{cosec}\theta = 2.236\,P\text{(tenison)}$$

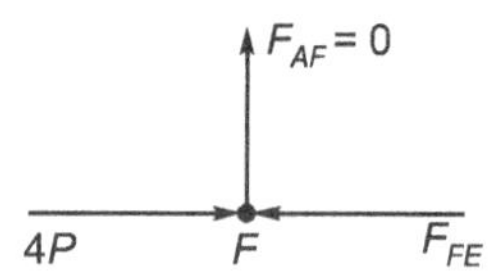

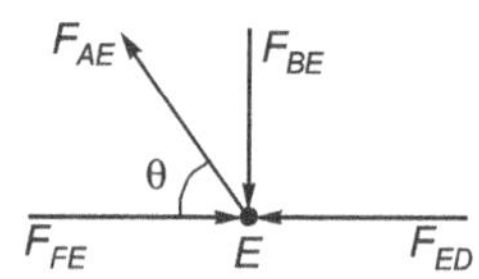

4.6 Method of Sections

Without going into the analysis of joint by joint for equilibrium, forces in a few chosen members can be determined by this method. A truss is cut into two parts with the help of a section and equilibrium of anyone part can be considered *i.e.*, equilibrium of forces due to (*i*) applied load or loads, (*ii*) reaction at the support and (*iii*) reactions of internal forces of the members cut.

The applied forces and support reactions are known, then using the conditions of equilibrium *i.e.*, $\Sigma F = 0$ and $\Sigma M = 0$, the magnitude and nature of the *forces in the broken-members are determined.* Let us consider a truss shown in **Fig. 4.19 (*a*).** To determine forces in various members, the truss can be cut in two parts by sections such as *mm*, *nn* and *pp*. If we calculate the reactions at the supports, then

$$R_A = R_1 = 1100 \text{ N and } R_E = R_2 = \mathbf{1100\ N}$$

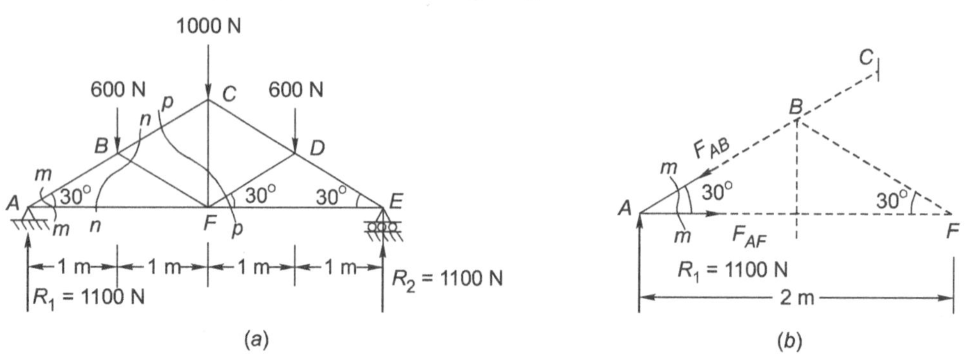

Fig. 4.19 (a) and (b)

Now the truss has been cut in two parts by a section *mm*. *Please take care that to avoid confusion, the cutting section drawn should be nearer the joints as section mm is closer to joint A.* Now on the first part of the truss shown by **Fig. 4.19 (*b*)** there are three forces (*i*) Reaction, R_1 (*ii*) Force in the member *AB*; F_{AB} (*iii*) Force in the member *AF*; F_{AF}. Let us take moments of the forces about the point *B*.

$$R_1 \times 1 - F_{AF} \times 1\tan 30° = 0$$

$$F_{AF} = \frac{1100}{\tan 30°} = \frac{1100}{0.57736} = 1905.2 \text{ N (Tension)}$$ **(pulling at the joint)**

Taking moments of the forces about the point F

$$R_1 \times 2 - F_{AB} \times 2\sin 30° = 0$$

$$F_{AB} = 1100 \times 2 = 2200 \text{ N (Compression)}$$ **pushing at the joint**

The truss can be cut in two parts by section nn also cutting members BC, BF and AF as shown in **Fig. 4.19 (*c*)**.

The first part of the cut truss has following forces (*i*) A vertical force of 600 N at B (*ii*) A vertical reaction of 1100 N at A (*iii*) Force, F_{BC} (*iv*) Force, F_{BF} (*v*) Force, F_{AF}.

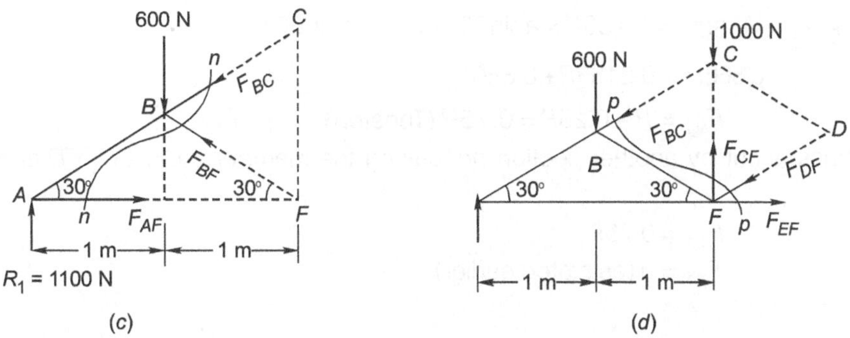

Fig. 4.19 (c) and (d)

Taking moments of the forces about the point F

$$1100 \times 2 = 600 \times 1 + F_{BC} \times 2 \sin 30°$$

$$F_{BC} = 1600 \text{ N (Compression)}$$

Taking moments about the point A

$$600 \times 1 = F_{BF} \times 2 \sin 30°$$

$$F_{BF} = 600 \text{ N (Compression)}$$

The truss is further cut by a section pp, cutting the members BC, FC, FD and FE as shown in **Fig. 4.19 (*d*)**. The truss is symmetrically loaded, therefore

$$F_{DF} = F_{BF} = 600 \text{ N (Compression)}$$

$$F_{EF} = F_{AF} = 1905.2 \text{ N (Tension)}$$

Taking moments of the forces about the point A

$$600 \times 1 + F_{DF} \times 2 \sin 30° = F_{CF} \times 2$$

$$600 + 600 = 2F_{CF}$$

$$F_{CF} = 600 \text{ N (Tension)}$$

It can be observed here that this method is time consuming and *many more sections are considered if we have to determine forces in all the members of the truss.* Therefore this method is only useful when the forces only in a few chosen members of the truss are to be determined.

Example 4.1 By the method of sections, determine the nature and magnitude of the force in the member BD of the truss shown in **Fig. 4.20**.

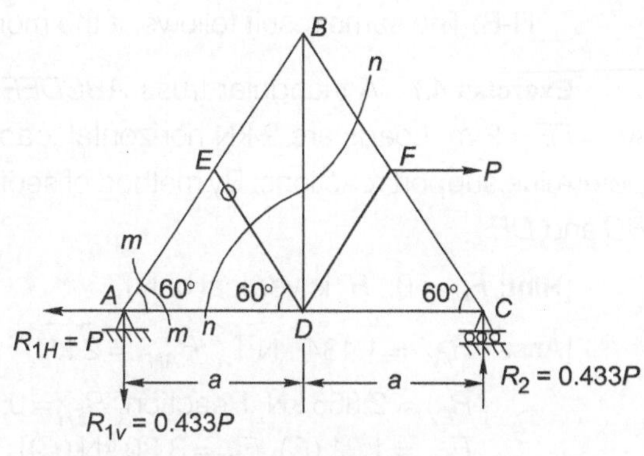

Fig. 4.20

Solution Let us first determine the support reactions. Taking moments about the point A

$$P\times a\sin 60^\circ = 2a\,.\,R_2$$

$$R_2 = 0.433P\uparrow$$

Reaction at A

For equilibrium, $R_{1V} = 0.433P\downarrow$

$$R_{1H} = \overleftarrow{P} \text{ (to balance } P \text{ at } F)$$

Now at the joint E, members AE and EB are collinear and member DE is joined at E. Therefore force in DE

i.e., $F_{DE} = 0$

Let us first take mm section as shown. Part I of the truss has forces (*i*) R_{1V} (*ii*) R_{1H} (*iii*) F_{EA} (*iv*) F_{AD} as shown in **Fig. 4.21**. Taking moments about the point E,

$$P\times a\times\sin 60^\circ = 0.433P\times a\sin 30^\circ + F_{AD}\times a\sin 60^\circ$$

$$0.866P = 0.2165P + 0.866F_{AD}$$

$$F_{AD} = P - 0.25P = \mathbf{0.75P\ (Tension)}$$

Now the truss is cut by another section nn, cutting the members, BF, BD, ED and AD as shown in **Fig. 4.21 (*b*).** Now

$$F_{AD} = 0.75P$$

$$F_{ED} = 0 \text{ (as stated earlier)}$$

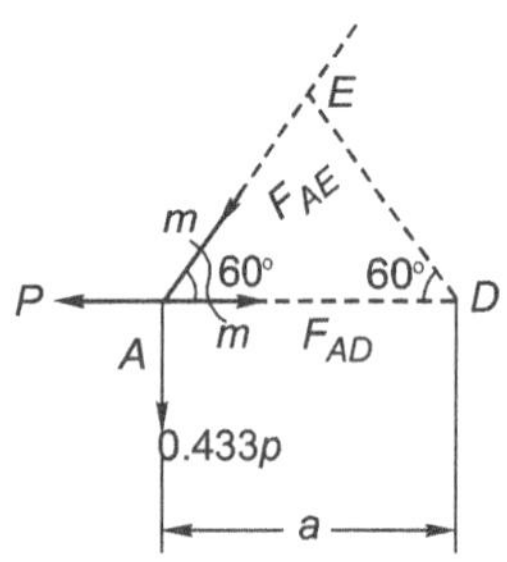

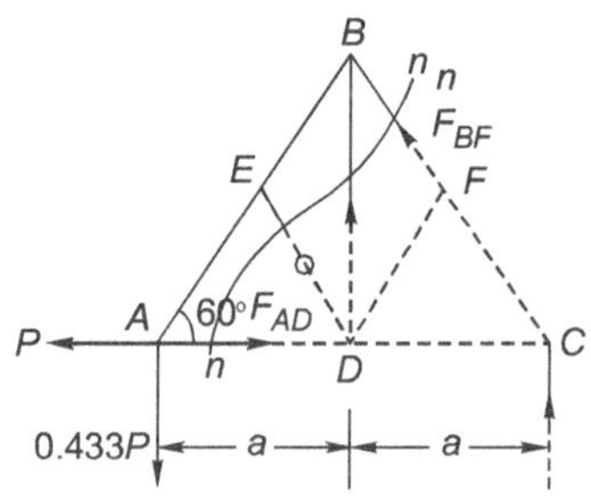

Fig. 4.22

Taking moments of the forces about the point F,

$$F_{AD}\times a\sin 60^\circ + 0.433P\times 1.5a - P\times a\sin 60^\circ - F_{BD}\times\frac{a}{2} = 0$$

$$0.75\times 0.866P + 0.433\times 1.5P - P\times 0.866 = 0.5F_{BD}$$

$$F_{BD} = 0.866P \text{ (Compression)}$$

N-B. The same result follows, if the moment centre is chosen at C.

Exercise 4.7 A triangular truss $ABCDEF$ is shown in **Fig. 4.22**. Distances $AF = FE = 2$ m. Loads are 2 kN horizontal load at D and 4 kN vertical load at F. Determine support reactions. By method of sections determine forces in members BC and DE.

[**Hint:** $F_{BF} = 0$, R_E is vertical only].

[**Ans:** $R_{AV} = 1.134$ kN $\uparrow$, $R_{AH} = \overleftarrow{2}$ kN,

$R_{EV} = 2.866$ kN, Reaction, $R_{EH} = 0$,

$F_{BC} = 1.31\ (C)$, $F_{DE} = 3.30$ kN (C)].

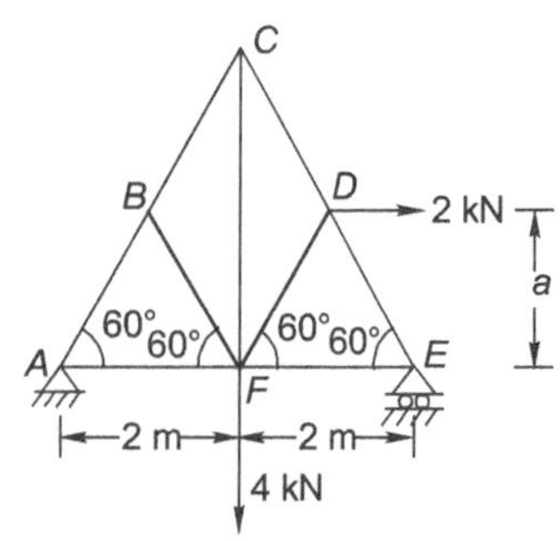

Fig. 4.22

Problem 4.1 A truss *ABCDE* is shown in **Fig. 4.23**. Truss is hinged at joint *C* and roller supported at joint *A*. Vertical and horizontal forces are applied on truss as shown. Determine support reactions what are zero force members?

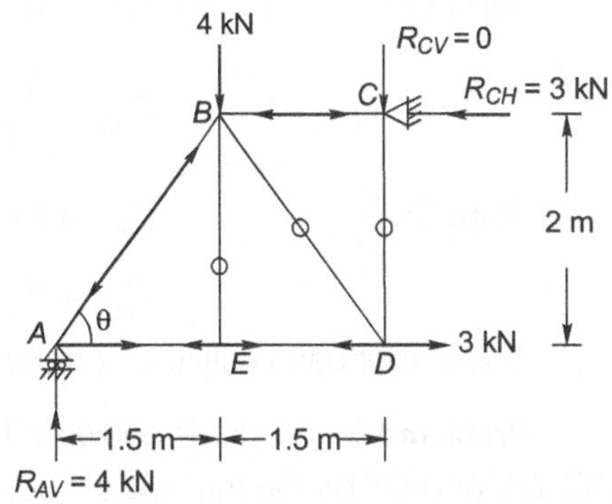

Fig. 4.23

Solution **Reactions:** Taking moments of forces about point *C*

$$3\times 2(ccw) + 4\times 1.5(ccw) - R_A \times 3(cw) = 0$$

$$R_{AV} = 4 \text{ kN}\uparrow,\ R_{AH} = 0$$

At *C* $R_{CV} = 4 - 4 = 0$

$$R_{CH} = \overleftarrow{3}\text{kN}$$

(to balance horizontal force at *D*)

Joint *E*: F_{AE} and F_{ED} are collinear, there $F_{BE} = 0$

$F_{AE} = F_{ED} = 3$ kN (*T*) Tensile

Joint *C* $R_{CV} = 0$

So $F_{CD} = 0$

$F_{BC} = R_{CH} = 3$ kN (*C*) (compressive).

Joint *D* $F_{CD} = 0$

Forces F_{ED} and 3 kN are collinear, force in member *BD*

i.e., $F_{BD} = 0$.

Problem 4.2 A cantilever truss *ABCDE* is shown in **Fig. 4.24**. It is subjected to loads *W* each at *C* and at *D*. Determine support reactions and force in member *BD* of the truss.

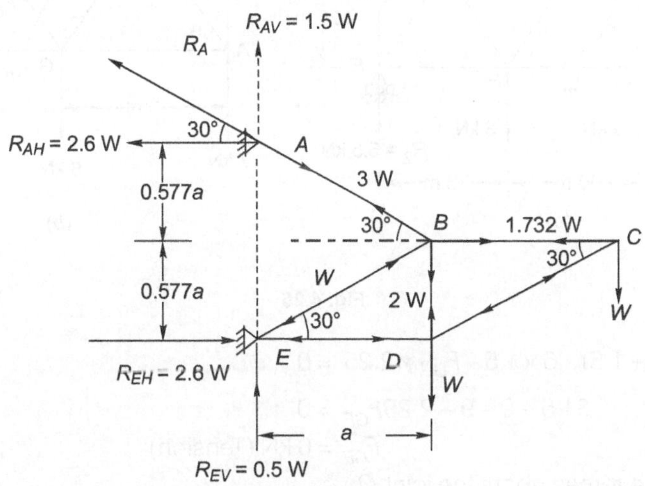

Fig. 4.24

Solution Distance $AE = 2\times a \times \tan 30° = 1.154a$

Reactions: Taking moments about *E*,

$R_{AH} \times 1.154a = W \times a + 2W \times a$ (Note that R_{AV} is passing through point *E*)

$$R_{AH} = \frac{3\text{ W}}{1.154} = 2.6 \text{ W}$$

$$R_{AV} = R_{AH}\tan 30° = 2.6 \times 0.577 = 1.5 \text{ W}$$

At *E*: Reaction $R_{EV} = 2\text{ W} - 1.5\text{ W} = 0.5\text{ W}\uparrow$

$R_{EH} = R_{AH} = 2.6$ W (for balancing)

Joint C: $F_{CD}\sin 30^\circ = W$

$$F_{CD} = \frac{W}{0.5} = 2W\,(C)$$

Joint D $F_{BD} = F_{CD}\sin 30^\circ + W = 2\,W \times 0.5 + W = 2\,W \uparrow (T)$

$$F_{ED} = F_{CD}\cos 30^\circ = 2\,W \times 0.866 = 1.732\,W\,(C)$$

The reader can verify the correctness of the solution.

Problem 4.3 A truss is shown in **Fig. 4.25 (*a*).** Determine the magnitude and nature of forces in members *BC*, *GC* and *GF* by the method of sections.

Solution Let us first determine the support reactions.

Taking moments of the forces about the joint *A*

$$6 \times 3 + 3 \times 6 + 1.5 \times 9 = 9 \times R_2$$

Reaction, $R_2 = 5.5$ kN

$$R_1 = 2 + 6 + 3 + 1.5 - 5.5 = 7 \text{ kN}$$

Now cutting the truss into two parts through the section *m–m*. This section is cutting the members *BC*, *GC* and *GF*. One part of the truss is shown in **Fig. 4.25 (*b*)**. Taking moments of the forces about the joint *C*.

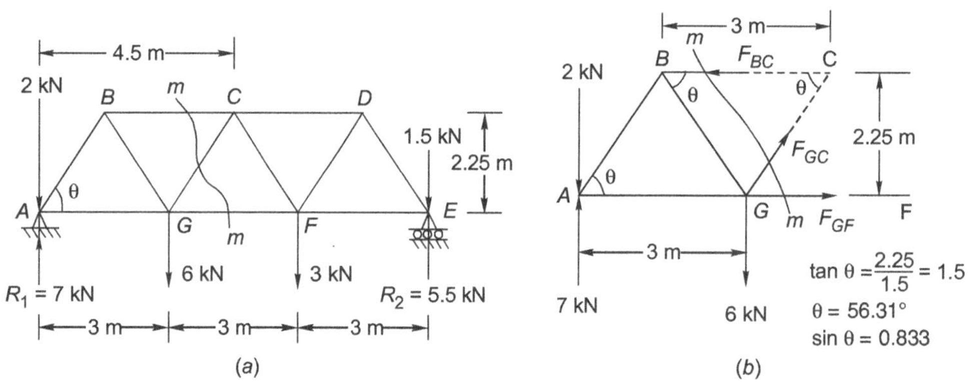

Fig. 4.25

$$7 \times (3 + 1.5) - 2 \times (3 + 1.5) - 6 \times 1.5 - F_{GF} \times 2.25 = 0$$

$$31.5 - 9 - 9 - 2.25F_{GF} = 0$$

$$F_{GF} = 6 \text{ kN (Tension)}$$

Taking moments of the forces about the joint *G*

$$7 \times 3 - 2 \times 3 - F_{BC} \times 2.25 = 0$$

$$F_{BC} = \frac{15}{2.25} = 6.667 \text{ kN (Compression)}$$

Taking moments of the forces about the joint *B*

$2 \times 1.5 - 7 \times 1.5 - 6 \times 1.5 + F_{GF} \times 2.25 + F_{GC} \times 3\sin\theta = 0$, putting the value of F_{GF}

$$-16.5 + 2.25 \times 6 + F_{GC} \times 3 \times 0.833 = 0$$

$$-3 + 2.5F_{GC} = 0$$

$$F_{GC} = 1.2 \text{ kN } (T).$$

PRACTICE PROBLEMS

4.1 A cantilever truss is shown in **Fig. 4.26**. Determine support reactions and forces in members *BE* and *DB* of the truss.

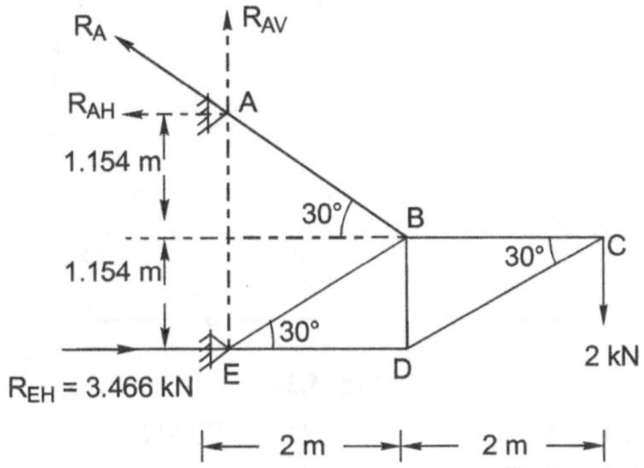

Fig. 4.26

[**Hint:** Calculate distance *AE*].

[**Ans:** $R_{AH} = 3.466$ kN, $R_{AV} = 2$ kN, $R_{EH} = \overrightarrow{3.466}$ kN,
$F_{BD} = 2$ kN, $F_{BE} = 0$

4.2 A truss *ABCDEF* is hinged at end *A* and roller supported at *D* as shown in **Fig. 4.27**. It is subjected to vertical and horizontal forces as shown. Calculate support reactions, and forces in members *DE* and *CE*.

[**Hint:** $F_{BF} = 0$].

[**Ans:** $R_D = 3$ kN ↑,

$R_{AH} = \overleftarrow{1 \text{ kN}}$, $R_{AV} = 1$ kN ↑, $F_{ED} = 3$ kN(*T*)
$F_{CE} = 3$ kN (*T*).

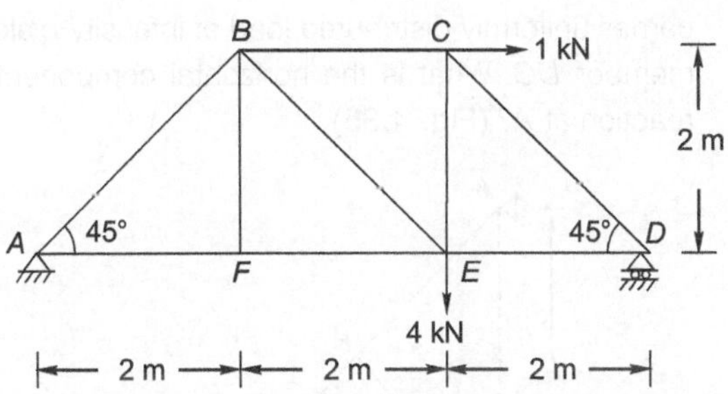

Fig. 4.27

4.3 A cantilever truss *ABCDEFG* is hinged at end *A* and roller supported at *G* as shown in Fig. 4.28. There are loads of 2 kN at *D* and 5 kN at *E*. By method of sections determine axial forces in members *AB*, *BF* and *EF*.

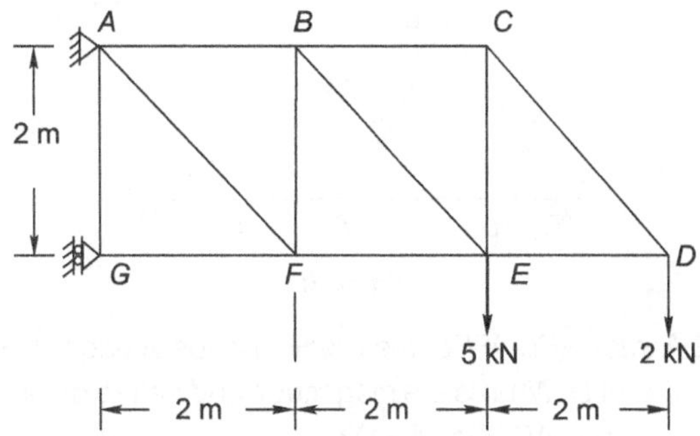

Fig. 4.28

[**Ans:** $F_{AB} = 9$ kN (*T*), $F_{BF} = 7$ kN (*C*),
$F_{FE} = 9$ kN (*C*)].

4.4 A truss *ABCDEF* is shown in **Fig. 4.29**, subjected to vertical loads at joints *F* and *E*. By method of sections, determine forces in members *BC*, *FE* and *CF*.

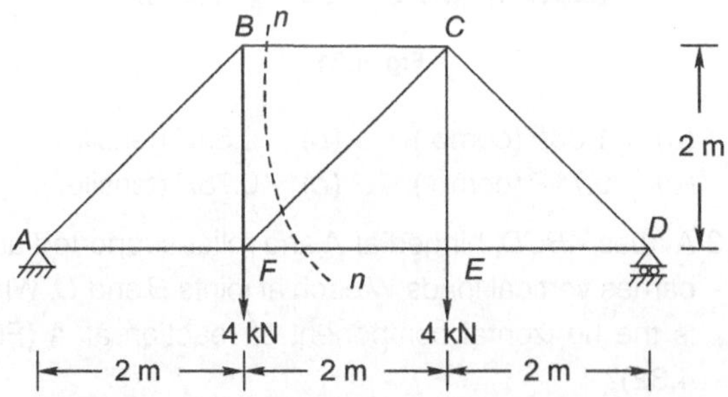

Fig. 4.29

[**Hint:** Determine reactions at supports and take a section to cut 3 members preferably near ends *B* and *F* as shown].

[**Ans:** $R_A = R_D = 4$ kN; $F_{BC} = 4$ kN (*C*),
$F_{FE} = 4$ kN (*T*), $F_{FC} = 0$].

MULTIPLE CHOICE QUESTIONS

4.1 A truss *ABCD*, is loaded by a vertical load *P* at joint *B* as shown in **Fig. 4.30**. Length of members *AB* = *BD* = *AD* = *DC* = *L*, what is the tensile force in member *AD*?

(*a*) *P* (*b*) 0.75*P*
(*c*) 0.5*P* (*d*) 0.433*P*.

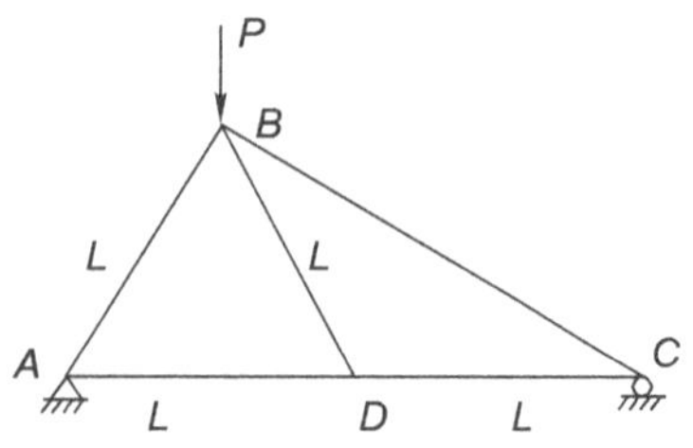

Fig. 4.30

4.2 A truss *ABCDEF* carries vertical loads *W* each at joints *C* and *D*. What is the magnitude and nature of force in member *AB* **(Fig. 4.31)**?

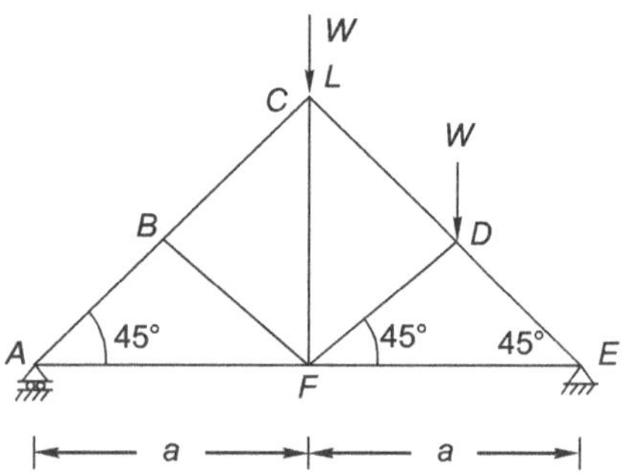

Fig. 4.31

(*a*) 1.06*P* (comp.) (*b*) 0.53*P* (tensile)
(*c*) 0.75*P* (comp.) (*d*) 0.75*P* (tensile)

4.3 A truss *ABCD*, hinged at *A* and roller supported at *D* carries vertical loads *W* each at joints *B* and *C*. What is the horizontal component of reaction at *A* **(Fig. 4.32)**?

(*a*) 2*W* (*b*) 1.5*W*
(*c*) *W* (*d*) 0.707*W*.

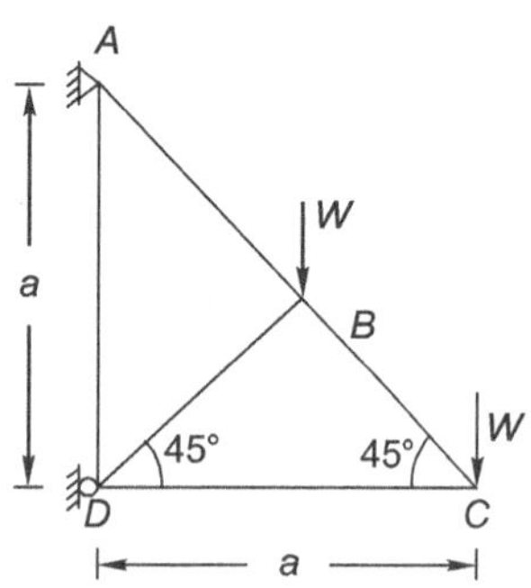

Fig. 4.32

4.4 A truss *ABCD* is shown in **Fig. 4.33**. It carries a load *W* at joint *D*. Area of cross-section of all the members is equal to 100 mm^2. If the maximum stress in any member is not to exceed 100 MPa, what is the maximum magnitude of *W*?

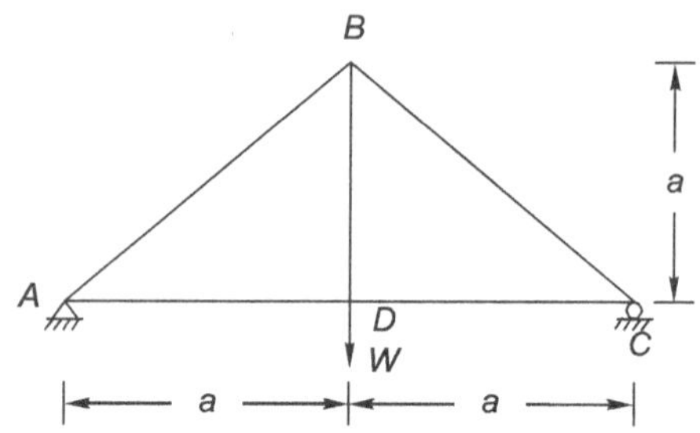

Fig. 4.33

(*a*) 10 kN (*b*) 8.66 kN
(*c*) 7.07 kN (*d*) 5 kN.

4.5 A truss *ABCDE* carries vertical load *W* each at joint *B* and *C*. What the magnitude of force in member *BD* **(Fig. 4.34)**?

(*a*) 0.66*W* (*b*) 0.75*W*
(*c*) *W* (*d*) None of these.

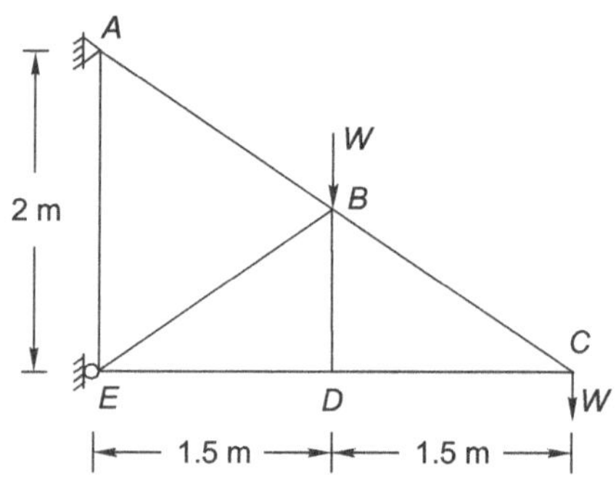

Fig. 4.34

4.6 A truss *ABCD*, hinged at *A*, roller supported at *D* carries uniformly distributed load of intensity *p* along member *DG*. What is the horizontal component of reaction at *A*? **(Fig. 4.35)**

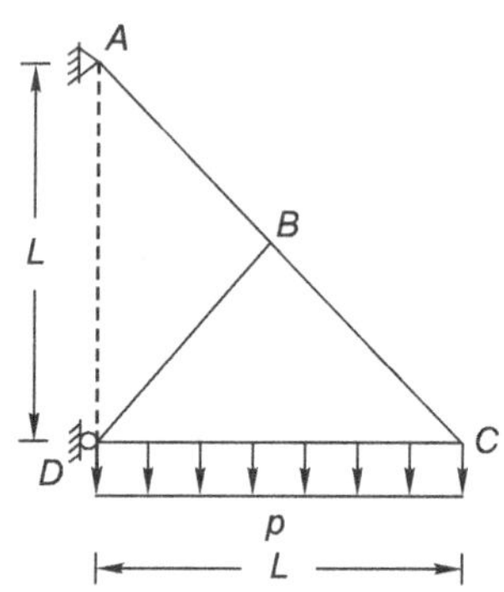

Fig. 4.35

(a) 0 (b) $pL/2$
(c) $3/4\ pL$ (d) pL.

4.7 A truss *ABCD* carries a vertical load at joint *A* as shown in **Fig. 4.36**. If force in member *AD* is 5 kN compressive, what is force in member *CB*?

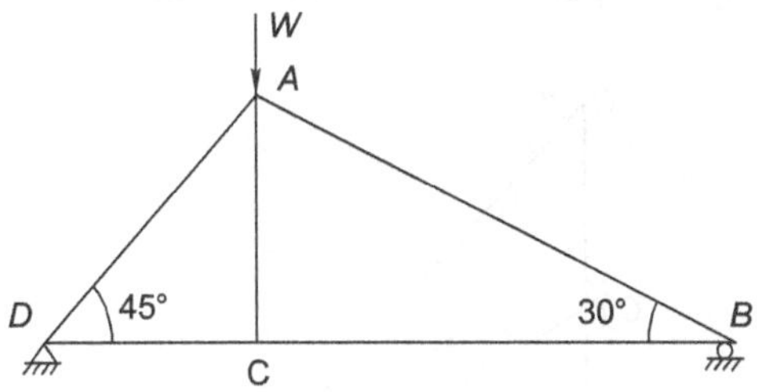

Fig. 4.36

(a) 5 kN (comp.) (b) 3.53 kN (tensile)
(c) 2.5 kN (tensile) (d) None of these.

4.8 A truss *ABCD*, hinged at end *A* is roller supported on inclined plane at *C*. Angle of inclination of plane is 45°. Load at joint *D* is *W* as shown in **Fig. 4.37**. What is the horizontal component of reaction at *C*?

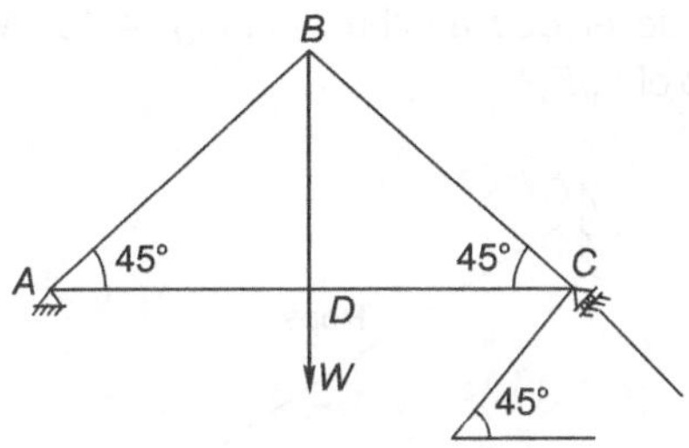

Fig. 4.37

(a) Zero (b) $0.5W$
(c) $0.707W$ (d) W

4.9 A truss *ABCDE* is shown in **Fig. 4.38**. A load *W* acts at joint *D* as shown in **Fig. 4.38**. What is the force in member *BC*?

(a) W (tension) (b) W (comp.)
(c) $\sqrt{2}W$ (tensile) (d) None of these.

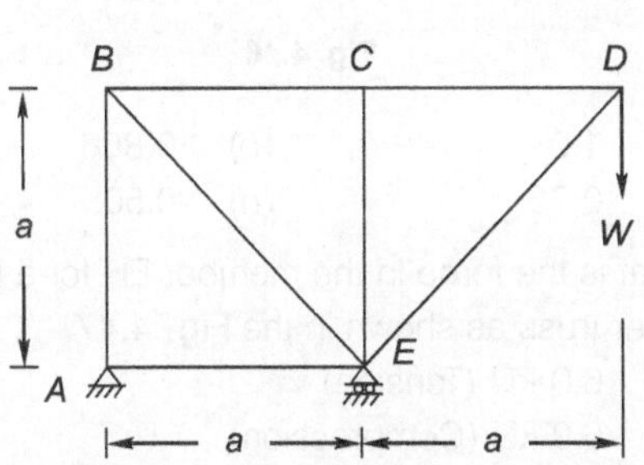

Fig. 4.38

4.10 A truss *ABCDEF*, shown in the **Fig. 4.39** carries loads *P* each at joints *B* and *C*. What is the value of *P* if force in member *AB* is 3 kN? Method of section may be used

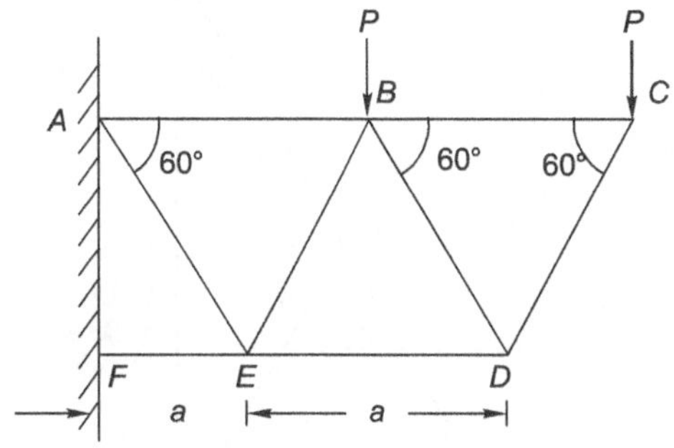

Fig. 4.39

(a) 1 kN (b) 1.299 kN
(c) 1.5 kN (d) None of these.

4.11 A truss *ABCD* hinged at *C*, roller supported at *D* is subjected to a vertical load 20 kN at joint *A*. What is the reaction at *C* **(Fig. 4.40)**?

(a) 20 kN (b) 15 kN
(c) 35 kN (d) None of these.

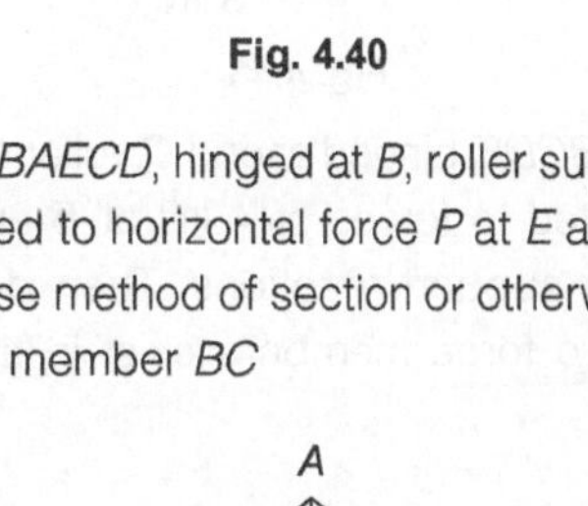

Fig. 4.40

4.12 A truss *BAECD*, hinged at *B*, roller supported at *D*, is subjected to horizontal force *P* at *E* as shown in **Fig. 4.41**. Use method of section or otherwise, determine force in member *BC*

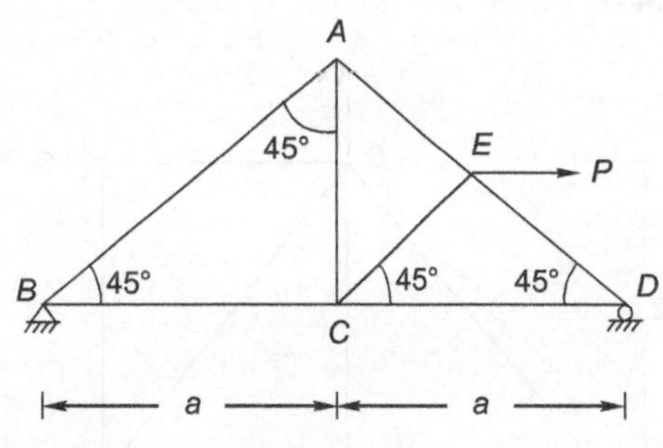

Fig. 4.41

(a) P (b) $0.75P$
(c) $0.5P$ (d) $0.25P$

4.13 A truss *ABCDE* is shown in **Fig. 4.42**. Truss is hinged at *D* and roller supported at *C*. How many zero force members are there in truss?

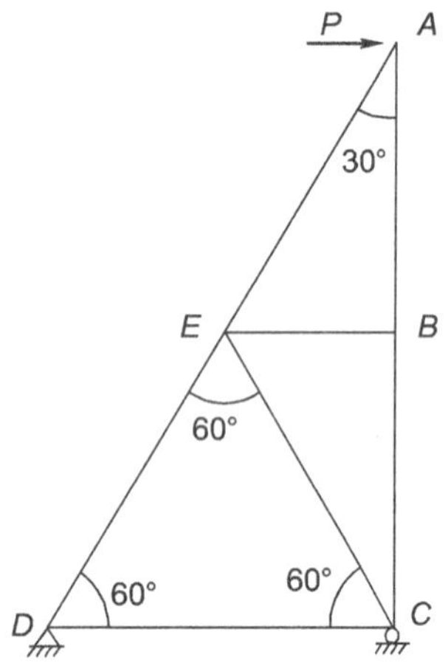

Fig. 4.42

(*a*) 4 (*b*) 3
(*c*) 2 (*d*) 1

4.14 A truss *ABCD* roller supported at *A* hinged at *C*, carries a vertical load of 3 kN at *D* as shown in **Fig. 4.43.** What is the magnitude and nature of force in member *AB*?

(*a*) 1 kN (comp.) (*b*) 2 kN (comp.)
(*c*) 2 kN (tensile) (*d*) 1.5 kN (comp.)

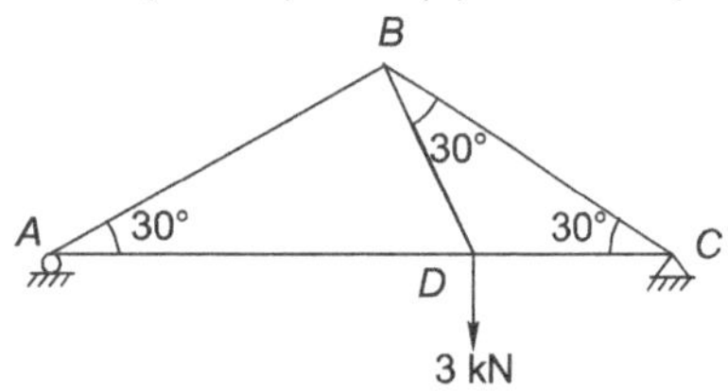

Fig. 4.43

4.15 A truss *ABCDE*, hinged at end *C*, roller supported at *A* is subjected to forces of 4 kN at *B* and 3 kN at *D*. Calculate support reactions. Then determine how many zero force members exist in the truss **(Fig. 4.44)**?

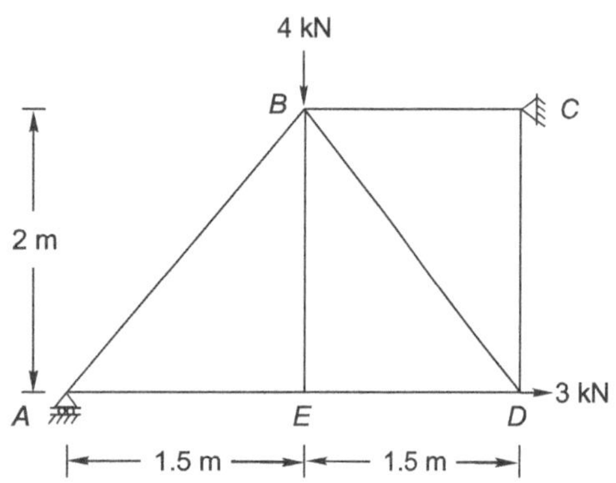

Fig. 4.44

(*a*) 1 (*b*) 2
(*c*) 3 (*d*) 4

4.16 A truss *ABC*, shown in **Fig. 4.45** carries a load *P* at joint *B*. What is the force in the member *AB*?

(*a*) 3.346*P* (*b*) 3*P*
(*c*) 1.732*P* (*d*) 0.866*P*

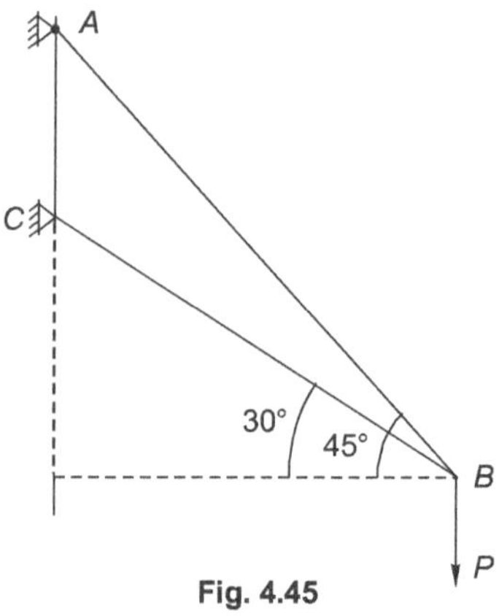

Fig. 4.45

4.17 A bar *AB* is held by a rope at the end of which load *W* is suspended. In two cases shown, forces F_1 and F_2 are developed as shown in **Fig. 4.46**. What is the ratio of F_1/F_2?

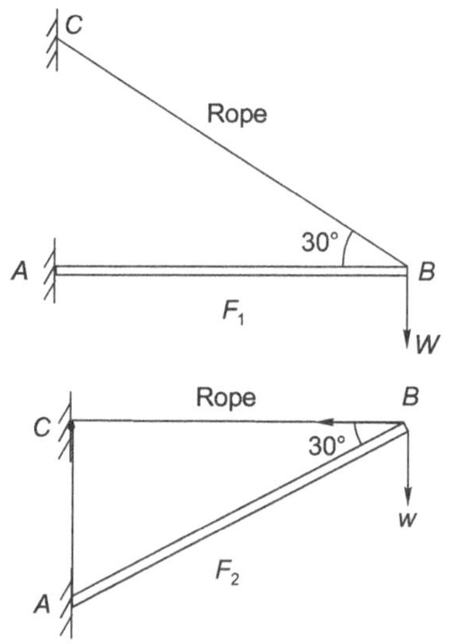

Fig. 4.46

(*a*) 1.0 (*b*) 0.866
(*c*) 0.75 (*d*) 0.50.

4.18 What is the force in the member EH for a pin jointed tower truss as shown in the **Fig. 4.47**.

(*a*) 6.0 kN (Tension)
(*b*) 6.0 kN (Compression)
(*c*) 7.5 kN (Compression)
(*d*) None of these

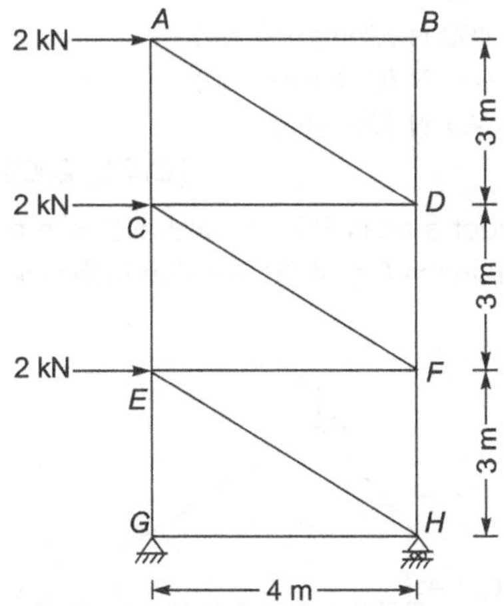

Fig. 4.47

[CSE, Prelim, CE : 2008]

4.19 In the plane truss shown in **Fig. 4.48**, how many members have zero force

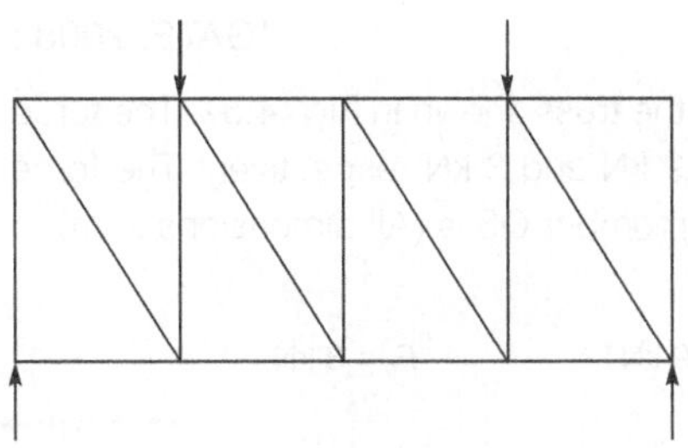

Fig. 4.48

(*a*) 3 (*b*) 5
(*c*) 7 (*d*) 9

[CSE, Prelim, CE : 2006]

4.20 In a truss work as shown in **Fig. 4.49**, what is the force in the member DE?

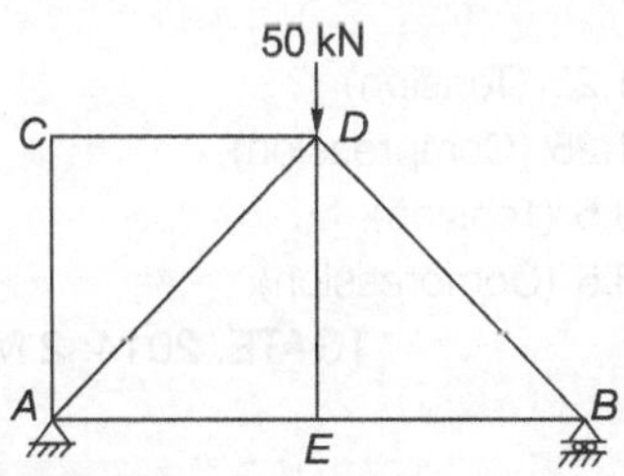

Fig. 4.49

(*a*) 50 kN (Tension)
(*b*) zero
(*c*) 50 kN (Compression)
(*d*) 25 kN (Compression)

[CSE, Prelim, CE : 2005]

4.21 Due to horizontal pull of 60 kN at C, what is the force induced in the member AB?

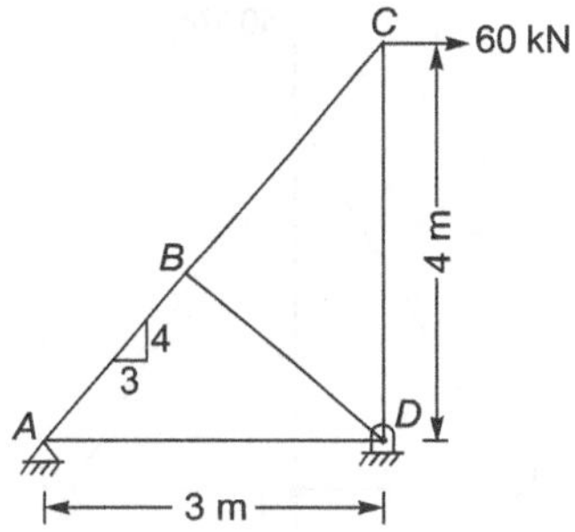

Fig. 4.50

(*a*) 0 (*b*) 40 kN
(*c*) 80 kN (*d*) 100 kN

[CSE, Prelim, CE : 2006]

4.22 A pin jointed lower truss is loaded as shown in **Fig. 4.51**. Force induced in member DF is

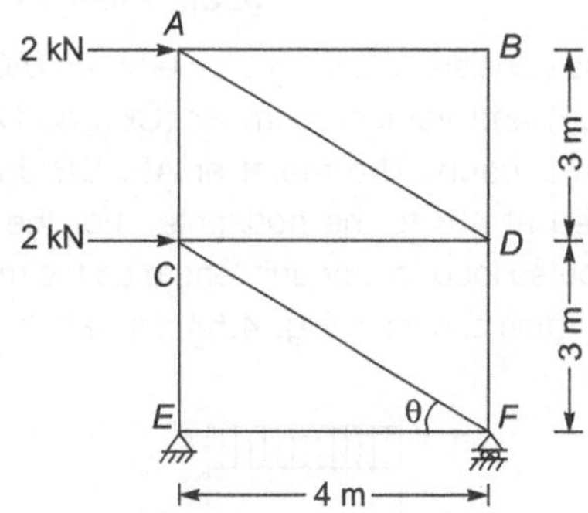

Fig. 4.51

(*a*) 1.5 kN (Tension)
(*b*) 4.5 kN (Tension)
(*c*) 1.5 kN (Compression)
(*d*) 4.5 kN (Compression)

[CSE, Prelim, CE : 2002]

4.23 For the truss shown in the **Fig. 4.52**, which one of the following members has zero force induced in it?

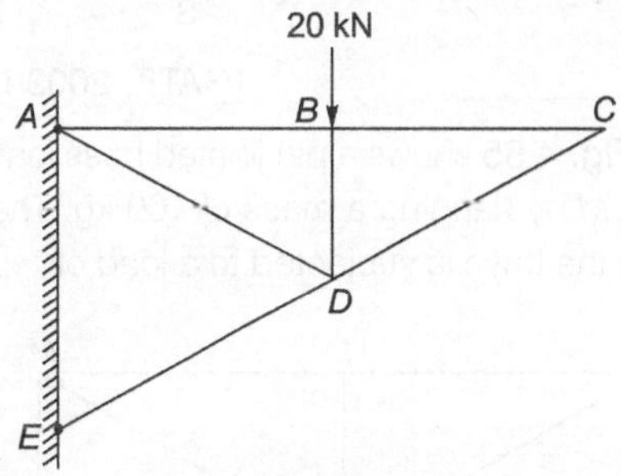

Fig. 4.52

(*a*) BC (*b*) AD
(*c*) DE (*d*) BD

[CSE, Prelim, CE : 2003]

4.24 The force induced in the vertical member CD of the symmetrical plane truss shown in **Fig. 4.53** is

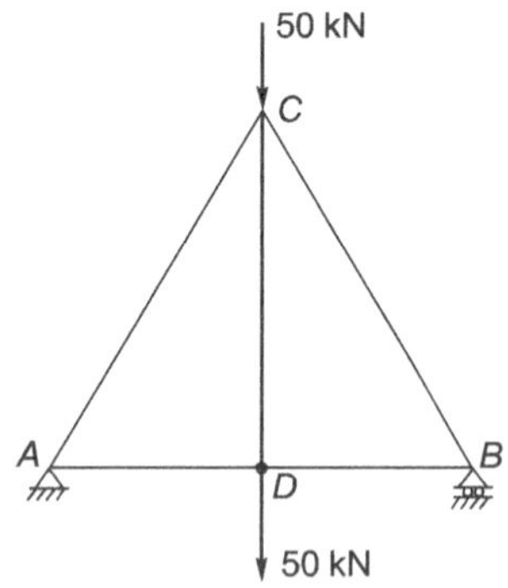

Fig. 4.53

(a) 50 kN (Tension)
(b) 100 kN (Compression)
(c) 50 kN (Compression)
(d) zero

[CSE, Prelim, CE : 2004]

4.25 A truss consists of horizontal members (AC, CD, DB and EF) and vertical member (CE and DF) having length L each. The member AE, DE and BF are inclined at 45° to the horizontal. For the uniformly distributes load 'p' per unit length as the member EF of the truss shown in **Fig. 4.54**, force in the member CD is

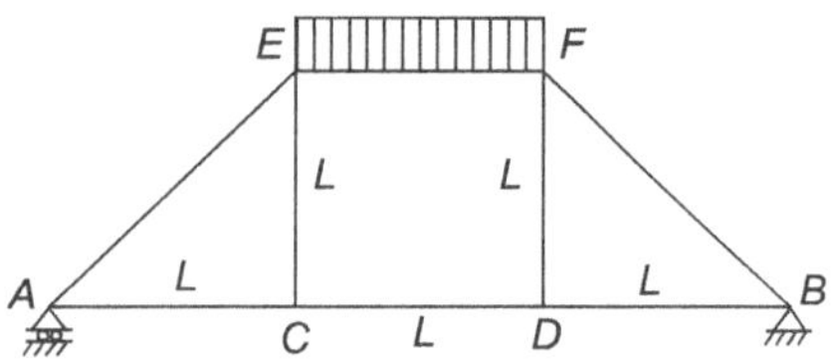

Fig. 4.54

(a) $\frac{pL}{2}$ (b) pL

(c) 0 (d) $\frac{2pL}{3}$

[GATE, 2003 : 1 Marks]

4.26 The **Fig. 4.55** shows a pin jointed truss loaded at the point *M* by hanging a mass of 100 kg. The member LN of the truss is subjected to a load of

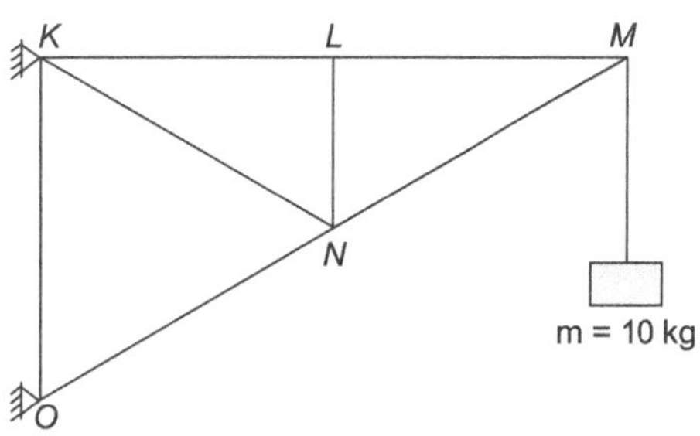

Fig. 4.55

(a) 0 N
(b) 490 N (Compression)
(c) 981 N (Compression)
(d) 981 N (Tension)

[GATE, 2003: 1 Marks]

4.27 Consider a truss *PQR* loaded at P with a force *F* as shown in the **Fig. 4.56**. Tension in the member *QR*

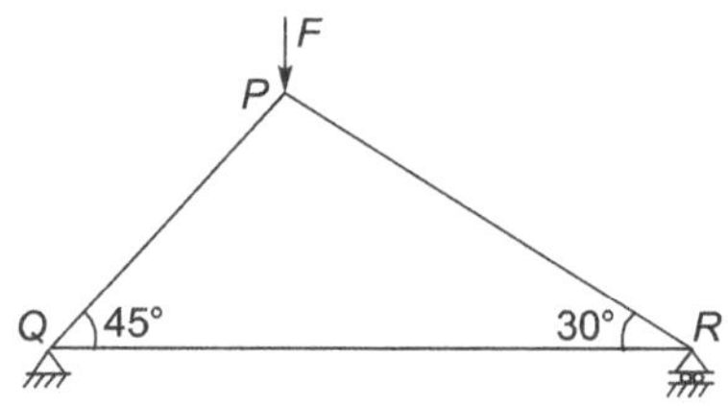

Fig. 4.56

(a) 0.5 F (b) 0.63 F
(c) 0.73 F (d) 0.87 F

[GATE, 2008 : 2 Marks]

4.28 For the truss shown in **Fig. 4.57**. The force F_1 and F_2 are 9 kN and 3 kN respectively. The force is (kN) in the member QS is (All dimensions in m)

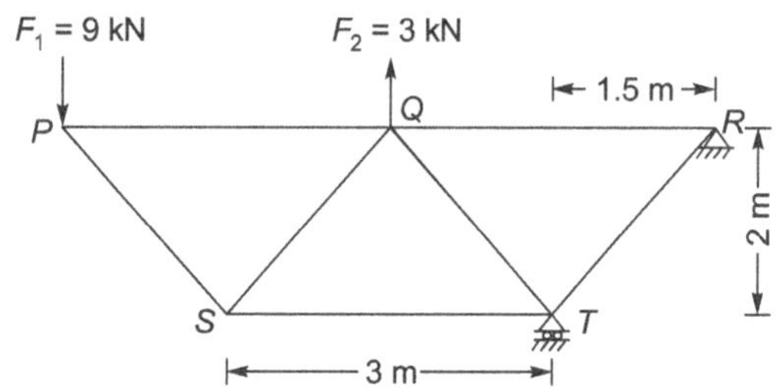

Fig. 4.57

(a) 11.25 (Tension)
(b) 11.25 (Compression)
(c) 13.5 (Tension)
(d) 13.5 (Compression)

[GATE, 2014: 2 Marks (sat 4)]

Answers

4.1 (d)	4.2 (a)	4.3 (b)	4.4 (a)	4.5 (d)
4.6 (b)	4.7 (b)	4.8 (b)	4.9 (a)	4.10 (b)
4.11 (d)	4.12 (b)	4.13 (b)	4.14 (b)	4.15 (c)
4.16 (a)	4.17 (c)	4.18 (c)	4.19 (d)	4.20 (b)
4.21 (d)	4.22 (c)	4.23 (a)	4.24 (a)	4.25 (a)
4.26 (a)	4.27 (b)	4.28 (a)		

EXPLANATIONS

4.1 (d)
Reaction at A (Fig. 4.58)

$$2L \,.\, R_A = P \times \frac{3L}{2}$$

$$R_A = \frac{3}{4}P$$

$$F_{AD} = R_A \times \tan 30^\circ$$

$$= \frac{3}{4}P \times 0.577 = 0.433P.$$

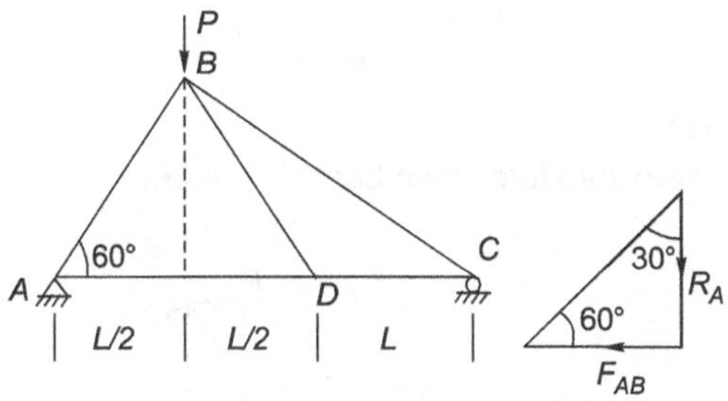

Fig. 4.58

4.2 (a)
Reaction at A (Fig. 4.59)

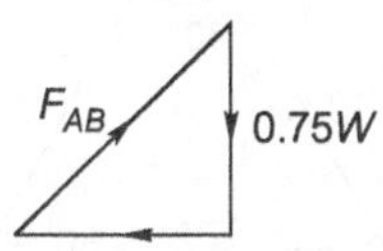

Fig. 4.59

$$W \times \frac{a}{2} + Wa = 2a \,.\, R_A$$

$$R_A = 0.75W$$

$$F_{AB} = \frac{R_A}{\cos 45^\circ} = \frac{0.705}{0.707} \times W = 1.06P \text{ (comp.)}$$

4.3 (b)

$$R_{DH} = \frac{1}{a}\left[W \times \frac{a}{2} + Wa\right] = \overrightarrow{1.5W}$$

$$R_{\overleftarrow{AH}} = \overrightarrow{R_{DH}}, \overleftarrow{R_{AH}} = 1.5W$$

4.4 (a)
Maximum force occurs in member BD

$$F_{BD} = W$$

$$\sigma_{BD} = 100 \text{ MPa} = \frac{W}{100}$$

$$W = 10{,}000 \text{ N} = 10 \text{ kN}$$

4.5 (d)
At joint D, forces F_{ED} and F_{DC} are collinear, forces are equal and opposite

$$F_{BD} = 0$$

4.6 (b)
Total load = pL
$C.G.$ of pl acts at $L/2$ from D or C.

$$R_{DH} = \frac{1}{2}\left(p\frac{l}{2} \times L\right) = \overrightarrow{\frac{pL}{2}}$$

$$R_{AH} = \overleftarrow{\frac{pL}{2}}$$

4.7 (b)

$$F_{CD} = \frac{F_{AD}}{\sqrt{2}}$$

$$= \frac{5}{\sqrt{2}} = 3.53 \text{ kN} = F_{CB}.$$

4.8 (b)

$$R_{CV} = \frac{W}{2} = R_{CH}. \text{ (Fig. 4.60).}$$

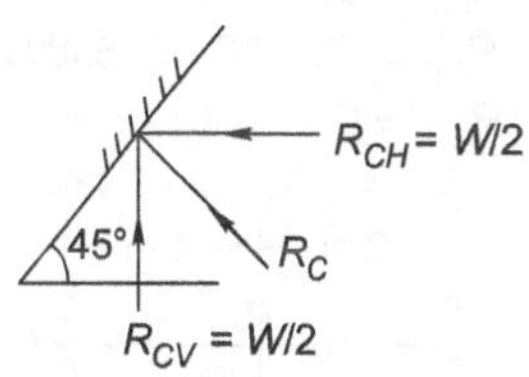

Fig. 4.60

4.9 (a)

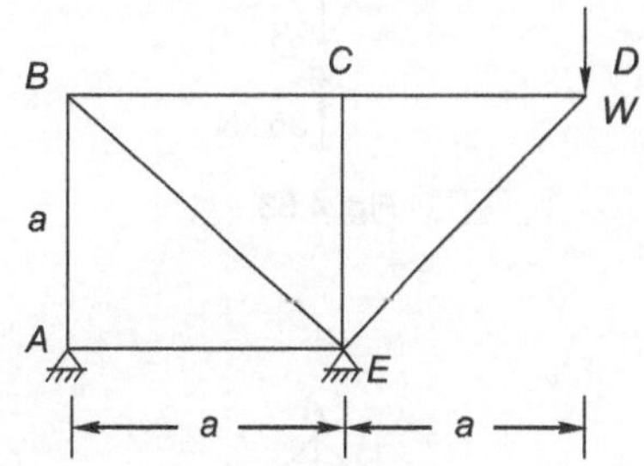

Fig. 4.61

At joint D (Fig. 4.61)

$$F_{CD} = W \text{ (tension)}$$

$$F_{CE} = 0$$

$$F_{BC} = W \text{ (tension)}$$

4.10 (b)
Method of section (**Fig. 4.62**)

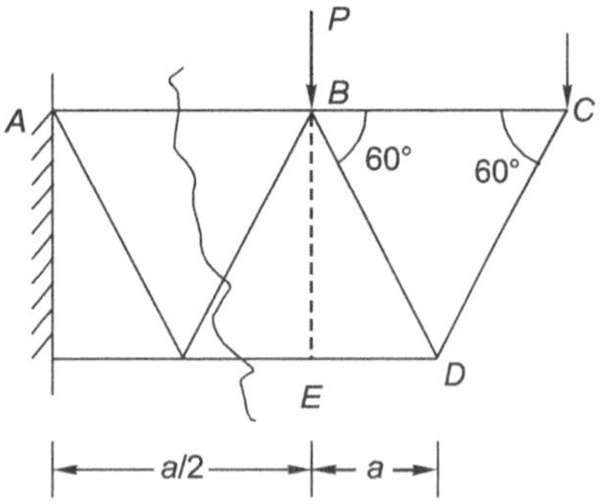

Fig. 4.62

Moment about E

$$F_{BA} \times 0.866a = \frac{Pa}{2} + \frac{3Pa}{2}$$

$$F_{BA} = \frac{2Pa}{0.866a}$$

$$= 2.31P = 3 \text{ kN}$$

$$P = 1.299 \text{ kN.}$$

4.11 (d)

Moments about C (**Fig. 4.63**)

$$27 \times 7 = R_{DV} + 4$$

$$R_{DV} = 35 \text{ kN}$$

$$R_{CV} = 35 - 20 = 15 \text{ kN}\downarrow$$

$$R_{CH} = 0.$$

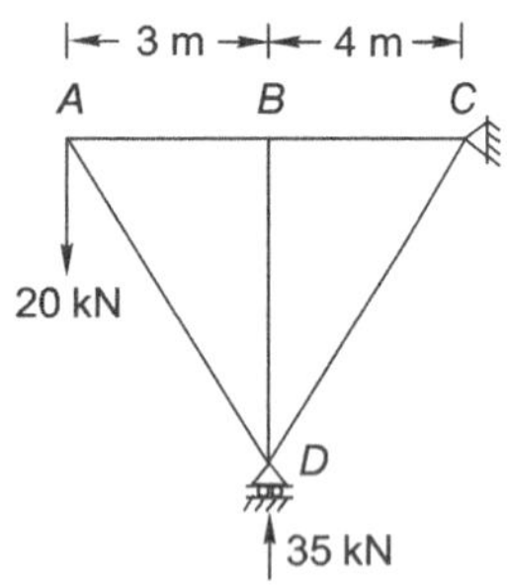

Fig. 4.63

4.12 (b)

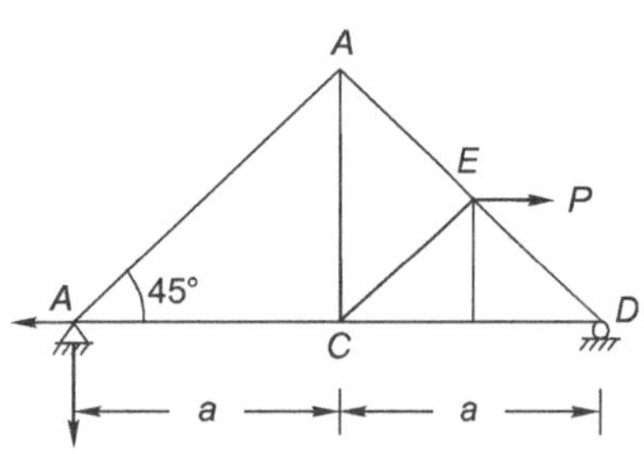

Fig. 4.64

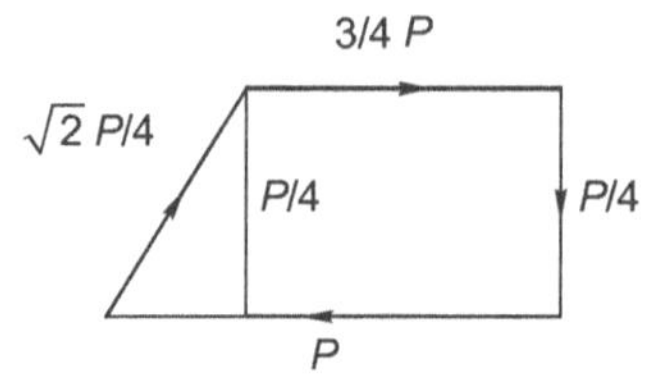

Fig. 4.65

Moments about D (**Fig. 4.64, 65**)

$$\frac{Pa}{2} = R_{AV} \times 2a,\ R_{AV} = 0.25P\downarrow$$

$$R_{AH} = \overleftarrow{P}$$

$$F_{BC} = 0.75P.$$

4.13 (b)

3 are zero force members (**Fig. 4.66**).

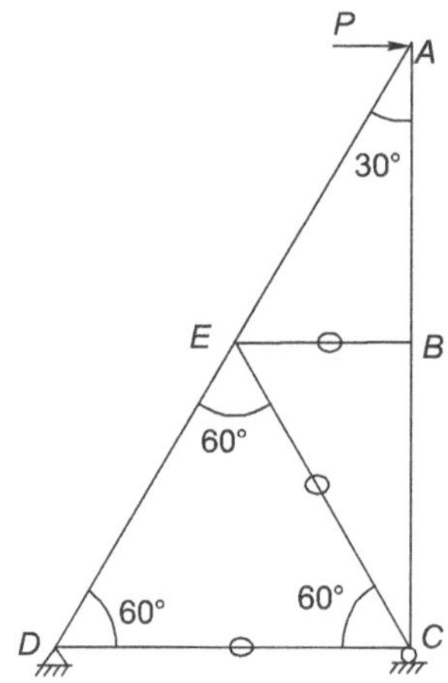

Fig. 4.66

4.14 (b)

(Fig. 4.67)

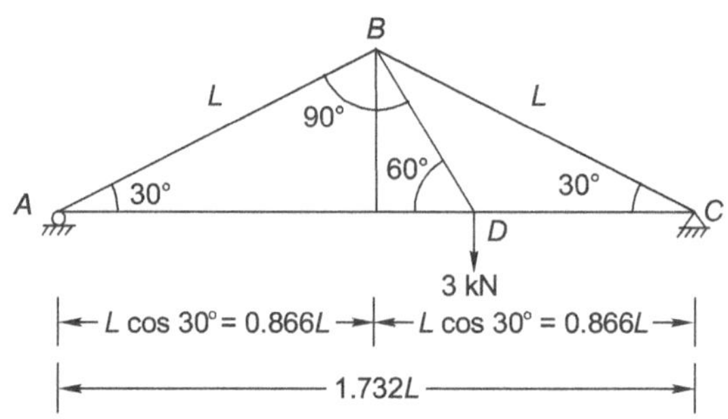

Fig. 4.67

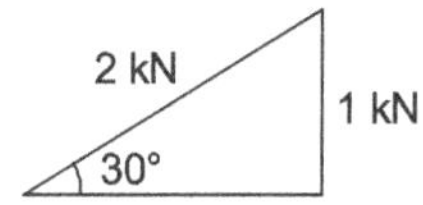

Fig. 4.68

$$AD = \frac{L}{\cos 30°} = \frac{L}{0.866} = 1.1547L$$

$$R_A = \frac{(1.732 - 1.1547)}{1.732} \times 3 \text{ kN} = 1 \text{ kN}$$

$$F_{AB} = \frac{R_A}{\sin 30°}$$

$$= \frac{1 \text{ kN}}{0.5}$$

$$= 2 \text{ kN (comp)}.$$

4.15 (c)

Moments about C (**Fig. 4.69**)

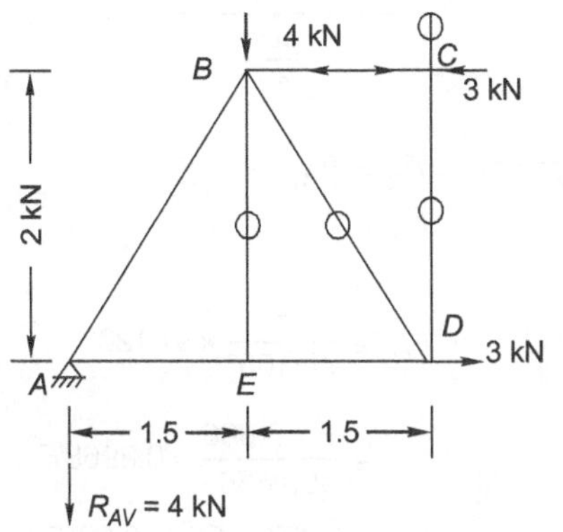

Fig. 4.69

$$4 \times 1.5 + 3 \times 4 = R_{AV} \times 3$$

$$R_{AV} = 3 \text{ kN}$$

$$R_{CV} = 0$$

$$R_{CH} = 3 \text{ kN}$$

There are three zero force members

$$F_{BE} = F_{BD} = F_{CD} = 0.$$

4.16 (a)

(**Fig. 4.70**)

$$F_{AB} \cos 45 = F_{CB} \cos 30°$$

$$F_{AB} = \frac{0.866}{0.707} = 1.225 F_{CB}$$

$$F_{CB} = 0.8164 F_{AB}$$

$$F_{AB} \sin 45° = F_{CB} \sin 30° + P$$

$$F_{AB} \times 0.707 = 0.8164 \times 0.5 \times F_{AB} + P \; 0.4082 F_{AB} + P$$

$$F_{AB} = \frac{P}{0.707 - 0.4082} = 3.346P.$$

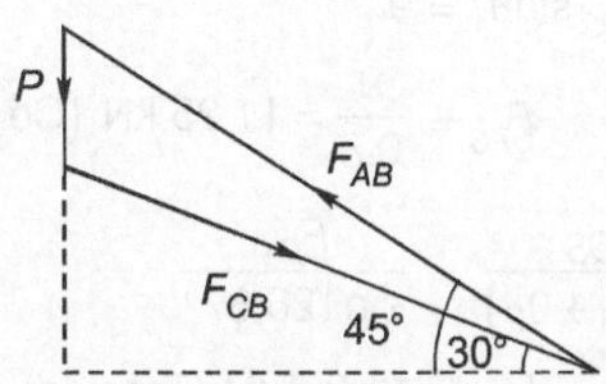

Fig. 4.70

4.17 (c)

(**Fig. 4.71**)

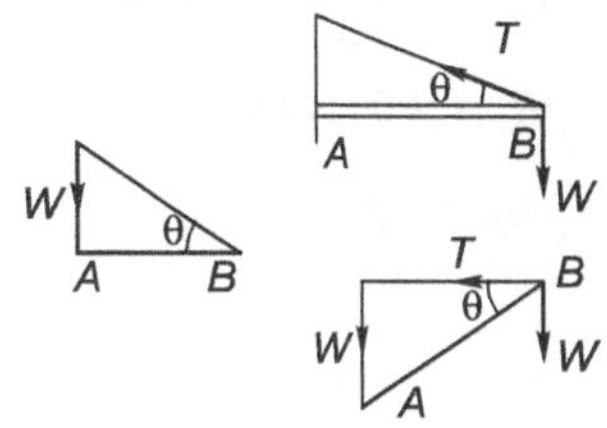

Fig. 4.71

$$AB = T \cos\theta$$

$$AB/\cos\theta = T$$

$$F_1 = T\cos\theta, \quad F_2 = \frac{T}{\cos\theta}$$

$$\frac{F_1}{F_2} = \cos^2\theta = 0.866^2 = 0.75$$

4.18 (c)

Reaction

$$R_{HV} = \frac{6 + 12 + 18}{4} = \frac{36}{4} = 9 \text{ kN}$$

$$R_{GH} = \overleftarrow{6 \text{ kN}}, R_{av} = 9 \text{ kN} (\downarrow)$$

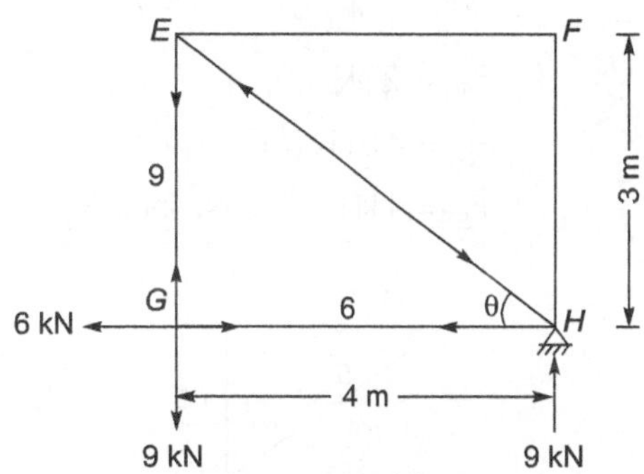

Fig. 4.72

$$\cos\theta = 0.8, \sin\theta = 0.6$$

$$EH \cos\theta = 6$$

$$EH = 7.5 \text{ kN (Compression)}$$

to balance GH force.

4.19 (d)

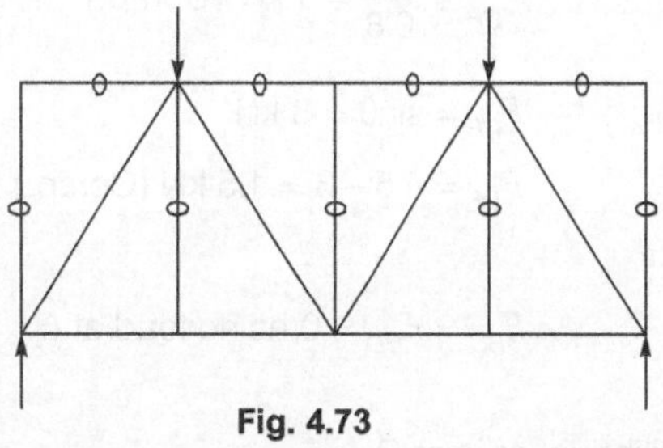

Fig. 4.73

9 zero force members.

4.20 (b)

Force in member DE is zero for equilibrium at point E.

4.21 (d)

$$R_{DV} = \frac{60 \times 4}{3} = 80 \text{ kN}$$

$$R_{AV} = 80 \text{ kN } (\downarrow)$$

$$R_{AH} = \overleftarrow{60 \text{ kN}}$$

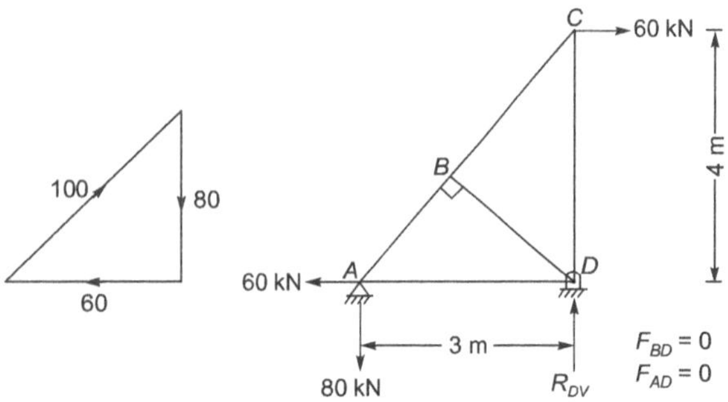

Fig. 4.74

Force in member

$$AB = 100 \text{ kN}$$

4.22 (c)

$$R_{FV} = \frac{12+6}{4} = 4.5 \text{ kN}(\uparrow)$$

$$R_{EH} = \overleftarrow{4 \text{ kN}}$$

$$R_{DV} = 4.5 \text{ kN } (\downarrow)$$

$F_{EF} = 4$ kN ((T) to balance R_{EH}

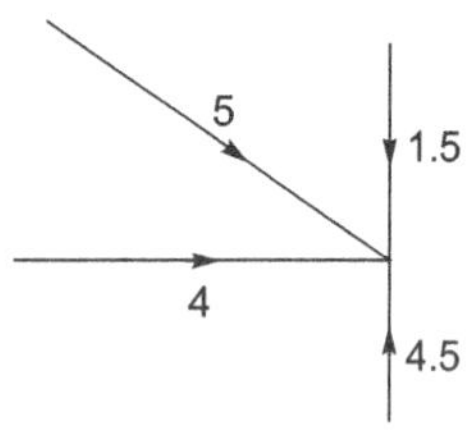

$$\cos\theta = 0.8$$

$$F_{CF} = \frac{4}{0.8} = 5 \text{ kN (Comp.)}$$

$$F_{CF} = \sin\theta = 3 \text{ kN}$$

$$F_{DF} = 4.5 - 3 = 1.5 \text{ kN (Comp.)}$$

4.23 (a)

$F_{BC} = F_{CD} = 0$ as no load at C.

4.24 (a)

Equilibrium of joint D

$F_{DC} = 50$ kN (Tension)

to balance 50 kN load

4.25 (a)

Reaction $R_A = R_C = \frac{pL}{2}(\uparrow)$

$$F_{EC} = 0$$

$$F_{AC} = F_{CD} = \frac{pL}{2}$$

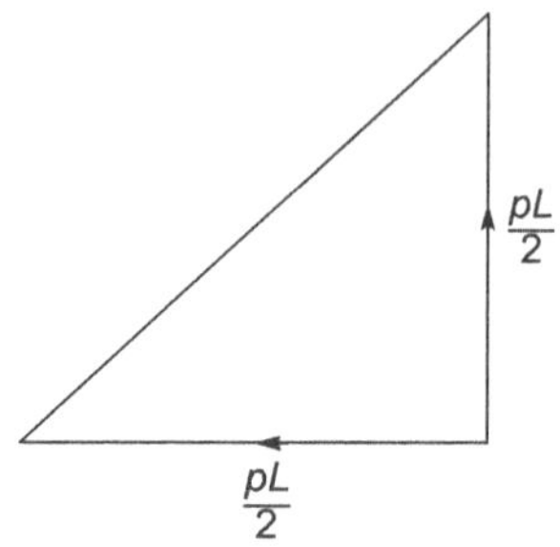

4.26 (a)

At point L, $F_{KL} = F_{LM}$

$$F_{LN} = 0$$

4.27 (b)

$$F_{PQ} = \frac{F}{\sin 105°} \times \sin 120°$$

$$= \frac{F \times 0.866}{0.9656} = 0.8968F$$

$$F_{QR} = 0.707 \times F_{PQ} = 0.63\, F$$

4.28 (a)

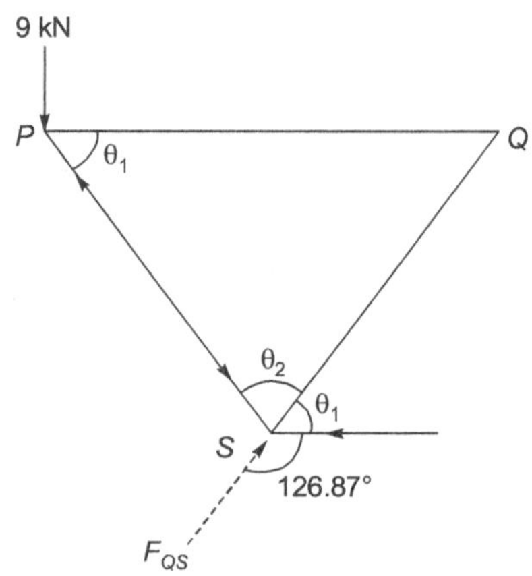

Fig. 4.75

$$\theta_1 = \tan^{-1}\frac{2}{1.5} = 53.13°$$

$$\theta_2 = 180 - 2\theta_1 = 73.74°$$

$$\theta_1 + \theta_2 = 126.87°$$

Force $F_{PS}\sin\theta_1 = 9$

$$F_{PS} = \frac{9}{0.8} = 11.25 \text{ kN (Comp.)}$$

$$\frac{F_{QS}}{\sin(\theta_1+\theta_2)} = \frac{F_{PS}}{\sin 126.87°}$$

$$F_{QS} = 11.25 \text{ kN (Tension)}$$

■■■

05

CHAPTER

Centroid and Center of Gravity

5.1 Introduction

Center of Gravity is that point of a body through which passes the resultant of distributed parallel gravitational forces. In the case of plane laminas, *CG* is termed as centroid.

While computing the stresses in a member, determination of centroid of a plane section is of paramount importance because stresses developed at a point in a beam or column section depend upon the location of centroid of section.

A team of engineers and scientists are engaged to determine the exact location of centre of gravity of a missile taking into account the location of weights and *CG* of all the components making a missile.

Location of *CG* of all sections of a rotor in a turbine is of much importance so as to avoid any unbalanced mass and resulting vibrations.

Whatever may be the position of a body in space, resultant of all parallel forces always passes through the point which is known as **Centre of gravity**.

Fig. 5.1 shows a body with its mass distributed through elementary volumes V_1, V_2, V_3, ..., V_n etc. The gravitational forces on these elementary masses are $\rho V_1 g$, $\rho V_2 g$, $\rho V_3 g$, ..., $\rho V_n g$ where ρ is the density of the body and *g* is the acceleration due to gravity. *These gravitational forces are always vertical* (directed towards the centre of the earth). F_R is the resultant of all these forces F_1, F_2, F_3, ..., F_n passing through *CG*, the centre of gravity. Many a times only a letter *G* denotes the location of centre of gravity.

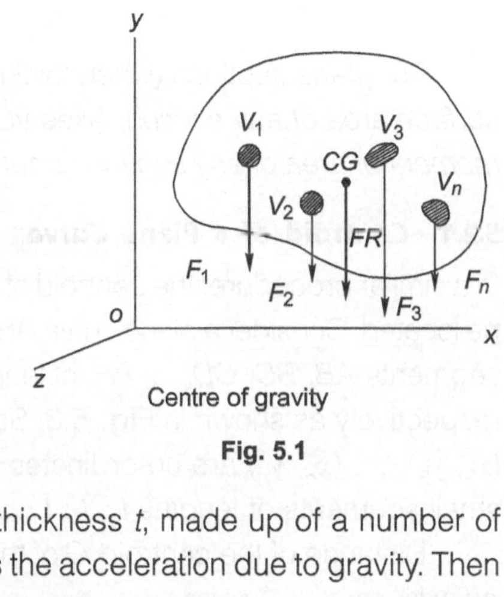

Centre of gravity

Fig. 5.1

5.2 Location of Centroid

Centroid is determined for a plane lamina. Consider a plate of uniform thickness *t*, made up of a number of elementary areas a_1, a_2, a_3, ..., a_n. Say ρ is the density of the plate and *g* is the acceleration due to gravity. Then vertical gravitational forces on these elements will be F_1, F_2, F_3, ..., F_n, where these forces are $\rho a_1 tg$, $\rho a_2 tg$, $\rho a_3 tg$, ..., $\rho a_n tg$. **(Fig. 5.2).**

To locate position of centroid *G*, let us take moments of the forces about the axis *oy* of the co-ordinate system. Then as per Vavignon's theorem

$$F_R \bar{x} = F_1 x_1 + F_2 x_2 + F_3 x_3 \ldots F_n x_n$$

$$M = \text{Total mass of the body} = A\rho t$$

A is the total surface area of the plate.

Then resultant force, $F_R = Mg$

Substituting these values

$$(A\rho t)g\bar{x} = \rho tg\,(a_1x_1 + a_2x_2 + a_3x_3 \dots a_nx_n)$$

or

$$\bar{x} = \frac{a_1x_1 + a_2x_2 + a_3x_3 \dots a_nx_n}{A}$$

$$= \frac{a_1x_1 + a_2x_2 + a_3x_3 \dots a_nx_n}{a_1 + a_1 + a_3 \dots a_n}$$

Therefore distance of G from vertical y-axis oy is

$$\bar{x} = \frac{\sum_{i=1}^{n} a_i x_i}{\sum_{i=1}^{n} a} = \frac{\sum_{i=1}^{n} a_i x_i}{A}$$

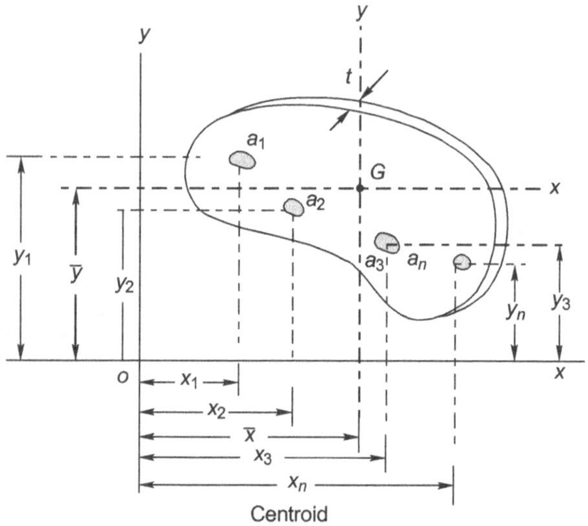

Fig. 5.2

Note that $\sum_{i=1}^{n} a_i x_i$, is the first moment of area of the surface of the plate about oy-axis. Axis Gy is passing through G, therefore from the above expression, it is obvious that *first moment of area of the surface area about G will be zero.*

Now if the plate is rotated in space by an angle of 90°, or axis oyx is changed to xoy, then point G remains the same. Distance of centroid G of the plate from horizontal axis can be written as

$$\bar{y} = \frac{a_1y_1 + a_2y_2 + a_3y_3 \dots a_ny_n}{a_1 + a_2 + a_3 \dots a_n} = \frac{\sum_{i=1}^{n} a_i y_i}{\sum_{i=1}^{n} a_i} = \frac{\sum_{i=1}^{n} a_i y_i}{A}$$

For plane sections (when thickness t of plate is very small) centre of gravity is termed as *centroid of the surface area of any section.* Axes yG and xG passing through centroid G are termed as *centroidal axes. First moment of area of any section about any centroidal axis is zero.*

5.2.1 Centroid of a Plane Curve

In a similar procedure, the centroid of a plane curve can be located. Consider a plane curve AH, made up of small segments AB, BC, CD, ..., FH, of lengths $l_1, l_2, l_3, \dots, l_n$ respectively as shown in **Fig. 5.3**. Say (x_1, y_1), (x_2, y_2), (x_3, y_3), ..., (x_n, y_n) are co-ordinates of of centroids of small segments of lengths $l_1, l_2, l_3, \dots, l_n$ respectively.

Distance of the centroid G of the curve AH about oy-axis

$$\bar{x} = \frac{\sum_{i=1}^{n} l_i x_i}{\sum_{i=1}^{n} l_i} = \frac{\sum_{i=1}^{n} l_i x_i}{L},$$

where L is the full length of the curve

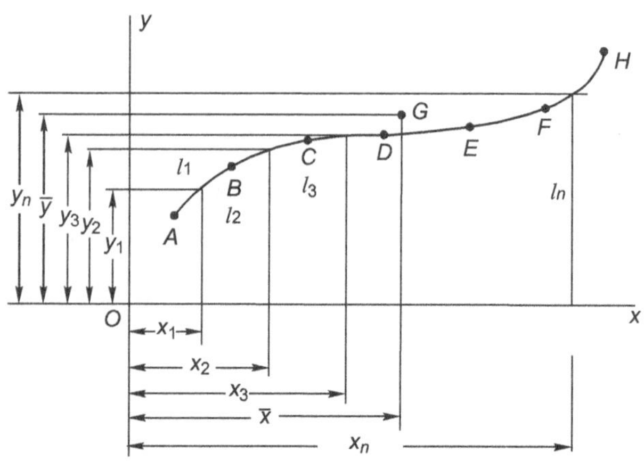

Centroid of a plane curve

Fig. 5.3

Rotating the axis by 90°, and G of the curve will remain the same, distance $\bar{y}$ can be calculated as follows:

$$\bar{y} = \frac{\sum_{i=1}^{n} l_i y_i}{\sum_{i=1}^{n} l_i}.$$

Example 5.1 A circular line segment AB subtends an angle 2α, at the centre as shown in **Fig. 5.4**. Determine the position of centroid of line segment.

Solution Line segment is symmetrical about the axis ox, so its centroid will be along x-axis or distance $\bar{y} = 0$ from axis yox. Now consider a small length $cc' = \delta l = Rd\theta$

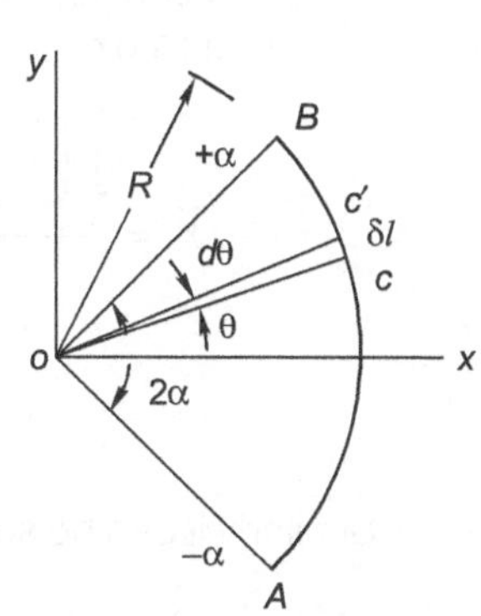

Fig. 5.4

Co-ordinates of mid point of cc' or length δl

$$x = R\cos\theta;\ y = R\sin\theta$$

$$\bar{x} = \frac{\int_{-\alpha}^{+\alpha} R\cos\theta.\delta l}{\int_{-\alpha}^{\alpha} dl}, \text{ where } \delta l = Rd\theta$$

$$= \frac{\int_{-\alpha}^{+\alpha} R^2 \cos\theta\, d\theta}{\int_{-\alpha}^{+\alpha} Rd\theta} = \frac{R^2\, |\sin\theta|_{-\alpha}^{+\alpha}}{|R\theta|_{-\alpha}^{+\alpha}}$$

$$= \frac{R^2\,(\sin\alpha + \sin\alpha)}{R.2\alpha}$$

$$= \frac{R^2\, 2.\sin\alpha}{R2\alpha} = \frac{2R\sin\alpha}{2\alpha} = \frac{R\sin\alpha}{\alpha}$$

where α is the half of angle subtended by the line segment at the centre.

Exercise 5.1 Determine the location of centroid of:

(*a*) quadrant of a circle of radius R as shown in **Fig. 5.5 (*a*)**.

(*b*) semi circle of radius R as shown in **Fig. 5.5 (*b*)**.

[**Hint:** For (*a*) take centre line at $\alpha = 45°$ from ox-axis, determine the centroid distance along the line, then find $\bar{x}$ and $\bar{y}$. For (*b*) $\bar{y} = 0$, $\alpha = \pm\frac{\pi}{2}$].

[**Ans:** (*a*) $\bar{x} = \bar{y} = 0.636R$ (*b*) $\bar{y} = 0$, $\bar{x} = \frac{2R}{\pi}$].

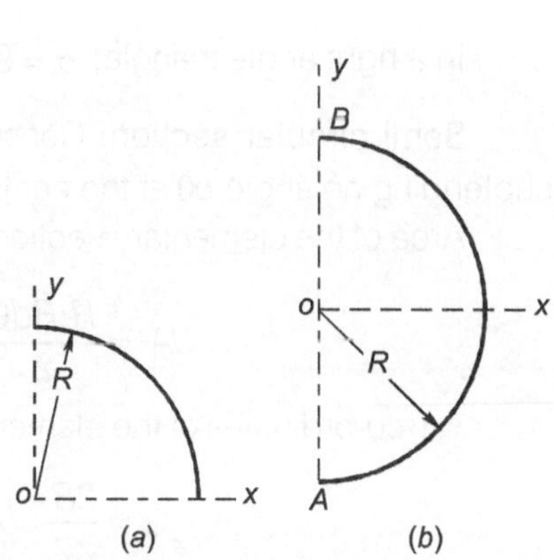

Fig. 5.5

5.3 Centroid of Simple Sections

Location of centroids of simple sections such as rectangle, square, triangle, semi circular etc. should be at the tips of engineering students. Centroid of these simple regular sections is determined by integration method.

Triangular section: Consider a triangle OBC, with base $OB = b$, altitude $CD = h$ as shown in the **Fig. 5.6**. Take a small strip of thickness dy, at a distance y from axis ox.

Breadth of elementary strip,

$$b_y = \left(\frac{h-y}{h}\right)\times b$$

Area of the elementary strip,

$$dA = b_y.dy = \left(\frac{h-y}{h}\right)b.dy$$

Area of the triangle OBC,

$$A = \frac{bh}{2}$$

Distance of axis mm passing through the centroid of triangle from axis ox

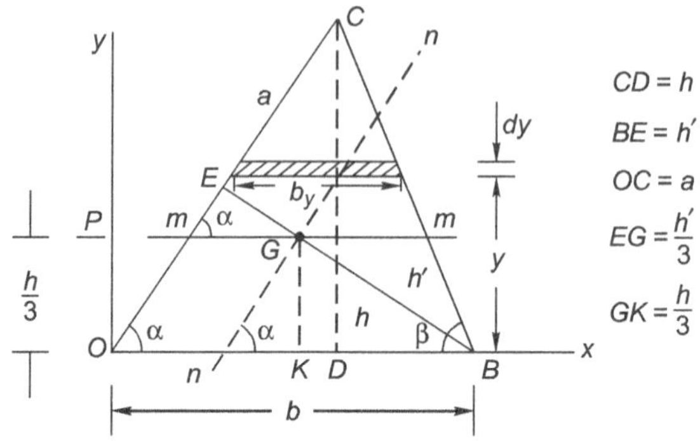

Fig. 5.6

$$\bar{y} = \frac{\int_0^h \left(\frac{h-y}{h}\right)by\,dy}{A} = \frac{2b}{bh}\int_0^h\left(y-\frac{y^2}{h}\right)dy$$

$$= \frac{2}{h}\left|\frac{y^2}{2}-\frac{y^3}{3h}\right|_0^h = \frac{2}{h}\left[\frac{h^2}{2}-\frac{h^3}{3h}\right] = \frac{2}{h}\left[\frac{h^2}{6}\right] = \frac{h}{3} = \text{distance of } G \text{ from } ox\text{-axis}$$

Similarly it can be shown that distance of axis n-n passing through the centroid of the triangle from the side OC is $\frac{h'}{3}$ where h' is the altitude of the triangle from B to side DC.

Axes mm and nn intersect at G which is the centroid of the triangle.

Distance of G from oy-axis

$$\bar{x} = PG = OK = On + nK = mG + nK$$

$$\bar{x} = \frac{h'}{3\sin\alpha}+\frac{h}{3\tan\alpha} = \frac{h'}{3\sin\alpha}+\frac{h}{3}\times\frac{\cos\alpha}{\sin\alpha}$$

But $h' = b\sin\alpha$, $h = \frac{a}{\sin\alpha}$, putting these values $\bar{x} = \frac{b}{3}+\frac{a}{3}\times\cos\alpha.$

In a right angle triangle, $\alpha = 90°$, $\bar{x} = \frac{b}{3}$.

Semi circular section: Consider a semi-circular section as shown in **Fig. 5.7**. Consider a small sector subtending an angle $d\theta$ at the centre O, and making an angle θ with the ox-axis.

Area of the elementary section oab,

$$dA = \frac{R.Rd\theta}{2} = \frac{R^2}{2}d\theta$$

x, y co-ordinates of the elementary area,

$$x = \frac{2R}{3}\cos\theta,\quad y = \frac{2R}{3}\sin\theta$$

For the semicircular section, symmetrical about oy-axis, $\bar{x} = 0$.

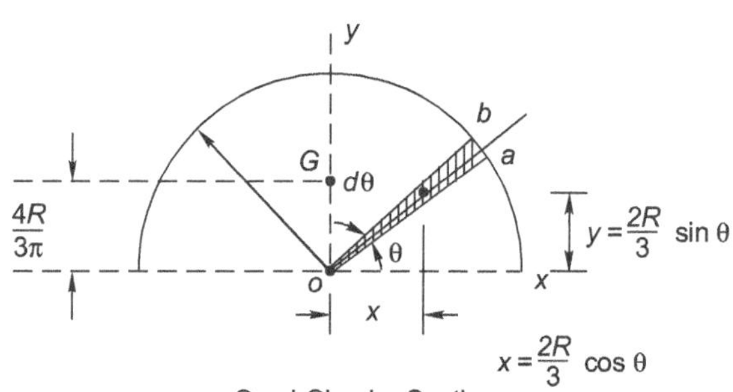

Semi-Circular Section

Fig. 5.7

Let us calculate $\bar{y}$,

$$\text{area, } A = \frac{\pi R^2}{2}$$

$$\bar{y} = \frac{1}{A}\int_0^\pi y\,dA = \frac{1}{A}\int_0^\pi \frac{2R}{3}\sin\theta \times \frac{R^2}{2}\,d\theta = \frac{1}{A}\int_0^\pi \frac{R^3}{3}\sin\theta = d\theta$$

$$= \frac{2}{\pi R^2}\times\frac{R^3}{3}\,|-\cos\theta|_0^\pi = \frac{2R}{3\pi}[1+1] = \frac{4R}{3\pi}$$

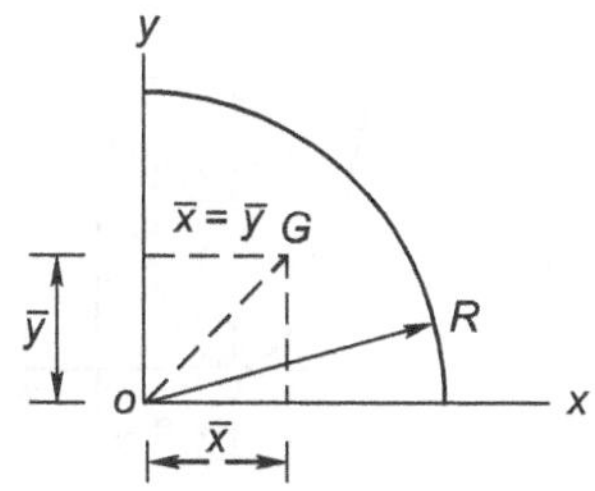

Fig. 5.8

Note that for a quadrant of radius R, G will be located at

$$\bar{x} = \bar{y} = \frac{4R}{3\pi},$$

this can be proved by the reader as shown in the **Fig. 5.8.**

Table 5.1 shows co-ordinates of centroid of plane sections and centre of gravity of solid bodies commonly used in engineering applications.

Table 5.1

Plane section	*Area*	$\bar{x}$	$\bar{y}$
Rectangle 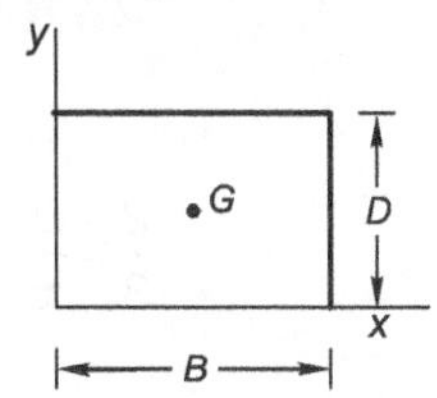	BD	$\frac{B}{2}$	$\frac{D}{2}$
Triangle 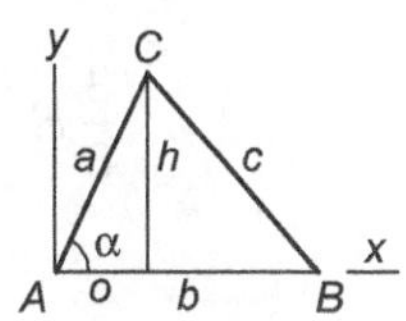	$\frac{bh}{2}$	$\frac{b}{3}+\frac{a\cos\alpha}{3}$	$\frac{h}{3}$
Semi circle 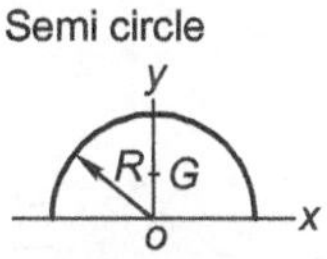	$\frac{\pi R^2}{2}$	0	$\frac{4R}{3\pi}$
Quarter circle 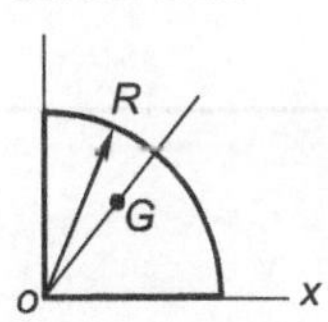	$\frac{\pi R^2}{4}$	$\frac{4R}{3\pi}$	$\frac{4R}{3\pi}$
Circular sector 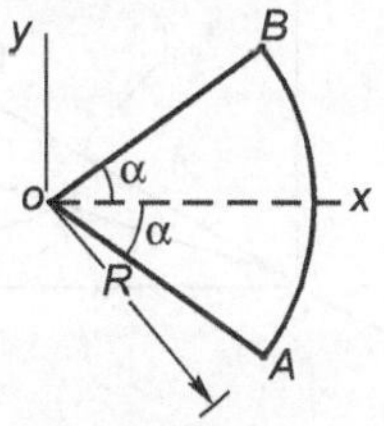	αR^2	$\frac{2R\sin\alpha}{3\alpha}$	0
Line segment 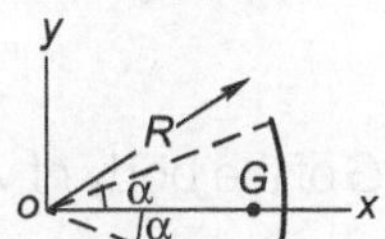	Length $2\alpha R$	$\frac{R\sin\alpha}{\alpha}$	0

Parabola 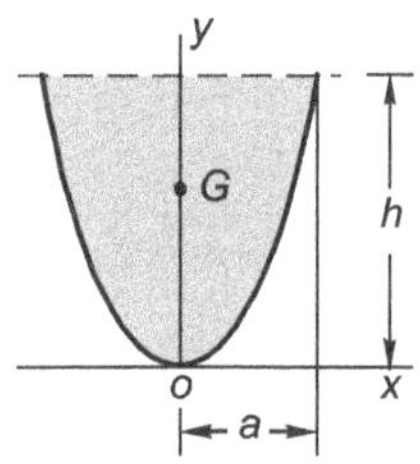	$\frac{4ah}{3}$	0	$\frac{3h}{5}$
General spander 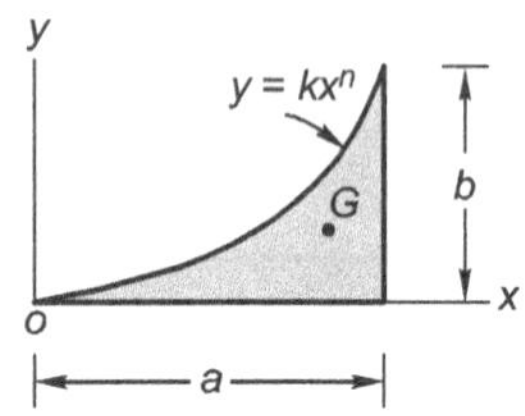	$\frac{ab}{n+1}$	$\left(\frac{n+1}{n+2}\right)a$	$\left(\frac{n+1}{2n+1}\right)\frac{b}{2}$
Ellipse 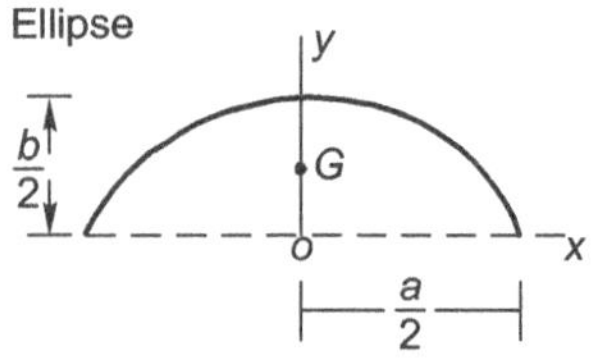	$\frac{\pi ab}{8}$	0	$\frac{2}{3}\times\frac{b}{\pi}$
Solids Hemisphere 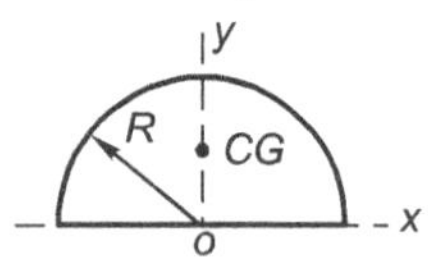	Volume $\frac{2}{3}\pi R^3$	0	$\frac{3}{8}R$
Right Circular Cone 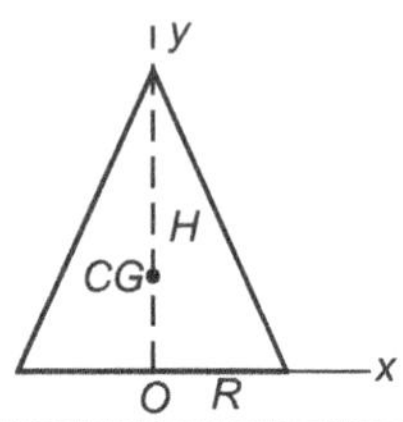	$\frac{\pi R^2 H}{3}$	0	$\frac{H}{4}$

5.4 Centre of Gravity of 3 Dimensional Bodies

The centre of gravity (*CG*) of a three dimensional body is determined by dividing the whole body into small elements ($\Delta V = \Delta x \times \Delta y \times \Delta z$) and expressing the weight *W* of the body attached at *CG* as equivalent to the system of distributed forces or $\delta W = \rho g\, \Delta V$, representing the weight of small elements **(Fig. 5.9)**.

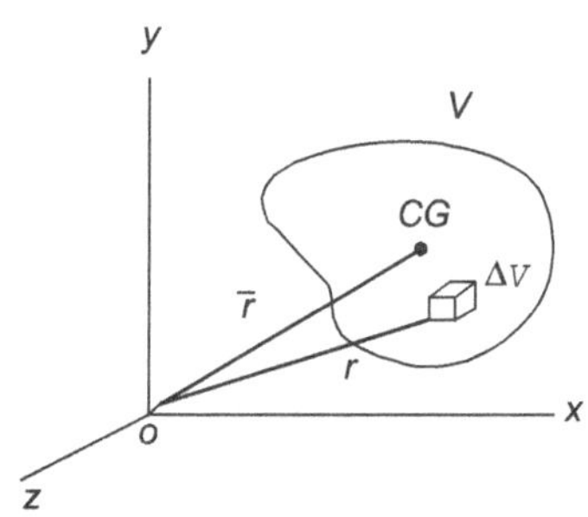

Fig. 5.9

Let us consider a body of volume *V*, the *CG* of the volume having $\bar{x}$, $\bar{y}$, $\bar{z}$ co-ordinates or

$$\bar{r} = \bar{x}i + \bar{y}j + \bar{z}k$$

or $$\bar{r}V = \int r\, dV,$$

where *r* is the position vector of small volume *dV* and $\bar{r}$ is the position vector of the *CG* of the body of volume *V*.

In the scalar form, $\bar{x}V = \int x\, dV,\ \bar{y}V = \int y\, dV$ and $\bar{z}V = \int z\, dV$

The integral $\int x dV$ is known as the *first moment of the volume V with respect to yz plane*. Similarly integrals $\int y dV$ and $\int z dV$ define the first moments of the volume about the planes xz and xy respectively.

If *CG* of a body is located in *any co-ordinate* plane, then *first moment of its volume about that plane will be zero.* If a volume *V* possesses a *plane of symmetry* then the first moment of *V* with respect to that plane of symmetry is zero and *CG* is located in that plane of symmetry. If a volume possesses two planes of symmetry then *CG* of the body is located at the intersection of the two planes of symmetry. If a volume possesses three planes of symmetry then *CG* of the body will lie at a well defined point *i.e.*, point of intersection of 3 planes of symmetry as in the case of a sphere, an ellipsoid, a cube etc.

Please note that the centre of gravity (CG) of a body generated as a volume of revolution *does not collaborate* with the centroid of the *generating section.*

As an example *CG* of a hemisphere is different from that the centroid (*G*) of a quarter circular section which generates the hemisphere, about the axis of revolution. Similarly *CG* of a cone is different from the centroid (*G*) of a triangle (which is section of a cone), even if the bodies are homogeneous.

Example 5.2 Determine the co-ordinates of the *CG* of a hemisphere of radius *R*. The body is homogeneous.

Solution Fig. 5.10 (*a*) shows a hemisphere of radius *R*, its base is in the *x-z* plane and hemisphere is symmetrical about *oy*-axis; therefore its centre of gravity will lie on the axis *oy*.

Consider an element of thickness *dy* at a height *y* from the *zx* plane.

Radius of small element,

$$r = R\sin\theta,$$

height, $y = R\cos\theta$

as shown in Fig. 5.10 (*b*)

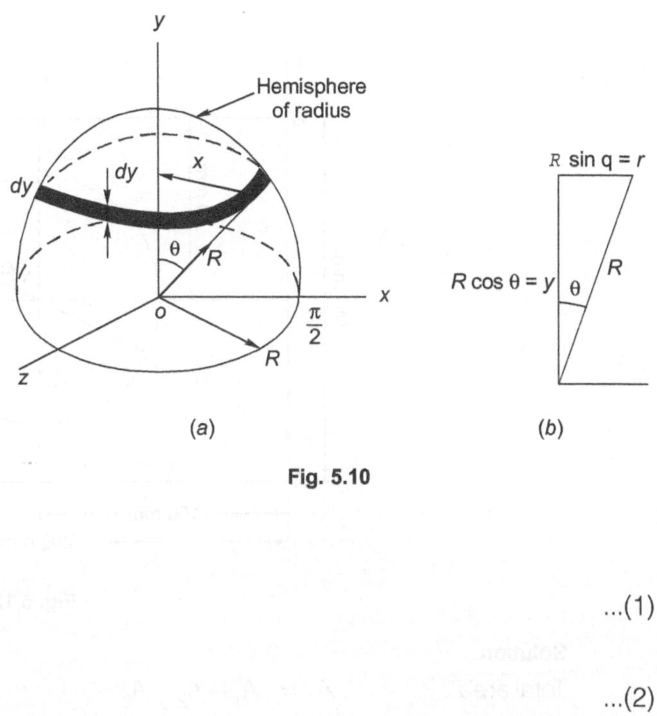

Fig. 5.10

Volume of the small element,

$$dV = \pi (R\sin\theta)^2 dy$$

$$= \pi R^2 \sin^2\theta dy \qquad ...(1)$$

Now $y = R\cos\theta$

So $dy = -R\sin\theta . d\theta \qquad ...(2)$

Putting the value in Equation (1)

$$dV = \pi R^2 \sin^2\theta (-R\sin\theta) d\theta = -\pi R^3 \sin^3\theta d\theta$$

First moment of the volume about *xz* plane

$$M_{xz} = \bar{y}V = \int y dV = \int_{\pi/2}^{0} (-\pi R^3 \sin^3\theta)(R\cos\theta) d\theta$$

$$= \int_{\pi/2}^{0} -\pi R^4 \sin^3\theta \cos\theta d\theta = -\pi R^4 \left|\frac{\sin^4\theta}{4}\right|_{\pi/2}^{0}$$

$$= +\frac{\pi R^4}{4}$$

Volume of the hemisphere, $V = \frac{2}{3}\pi R^3$

Distance of *CG* from *xz* plane,

$$\bar{y} = \frac{\pi R^4}{4} \times \frac{3}{2\pi R^3} = \frac{3}{8}R$$

Other distances $\bar{x} = \bar{z} = 0.$

Exercise 5.2 Determine the co-ordinates of *CG* of a semi right circular cone of base radius *R* and axial length *H* as shown in **Fig. 5.11.**

[**Hint:** Take small element at *x*, $r = \frac{xR}{H}$, for $\bar{y}$, consider $\frac{4r}{3\pi}$ as *CG* distance from *ox*].

[**Ans:** $\bar{z} = 0$, $\bar{y} = \frac{R}{\pi}$, $\bar{x} = \frac{3H}{4}$].

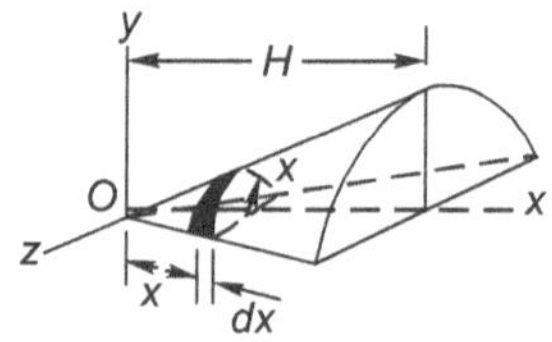

Fig. 5.11

Example 5.3 Find the centroid of the element shown in **Fig. 5.12**. Let as take x-axis along AF and y-axis along AB.

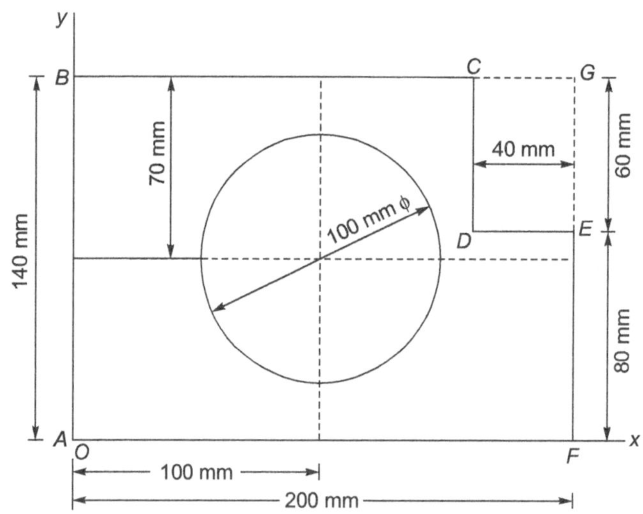

Fig. 5.12

Solution:

Total area

$$A = A_1 - A_2 - A_3$$

$$A_1 = \text{rectangle } 140 \times 200 = 2800 \text{ mm}^2$$

$$A_2 = \text{circle, radius 50 mm} = 7854 \text{ mm}^2$$

$$A_3 = \text{rectangle } 40 \times 60 \text{ mm} = 2400 \text{ mm}^2$$

$$A = 28000 - 7854 - 2400 = 17746 \text{ mm}^2$$

$$\bar{x} = \frac{28000 \times 100 - 7854 \times 100 - 2400 \times 180}{17746}$$

$$= \frac{2800000 - 785400 - 432000}{17746} = \frac{1582600}{17746}$$

$$= 89.18 \text{ mm from } A$$

$$\bar{y} = \frac{2800 \times 70 - 7854 \times 70 - 2400 \times 110}{17746}$$

$$= \frac{1960000 - 549780 - 264000}{17746}$$

$$= \frac{1146220}{17746} = 64.59 \text{ mm from } A$$

5.5 Theorems of Pappus-Guldinus

These theorems provide relationship between:

(*i*) Surface of revolution of the length of a generating curve

(*ii*) Volume of revolution by rotating a plane surface and the *circumference of the circle formed by the centroid of the plane surface.*

Note that generating curve can touch the axis of revolution but it must not cross the axis of rotation.

Theorem I *states The surface of revolution developed by revolving the generating curve about the axis of revolution has an area equal to the product of this length of the generating curve times the circumference of the circle formed by the centroid of the generating curve in the process of generating the surface of revolution.*

Theorem II *states that the volume of the body of revolution developed by rotating the plane surface about the axis of revolution equals the product of the area of the surface times the circumference of the circle formed by the centroid of the surface in the process of generating the body of revolution.*

Example 5.4 The volume of ellipsoidal body of revolution is known from the calculus, to be equal to $\frac{\pi ab^2}{6}$, where *a* and *b* are major and minor axis of the ellipse. If the area of the ellipse is $\frac{\pi ab}{4}$, find the centroid of area of semi ellipse as shown in **Fig. 5.13.**

Solution Area of semi ellipse surface $= \frac{\pi ab}{2\times 4} = \frac{\pi ab}{8}$

Volume of ellipsoid $= \frac{\pi ab^2}{6}$

Say *G* is centroid of semi ellipse.

Circumference made by *OG* during one revolution $= 2\pi.OG$

Then $(2\pi.OG)\times\left(\frac{\pi ab}{8}\right) = \frac{\pi ab^2}{6}$

or $2\pi.OG = \frac{4}{3}\times b$

$$\bar{y} = OG = \frac{2}{3\pi}\times b.$$

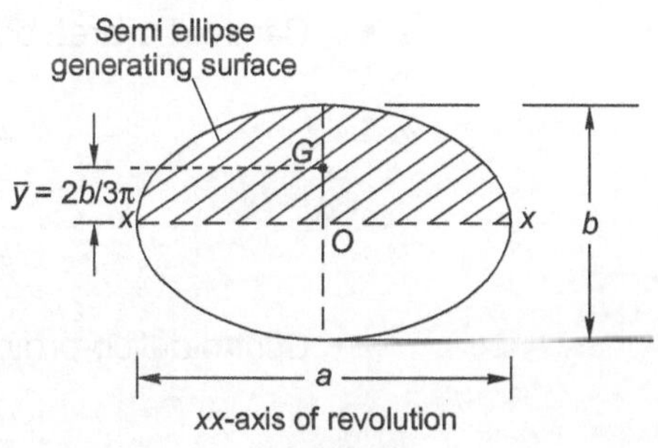

Fig. 5.13

Exercise 5.3 **Fig. 5.14** shows a cone of base radius *R* and axial length *H*. Surface of the cone is generated by revolving a line *OA* (generating curve) about axis *ox*. Determine area of surface of revolution.

[**Hint:** $O'G = \frac{L}{2}\sin\alpha$].

[**Ans:** πRL, where $L = OA$].

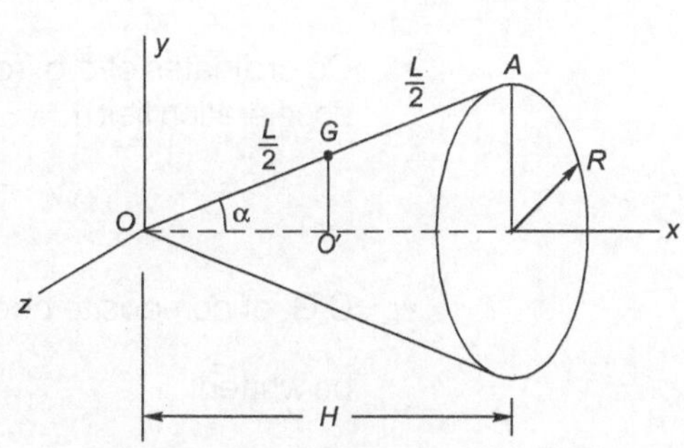

Fig. 5.14

Remember

- Centre of gravity of a body is the point through which passes the resultant of the distributed gravitational parallel forces.
- If an area can be considered as composed of a number of small areas $a_1, a_2, \ldots, a_n$ with their co-ordinates as $x_1, x_2, \ldots, x_n$ and $y_1, y_2, \ldots, y_n$ then co-ordinates of the centroid of the area will be

$$\bar{x} = \frac{\sum_{i=1}^{n} a_i x_i}{\sum_{i=1}^{n} a_i}; \quad \bar{y} = \frac{\sum_{i=1}^{n} a_i y_i}{\sum_{i=1}^{n} a_i}$$

Similarly the co-ordinates of the centroid of a line can be written as

$$\bar{x} = \frac{\sum_{i=1}^{n} l_i x_i}{\sum_{i=1}^{n} l_i}; \quad \bar{y} = \frac{\sum_{i=1}^{n} l_i y_i}{\sum_{i=1}^{n} l_i}$$

- Centroid of a circular line segment of radius R, subtending angle 2α at the centre.

If $\bar{y} = 0, \ \bar{x} = \frac{R\sin\alpha}{\alpha}$.

- Centroidal co-ordinate for a semi circular area $\bar{y} = \frac{4R}{3\pi}; \ \bar{x} = 0$.
- Centroid of a sector of a circular area, subtending an angle 2α at the centre, $\bar{x} = \frac{2R\sin\alpha}{3\alpha}, \ \bar{y} = 0$.
- Centroid of area bounded by a curve can be obtained by integration

$$A = \int y\,dx, \quad M_y = \int xy\,dx, \quad M_x = \int \frac{y^2}{2}\,dx$$

$$\bar{x} = \frac{M_y}{A}; \quad \bar{y} = \frac{M_x}{A}.$$

- Centroidal co-ordinates of a composite line

$$\bar{x} = \frac{l_1 x_1 + l_2 x_2 + \ldots l_n x_n}{l_1 + l_2 + \ldots l_n}; \quad \bar{y} = \frac{l_1 y_1 + l_2 y_2 + \ldots l_n y_n}{l_1 + l_2 + \ldots l_n}.$$

- Centroidal co-ordinates of a composite section

$$\bar{x} = \frac{A_1 x_1 + A_2 x_2 \ldots A_n x_n}{A_1 + A_2 \ldots A_n}; \quad \bar{y} = \frac{A_1 y_1 + A_2 y_2 \ldots A_n y_n}{A_1 + A_2 \ldots A_n}.$$

- Co-ordinates of c.g. (centre of mass) of a three dimensional body [uniform body, uniform acceleration field]

$$\bar{x}V = \int x\,dV; \quad \bar{y}V = \int y\,dV, \quad \bar{z}V = \int z\,dV.$$

- C.G. of composite bodies $\bar{x} = \frac{x_1 V_1 + x_2 V_2 \ldots x_n V_n}{V_1 + V_2 + \ldots V_n}$. Similarly expression for $\bar{y}, \bar{z}$ can be written.

- Theorems of Pappus Guldinus
  I. Surface of revolution developed by revolving generating curve about the axis of revolution has an area equal to the product of the length of generating curve times the circumference of the circle formed by the centroid of generating curve in the process of generating the surface.
  II. Volume of the body of revolution developed by rotating the plane surface about the axis of revolution equals the product of the area of surface times the circumference of the circle formed by the centroid.

MULTIPLE CHOICE QUESTIONS

5.1 A triangle *ABC*, has attitude $h = 6$ cm. What moment is of inertia about centroidal axis which is parallel to base *AC*?

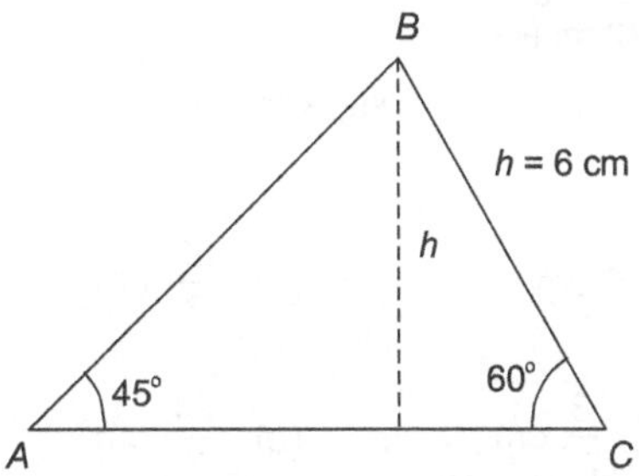

Fig. 5.15

(*a*) 56.77 cm^4 (*b*) 113.54 cm^4
(*c*) 170.35 cm^4 (*d*) None of these

5.2 What is moment of inertia of triangle *ABC* as shown in **Fig. 5.16** about base *BC*. ?

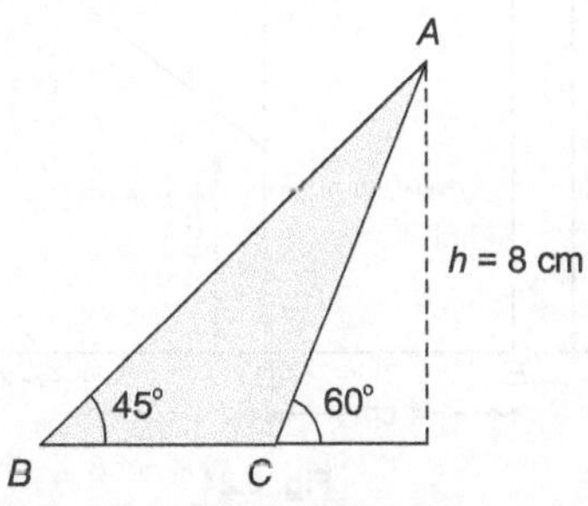

Fig. 5.16

(*a*) 48.13 cm^4 (*b*) 144.26 cm^4
(*d*) 341.3 cm^4 (*d*) None of these.

5.3 What is $\bar{x}$ of quarter of a circle of radius *R* as shown in **Fig. 5.17**

(*a*) 0.8*R* (*b*) 0.75*R*
(*c*) 0.636*R* (*d*) 0.577*R*

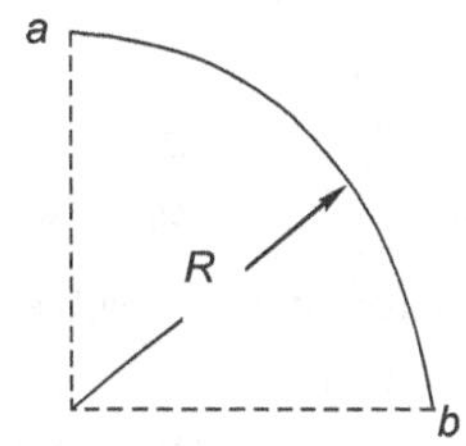

Fig. 5.17

5.4 A solid hemisphere of radius *R* is shown in **Fig. 5.18**. What is $\bar{y}$ of this solid hemisphere about *oy*-axis

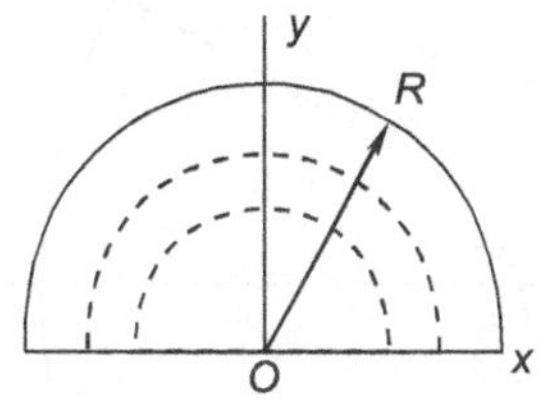

Sold hemisphere

Fig. 5.18

(*a*) 0.625*R* (*b*) 0.636*R*
(*c*) 0.375*R* (*d*) 0.25*R*

5.5 What is $\bar{x}$ of shaded area of a square **(Fig. 5.19)**?

(*a*) $-\frac{a}{4}$ (*b*) $-\frac{a}{6}$
(*c*) $-\frac{a}{8}$ (*d*) None of these.

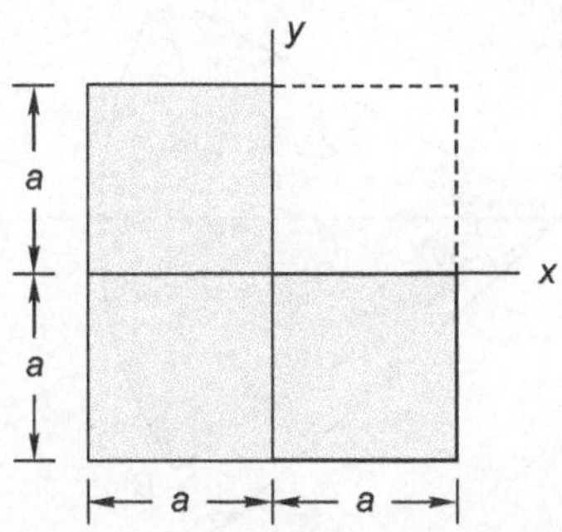

Fig. 5.19

5.6 What is $\bar{y}$ of shaded area of a triangle **(Fig. 5.20)**?

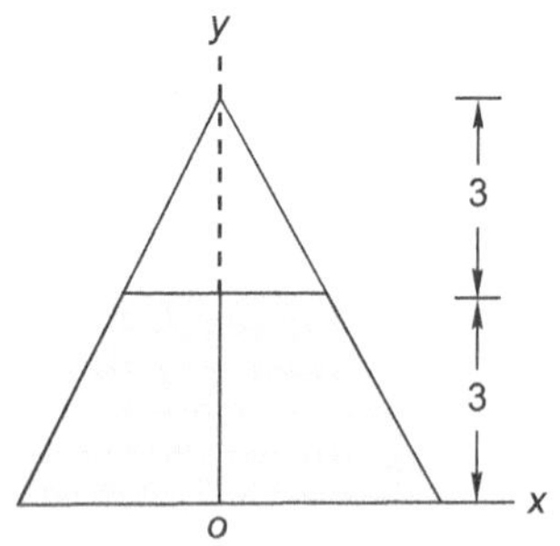

Fig. 5.20

(*a*) 2.4 cm (*b*) 1.5 cm
(*c*) 1.33 cm (*d*) None of these.

5.7 What is $\bar{x}$ of circular arc *A*, subtending an angle of 60° at *O* **(Fig. 5.21)**?
(*a*) 0.985*R* (*b*) 0.955*R*
(*c*) 0.866*R* (*d*) 0.75*R*

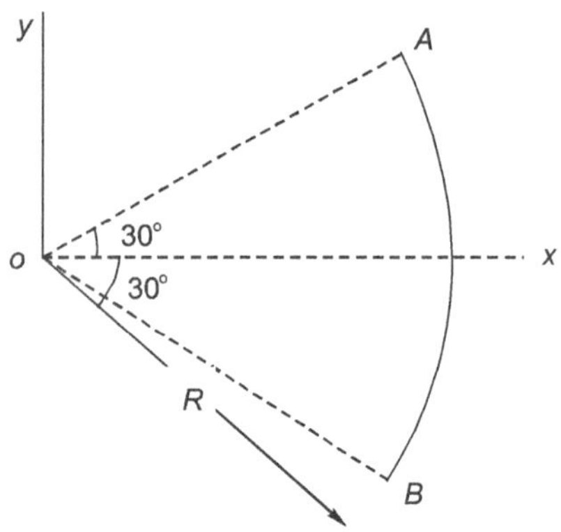

Fig. 5.21

5.8 What is $\bar{x}$ of a sector of a circular area of radius *R* subtending an angle of 60° at the centre **(Fig. 5.22)**?

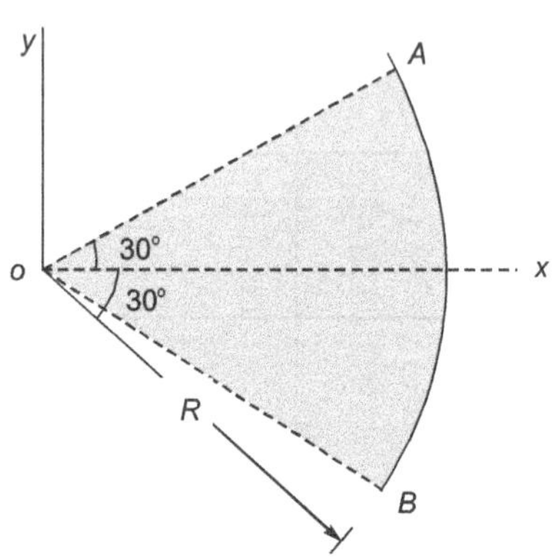

Fig. 5.22

(*a*) 0.866*R* (*b*) 0.75*R*
(*c*) 0.6366*R* (*d*) 0.619*R*

5.9 In a thin uniform lamina having symmetrical central axis as shown in **Fig. 5.23**. The distance of centre of gravity from AD is

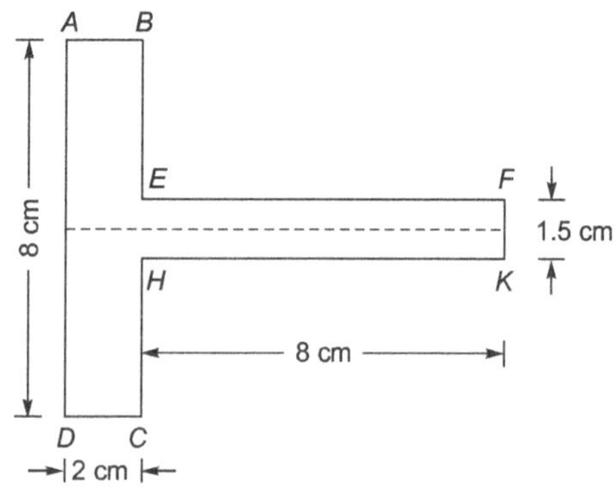

Fig. 5.23

(*a*) 3 cm (*b*) $\frac{22}{7}$ cm
(*c*) $\frac{23}{7}$ cm (*d*) $\frac{24}{7}$ cm

[CSE, Prelim, CE : 2004]

5.10 What is the y-coordinate of the centroid of the area a ABCDE, as shown in **Fig. 5.24.**

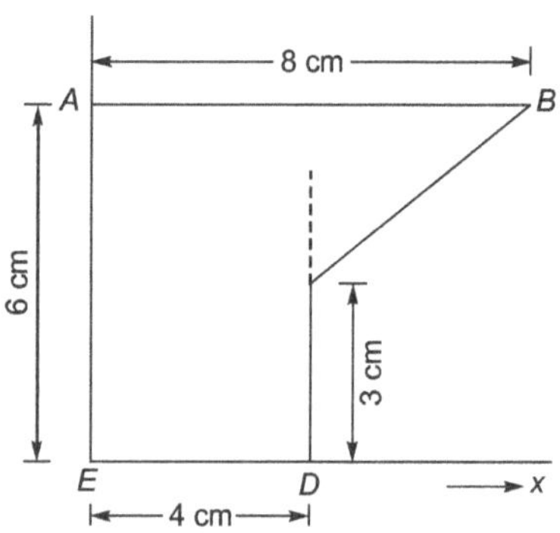

Fig. 5.24

(*a*) 3.2 (*b*) 3.3
(*c*) 3.4 (*d*) 3.5

[CSE, Prelim, CE : 2006]

Answers

5.1 (a)	5.2 (b)	5.3 (c)	5.4 (c)	5.5 (b)
5.6 (c)	5.7 (b)	5.8 (c)	5.9 (b)	5.10 (c)

EXPLANATIONS

5.1 (a)
Base,

$b = 6 + 6 \times \tan 30^o = 9.464$ cm

$$I_{xx} = \frac{bh^3}{36} = \frac{9.464 \times 6^3}{36} = 56.77 \text{ cm}^4.$$

5.2 (b)
Base,

$$b = 8 - \frac{8}{1.732} = 3.381 \text{ cm},$$

$$I = \frac{bh^3}{12} = \frac{3.381 \times 8^3}{12} = 144.26 \text{ cm}^4$$

5.3 (c)

$\bar{x} = 0.636R.$

5.4 (c)
Hemisphere $\bar{y} = 0.375R.$

5.5 (c)

$$\bar{x} = \frac{0 - a^2 \times a/2}{3a^2} = -\frac{a^3}{2} \times \frac{1}{3a^2} = -\frac{a}{6}$$

5.6 (c)

$$\bar{y} = \frac{A.2 - \frac{A}{4} \times 4}{\frac{3A}{4}} = \frac{4A}{3A} = \frac{4}{3}, \quad A = \text{area of triangle}$$

5.7 (b)
Arc

$\theta = 30^o = \pi/6$

$$\bar{x} = \frac{R \sin\theta}{\theta} = \frac{R \times 0.5}{\pi} \times 6 = \frac{3}{\pi} R = 0.955R.$$

5.8 (c)

$$\bar{x} = \frac{2R \sin\theta}{3\theta}, \quad \theta = \frac{\pi}{6}$$

$$\bar{x} = \frac{2R \times \sin 30^o}{3 \times \frac{\pi}{6}} = \frac{2R}{\pi} = 0.6366R.$$

5.9 (b)

$$\bar{x} \text{ from AD} = \frac{16 \times 1 + 12 \times 6}{28} = \frac{88}{28} = \frac{22}{7}$$

5.10 (c)

$$\bar{y} = \frac{24 \times 3 + 6 \times 5}{24 + 6} = \frac{72 + 30}{30} = \frac{102}{30} = 3.4$$

■■■■

06
CHAPTER

Friction

6.1 Introduction

Friction is a necessary evil. Many actions in daily life can not be performed without the aid of frictional force. A man cannot walk on road, if there is no friction between the feet of the man and the road. Friction takes place between two contacting surfaces.

- Contacting surfaces between two bodies can support normal as well as tangential forces.
- When one surface of a body tends to slide over another surface, frictional force (a tangential force) opposes the motion.
- The contacting surfaces resist the applied tangential force by frictional resistance between the surfaces.
- As the applied force increases, the frictional resistance also increases but to a limit.
- However, when the applied force becomes greater than the limiting value of frictional force, one body slides over another body.
- In a machine, there is loss of energy due to friction and efficiency of the machine is reduced.
- Due to friction, heat is generated between two surfaces, and this heat is dissipated away by lubrication.
- However, frictional force is used to advantage in friction drive as belt, rope, clutch drive.
- In the case of wheeled vehicles, friction is necessary for starting, moving and stopping the vehicle.

Block *A* of weight *W* rests on another block *B* as shown in **Fig. 6.1**. A force *P* is applied on block *A* and frictional force *F* is generated between the two contacting surfaces as shown. If the block *A* does not move, then $P = F$, Normal reaction from block *B* on block *A* is *N*, such that $W = N$. Forces *P* and *F* constitute a couple (*cw*) and to balance this *W* and *N* produce a couple (*ccw*) as shown. However in making calculations for frictional force and normal reactions, the effect of these couples is neglected and normal reaction is shown in line with the vertical load.

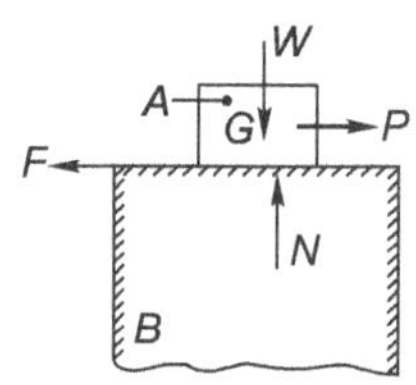

Fig. 6.1

In reality the two contacting surfaces are neither smooth nor plane, they are rough and contain irregularities as ridges and valleys as shown in **Fig. 6.2**. At contact points, there are frictional forces $F_1, F_2, F_3, ..., F_n$ and normal reactions $N_1, N_2, N_3, ..., N_n$. Resultant of normal reactions and forces of friction *i.e.*, $R_1, R_2, R_3, ..., R_n$ are also shown.

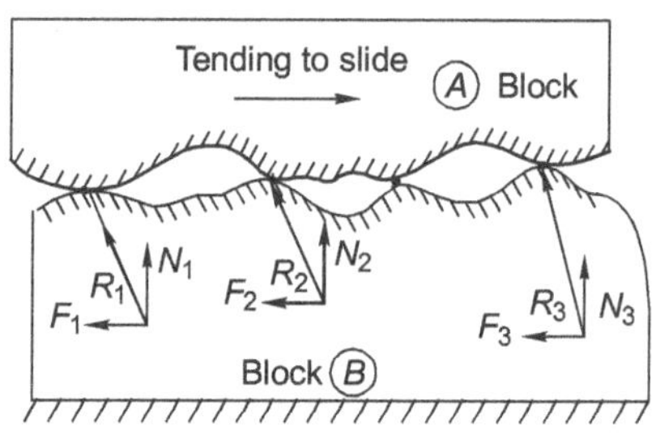

Fig. 6.2

Total normal reaction, $N = \sum_{i=1}^{n} N_i$

Total frictional force, $F = \sum_{i=1}^{n} F_i$

At the contact points, yielding, crushing and tearing of the material take place and force of friction depends upon:

(*i*) generation of local high temperature
(*ii*) adhesion at contact points
(*iii*) relative hardness of contacting surfaces
(*iv*) presence of thin oxide films, oil, dirt or dust etc. on surfaces.

As the applied force P is increased gradually, frictional force also increases gradually but upto a limit F_{max}, beyond which there is no increase in frictional force and body starts slipping *i.e.*, in this example, block A starts sliding over block B. As the motion starts, there is slight reduction in frictional force and F_{max} is reduced to F_k, kinetic frictional force which remains constant as shown in **Fig. 6.3.**

Friction between two dry or unlubricated surfaces is called *Coulomb's Friction.*

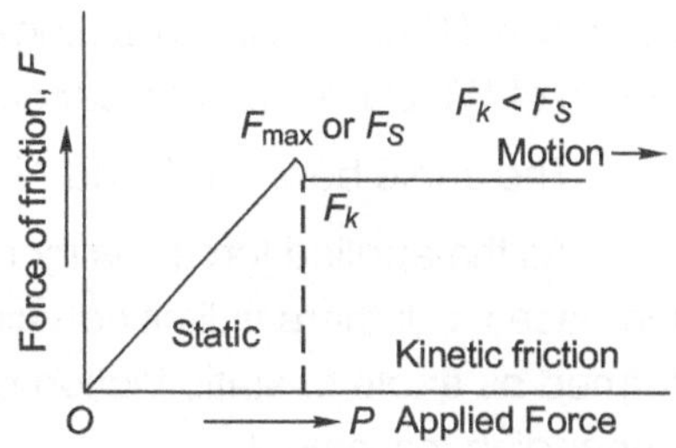

Fig. 6.3

6.2 Laws of Dry Friction or Coulomb's Friction

Frictional force theory or impending slippage or impending motion depends on following laws termed as Coulomb's laws of dry friction, described as follows:

1. **Total frictional force is independent of the magnitude of area of contact between two surfaces.**
2. **Frictional force always opposes the motion and acts in a direction opposite to direction of slippage.**
3. For low sliding velocities, frictional force is practically independent of the magnitude of velocity, however frictional force developed at the instant of impending sliding motion is slightly reduced, and when motion has already started $F_k < F_s$; kinetic frictional force is less than static force of friction.
4. **Total frictional force developed is proportional to the normal reaction provided by mating surfaces.**

Mathematically static frictional force, $F_s \propto N$ (normal reaction)

Kinetic frictional force, $F_k \propto N$

or $F_s = \mu_s \times N$

$F_k = \mu_k \times N,$

where μ_s and μ_k are coefficients of static and kinetic friction respectively between two contacting surfaces. Table 6.1 gives values of μ_s for contacting surfaces.

Table 6.1 Approximate Values of μ_s

Contacting surface materials	μ_s	*Contacting surface materials*	μ_s
Metal on metal	0.15–0.60	Wood on leather	0.25–0.50
Metal on wood	0.20–0.60	Stone on stone	0.4–0.70
Metal on stone	0.30–0.70	Earth on earth	0.2–1.0
Metal on leather	0.30–0.60	Rubber on concrete	0.60–0.95
Wood on wood	0.25–0.50	Rope on wood	0.50–0.80

Corresponding values of μ_k would be about 20–25 per cent smaller than μ_s

6.3 Angle of Friction

Consider a block of weight W resting on a horizontal plane as shown in **Fig. 6.4**. Weight acts through centre of gravity, G of the block. Now a horizontal force P is applied on block and a horizontal frictional force F is developed

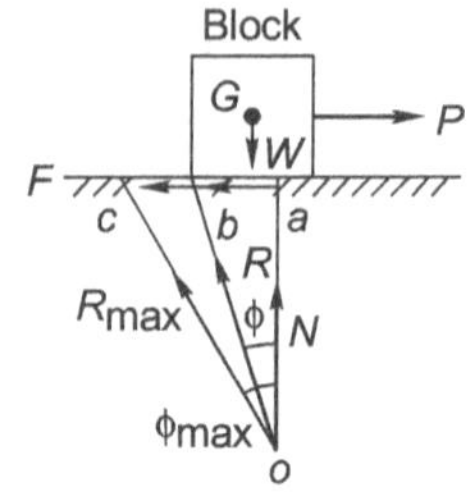

Fig. 6.4

between the contacting surfaces of the block. If N is the reaction of the plane on the block, represented by oa and force of friction F is represented by ab. Then $\overrightarrow{ob}$ is the total reaction, R.

Applied force, $P = F$, force of friction.

Weight of block $W = N$, normal reaction on block. The forces P and F constitute a clockwise couple, and to balance this couple, W and N constitute an anticlockwise couple as shown. (Therefore line of action of normal reaction N, is slightly different than the line of action of W). *But in actual practice, W and N are shown always along the same line.*

The angle between R and N, *i.e.*, $\angle aob = \varphi$, friction angle at this particular instant.

As the applied force P is increased, then friction force F also increases and consequently friction angle φ increases, but there is limiting value of force of friction, *i.e.*, F_{max}, known as *static force of friction* and φ_{max} is termed as angle of static friction φ_s, *i.e.*, angle between total reaction R and normal reaction N, *at the time of impending movement.*

$$\text{Ratio,} \qquad \frac{F_{max}}{N} = \frac{F_s}{N} = \mu_s, \text{ coefficient of static friction}$$

$$\frac{F_s}{N} = \tan\varphi_{max} = \tan\varphi_s = \mu_s$$

However many a times, μ is written as coefficient of static friction and ϕ is written as angle of friction at the time of impending slipping or sliding motion.

When the block has actually started sliding, friction angle ϕ_s is reduced to ϕ_k (angle of kinetic friction).

Direction of applied force on the block can be varied. **If the block is rotated through 360°, then total reaction R_{max} will generate a cone of apex angle $2\varphi_s$ as shown in** Fig. 6.5. **Base of cone of radius F_{max} is called *static friction circle* as shown by radius *oa*.** Similarly cone of kinetic friction with apex angle $2\varphi_k$ and kinetic friction circle of radius F_k are produced as shown by radius oc.

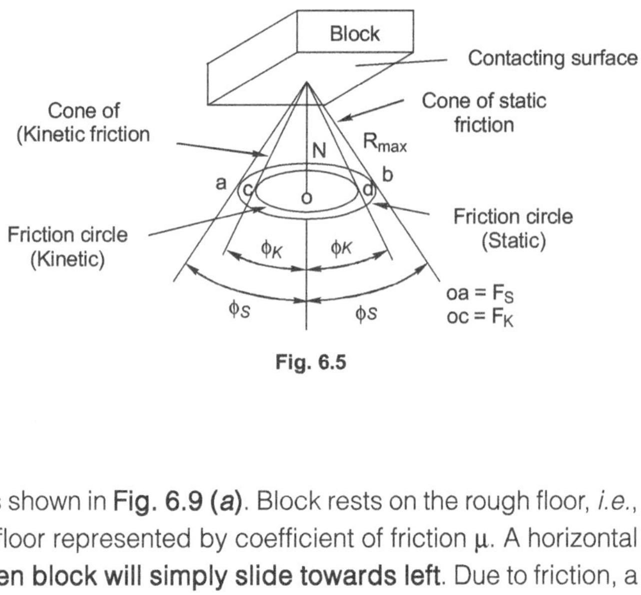

Fig. 6.5

6.4 Block on a Horizontal Plane

A block of height h, width b, weight W acting at its G is shown in **Fig. 6.9 (*a*)**. Block rests on the rough floor, *i.e.*, friction between the contacting surface of block and floor represented by coefficient of friction μ. A horizontal force P is applied at edge B. **If there is no friction, then block will simply slide towards left**. Due to friction, a frictional force F is developed at the contacting surface towards right as shown. Force P produces a moment $P \times h$ (ccw) at edge D and the block may try to tip at edge D. **To prevent tipping, a stabilising moment** $W \times \frac{b}{2}(cw)$ **acts at D.** A normal reaction, N will develop at edge D, as shown in **Fig. 6.6 (*b*).**

Moments about the edge D

$$P \times h\,(ccw) - W \times \frac{b}{2}(cw) = 0$$

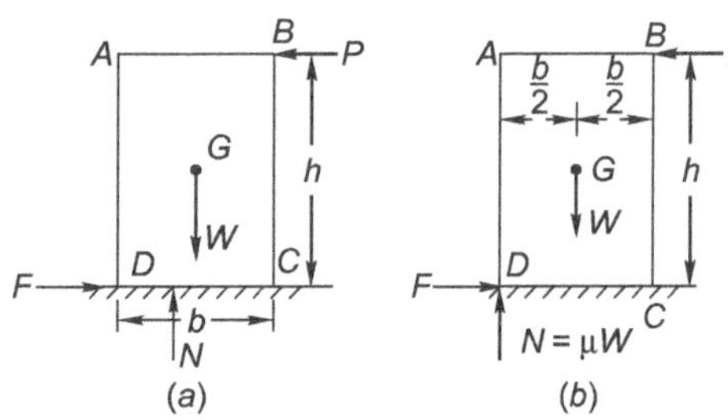

Fig. 6.6

Overturning moment – stabilishing moment = 0

To prevent overturning, $Ph < \frac{Wb}{2}$

$$h < \frac{Wb}{2P}$$

If the block is to tip at edge *A*, then

$$h > \frac{Wb}{2P}$$

Since the forces *F* and *N* are passing through edge *D*, there is no effect of these forces on tipping of the block at the edge.

If the block is *to slide without tipping*, then

$$P > F \text{(force of friction)}$$
$$> \mu \times \text{normal reaction}$$

$$P > \mu W \quad \text{and} \quad h < \frac{Wb}{2P}$$

or $$h < \frac{Wb}{2\mu W} \quad \text{or} \quad h < \frac{b}{2\mu}.$$

6.5 Two Mating Blocks

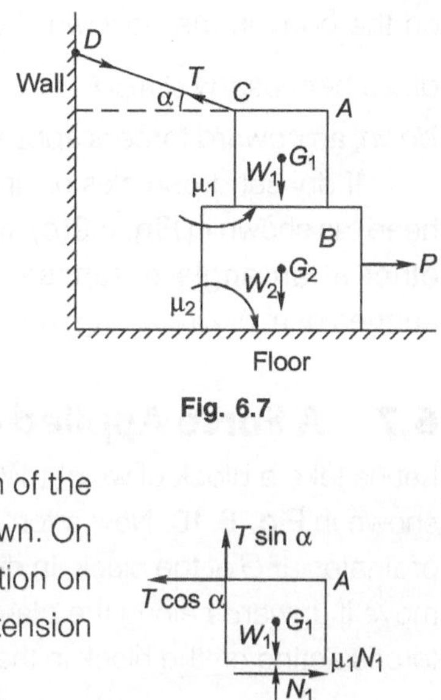

Fig. 6.7

Block *A* of weight W_1 rests on a block *B* of weight W_2, which is lying on floor, as shown in the **Fig. 6.7**. Upper block *A* is connected to the wall through a string *CD* and therefore restricted from movement. Let us take μ_1 as coefficient of friction between blocks *A* and *B* and μ_2, as coefficient of friction between block *B* and floor.

A force *P* is applied on block *B* and let us find P_{min} which causes movement of block *B* in between floor and block *A*. So the block *B* is subjected to frictional force on top and bottom surfaces. Let us draw *free body diagrams for both blocks as shown in* **Fig. 6.8.** On block *B*, there are normal reactions of block *A* and floor *i.e.*, N_1 and N_2. Block *B* tends to move towards right under the action of the force *P*, so forces of friction will act in opposite direction *i.e.*, *towards left* as shown. On block *A*, force of friction $\mu_1 N_1$ will act in the opposite direction to force of friction on block *B*, *i.e.*, $\mu_1 N_1$ will act towards right. Moreover frictional force $\mu_1 N_1$ will cause tension *T* in the string in the opposite direction as shown.

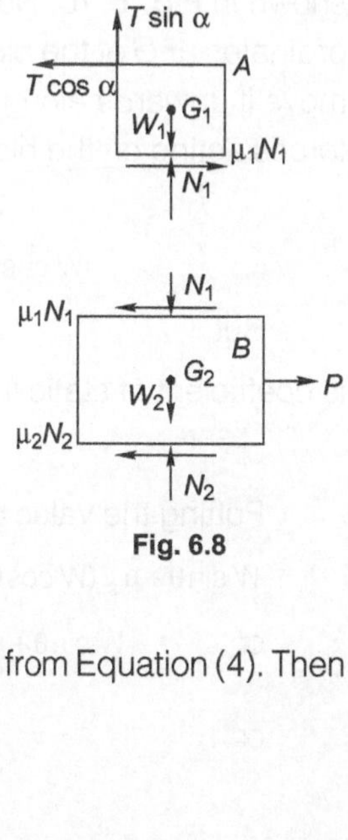

Fig. 6.8

Block A, For equilibrium

$$T \cos\alpha = \mu_1 N_1 \quad ...(1)$$
$$T \sin\alpha + N_1 = W_1 \quad ...(2)$$

From these two equations values of *T* (tension in string) and normal reaction N_1 can be determined.

Block B. For equilibrium

$$P = \mu_1 N_1 + \mu_2 N_2 \quad ...(3)$$
$$N_1 + W_2 = N_2 \quad ...(4)$$

Knowing the value of N_1 from Equations (1) and (2), we can find value of reaction, N_2 from Equation (4). Then finally the magnitude of minimum force *P* required can be determined

6.6 Angle of Repose

Consider a block of weight W, resting on an inclined plane AB as shown in **Fig. 6.9 (*a*)**. **Fig. 6.18 (*b*)** shows the free body diagram of block, in which,

Normal reaction on block, $N = W\cos\theta$

Friction force $F = W\sin\theta$

Under this condition, the block is at rest and there is *no downward motion* of the block.

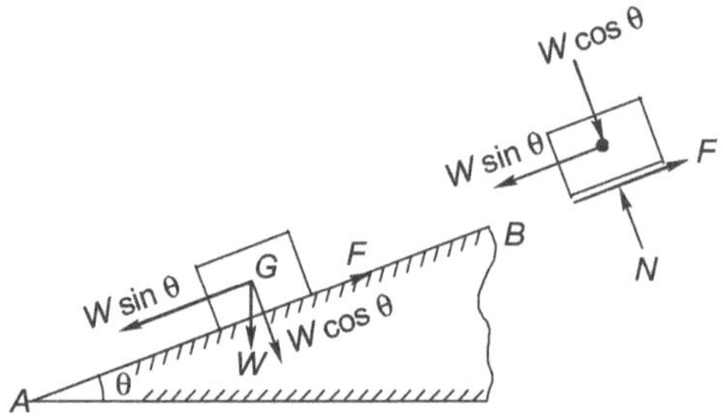

(a) (b)

Fig. 6.9

Now the angle of inclination of the block is gradually increased till the block *tends* to slide down. The angle of inclined plane, at which the block tends to slide down is known as *angle of repose*. Say this angle is φ and coefficient of static friction is μ (between block and inclined plane). Then

At, angle $\theta = \phi$.

$$W\sin\phi > F = \mu W\cos\varphi \text{ for impending downward motion}$$

or in the limiting case $\mu = \tan\varphi$, where φ is angle of friction or angle of repose.

Note that if angle of inclination of the plane $\theta < \varphi$, the block will remain at rest on inclined plane and if the body is to slide down, then a force has to be applied on the body in the downward direction. However if $\theta > \varphi$, then block will slide down because $W\sin\theta > F$, force of friction and to *prevent the body from sliding down*, an upward force is applied on the block.

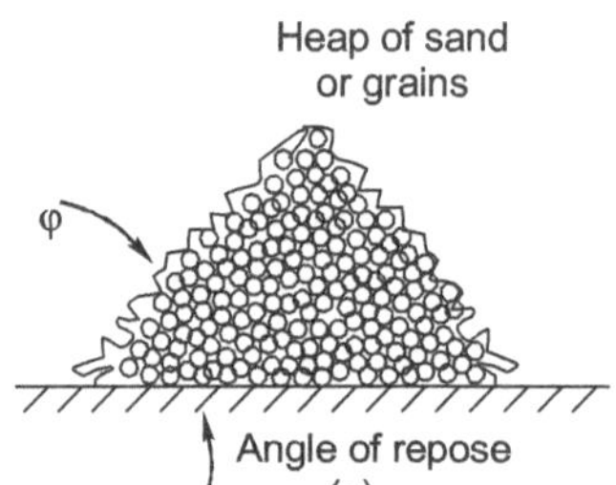

(c)

Fig. 6.9

If dry sand particles or grains are dropped on the ground in the form of a heap, as shown in **Fig. 6.9(c)**, **then sand particles or grains *rest against each* other at an angle of repose**. Use of angle of repose is made in the case of earthen dam or a retaining wall.

6.7 A Force Applied on a Body Lying on an Inclined Plane

Let us take a block of weight W, lying on an inclined plane AB, with angle of inclination θ with the horizontal as shown in **Fig. 6.10**. Now a force P inclined at an angle α with the inclined plane (or with x-axis, taking xy co-ordinates at G of the block, in directions parallel and perpendicular to inclined plane) is applied on the block to move it upwards along the plane. The force of friction on block will act downwards as shown. Considering the forces acting on the block in the directions x and y,

$$W\sin\theta + F = P\cos\alpha \quad \text{...(1)}$$

$$W\cos\theta + P\sin\alpha = N, \text{ normal reaction} \quad \text{...(2)}$$

But $F = \mu_s N$, where μ_s is coefficient of static friction

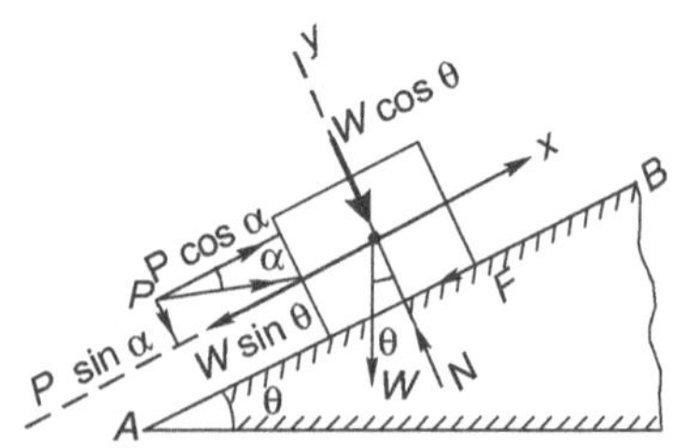

Fig. 6.10

or $F = \mu_s (W\cos\theta + P\sin\alpha)$ ...(3)

Putting the value of F in Equation (1), we get

$$W\sin\theta + \mu_s (W\cos\theta + P\sin\alpha) = P\cos\alpha$$

or $W\sin\theta + \mu_s W\cos\theta = P\cos\alpha - \mu_s P\sin\alpha$

or $$P = W\left[\frac{\sin\theta + \mu_s\cos\theta}{\cos\alpha - \mu_s\sin\alpha}\right]$$

If $\theta = \alpha$, then $$P = W\left[\frac{\sin\theta + \mu_s\cos\theta}{\cos\theta - \mu_s\sin\theta}\right] = W\left[\frac{\tan\theta + \mu_s}{1 - \mu_s\tan\theta}\right]$$

If $\theta < \varphi$, and the body is to move up the plane, with the application of the applied force P, the expression for P and F and N will remain the same as described above.

$\theta < \varphi$ and body is to slide down

A force P is required to be applied on block to move the block down the plane as shown in **Fig. 6.11**. Then

Friction force, $F = W\sin\theta + P$

Normal reaction, $N = W\cos\theta$

or $F = \mu_s N = \mu_s N$

$= \mu_s W\cos\theta$

or $\mu_s W\cos\theta = W\sin\theta + P$

or $P = \mu_s W\cos\theta - W\sin\theta.$

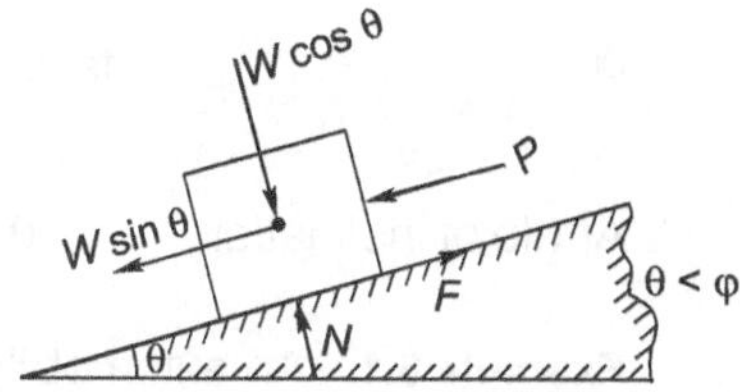

Fig. 6.11

6.8 Bodies on Inclined Plane Connected Through a Link

Consider two bodies of weights W_1 and W_2 respectively, connected through a link AB, coefficients of friction between block A and inclined plane is μ_1 and that between block B and inclined plane is μ_2 as shown in **Fig. 6.12(*a*)**. Let us say coefficient of friction $\mu_1 < \mu_2$. Let us determine the angle θ of the plane at which blocks will tend to slide down. As μ_1 is less than μ_2, block A of weight W_1 will exert force on link AB and link will be in compression. Take the force in link as P as shown. **Fig. 6.12 (*b*)** shows the free body diagrams of both blocks A and B. Resolving the forces on the block in the direction parallel and perpendicular to the inclined plane.

Block *A*:

Normal reaction, $N_1 = W_1\cos\theta$

Force of friction $F_1 = \mu_1 N_1 = \mu_1 W_1\cos\theta$

(blocks tend to slide down, force of friction will be upward along the plane)

Then $W_1\sin\theta = P + F_1 = P + \mu_1 W_1\cos\theta$

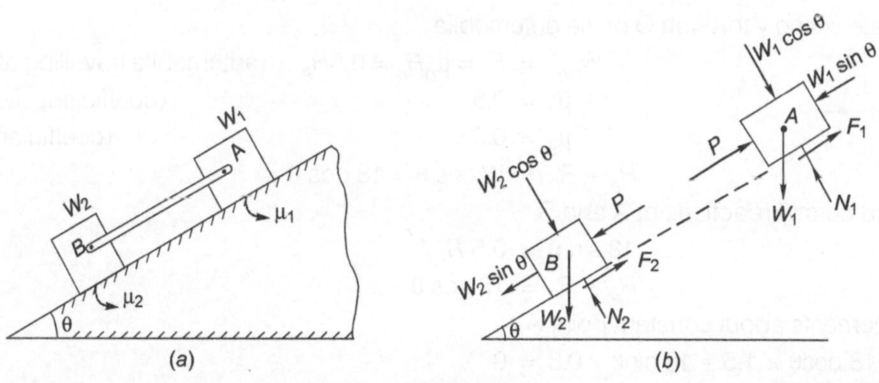

Fig. 6.12

or $P = W_1\sin\theta - \mu_1 W_1\cos\theta$ ...(1)

Block *B*:

Normal reaction, $N_2 = W_2\cos\theta$

Force of friction, $F_2 = \mu_2 N_2 = \mu_2 W_2 \cos\theta$

$$W_2 \sin\theta + P = F_2 = \mu_2 W_2 \cos\theta$$

or $P = \mu_2 W_2 \cos\theta - W_2 \sin\theta$ ...(2)

From Equations (1) and (2)

$$W_1 \sin\theta - \mu_1 W_1 \cos\theta = \mu_2 W_2 \cos\theta - W_2 \sin\theta$$

or $(W_1 + W_2)\sin\theta = (\mu_1 W_1 + \mu_2 W_2)\cos\theta$

or $\tan\theta = \dfrac{\mu_1 W_1 + \mu_2 W_2}{W_1 + W_2}$

Angle of inclination, $\theta = \tan^{-1}\left(\dfrac{\mu_1 W_1 + \mu_2 W_2}{W_1 + W_2}\right)$.

Example 6.1 An automobile with the distance between the front and rear wheels 3.0 m and height of the centre of gravity 300 mm is travelling over a road way inclined at an angle θ with the horizontal. The coefficient of static and dynamic friction are 0.6 and 0.5 respectively. Automobile is a rear wheel drive and has a load of 18 kN. What is the maximum inclination θ_{max} that the automobile can climb at uniform speed?

Solution:

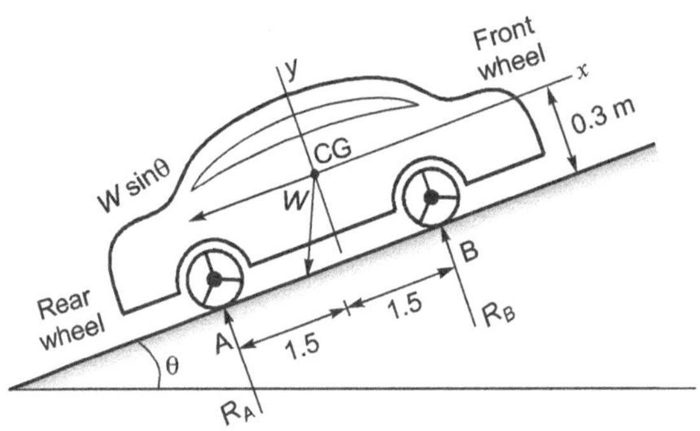

Fig. 6.13

Rear wheel drive rear wheel at *A*, front wheel at *B* relative velocity, between road and tyre surface is now at the point of contact at *A*.

Coordinate, x and y through G of the automobile

$W_{\sin\theta} = F_A = \mu_d R_A = 0.5 R_A$ (automobile travelling at uniform speed)

$\mu_d = 0.5$ (coefficient of dynamic friction)

$\mu_s = 0.6$ (coefficient of static force)

$R_A + R_B = W\cos\theta = 18\cos\theta$

R_A, R_B are normal reactions at A and B.

$18\sin\theta = 0.5 R_A$ ...(i)

$R_A + R_B = 18\cos\theta$ ...(ii)

Taking moments about constant point *A*

$3R_B - 18\cos\theta \times 1.5 + 18\sin\theta \times 0.3 = 0$ ...(iii)

$R_A = 36\sin\theta$ ...(iv)

$R_B = 18\cos\theta - 36\sin\theta$ ...(v)

Substituting for R_B in equation (iii)

$3(18\cos\theta - 36\sin\theta) - 27\cos\theta + 5.4\sin\theta = 0$

$54\cos\theta - 108\sin\theta - 27\,\text{co}\theta + 5.4\,\text{sn}\theta = 0$

$$27\cos\theta - 102.6\sin\theta = 0$$
$$27\cos\theta = 102.6\sin\theta$$
$$\tan\theta = 0.263,\ \theta = 14.73°$$

6.9 Ladder

Ladder cannot be functional without the help of friction between floor and end of the ladder in contact with floor. Frictional force on ladder at the ground maintains the equilibrium of ladder. Consider a uniform ladder *AB* of length *L*, weight of the ladder W_L supporting a man of weight W_m at a horizontal distance nL from end *A* of ladder. The lower end of the ladder is on a rough floor and upper end of the ladder rests against a rough vertical wall as shown in **Fig. 6.14**. Say coefficient of friction between ladder and floor is μ_1 and that between ladder and rough vertical wall is μ_2. Let us determine the minimum angle of inclination θ of the ladder with the horizontal to maintain equilibrium. Say the normal reactions at ends *A* and *B* are N_1 and N_2 respectively. The ladder will try to slip down, or end *A* tries to move towards left, then

Force of friction F_1 at *A* will act towards right.

Force of friction F_2 at *B* will act upwards as shown in the force.

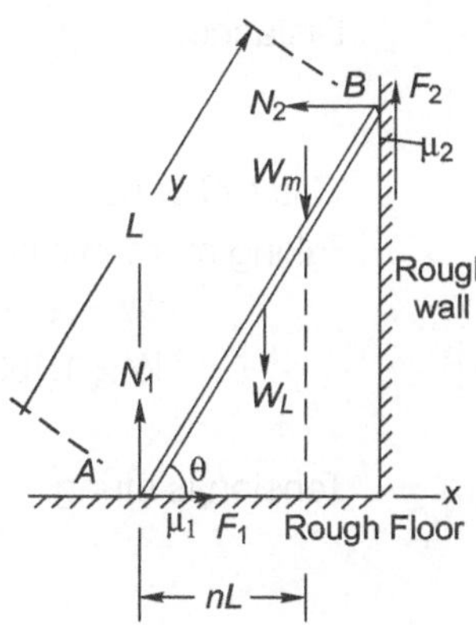

Fig. 6.14

Force of friction, $F_1 = \mu_1 N_1$

$$F_2 = \mu_2 N_2$$

Considering the equilibrium of forces

$$\sum F_x = 0, \quad F_1 = N_2 \quad ...(1)$$

$$\sum F_y = 0, \quad W_L + W_m = N_1 + F_2 \quad ...(2)$$

But $F_1 = \mu_1 N_1$

$$F_2 = \mu_2 N_2,$$

putting this values in Equations (1) and (2)

$$\mu_1 N_1 = N_2 \quad ...(3)$$

$$W_L + W_m = N_1 + \mu_2 N_2 \quad ...(4)$$

From these equations normal reactions N_1 and N_2 can be determined. All the forces shown in the ladder are not concurrent. So to determine minimum angle for equilibrium, we have to consider one more equation of equilibrium

$\sum M_A = 0$, (moments about any point is zero)

= Moments about *A*

$$W_L \times \frac{L}{2}\cos\theta + nL\,W_m = F_2 . L\cos\theta + N_2 \times L\sin\theta$$

or
$$\frac{W_L}{2}\cos\theta + nW_m = F_2\cos\theta + N_2\sin\theta \quad ...(5)$$

From Equation (4)

$$W_L + W_m = N_1 + \mu_2 N_2, \text{ but } N_2 = \mu_1 N_1 \text{ from Equation (3),}$$

So
$$W_L + W_m = N_1 + \mu_2\mu_1 N_1$$

or reaction
$$N_1 = \frac{W_L + W_m}{1 + \mu_1\mu_2}$$

$$N_2 = \mu_1 N_1 = \frac{\mu_1 [W_L + W_m]}{1 + \mu_1\mu_2} \quad ...(6)$$

Value of N_1 and N_2 can be substituted to know the minimum value of angle θ, *i.e.*, inclination of ladder.

$$\frac{W_L}{2}\cos\theta + nW_m = \mu_2 N_2 \cos\theta + N_2 \sin\theta$$

$$= \frac{\mu_1\mu_2[W_L + W_m]}{1+\mu_1\mu_2}\times\cos\theta + \frac{\mu_1[W_L+W_m]}{1+\mu_1\mu_2}\cdot\sin\theta \quad ...(7)$$

Equation (7) can be solved for minimum value of angle of inclination θ of the ladder.

Example 6.2 A rod *AB*, 6 m long is held against sliding by a string *AD*. The rod weights 10 kN, determine the tension in the string *AD* assuming that all surfaces are smooth.

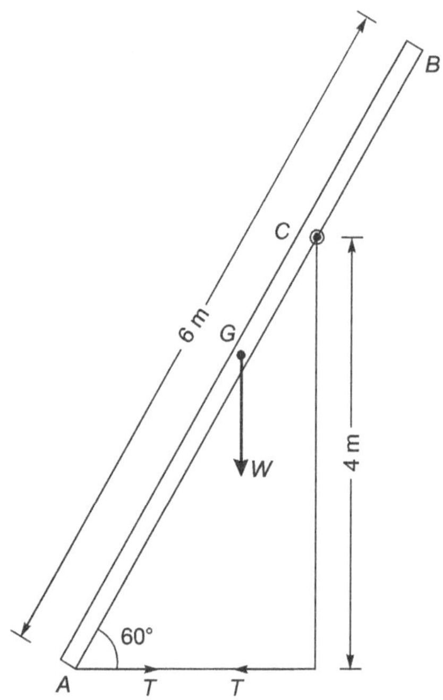

Fig. 6.15

Solution:

$$W = 10 \text{ kN, weight of rod}$$

Distance, $$GC = \frac{4}{\sin 60^\circ} - 3$$

$$= 4.169 - 3 = 1.169 \text{ m}$$

G is *CG* of bar.

Taking moments about constact point *C* of bar

$$W \times GC\cos 60^\circ = T \times 4$$

$$W \times 1.169 \times 0.5 = 4T$$

Tension is string, $$T = \frac{10\times 1.169\times 0.5}{4}$$

$$= 1.461 \text{ kN}$$

6.10 Friction in Wheels

There are two types of driven wheels as follows:

(*i*) Wheels being pulled by a force applied on the axle.

(*ii*) Wheels driven by an engine of an automobile *i.e.*, power is transmitted to the wheels of an automobile, which become the driving wheels and the other pair becomes the driven wheels.

(*i*) **Wheels being pulled by a force:** Say a wheel of radius *R* is being pulled by a force *P* over a rough surface, as shown in **Fig. 6.16**. If coefficient of friction between wheel and surface is μ, then frictional force $F = \mu N$, where N = normal reaction = W

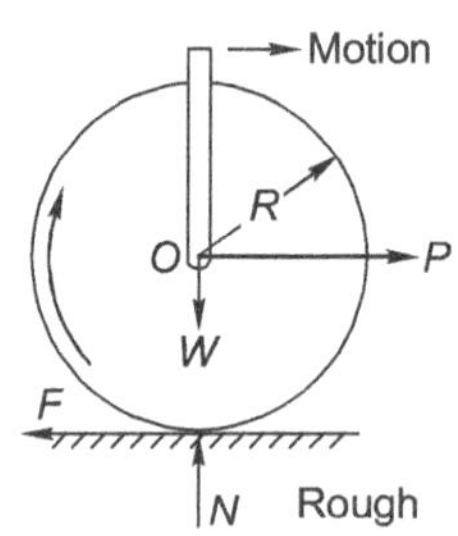

Fig. 6.16

Turning moment at the centre of the wheel

Turning moment, $T = FR = \mu WR$

If surface is smooth, $\mu = 0$, $T = 0$, the wheel will not rotate.

The wheel moves in the clockwise direction and wheel moves towards right. Magnitude of frictional force *F* will be maximum equal to μN. *Force P applied at axis O cannot provide turning moment about axis O.* Frictional force provides clockwise turning moment about axis *O*.

(*ii*) **Driving and Driven wheels of an automobile:** Consider a *rear wheel drive* vehicle as shown in **Fig. 6.17**, travelling towards right as shown. Driving torque on rear wheels is *T*. This torque is provided by the engine through the propeller shaft. **The engine torque *T*, tends to rotate the wheels in the clockwise direction.** (*i*) If the road is smooth,

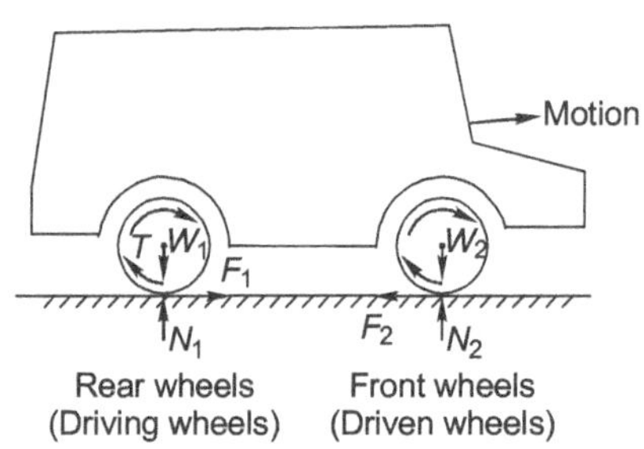

Fig. 6.17

$\mu = 0$, *the wheels will simply rotate clockwise about their axis but the axle does not move forward*, remains stationery and the vehicle does not move. This phenomenon is sometimes observed when the car is started on a *slushy ground*.

However the road is generally rough and a frictional force F_1 acts on the outer surface of wheel, if R is the radius of the wheel, then frictional torque, $T_f = F_1 \times R$, (where F_1 **is in the opposite direction at the point of contact, wheel tends to rotate and tends to slip towards left at contact point)**. This frictional force F_1, becomes the propelling force on the **vehicle in the direction of motion. Frictional force** F_1 at contact point A can be replaced by a anticlockwise couple $F_1 \times R$ and a force F_1 at axle axis O as shown in **Fig. 6.18(a)**. *So this is the frictional force on the driving wheel which is responsible for producing motion in vehicle.*

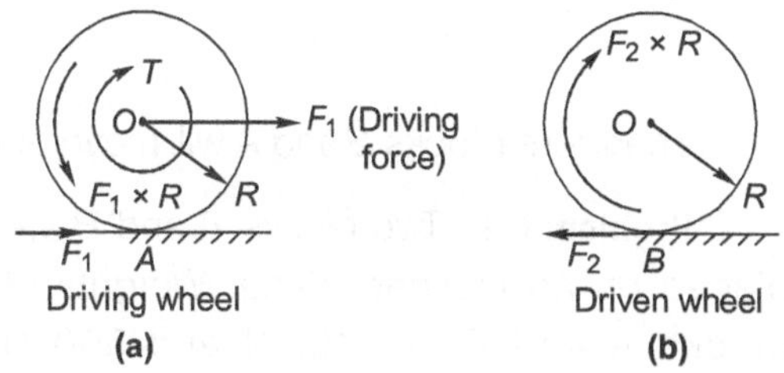

Fig. 6.18

Now the *driven wheels* (front wheels) *are not driven by the engine but are driven by the vehicle as the vehicle is propelled by the engine through the driving wheels* (rear wheels). Friction on driven wheels will be towards left as shown in **Fig. 6.18(b)**, and this provides clockwise turning moment $F_2 \times R$ on the front wheels. Normally $F_2 < F_1$. Following points must be remembered in an engine powered vehicle:

(*a*) On a smooth surface driving wheels will rotate about axis of axle but vehicle will not move.
(*b*) Frictional torque on driving wheels is overcome by engine torque.
(*c*) Friction on driven wheels provides motion to driven wheels.
(*d*) Friction on driving wheels provides driving force or tractive force to the vehicle.

PROBLEMS

Problem 6.1 Blocks A and B connected through a cord rest on inclined planes as shown in **Fig. 6.19**. Coefficient of friction between plane and block A is 0.2 and that between block B and plane is 0.15. What is the tension in the cord if the friction at block A reaches the maximum value?

Solution Say tension in the cord is T. Free body diagrams of blocks A and B are shown in the figure.

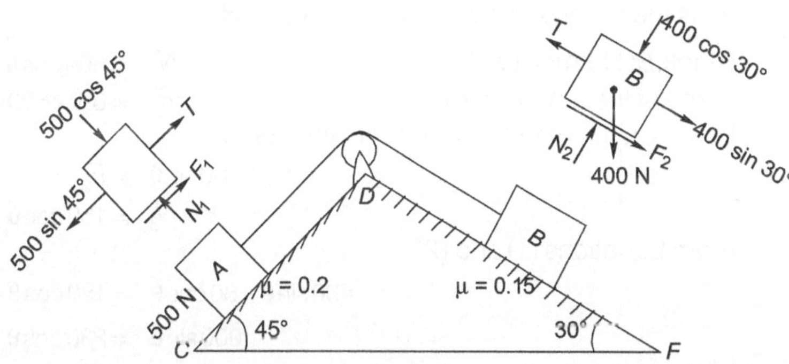

Fig. 6.19

Block A: Angle of inclination of plane CD is more than the angle of inclination of plane DF. Block A tends to slide down, force of friction F_1 will act upwards.

$$T + F_1 = 500 \sin 45° = 500 \times 0.707 = 353.5 \text{ N}$$

$$F_1 = \text{maximum value of frictional force}$$

$$= \mu_1 N_1$$

$$= 0.2\,(500 \cos 45°) = 70.7 \text{ N}$$

Tension in cord, $$T = 353.5 - 42.42 = 282.8 \text{ N}$$

Block B: $$T = 400 \sin 30° + F_2$$

$$282.8 = 400 \times 0.5 + F_2$$

Force of friction $$F_2 = 82.8 \text{ N}$$

Limiting force of friction on block B

$$= \mu_2 N_2$$

$$= 0.15\,(400\cos 30°) = 51.96 \text{ N}$$

$$m_2 N_2 < F_2$$

$$T > 200 = 51.96$$

$$T > 251.96$$

Therefore Blocks B and A will move but with acceleration.

Problem 6.2 Two blocks A and B are placed on an inclined plane. Weights of blocks A and B are 400 N and 500 N respectively. Coefficients of friction between block A and plane is 0.2 and between block B and plane is 0.3 as shown in **Fig. 6.20 (*a*)**. To what angle θ, the plane should be raised so that bodies start slipping down the plane?

Solution Say the blocks are at the point of sliding down the plane. Free body diagrams of blocks A and B are given in **Fig. 6.20 (*b*)**.

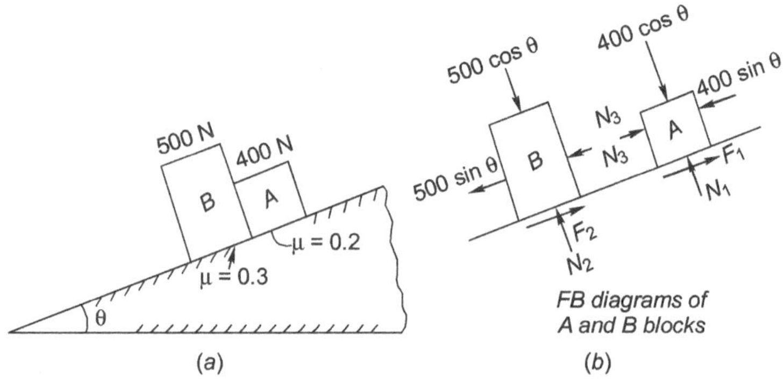

Fig. 6.20

Frictional force on block A is less than frictional force on block B, so block A will exert push on block B.

Block *A*: Normal Reaction, $N_1 = 400\cos\theta$

Friction force, F_1 (upwards) $= 0.2 \times 400\cos\theta = 80\cos\theta$

Forces in the direction parallel to the plane $N_3 = 400\sin\theta - F_1 = 400\sin\theta - 80\cos\theta$ ...(1)

N_3 is the reaction between blocks A and B

Block *B*: Normal reaction, $N_2 = 500\cos\theta$

Force of friction (upwards), $F_2 = 0.3 \times 500\cos\theta = 150\cos\theta$

Forces in the direction parallel to the plane

$$N_3 + 500\sin\theta = F_2$$

$$N_3 = 150\cos\theta - 500\sin\theta \quad ...(2)$$

From Equations (1) and (2)

$$400\sin\theta - 80\cos\theta = 150\cos\theta - 500\sin\theta$$

$$900\sin\theta = 230\cos\theta$$

$$\tan\theta = \frac{230}{900} = 0.2555$$

$$\theta = \tan^{-1} 0.2555 = 14.335° = 14°20'.$$

Remember

- Tangential force generated between two contacting surfaces which tend to slide over one another due to surface irregularities is called *force of friction*.
- Force of friction on a body always opposes the motion of the body.
- Coefficient of kinetic friction is less than the coefficient of static friction *i.e.*, $\mu_k < \mu_s$.
- Angle made by the limiting force of friction with the normal reaction is called *angle of friction*.
- Force of friction, $F = \mu \times N$ (normal reaction), normal reaction is perpendicular to the surface on which a body rests.
- Angle of repose = angle of friction.

- A wedge is a simple machine to lift heavy loads or heavy stone blocks. A wedge is generally used for small adjustments.
- An inclined plane is also a lifting machine since it provides mechanical advantage.
- For an engine powered vehicle remember that:
  (*a*) on a smooth surface, driving wheels will rotate about their axis but vehicle will not move.
  (*b*) frictional torque on driving wheels is overcome by engine torque.
  (*c*) friction on driven wheels provides moving torque for the vehicle.
  (*d*) friction on driving wheels provides driving force for the vehicle.

PRACTICE PROBLEMS

6.1 A packing crate of mass 45 kg is pulled by a rope as shown in **Fig. 6.21**. If the coefficient of friction between crate and rope is 0.3, determine (*a*) the magnitude of pull *P* required to move the crate, (*b*) whether the crate will slide or tip.

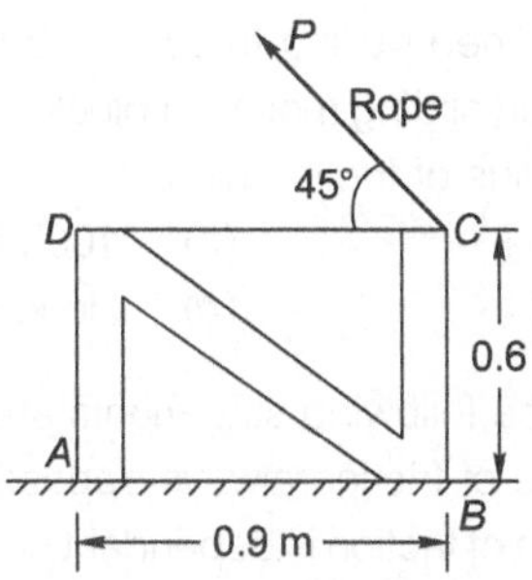

Fig. 6.21

[**Hint:** Take components of pull *P* in rope, and consider edge *A* for tipping or no tipping].
[**Ans:** (*a*) P = 144.1 N, (*b*) Crate will slide].

6.2 A bar of weight 800 N, 8 m long rests on a stationary support *A* and on a roller support at *B*, as shown in **Fig. 6.22**. The static coefficient of friction at *A* is 0.4, while the dynamic coefficient of friction of roller support *B* is 0.20. *If support B is moved at constant speed towards left,* how far does it move before the bar begins to move.

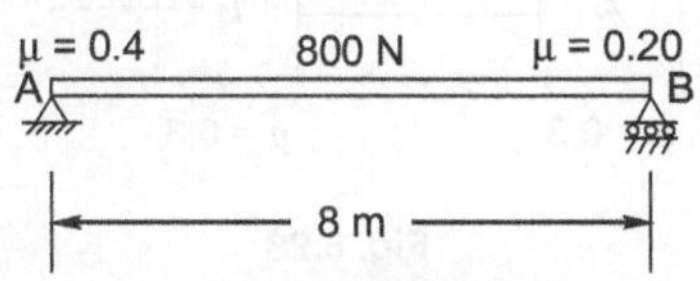

Fig. 6.22

[**Hint:** *Frictional force on bar at B will act in left direction, because while moving roller support, towards left, force of friction on roller support will act in right direction,* as roller support moves towards left, the normal reaction at *B* goes on increasing. Say the roller support moves by distance *x* m].
[**Ans:** x = 1.33 m, roller support moves by 1.33 m].

6.3 Determine the range of magnitude of weight *W* for which the block of 500 N weight will either slide down the plane or slide up the plane. At what value of *W*, the friction force will become zero. Refer to **Fig. 6.23**.

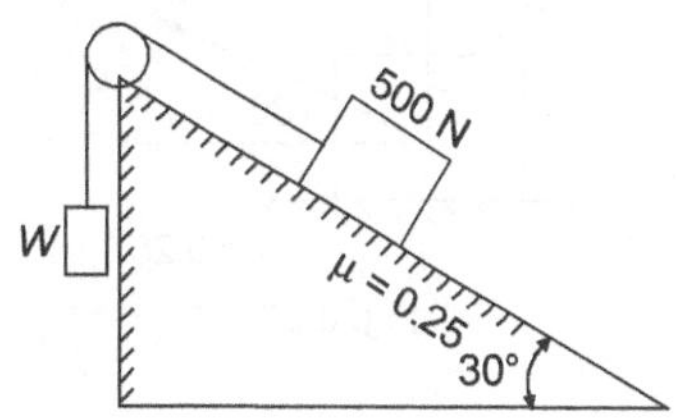

Fig. 6.23

[**Ans:** 141.75 N; 358.25 N; W = 250 N, friction force will be zero].

6.4 Two blocks *A* and *B* of weights 400 N and 300 N respectively rest on an inclined plane. The coefficient of friction between block *A* and plane is 0.35 and that between block *B* and plane is 0.2. The angle of inclination of the plane is increased gradually. At what angel θ, blocks *A* or *B* or both start sliding down the plane. Refer to **Fig. 6.24**.

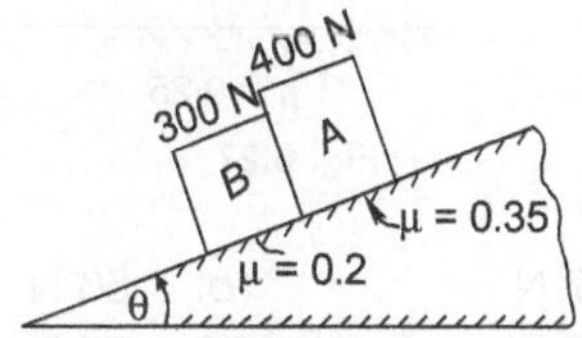

Fig. 6.24

[**Hint:** Coefficient of friction for upper block *A* is more than for lower block, therefore lower block will start sliding as angle θ increases].
[**Ans:** 11.31°].

6.5 A block weighing 1000 N rests on a horizontal plane as shown in **Fig. 6.25**. The coefficient of static friction between block and plane is 0.25. Over what range of angle θ, there will be no movement of the block if a force of 300 N acts on the block as shown.

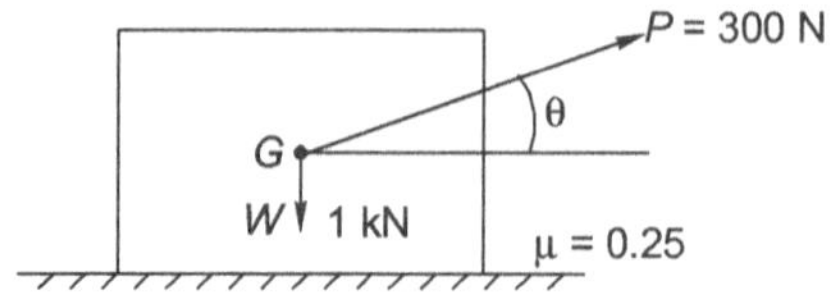

Fig. 6.25

[**Hint:** $P\cos\theta < \mu\,(W - P\sin\theta)$].

[**Ans:** $\theta = 50°$ to $130°$].

MULTIPLE CHOICE QUESTIONS

6.1 A block weighing 200 N is lying on a rough surface. An inclined force of 100 N acts on the block. If coefficient of friction between block and surface is 0.3, what is force of friction on block when block tends to slide under the action of force *P* **(Fig. 6.26)**?

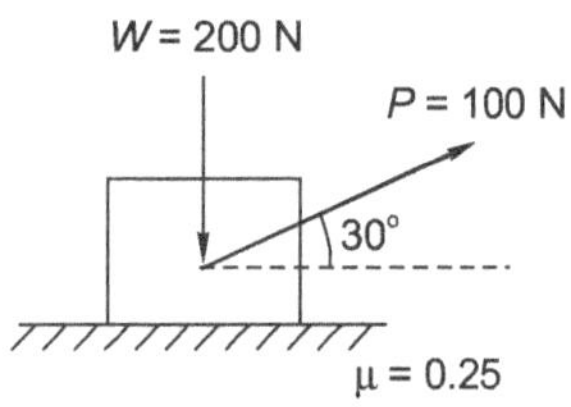

Fig. 6.26

(*a*) 75 N (*b*) 50 N
(*c*) 45 N (*d*) None of these

6.2 A block weighing 200 N is lying on floor. Coefficient of friction between floor and block is 0.25. A pushing force *P* is applied on block as shown in **Fig. 6.27**. What is maximum value of *P* so that block starts sliding?

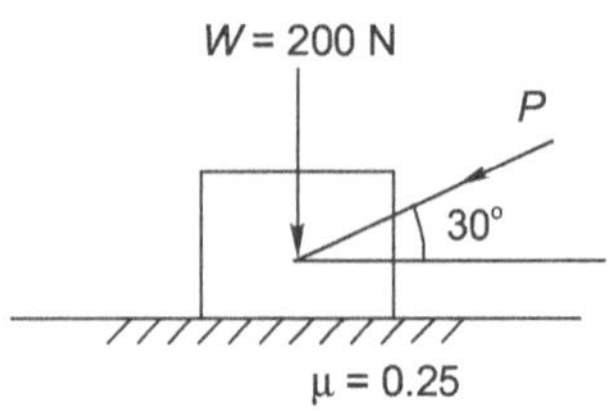

Fig. 6.27

(*a*) 57.7 N (*b*) 60 N
(*c*) 67.5 N (*d*) 75 N

6.3 Arrange the following contacting surfaces in increasing order of coefficient of static friction

(*i*) metal on metal–I
(*ii*) rubber on concrete–II
(*iii*) metal on stone–III
(*iv*) stone on stone–IV.

(*a*) I III IV II
(*b*) I II III IV
(*c*) II I III IV
(*d*) III I II IV

6.4 A block of weight 500 N rests on rough surface. The coefficient of friction between block and surface is 0.2. An inclined pulling force, $\alpha = 30°$ is applied on block to start sliding motion of block. If force is 200 N, what is radius of friction circle?

(*a*) 120 N (*b*) 100 N
(*c*) 80 N (*d*) None of these

6.5 Which of the following statements are correct?

(*i*) Force of friction always opposes motion
(*ii*) Force of friction is dependent on area of contact
(*iii*) Rolling friction is much less than sliding friction
(*iv*) Force of friction depends upon surface roughness.

(*a*) (*i*), (*ii*), (*iii*) (*b*) (*i*), (*iii*), (*iv*)
(*c*) (*ii*) and (*iv*) only (*d*) (*i*) and (*iii*) only

6.6 Two blocks *A* and *B* are connected through a string. Weight of each block is 50 N. Coefficient of friction between each block and floor is 0.3. An inclined force *P* is applied to start sliding of the system. What is the minimum value of *P* required **(Fig. 6.28)**?

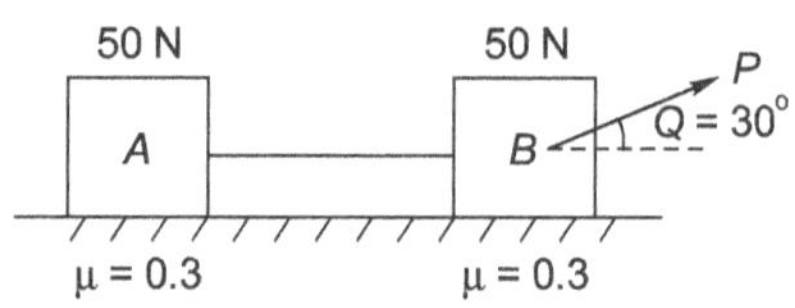

Fig. 6.28

(*a*) 34.64 N (*b*) 31.21 N
(*c*) 29.53 N (*d*) None of these

6.7 **Fig. 6.29** shows a block of width *b* and weight *mg*. Force *P* is applied to block at height *h*. Coefficient of friction between block and floor is 0.3. At what height h force can be applied so that block slips without tipping?

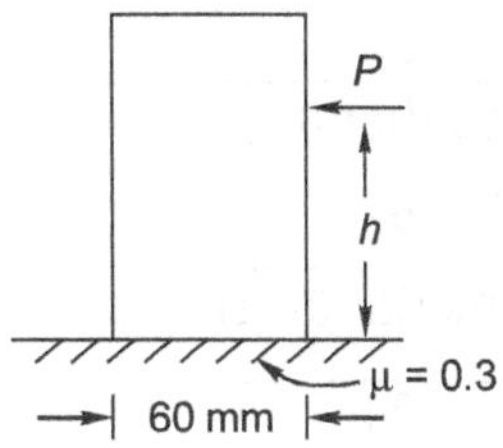

Fig. 6.29

(a) 120 mm (b) 110 mm
(c) 100 mm (d) None of these

6.8 Angle of repose α holds the following relations with the angle of friction ϕ in the condition of limiting equilibrium
(a) $\alpha = \phi$ (b) $\alpha < \phi$
(c) $\alpha > \phi$ (d) $\alpha = 0.8\phi$

6.9 The coefficient of friction between two surfaces is the constant of proportionality between the tangential force and normal reaction at the instant of
(a) application of force
(b) impending motion
(c) body is in motion
(d) body is at rest

6.10 If a ladder is not in equilibrium on a rough floor and against a smooth vertical wall, then it can be made in equilibrium by
(a) increasing the angle of inclination of ladder
(b) slightly decreasing the angle of inclination of ladder
(c) increasing the weidth of the ladder
(d) None of these

6.11 A body of weight W is placed on an inclined plane. Angle of inclination of the plane is gradually increased till the body starts sliding down, then this angle is called
(a) angle of friction
(b) angle of repose
(c) angle of limiting inclination
(d) None of these

6.12 A flat stone rests on an inclined skid way which makes an angle α with the horizontal. If angle of friction is ϕ what is the condition of equilibrium?
(a) $\alpha \leq \phi$ (b) $\alpha = \phi$
(c) $\alpha > \phi$ (d) $\alpha = 0.5\phi$

6.13 What is the necessary coefficient of friction between tyres and roadway to enable a four-wheel drive automobile on a climb a 30% grad?
(a) $\mu > 0.25$ (b) $\mu > 0.30$
(c) $\mu > 0.35$ (d) None of these

6.14 A body of mass m is pushed up the inclined plane by a force P parallel to the plane. If ϕ is angle of friction between body and plane and θ is the angle of inclination of the plane, then what is minimum value of $P (\mu = \tan\phi)$?
(a) $mg \sin\theta$
(b) $mg \sin\theta + \mu\, mg \cos\theta$
(c) $mg \sin\theta - \mu\, mg \cos\theta$
(d) $\mu\, mg \cos\theta$

[CSE, Prelim, CE, 2010]

6.15 A body is lying on a plane with its angle of inclination θ more than the angle of friction. By the application of a force P parallel to the plane it is prevented from slipping down. If $\mu = \tan\phi$, m is mass of the body, then what is minimum value of P?
(a) $mg \sin\theta$
(b) $mg \sin\theta - \mu\, mg \cos\theta$
(c) $mg \sin\theta + \mu\, mg \cos\theta$
(d) $\mu\, mg \cos\theta$

6.16 If a body of weight 100 N is hauled along a rough horizontal plane by a pull of 50 N acting at an angle of 30° with the horizontal, the coefficient of friction is
(a) $\frac{1}{\sqrt{3}}$ (b) $\frac{2}{\sqrt{3}}$
(c) $\sqrt{3}$ (d) $\frac{4}{\sqrt{3}}$

[CSE, Prelim, CE : 2001]

6.17 A vehicle is moving up an inclined plane. The vehicle has rear wheels drive. Force of friction on the rear wheels will be in a direction
(a) along the plane but backward
(b) backward at angle ϕ with the plane, where ϕ is the friction angle
(c) along the plane but forward
(d) forward at an angle ϕ with the plane, where ϕ is the friction angle

[CSE, Prelim, CE : 2003]

6.18 A 12 kg mass rests an a surface for wheels. The coefficient of friction is $\mu = 0.15$. What is the smallest force that can give the mass an acceleration of 3 m/s^2. (Take $g = 10$ m/s^2)
(a) 120 N (b) 100 N
(c) 65 N (d) 54 N

[CSE, Prelim, CE : 2007]

6.19 A block of mass m is in equilibrium on a rough inclined plane inclined at angle α to the horizontal. If μ is the coefficient of friction between the plane and the block, what is the frictional force between the block and the plane?

(a) μ mg sinα (b) μ mg cosα
(c) mg sinα (d) mg cosα

[CSE, Prelim, CE : 2008]

6.20 Two books of mass 1 kg each are kept on a table, one over the other. The coefficient of friction on every pair of contacting surfaces is 0.3. The lower book is pulled with a horizontal force F. The minimum value of F for which slip occurs between the two books is

(a) Zero (b) 1.06 N
(c) 5.74 N (d) 8.83 N

[GATE 2005 : 2 Marks]

6.21 The figure shows a pair of pin-jointed gripper-tongs holding an object weighting 2000 N. The coefficient of friction (m) at the gripping surface is 0.1 XX is the line of action of the input force and YY is the line of application of gripping force. If the pin-joint is assumed to be frictionless, the magnitude of force F required to hold the weight is

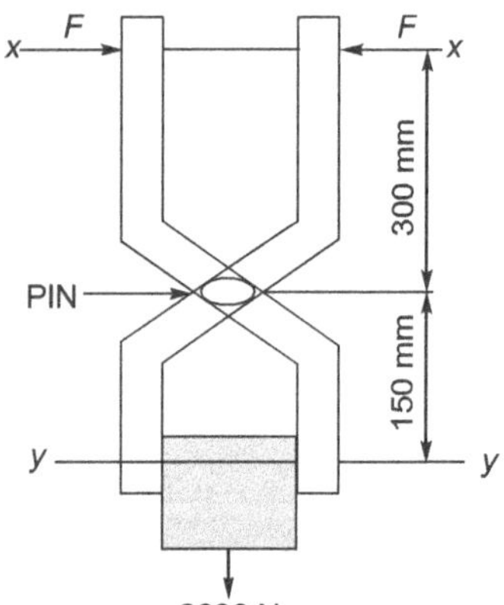

Fig. 6.30

(a) 1000 N (b) 2000 N
(c) 2500 N (d) 5000 N

[GATE 2004 : 2 Marks]

6.22 A block of mass *M* is released from point *P* on a rough inclined plane with inclination angle q, shown in the figure below. The coefficient of friction is *m*. If *m* < tanθ, then the time taken by the block to reach another point *Q* on the inclined plane, where *PQ* = s, is

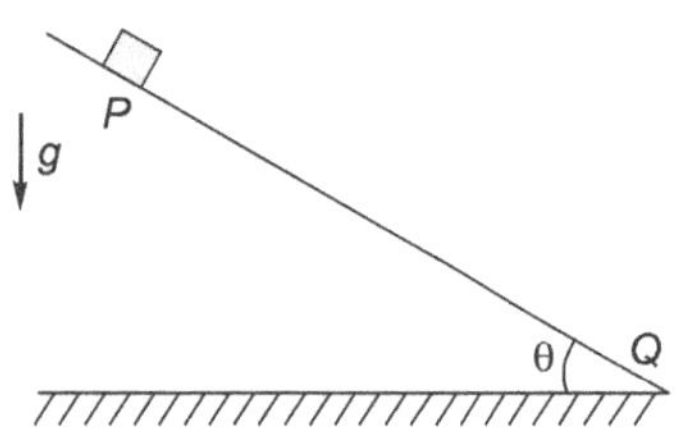

Fig. 6.31

(a) $\sqrt{\dfrac{2s}{g\cos\theta\,(\tan\theta-\mu)}}$

(b) $\sqrt{\dfrac{2s}{g\cos\theta(\tan\theta+\mu)}}$

(c) $\sqrt{\dfrac{2s}{g\sin\theta\,(\tan\theta-\mu)}}$

(d) $\sqrt{\dfrac{2s}{g\sin\theta\,(\tan\theta+\mu)}}$

[GATE 2007 : 2 Marks]

6.23 A block weighing 981 N is resting on a horizontal surface. The coefficient of friction between the block and the horizontal surface is *m* = 0.2. A vertical cable attached to the block provides partial support as shown. A man can pull horizontally with a force of 100 N. What will be the tension, *T* (in N) in the cable if the man is just able to move the block to the right?

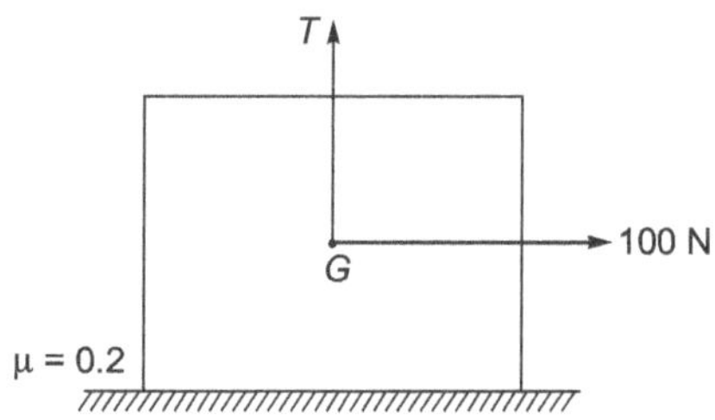

Fig. 6.32

(a) 176.2 (b) 196.0
(c) 481.0 (d) 981.0

[GATE 2009 : 1 Mark]

6.24 A 1 kg block is resting on a surface with coefficient of friction μ = 0.1. A force of 0.8 N is applied to the block as shown in the figure. The friction force is

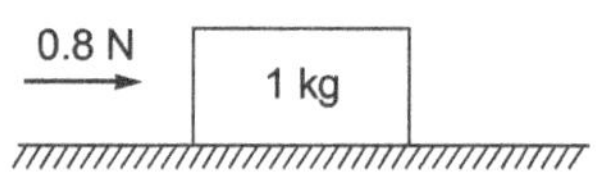

Fig. 6.33

(a) 0 (b) 0.8 N
(c) 0.98 N (d) 1.2 N

[GATE 2011 : 2 Marks]

6.25 A block weighing 200 N is in contact with a level plane whose coefficients of static and kinetic friction are 0.4 and 0.2, respectively. The block is acted upon by a horizontal force (in newton) *P* =10 t, where t denotes the time in seconds. The velocity (in m/s) of the block attained after 10 seconds is ______.

[GATE 2014 : 2 Marks, Set-1]

6.26 A block *R* of mass 100 kg is placed on a block *S* of mass 150 kg as shown in the figure. Block *R* is tied to the wall by a massless and inextensible string *PQ*. If the coefficient of static friction for all surfaces is 0.4, the minimum force F (in kN) needed to move the block *S* is

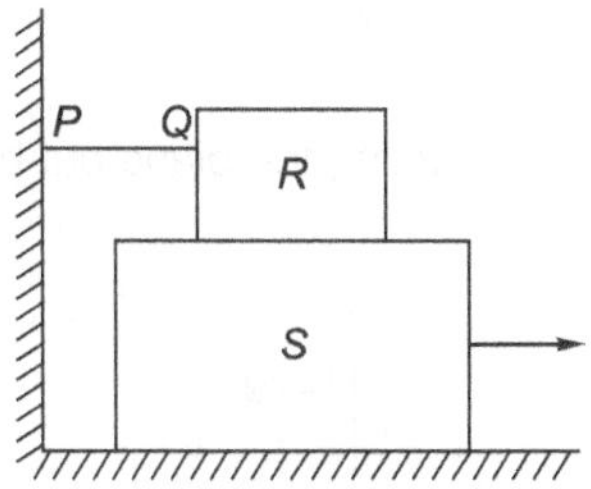

Fig. 6.34

(a) 0.69 (B) 0.88
(c) 0.98 (d) 1.37

[GATE 2014 : 2 Marks, Set-1]

6.27 A truck accelerates up a 10° incline with a crate of 100 kg. Value of static coefficient of friction between the crate and the truck surface is 0.3. The maximum value of acceleration (in m/s^2) of the truck such that the crate does not slide down is ______.

[GATE 2014 : 2 Marks, Set-2]

6.28 A body of mass (M) 10 kg is initially stationary on a 45° inclined plane as shown in figure. The coefficient of dynamic friction between the body and the plane is 0.5. The body slides down the plane and attains a velocity of 20 m/s. The distance travelled (in meter) by the body along the plane is ______.

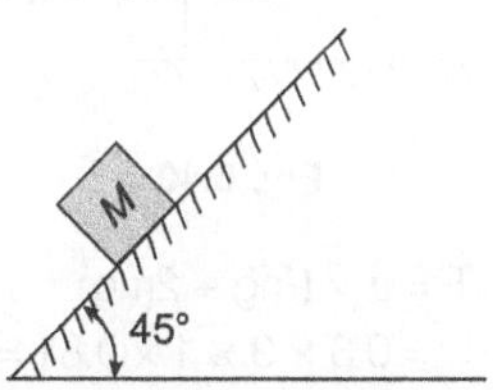

Fig. 6.35

[GATE 2014 : 2 Marks, Set-3]

6.29 A wardrobe (mass 100 kg, height 4 m, width 2 m, depth 1 m), symmetric about the Y-Y axis, stands on a rough level floor as shown in the figure. A force P is applied at mid-height on the wardrobe so as to tip it about point Q without slipping. What are the minimum values of the force (in newton) and the static coefficient of friction m between the floor and the wardrobe, respectively?

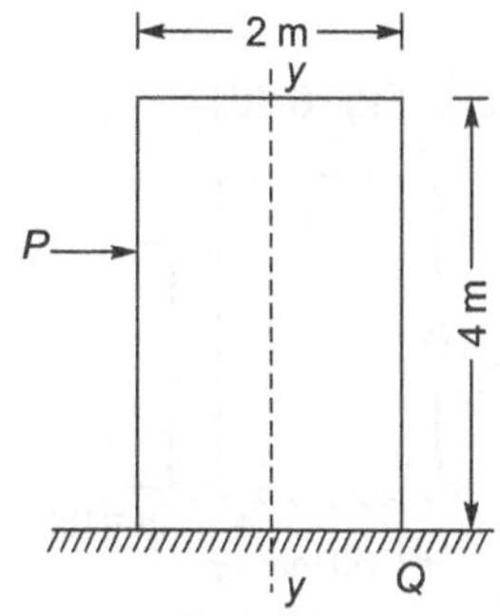

Fig. 6.36

(a) 490.5 and 0.5 (b) 981 and 0.5
(c) 1000.5 and 0.15 (d) 1000.5 and 0.25

[GATE 2014 : 2 Marks, Set-4]

Answers

6.1 (c)	6.2 (c)	6.3 (a)	6.4 (c)	6.5 (b)
6.6 (c)	6.7 (c)	6.8 (a)	6.9 (b)	6.10 (a)
6.11 (b)	6.12 (a)	6.13 (b)	6.14 (b)	6.15 (b)
6.16 (a)	6.17 (a)	6.18 (d)	6.19 (b)	6.20 (d)
6.21 (d)	6.22 (a)	6.23 (c)	6.24 (b)	
6.25 (4.9 m/s)		6.26 (d)	6.27 (4.6 m/s^2)	
6.28 (57.63 m)		6.29 (a)		

EXPLANATIONS

6.1 (c)

$$N = 200 - P\sin 30^\circ$$
$$= 200 - 100 \times 0.5 = 150 \text{ N}$$
$$F = \mu N = 0.3 \times 150 = 45 \text{ N.}$$

6.2 (c)

$$N = 200 + P\sin 30^\circ = 200 + 0.5P$$
$$= 200 - 100 \times 0.5 = 150 \text{ N}$$
$$F = \mu N = 0.25\,(200 + 0.5P)$$
$$= 50 + 0.125P$$

$$P\cos 30^\circ = 0.866P$$
$$0.866P > 50 + 0.125P$$
$$P \geq \frac{50}{0.741} \text{ or } P > 67.5 \text{ N.}$$

6.3 (a)

Metal on metal	0.15–0.6
Rubber on concrete	0.6–0.95
Metal on stone	0.3–0.7
Stone on stone	0.4–0.7

6.4 (c)

$N = W - P\sin 30^\circ = 500 - 200 \times 0.5 = 400$ N

$\mu N = 0.2 \times 400 = 80$ N = Radius of friction circle.

6.6 (c)

$$15 + 15 - 0.15P = 0.866P$$
$$0.866 + 0.15P = 30$$
$$P = \frac{30}{1.016} = 29.53 \text{ N.}$$

6.7 (c)

For no tipping **(Fig. 6.35)**

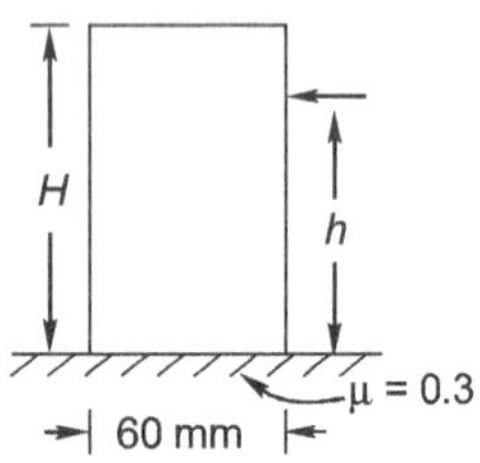

Fig. 6.37

$$h < \frac{b}{2\mu}$$
$$< \frac{60}{0.6}$$
$$h < 100 \text{ mm.}$$

6.8 (a)

$\alpha = \phi$.

6.9 (b)

Impending motion.

6.10 (a)

Increasing angle of inclination of ladder.

6.11 (b)

Angle of repose.

6.12 (a)

$\alpha \le \phi$.

6.13 (b)

$\mu > 0.3$.

6.14 (b)

$mg\sin\theta + \mu\, mg\cos\theta$.

6.15 (b

$mg\sin\theta - \mu\, mg\cos\theta$.

6.16 (a)

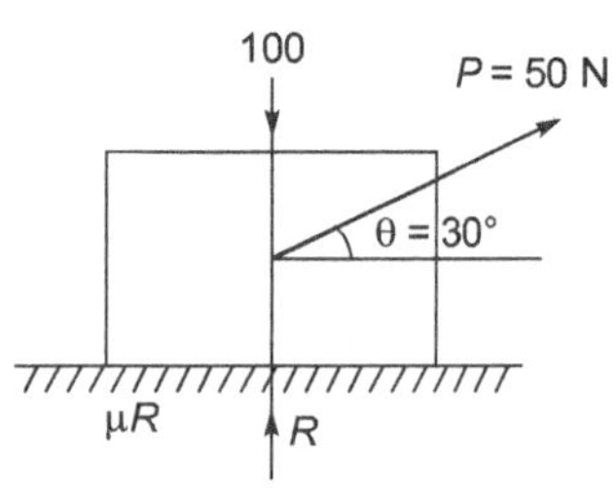

Fig. 6.38

$$R = 100 - 50\sin 30^\circ = 75 \text{ N}$$
$$P\cos 30^\circ = 50 \times 0.866 = 43.3$$
$$\mu = \frac{43.3}{75} = 0.577 = \frac{1}{\sqrt{3}}$$

6.17 (a)

Force of friction acting the plane but backward.

6.18 (d)

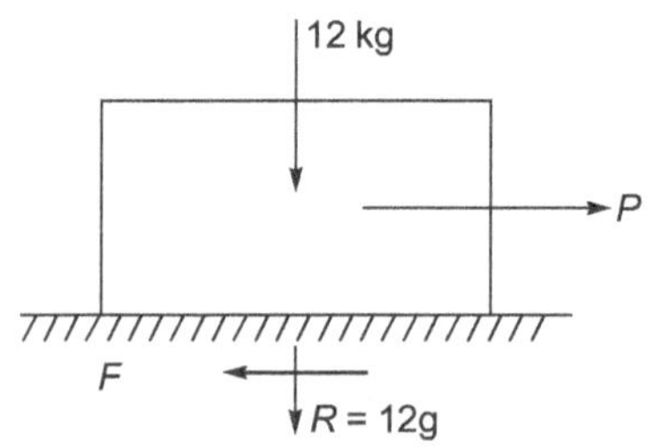

Fig. 6.39

$$P - 129 \times \mu = m \times 3$$
$$P - 12 \times 10 \times 0.15 = 12 \times 3$$
$$P = 54 \text{ N}$$

6.19 (b)

$F = \mu\, mg\cos\alpha = \mu$ (normal reactions)

6.20 (d)

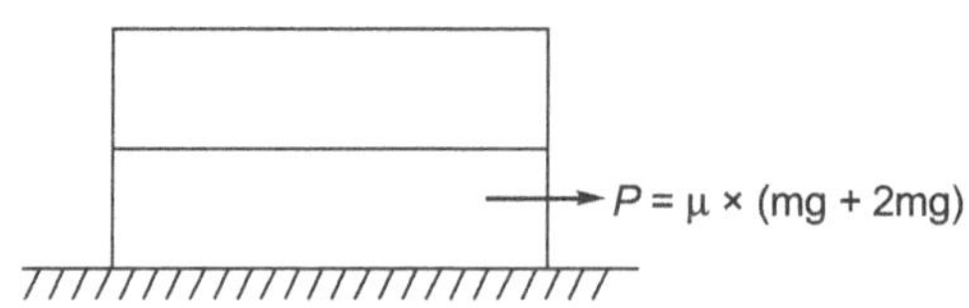

Fig. 6.40

$$P = \mu \times (mg + 2mg)$$
$$= 0.3 \times 3 \times 1 \times 9.81 = 8.83 \text{ N}$$

6.21 (d)

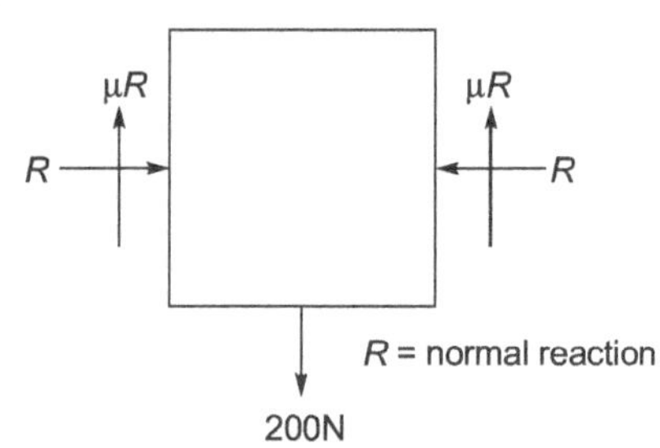

Fig. 6.41

$$2000 \times 1 = 2\mu R = 0.2\,R$$
$$R = 10{,}000 \text{ Newton}$$

Moment about pin,

$$R \times 150.0 = F \times 300$$
$$F = \frac{R}{2} = 5000 \text{ N}$$

6.22 (a)

$$a = g\,(\sin\theta - \mu\cos\theta)$$
$$= g\cos\theta(\tan\theta - \mu)$$
$$s = \frac{1}{2}g\cos\theta(\tan\theta - \mu)t^2$$
$$t = \sqrt{\frac{2s}{g\cos\theta(\tan\theta - \mu)}}$$

6.23 (c)

$$(981 - T) \times 0.2 = 100$$
$$T = 981 - 500 = 481\text{ N}$$

6.24 (b)

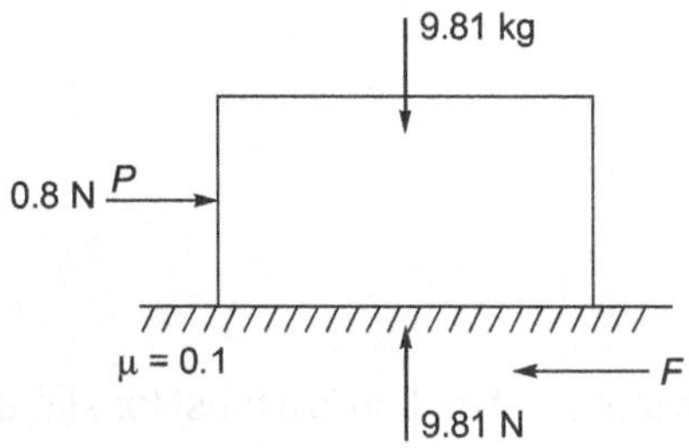

Fig. 6.42

$$F = \mu R = 0.1 \times 9.81 = 0.98\text{ N limiting}$$
$$\text{Factory} = 0.8\text{ N} < 0.98\text{ (friction force)}$$

6.25 (4.9 m/s)

$$P = 10t$$
$$\mu_R = 0.2$$
$$\text{Friction force} = 0.2 \times 200 = 400$$
$$(10t - 40) = \frac{200}{g}\times a = \frac{200}{9.81}\times\frac{dV}{dt}$$
$$\int_0^{10}(10t - 40)dt = 20.388\int dV$$
$$\left(5t^2 - 40t\right)\Big|_0^{10} = 20.388 \times V_{10}$$
$$\text{Velocity after 10 sec.} = \frac{1}{20.388}(5\times100 - 40\times10)$$
$$= 4.9\text{ m/s}$$

6.26 (d)

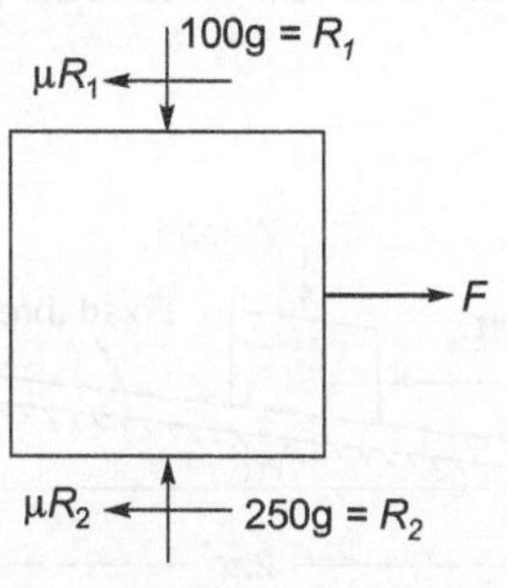

Fig. 6.43

$$F = 0.4[100\,g + 2\,N\,g] = 140 \times 9.81$$
$$= 1373.4\text{ N} = 1.37\text{ kN}$$

6.27 (4.6 m/s^2)

$$a = \sin\theta + \mu g\cos\theta$$
$$= g(\sin 10° + 0.3 \times \cos 10°)$$
$$= g(0.1736 + 0.2954)$$
$$= 9.81 \times 0.469 = 4.6\text{ m/s}^2$$

6.28 (56.67 m)

Acceleration,

$$a = g\sin\theta - \mu g\cos\theta$$
$$= 9[0.707 - 0.5 \times 0.707]$$
$$= 9.81 \times 0.707 \times 0.5 = 3.4678\text{ m/s}^2$$
$$V^2 = 2a.S$$
$$20^2 = 2 \times 3.4678 \times S$$
$$S = 57.67\text{ m}$$

6.29 (a)

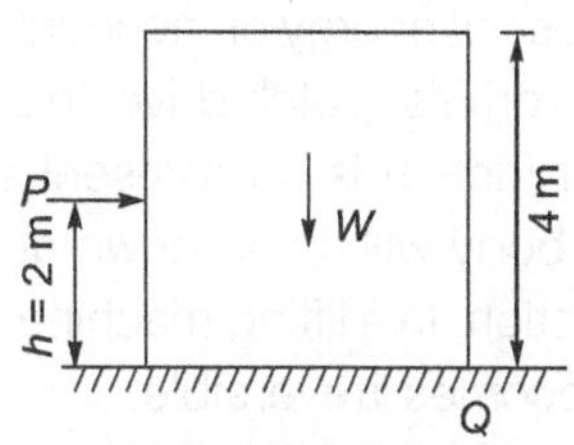

Fig. 6.44

To tip about the edge Q

$$P \times 2M > W \times \frac{b}{2}$$
$$P > \frac{W \times 1}{2}$$
$$> \frac{981}{2} = 490.5\text{ N}$$

But $P > \frac{W}{2}$

or $\mu W > \frac{W}{2}$

$$\mu > 0.5$$

■■■■

07

CHAPTER

Application of Friction in Machines

7.1 Introduction

While there is loss of energy on account of friction in a mechanical system, yet friction is a must for **slip drives** as in belt and rope drives, clutch drive. In a **lifting machine**, though there is loss of work due to friction in a machine but if sufficient friction is not present and **if applied effort is removed after lifting the body to a higher position**, then body will come down at its own, because the force necessary to maintain the lifted position is provided by friction. In a lifting machine, it is necessary that its efficiency should be less than 50 percent, so that the machine becomes irreversible.

Sometimes in slip drive, it is necessary to increase the effects of friction by passing the belt through a grooved pulley, as in the case of *V*-belts.

However, in bearings, friction between journal and bearing should be minimised by providing lubrication, is in power generation, friction between rotor and bearings is minimised by lubrication.

Rolling friction is made less than sliding friction. Roller bearings are termed as antifriction bearings, and energy loss due to friction is much less.

7.2 Screw Threads

Screw threads are frequently used in screw jacks, fly presses, lead screw of a lathe. Moreover, square threads are more efficient than *V*-threads for power transmission. Friction along the threads determines the amount of effort required to lift a load.

In a screw jack, the *threads of the screw* slide around in the fixed threads of a nut, fitted in a frame.

A screw thread acts as an inclined plane and angle of inclination of plane is equal to the helix angle of thread as shown in **Fig. 7.1**. **Fig. 7.1** shows a fixed thread of a nut, in which thread of the screw slides up or down depending upon whether the load is to be lifted or lowered. Effort *P* required to raise the load *W* shown in the figure, is the effort applied at the mean radius of screw.

Helix angle of the thread,

$$\alpha = \tan^{-1}\frac{np}{2\pi r},$$

where n is the number of start of threads, p pitch of the threads, np = lead of the screw and r is the mean radius of the screw. The threads can be single start ($n = 1$) or multi-start ($n > 1$).

If φ = Angle of friction between threads of screw and nut

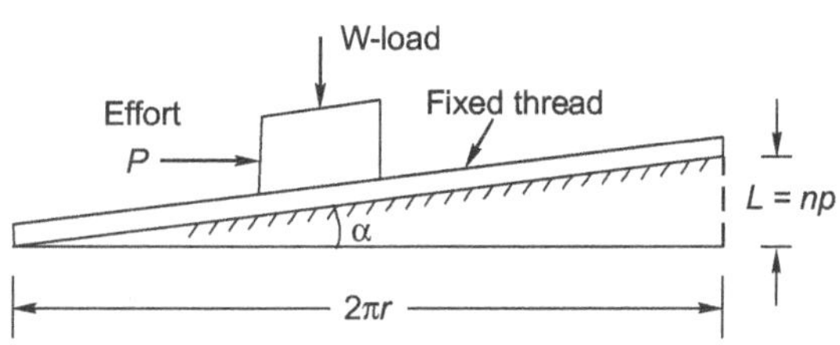

Fig. 7.1

$\varphi = \tan^{-1}\mu$, where μ is coefficient of the friction between threads of screw and the nut

Say helix angle $\alpha > \varphi$, friction angle. Once the load is lifted and effort removed from handle, then screw thread will roll down and the load is lowered. In this condition, *screw jack becomes reversible, which is undesirable.* Therefore for a screw jack $\varphi > \alpha$. **Fig. 7.2** shows a sketch of a screw jack. A load *W* placed on screw cap is lifted by an effort P' applied at the end of a handle of length *l*. Mean radius of screw is *r*. Effort P' applied at handle is in horizontal direction. **Fig. 7.3** shows unwrapped thread of the nut, with angle of inclination α.

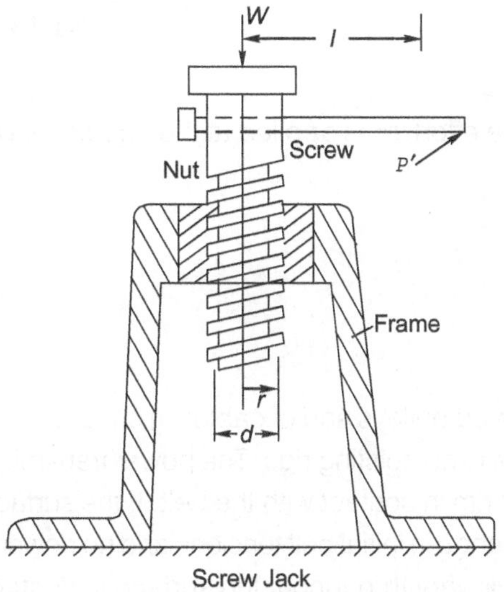

Fig. 7.2

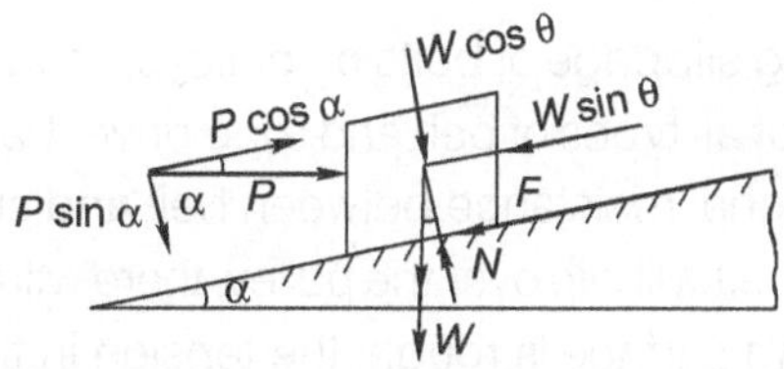

Fig. 7.3

A weight *W* acting on the inclined plane, is to be lifted by an effort *P* applied on the block of weight *W* (or in other words applied at radius *r* of the screw).

Force of friction on block will act downwards, as the load is lifted upwards. Resolving the forces *P*, *W* and *F* in the directions parallel and perpendicular to the inclined plane.

Normal reaction of plane, $N = W\cos\alpha + P\sin\alpha$

Force of friction, $F = \mu N = \mu\,[W\cos\alpha + P\sin\alpha]$

Moreover $P\cos\alpha = W\sin\alpha + F$

$$= W\sin\alpha + \mu W\cos\alpha + \mu P\sin\alpha$$

or Effort, $$P = W\left[\frac{\mu + \tan\alpha}{1 - \mu\tan\alpha}\right]$$

Putting $\mu = \tan\varphi$, where φ is friction angle

Effort, $$P = W\left[\frac{\tan\varphi + \tan\alpha}{1 - \tan\varphi\tan\alpha}\right] = W\tan[\alpha + \varphi]$$

Helix angle, $$\alpha = \frac{\pi}{4} - \frac{\varphi}{2}, \text{ for maximum efficiency} \quad \ldots(1)$$

So maximum efficiency of the screw jack will be achieved when helix angle α is given by Equation (1) as above.

But, if α becomes greater than φ, screw jack will become reversible and it will not be self locking.

Generally, $\varphi = 6° - 8°$, and $\alpha < \varphi$ for self locking.

Self locking screw jack: Helix angle, $\alpha < \varphi$, friction angle

Effort required to lift the load $P = W\tan(\alpha + \varphi)$.

To lower down the load, some effort has to be applied because screw jack has become self locking and effort P is applied in downward direction and force of friction F will act upwards as shown in the Fig. 7.4.

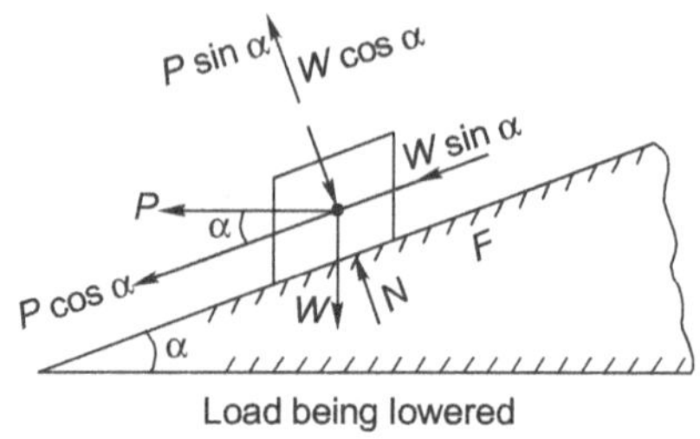

Fig. 7.4

Effort to lower down the load,

$$P = W\left[\frac{\mu - \tan\alpha}{1+\mu\tan\alpha}\right] = W\left[\frac{\tan\varphi - \tan\alpha}{1+\tan\varphi\tan\alpha}\right]$$

as $\mu = \tan\varphi$, where φ is friction angle

$$P = W\tan(\varphi - \alpha)$$

This effort is applied at mean radius, r of screw, while effort P' is applied at the end of handle of length, l,

$$P.r = P' \times l$$

or effort at handle

$$P' = \frac{Pr}{l} = \frac{Wr\tan(\varphi-\alpha)}{l}.$$

7.3 Flat Belt Drive

The impending slippage of belts on pulleys, ropes in grooved pulleys and of cables in sheaves is of importance in the design of all types of belt and rope drive, band brakes and hoisting rigs. The power transmission depends upon the frictional resistance between belt and surface of rim in contact with the belt. If the surface of the rim is smooth, then belt will slip over the pulley, there will be no change in initial belt tensions and no power transmission. If the pulley rim surface is rough, the tension in the belt will vary throughout the surface in contact. Due to this incremental sum of frictional resistance, difference in belt tensions is developed which is responsible for power transmission.

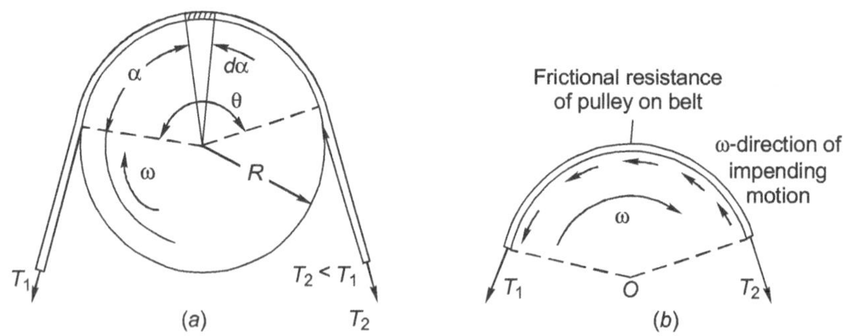

Fig. 7.5

Consider a pulley of radius R, on which a belt is in contact over an angle θ, as shown in **Fig. 7.5 (*a*)** T_1 is the tension on tight side and T_2 is the tension on slack side belt. Tension T_1 is gradually reduced to T_2 to overcome the resisting frictional moment of the pulley. **Fig. 7.5 (*b*)** shows the frictional forces cumulatively make up the difference between T_2 and T_1.

Ratio of belt tension

$$\frac{T_1}{T_2} = e^{\mu\theta}$$

This expression can also be written as

$$2.3\log_{10}\frac{T_1}{T_2} = \mu\theta$$

Initial Tension in the belt,

$$T_0 = \frac{T_1+T_2}{2}$$

When the motion starts, T_0 is increased by $T_0 + \Delta T = T_1$ on tight side and T_0 is decreased to $T_0 - \Delta T = T_2$ on the slack side of the belt.

Moreover formula can be further modified as

$$\frac{T_1 - T_c}{T_2 - T_c} = e^{\mu\theta},$$

where T_c is the centrifugal tension in belt due to centrifugal force acting on the belt throughout the operation, where $T_c = mV^2$, m is mass of the belt per unit length

Velocity, $\quad v = \sqrt{\frac{T_{max}}{3\,m}}$, power transmission is maximum

where centrifugal tension in the belt is one third of T_{max}, but

$$T_{max} = T_1 = 3T_c$$

or $\quad T_c = \frac{T_1}{3}$

Maximum permissible tension in the belt depends upon the strength of the belt. For leather belt, allowable stress is 2 – 4 N/mm^2.

7.4 V-Belt and Rope Drive

V-belts and ropes run in *V*-grooved pulleys, *i.e.*, groove provided in the rim of the pulley. In these cases ratio of belt tensions is greatly enhanced and more power can be transmitted. In *V*-groove, the actual coefficient of friction is more than μ, it is affected by the groove angle.

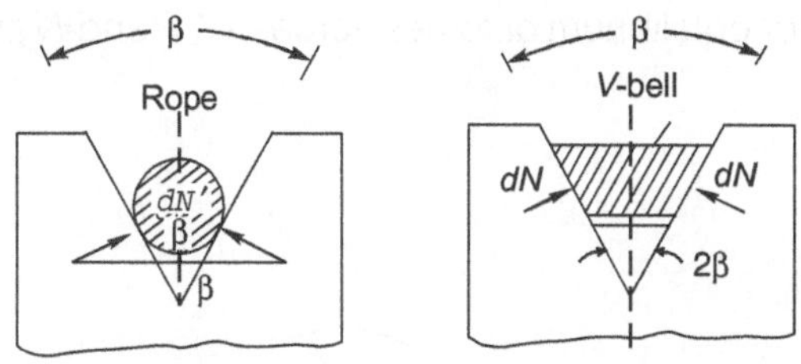

(*a*) Rope in *V*-groove (*b*) *V*-belt in groove

Fig. 7.6

Ratio of belt tensions, $\quad \frac{T_1}{T_2} = e^{\mu \text{ cosec } \beta . \theta}$

where θ = angle of lap or angle of contact and $e^{\mu \text{ cosec } \beta.\theta} > e^{\mu\theta}$, (where β is semi groove angle)

Example 7.1 A rope drive is used for transmitting power through a pulley of 0.9 m diameter. The angle of lap is 160° and groove angle of pulley section is 45°. Maximum allowable tension in rope is 1200 N and coefficient of friction between rope and groove is 0.3. Determine the torque provided by the rope on the pulley.

Solution Angle of lap, $\quad \theta = 160° = \frac{160 \times \pi}{180} = 2.792$ radians

Groove angle, $\quad 2\beta = 45°$

$\beta = 22.5°$

$\text{cosec } \beta = 2.613$

Coefficient of friction, $\quad \mu = 0.3$

$$\mu\,(\text{cosec } \beta)\,\theta = 0.3 \times 2.613 \times 2.792 = 2.1886$$

$$e^{\mu\,(\text{cosec } \beta)\,\theta} = e^{2.1886} = 8.92$$

Maximum allowable tension, $\quad T_1 = 1200$ N

Tension on slack side, $\quad T_2 = \frac{T_1}{8.92} = \frac{1200}{8.92} = 134.53$ N

Pulley Radius, $\quad R = 0.45$ m

Torque provided by the rope $\quad = R(T_1 - T_2)$

$= 0.45\,(1200 - 134.53) = 479.46$ Nm.

Exercise 7.1 A *V*-belt drive is used to transmit power from a smaller pulley to a bigger pulley in an open belt system, the angle of lap for bigger pulley is 200°. The coefficient of friction between *V*-belt and pulley groove is 0.3. If the ratio of belt tensions on tight side and slack side is 3.2, what is the groove angle?

[**Hint:** $\theta_1 = 360 - 200 = 160°$].

[**Ans:** Groove angle 46°].

7.5 Rolling Resistance

At the point of contact of wheel on the ground, the point on **wheel is instantaneously at rest and rolling takes place without slipping and lot of frictional resistance is eliminated** (But in sliding, due to relative motion between the two surfaces, lot of frictional resistance occurs). Theoretically moment of forces at the point of contact is zero, *i.e.*, at *A*. $M_A = 0$, as shown in **Fig. 7.7 (*a*)**, and wheel may move indefinitely towards right once the motion has started, however the wheel stops after some distance due to frictional moment and for continued motion, an effort *P* opposite and equal to the rolling friction force has to be applied at the wheel centre. Practically either the wheel deforms as shown in **Fig. 7.7(*b*)**, deformation of tyre of a bus on a road or the ground deforms as shown. Consider the **Fig. 7.7 (*c*)**, in which ground is shown as deformed, line of action of load *W* of wheel is vertically passing through centre of gravity *G* of the wheel, ground deforms and reaction from ground *i.e.*, *N* is ahead of the line of action of weight *W* by a distance '*a*' as shown. As *P* and *W* are non-parallel forces, therefore for equilibrium of forces, forces *W*, *P* and *N* are concurrent and pass through centre of gravity *G* of the wheel.

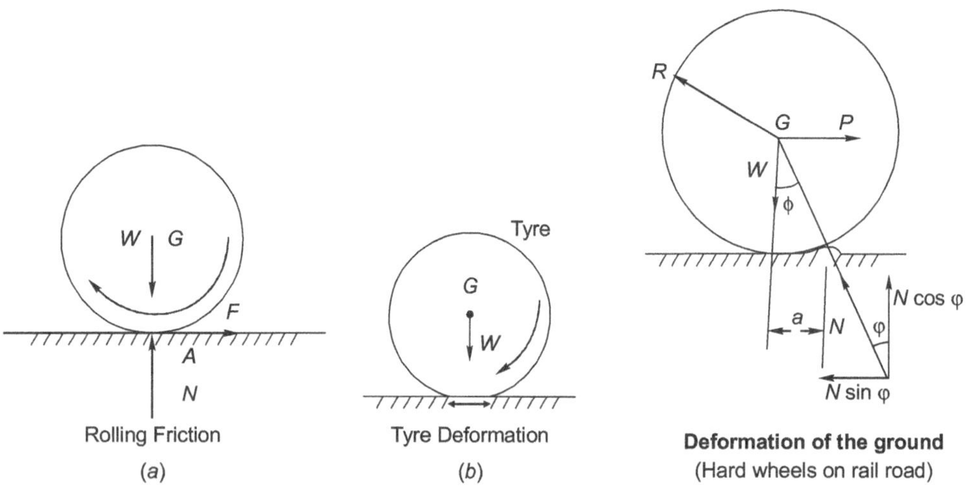

Fig. 7.7

Reaction from ground, *N* makes an angle φ with the vertical. In the figure angle φ is highly exaggerated, otherwise in rolling, frictional resistance is small and *angle φ is also very small.* For equilibrium

$$P = N\sin\varphi$$

$$W = N\cos\varphi = \text{Normal reaction}$$

$$\frac{P}{W} = \tan\varphi$$

$$P = W\tan\varphi \simeq W\sin\varphi \text{ (an angle } \phi \text{ is very small)}$$

$$= W\cdot\frac{a}{R}, \text{ since the angle } \varphi \text{ is very very small, } \tan\varphi \simeq \sin\varphi$$

R is the radius of the wheel.

$$P = \frac{W}{R}(a) \text{ (Rolling friction force)}$$

Coefficient '*a*' is termed as coefficient of rolling resistance. Ratio of *P*/*W* varies inversely with the radius of the wheel. For a given material of the wheel, on a given surface (on ground) *distance 'a' is constant, known as coefficient of rolling resistance.*

For some wheel and ground materials, coefficient '*a*' is given in **Table 7.1.**

Table 7.1 Coefficients of Rolling resistance

Materials	a *in* (*mm*)
Steel on steel	0.175–0.225
Steel on wood	0.15–0.25
Pneumatic tyres on smooth road	0.5–0.75
Pneumatic tyres on mud road	1.0–1.5
Hardened steel on hardened steel (as in the case of ball bearings, balls on races)	0.0050–0.0125

Example 7.2 A boat exerts a pull of 20 kN on its hawser which is wrapped around a capstan on the pier. If the coefficient of friction is 0.3, how many turns must the hawser make around the capstan so that the pull at the other end does not exceed 250 N.

Solution Pull exerted by boat, $= 20 \text{ kN} = 20{,}000 \text{ N} = T_1$

Pull at the other end of hawser $= 250 \text{ N} = T_2$

Ratio of hawser tensions $= \dfrac{T_1}{T_2} = \dfrac{20{,}000}{250} = 80 = e^{\mu\theta}$

$$\mu\theta = 4.382$$

Coefficient of friction, $\mu = 0.3$

Therefore lap angle, $\theta = \dfrac{4.382}{0.3} = 14.60 \text{ radians} = 2\pi n$

where n = number of turns of hawser around capstan.

So number of turns, $n = \dfrac{14.60}{2\pi} = 2.32$ turns.

Remember

- Screw jack, bench vice, belt and rope drive, journal bearing, thrust bearing, disc brakes etc. are applications of friction.
- For a screw jack, helix angle,

$$\alpha = \tan^{-1}\frac{\text{lead}}{\pi d}$$

where $d = 2r = 2 \times$ mean radius of screw

φ = friction angle between threads of screw and nut.

Effort required to lift the load, $P = W\tan(\alpha + \varphi)$.

- If $\varphi > \alpha$, then effort required to lower the load

$$P' = W\tan(\varphi - \alpha).$$

- For maximum efficiency of screw jack, $\alpha = \dfrac{\pi}{4} - \dfrac{\varphi}{2}$.
- If $\alpha < \varphi$, screw jack is irreversible.
- Belt and rope drive $\dfrac{T_1}{T_2} = e^{\mu\theta}$,

where T_1 and T_2 are tensions on tight side and slack side respectively on belt, μ is the coefficient of static friction, θ is the angle of lap in radians.

- V-belt and rope drive, $\frac{T_1}{T_2} = e^{\mu\theta \text{ cosec } \beta}$

  where β is the semi groove angle of pulley.
- Coefficient of rolling resistance, $a = R\sin\varphi$

  φ is the angle of rolling friction, R = wheel radius.

PRACTICE PROBLEMS

7.1 A dock worker adjusts a spring line rope which keeps a ship from drifting alongside a wharf. If he exerts a pull of 300 N on the rope which has 1.25 turns around the mooring bit, how much force T can he support? The coefficient of friction between the rope and cast steel mooring bit is 0.30. Refer to Fig. 7.8].

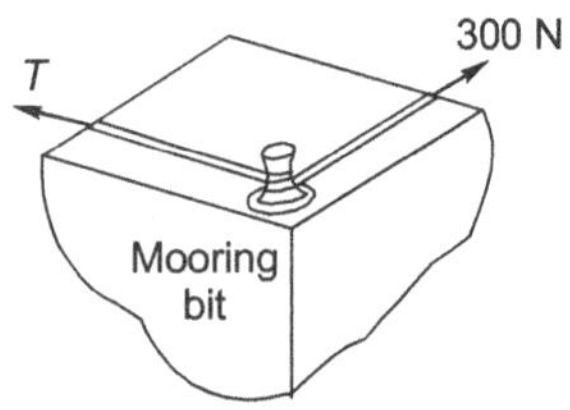

Fig. 7.8

[**Ans:** 3164 N].

7.2 A hand brake is used to control the speed of a flywheel as shown in Fig. 7.9. The coefficients of friction are $\mu_s = 0.30$ and $\mu_k = 0.25$. What couple should be applied to the flywheel to keep it rotating counter-clockwise at a constant speed, when $P = 60$ N?

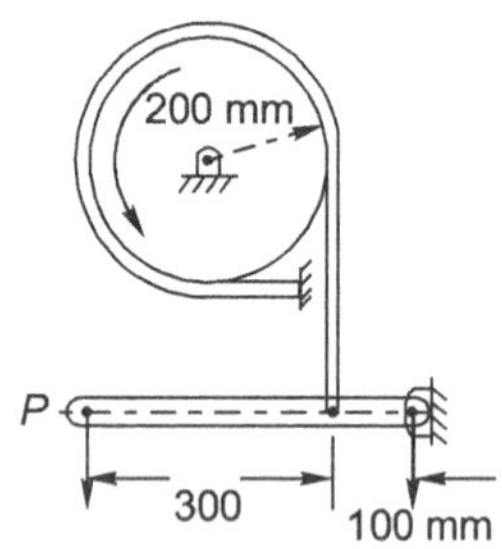

Fig. 7.9

[**Hint:** $\theta = 270°$, $100T_1 = 400P$].

[**Ans:** 33.225 Nm].

7.3 A 100 kg package is attached to a rope which passes over an irregularly shaped boulder with uniform surface texture as shown in Fig. 7.10. If the downward pull P = 140 N is required to lower the package at constant speed, determine: (*a*) coefficient of friction between rope and boulder, (*b*) force P required to raise the package at a constant speed.

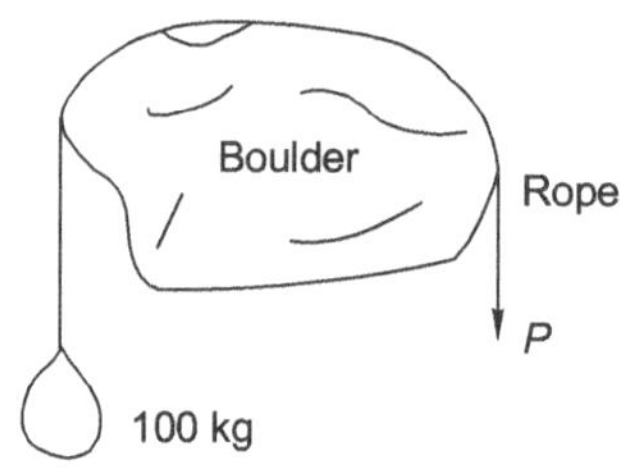

Fig. 7.10

[**Ans:** $\mu = 0.62$, 6.867 kN].

7.7 A brake drum of radius 150 mm is rotating in anticlockwise direction when a force P = 60 N is applied as shown in Fig. 7.11. If the coefficient of kinetic friction is 0.4, determine the moment of friction forces applied to the drum, when a = 250 mm and b = 300 mm.

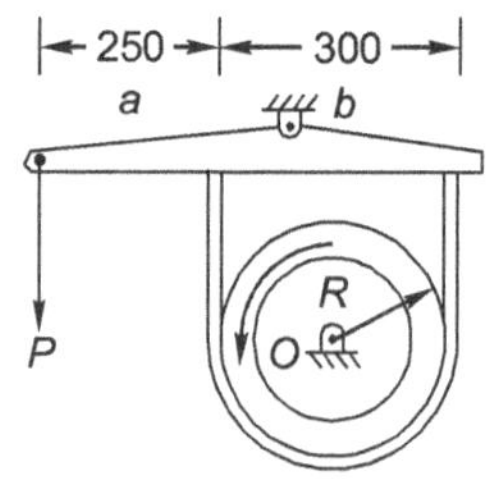

Fig. 7.11

[**Ans:** 91.2 Nm].

MULTIPLE CHOICE QUESTIONS

7.1 In the screw of a screw jack, what type of threads are used?

(*a*) V (*b*) Buttress
(*c*) Square (*d*) Trapezoidal

7.2 In a screw jack, friction angle between screw and nut is 12°. For maximum efficiency of screw jack, helix angle of thread should be

(*a*) 51° (*b*) 45°
(*c*) 39° (*d*) 33°

7.3 In a screw jack, angle of friction φ = 15°, helix angle α = 9°, to lower down a load W, how much effort is required in a horizontal direction

(*a*) $W\tan 24°$ (*b*) $W\tan 15°$
(*c*) $W\tan 10.5°$ (*d*) $W\tan 6°$

7.4 A belt passes over a driving pulley, making angle of lap of 3 radians. What is the ratio of belt tensions on tight side/belt tension on slack side coefficient of friction between belt and pulley is 0.3
(*a*) 3 (*b*) 2.46
(*c*) 2.16 (*d*) None of these

7.5 In a belt drive T_1 = tension on tight side, T_2 = tension on slack side, T_C = centrifugal tension, for maximum power transmission, what should be the value of T_C?
(*a*) $\frac{T_1}{3}$ (*b*) $\frac{T_2}{3}$
(*c*) $\frac{T_1+T_2}{3}$ (*d*) $\frac{T_1-T_2}{3}$

7.6 A boat exerts a pull of 20 kN on its hawser which is wrapped around a capstan on the pier. If the coefficient of friction is 0.25, how many turns must the hawser make around the capstan so that pull on the other end does not exceed 250 N.
(*a*) 2.8 (*b*) 2.32
(*c*) 2.16 (*d*) 1.98

7.7 A flat belt passes over a pulley making an angle of lap of 180°. If ratio of belt tensions is 2.72, what is coefficient of friction between belt and pulley?
(*a*) 0.34 (*b*) 0.32
(*c*) 0.30 (*d*) 0.25

7.8 A steel wheel of 600 mm diameter on a horizontal steel rail. It carries a load of 500 N. The coefficient of rolling resistance is 1.3 mm. The force in Newton, necessary to roll the wheel along the rail is
(a) 0.5 (b) 5
(c) 1.5 (d) 150
[GATE, 2000 : 1 Marks]

7.9 A force of 400 N is applied to the brake drum of 0.5 m diameter in a band brake system as shown in figure, where the wrapping angle is 180°. If the coefficient of friction between the drum and the band is 0.25, the braking torque applied, in Nm is

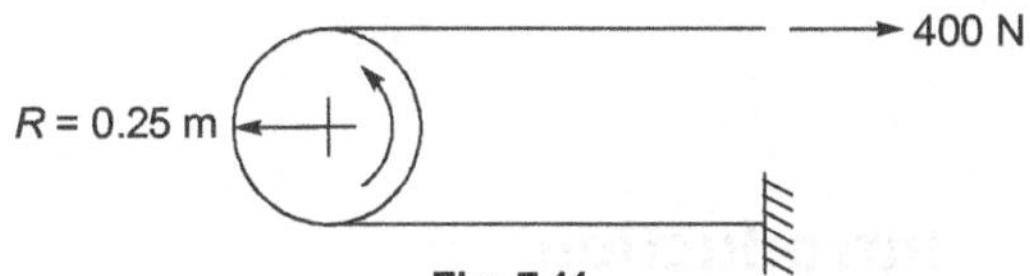

Fig. 7.11

(a) 100.6 (b) 54.4
(c) 22.1 (d) 15.7
[GATE, 2012 : 2 Marks]

Answers

7.1 (c)	**7.2 (c)**	**7.3 (d)**	**7.4 (b)**	**7.5 (a)**
7.6 (a)	**7.7 (b)**	**7.8 (a)**	**7.9 (b)**	

EXPLANATIONS

7.1 (c)
Square threads.

7.2 (c) $\alpha = \frac{\pi}{4} - \frac{\varphi}{2} = 45 - 6 = 39°$.

7.3 (d) $P = W\tan(\varphi - \alpha) = W\tan 6°$.

7.4 (b) $e^{0.9} = 2.46$.

7.5 (a) $T_c = \frac{T_1}{3}$, for maximum power transmission.

7.6 (a)
$$e^{\mu\theta} = \frac{20{,}00}{250} = 80$$
$$\mu\theta = 4.38$$
$$\mu = 0.25$$
$$\theta = 17.528 \text{ rad} = 2\pi n$$
$$n = 2.8$$

7.7 (b)
$$e^{\mu\theta} = 2.72,\ \mu\theta = 1.0,$$
$$\theta = \pi,\ \mu = 0.32.$$

7.8 (a)
$$P = \frac{W}{R} \times a, \text{ where } a = 0.3 = \frac{500 \times 0.3}{300} = 0.5 \text{ N}$$

7.9 (b)
Braking torque, B_T

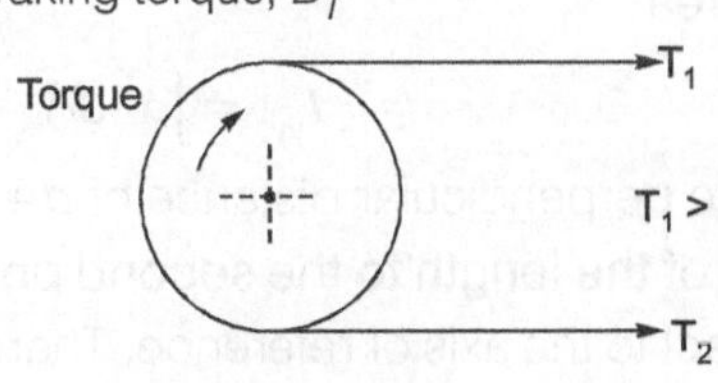

$$= (T_1 - T_2)R$$
$$T_1 = 400$$
$$\frac{T_1}{T_2} = e^{\mu\theta}$$
$$\mu = 0.25$$
$$\theta = \pi$$
$$\therefore\ T_2 = 182.375$$
$$\therefore\ BT = (400 - 182.375) \times \frac{0.5}{2}$$
$$BT = 54.4 \text{ Nm}$$

■■■■

08

CHAPTER

Second Moment of Area (Moment of Inertia)

8.1 Introduction

A quantitative measure of the resistance of a beam is its *second moment of area* or so called *moment of inertia.*

In strength of materials, the expressions for stresses, slopes and deflections in beams and columns require the use of a term known as *Moment of inertia*, which is equal to $I = \int r^2 dA$, *i.e.*, an area A, is divided into small parts such as dA and **each small area dA is multiplied by the square of its *moments arm r*, about the reference axes.**

Let us consider an area A as shown in **Fig. 8.1**. This area can be divided into small parts such as dA shown in the figure. Let us consider ox as the reference axis.

First moment of this area about ox axis = $y \, . \, dA$.

Second moment of this area about ox-axis = $y^2 \, dA$.

Then the mathematical expression $\int y^2 dA$ is known as the moment of inertia of the section about axis ox

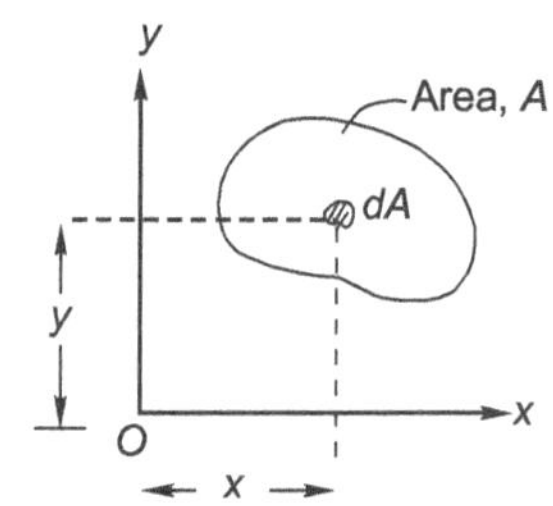

Fig. 8.1

or $$I_{xx} = \int y^2 dA \text{ (about } x\text{-}x \text{ axis)}$$

Similarly choosing oy-axis as the reference axis, moment of inertia or second moment of area

$$I_{yy} = \int x^2 dA,$$

where x is the perpendicular distance of dA from the axis oy. **The ratio of moment of inertia I and area A has dimensions of the length to the second power**. This length is called the *radius of gyration of the plane section A*, with respect to the axis of reference. Therefore,

Radius of gyration about ox-axis, $$\rho_x = \sqrt{\frac{I_{xx}}{A}}$$

Radius of gyration about oy-axis, $$\rho_y = \sqrt{\frac{I_{yy}}{A}}.$$

8.2 Rectangular Section

Consider a rectangular section of breadth b and depth d as shown in **Fig. 8.2.** Centroid of the section is located at G, at distances of $b/2$ and $d/2$ from vertical and horizontal axes OY' and OX'. Say the reference axis GX and GY pass through the centroid G of the section. Now take a small area $dA = b.dy$, of breadth b and thickness dy.

Second moment of area dA about GX-axis

$$= y^2\,dA = by^2\,dy$$

Limits on y are $-\frac{d}{2}$ to $+\frac{d}{2}$ as shown.

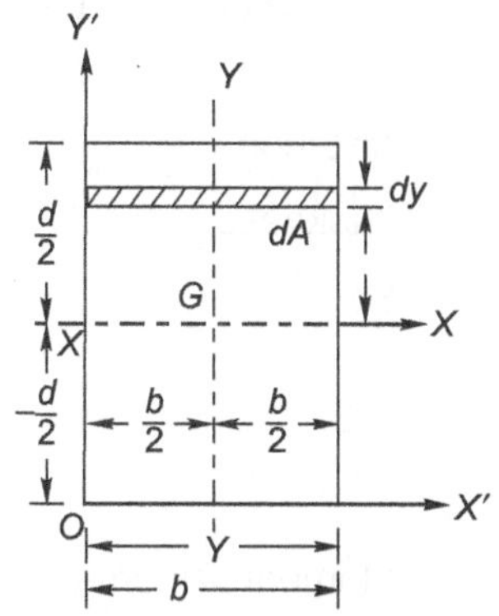

Fig. 8.2

Therefore $$I_{xx} = \int_{-d/2}^{+d/2} by^2 dy = b\left|\frac{y^3}{3}\right|_{-d/2}^{d/2}$$

$$= \frac{b}{3}\left[\frac{d^3}{8}+\frac{d^3}{8}\right] = \frac{bd^3}{12}$$

Area of the section, $A = bd$

Radius of gyration, $$\rho_x = \sqrt{\frac{I_{xx}}{A}} = \sqrt{\frac{bd^3}{12bd}} = \frac{d}{2\sqrt{3}}$$

Similarly considering a section $d.dx$ at a distance of x width of section dx, from axis YGY, we can determine

Moment of Inertia, $$I_{yy} = \frac{db^3}{12}$$

Radius of gyration, $$\rho_y = \frac{b}{2\sqrt{3}}.$$

8.3 Triangular Section

Fig. 8.3 shows a triangular section of base b and altitude h and say the reference axis is passing along the base of the triangular section. Take a small area $dA = b_x \,.\, dy$, at a distance of y from the reference axis ox.

Breadth of small strip $$= b_x = \left(\frac{h-y}{h}\right)b$$

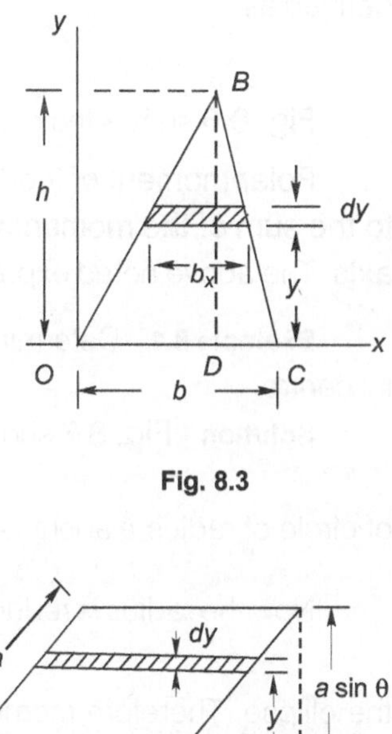

Fig. 8.3

Therefore $$dA = \left(\frac{h-y}{h}\right)b.dy$$

Moment of inertia about ox-axis,

$$I_{xx} = \int y^2 dA = \int_0^h \left(\frac{h-y}{h}\right) b.y^2\,dy$$

Limits of y are from 0 to h, as shown in Fig. 8.3

or $$I_{xx} = \int_o^h by^2\,dy - \int_o^h \frac{by^3}{h}\,dy$$

$$= \left|\frac{by^3}{3}\right|_o^h - \left|\frac{b}{h}\times\frac{y^4}{4}\right|_o^h = \frac{bh^3}{3}-\frac{bh^3}{4} = \frac{bh^3}{12}.$$

Exercise 8.1 Determine moment of inertia I_{xx} about OX-axis of the parallelogram of sides a and b with included angle θ as shown in the **Fig. 8.4**.

[Ans: $\frac{ba^3 \sin^3\theta}{3}$].

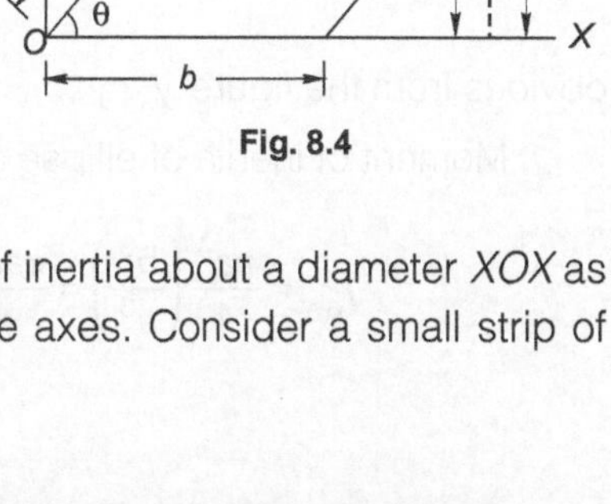

Fig. 8.4

8.4 Circular Section

Fig. 8.5 shows a circular section of radius R. Let us determine its moment of inertia about a diameter XOX as shown in the figure. Centre of the circle lies at the origin of XY co-ordinate axes. Consider a small strip of thickness dy at an angle θ as shown.

Then breadth of the strip

$$b = 2R\cos\theta$$

distance $y = R\sin\theta$

thickness, $dy = Rd\theta.\cos\theta$ (as shown in **Fig. 8.6**).

So $dA = 2R\cos\theta.R\cos\theta.d\theta$

$$= 2R^2\cos^2\theta.d\theta$$

$$y^2 = R^2\sin^2\theta$$

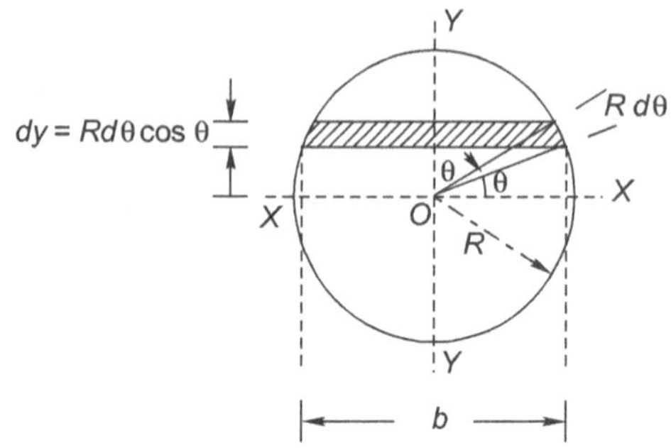

Fig. 8.5

Moment of inertia, $I_{xx} = \int y^2dA = \int 2R^2\cos^2\theta.R^2\sin^2\theta.d\theta$

$$= 2R^4\int_{-\pi/2}^{+\pi/2}\sin^2\theta\cos^2\theta d\theta$$

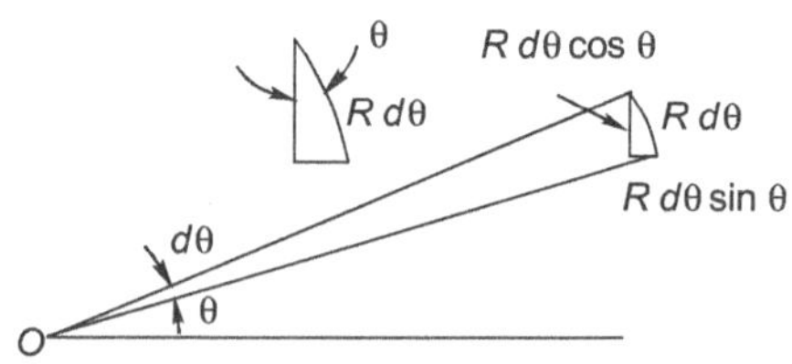

Fig. 8.6

Limits on θ are from $-\frac{\pi}{2}$ to $+\frac{\pi}{2}$

So $I_{xx} = 2R^4\int_{-\pi/2}^{\pi/2}\frac{(\sin 2\theta)^2}{4}d\theta = \frac{R^4}{2}\int_{-\pi/2}^{+\pi/2}\sin^2 2\theta d\theta$

$$= \frac{R^4}{2}\int_{-\pi/2}^{\pi/2}\frac{(1-\cos 4\theta)}{2}d\theta = \frac{\pi R^4}{4} = \frac{\pi D^4}{64}$$

8.5 Polar Moment of Inertia

Polar moment of inertia of a section about a line or about an axis perpendicular to the plane of the section is defined as

$$I_{00} = \int r^2dA = \int(x^2+y^2)\,dA = \int x^2dA + \int y^2dA = I_{yy} + I_{xx}$$

Fig. 8.6 shows the components *x* and *y* of radius vector *r*.

Polar moment of inertia of a section about an axis perpendicular to the plane of the section is equal to the sum of the moments of inertia about any two mutually perpendicular axes intersecting at the polar axis. The above noted expression is also known as the *perpendicular axes theorem*.

Example 8.1 Determine polar moment of inertia of an ellipse of major axis = 2*a* and minor axis = 2*b* about its centre.

Solution **Fig. 8.7** shows an ellipse with semi major axis = *a* and semi minor axis = *b*. The moment of inertia of circle of radius *a* about *x*-*x*-axis $= \frac{\pi a^4}{4}$.

Now the radius is reduced to *b* in the direction *y*, when we consider the ellipse. Therefore moment of inertia can be reduced by $\frac{b^3}{a^3}$ as is obvious from the figure $y < y'$.

Moment of inertia of ellipse about axis *xx*,

$$I_{xx} = \frac{\pi a^4}{4}\left(\frac{b^3}{a^3}\right) = \frac{\pi ab^3}{4}$$

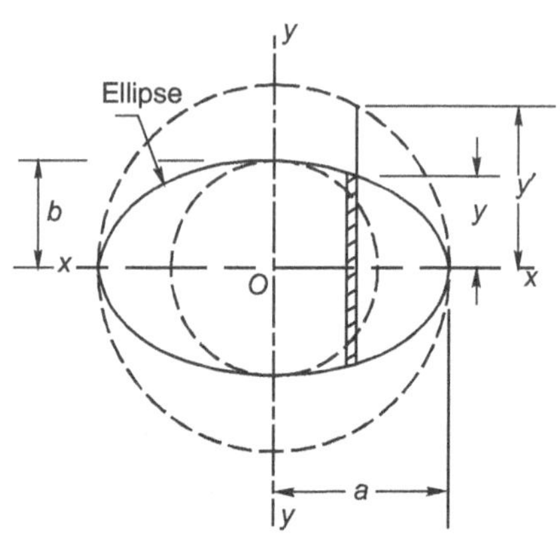

Fig. 8.7

Similarly moment of inertia of ellipse about axis *yy*,

$$I_{yy} = \frac{\pi b^4}{4}\left(\frac{a^3}{b^3}\right) = \frac{\pi a^3 b}{4}$$

Polar moment of inertia of ellipse,

$$I_{00} = I_{xx} + I_{yy} = \frac{\pi ab}{4}(a^2 + b^2).$$

Exercise 8.2 An ellipse has major axis 10 cm and minor axis 6 cm. Calculate (*i*) moment of inertia about major axis, (*ii*) moment of inertia about minor axis, (*iii*) polar moment of inertia.

[**Ans:** 106.03 cm^4, 294.52 cm^4, 400.55 cm^4].

8.6 Parallel Axes Theorem

Many a times it is necessary to transfer the moment of inertia from one axis to another parallel axis. Consider a section with *G* as centroid and an axis *xx* passing through the centroid. Say there is another axis $o'x'$ parallel to *xx* at a perpendicular distance *h* from *G*, as shown in the **Fig. 8.8**. Now take a small area *dA* at a distance of y' from the axis $x'x'$, then

Moment of inertia, $\quad I_{x'x'} = \int y'^2 dA$

but $\quad y' = y + h$

where *y* is the distance of *dA* from centroidal axis *xx*.

So
$$I_{x'x'} = \int (y+h)^2\, dA = \int y^2 dA + \int 2yh\, dA + \int h^2 dA$$
$$= I_{xx} + 2h \int y\, dA + Ah^2$$

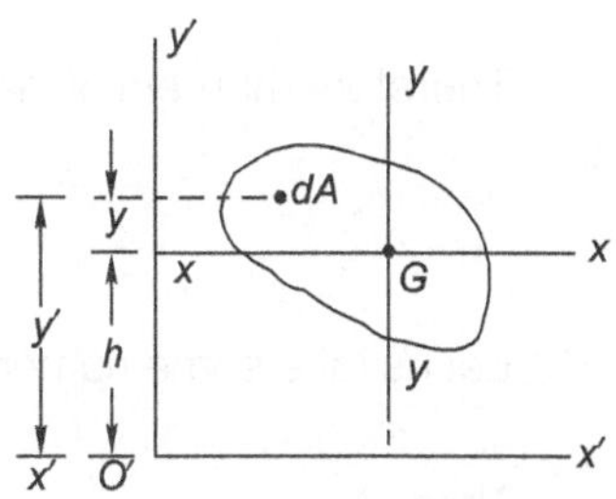

Fig. 8.8

But $\quad \int y dA$ = first moment of area about the centroidal axis = 0 (student can prove this)

Therefore $\quad I_{x'x'} = I_{xx} + Ah^2$

This is known as the *parallel axes theorem* which can be stated as that *moment of inertia of any section about an axis is the sum of the moment of inertia about an axis parallel to the given axis and passing through the centroid of the section plus the product of area and the square of the perpendicular distance between the two axes.*

This theorem is most commonly used to determine the moment of inertia of any composite section about any particular axis.

Example 8.2 Fig. 8.9 shows a *T*-section with a flange of area $B \times t$ and a web of area $b \times d$. Determine moment of inertia of this *T*-section about the centroidal axis *xx* passing through the *G*.

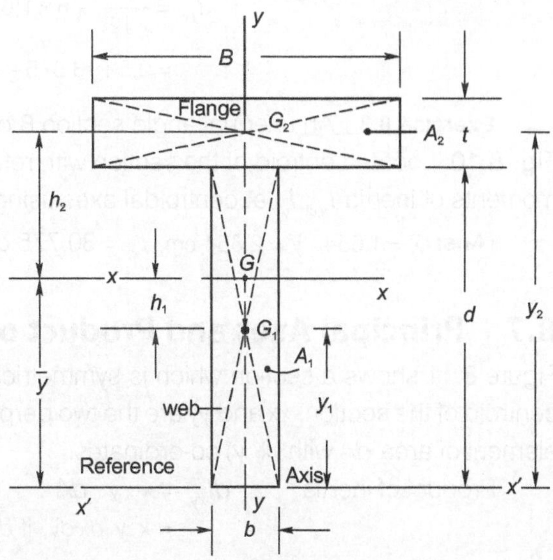

Fig. 8.9

Solution Let us locate the centroid of the section. The section is symmetrical about *yy*-axis which passes through the centroid of the section. Consider a reference axis $x'x'$ at the lower edge of the web. The section can be divided into two areas

$A_1 = b \times d$; $A_2 = B \times t$ and total area $A = A_1 + A_2$

Then distance of G from lower edge $(x'x')$

$$\bar{y} = \frac{A_1y_1 + A_2y_2}{A_1 + A_2}, \text{ where } y_1 = \frac{d}{2} \text{ and } y_2 = d + \frac{t}{2}$$

$$= \frac{bd\left(\frac{d}{2}\right) + B.t\left(d + \frac{t}{2}\right)}{bd + Bt}$$

Moment of inertia of the flange about xx-axis

$$= \frac{Bt^3}{12} + Bt \times h_2^2 = \frac{Bt^3}{12} + Bt\left(d + \frac{t}{2} - \bar{y}\right)^2$$

using parallel axes theorem.

Moment of inertia of the web about xx-axis

$$= \frac{bd^3}{12} + bd \times h_1^2 = \frac{bd^3}{12} + bd\left(\bar{y} - \frac{d}{2}\right)^2,$$

using parallel axis theorem because

$$h_1 = \bar{y} - \frac{d}{2} \text{ and } h_2 = d + \frac{t}{2} - \bar{y}$$

Therefore moment of inertia of T-section about centroidal axis xx

$$I_{xx} = \frac{Bt^3}{12} + Bt\left(d + \frac{t}{2} - \bar{y}\right)^2 + \frac{bd^3}{12} + bd\left(\bar{y} - \frac{d}{2}\right)^2$$

Let us take some numerical values

$$B = 6 \text{ cm}, \quad t = 1 \text{ cm}; \quad d = 6 \text{ cm and } b = 1 \text{ cm}$$

Then $$A_1 = A_2 = 6 \text{ cm}^2$$

$$y_1 = 3 \text{ cm}; \quad y_2 = 6 + \frac{1}{2} = 6.5 \text{ cm}$$

$$\bar{y} = \frac{6 \times 3 + 6 \times 6.5}{6 + 6} = \frac{18 + 39}{12} = 4.75 \text{ cm}$$

Moment of inertia,

$$I_{xx} = \frac{6 \times 1^3}{12} + 6 \times 1(6.5 - 4.75)^2 + \frac{1 \times 6^3}{12} + 6\,(4.75 - 3)^2$$

$$= 0.5 + 18.375 + 18 + 18.375 = 55.25 \text{ cm}^4.$$

Exercise 8.3 An unequal angle section 8 cm × 6 cm × 1 cm is shown in Fig. 8.10. Locate centroid of the section with reference to yox-axes. Calculate moments of inertia I_{xx}, I_{yy} at centroidal axes using parallel axes theorem.

[**Ans:** $\bar{x} = 1.654$, $\bar{y} = 2.654$ cm, $I_{xx} = 80.775$ cm^4, $I_{yy} = 38.775$ cm^4].

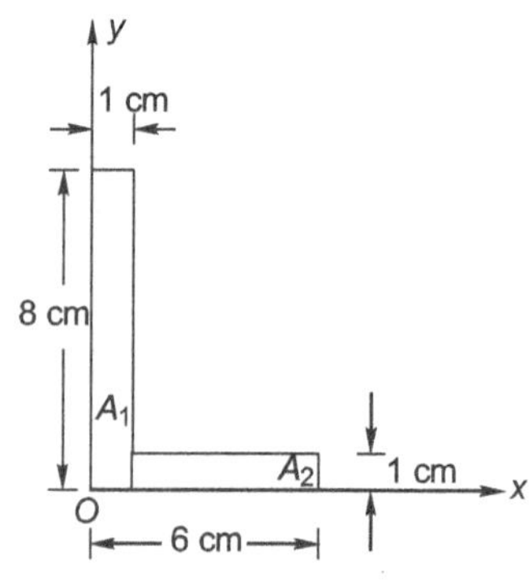

Fig. 8.10

8.7 Principal Axes and Product of Inertia

Figure 8.11 shows a section which is symmetrical about the axis y-y, G is the centroid of the section. xx and yy are the two perpendicular axes passing through the centroid. Consider a small element of area dA with (x, y) co-ordinates.

Product of inertia $$dI_{xy} = x \,.\, y \,.\, dA$$

$$= x.y.dx\,dy, \text{ if } dA = dx.dy$$

In this case $$\int xy\, dA = 0, \text{ because of symmetry about } yy\text{-axis.}$$

The expression $\int xy\,dA$ is called the *product of inertia* of the area about *xx* and *yy*-axes, represented by I_{xy}.

The product of inertia can be zero about the two co-ordinate axes passing through the centroid. *Such axes (about which product of inertia is zero) are called Principal axes of the section and moment of inertia about the principal axes are called Principal moments of inertia.*

The product of inertia may be positive, negative or zero depending upon the section and co-ordinate axes. The product of inertia of a section with respect to two perpendicular axes is zero *if either of the two axes is an axis of symmetry.*

The section shown in the **Fig. 8.11** is symmetrical about *yy*-axis, because of every small area *dA* considered, there is an area *dA* on the other side of *yy*-axis with negative value of *x*. There are many sections such as circular, rectangular, *I* section which have two axes of symmetry passing through their centroid.

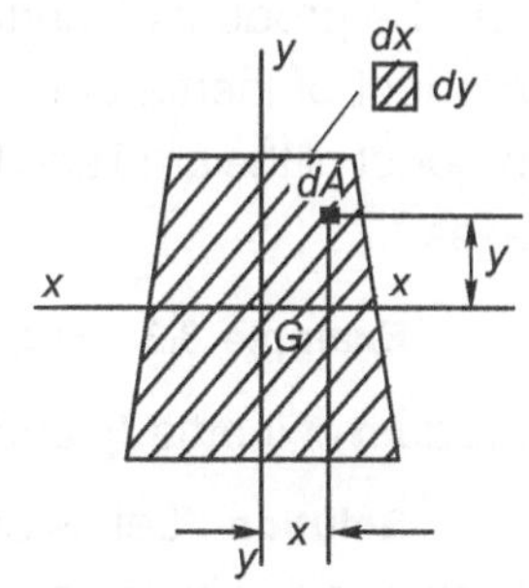

Fig. 8.11

In all such cases product of inertia *i.e.*, $\int xy\,dA = 0$. Then there are sections which have only one axis of symmetry passing through their centroids such as *T*, channel section, trapezoidal section, semicircular section. In these cases also product of inertia *i.e.*, $\int xy\,dA = 0$ about centroidal axes.

There are sections such as *L* section, *L* equal angle section, *Z* section, unequal channel section, which are neither symmetrical about *x*-axis nor symmetrical about *y*-axis. In all such cases $\int xy\,dA \neq 0$, about centroidal axes as shown in **Fig. 8.12.**

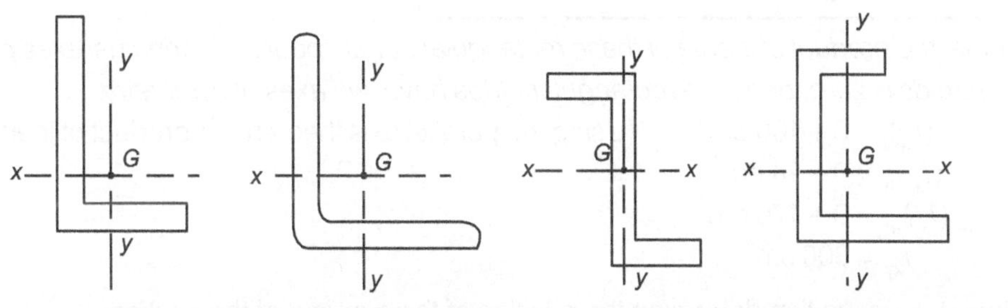

Fig. 8.12

8.8 Parallel Axes Theorem for Product of Inertia

Fig. 8.13 shows a section with its centroid at *G* and Gx', Gy' are the two rectangular co-ordinates passing through *G*. Say the product of inertia about $x'y'$ is $I\bar{x}\bar{y}$. Let us determine the product of inertia about the axes *Ox* and *Oy* *i.e.*, I_{xy}.

Say distance of *G* from *Ox*-axis $= \bar{y}$ and distance of *G* from *Oy*-axis $= \bar{y}$ as shown.

Consider a small element of area *dx.dy*.

Say co-ordinates of the element about the centroidal axis Gx', Gy' are x', y'.

Then co-ordinates of the element about *x-y*-axes are

$$x = \bar{x} + x' \text{ and } y = \bar{y} + y'$$

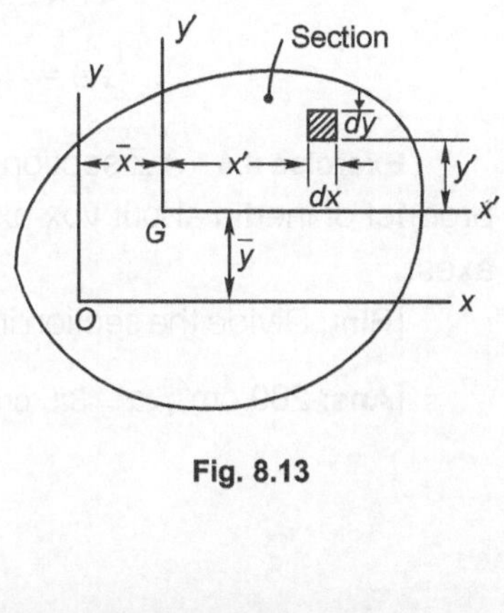

Fig. 8.13

So product of inertia,

$$I_{xy} = \int xy\,dA = \int (\bar{x}+x')(\bar{y}+y')\,dx\,dy$$
$$= \int x'y'\,dA + \bar{x}\bar{y}\int dA + \bar{y}\int x'\,dA + \bar{x}\int y'\,dA$$
$$= I_{\bar{x}\bar{y}} + \bar{x}\bar{y}A + 0 + 0$$

(because first moments of area, $\int x'dA = \int y'dA = 0$ about centroidal axes)

i.e., the product of inertia of any section with respect to any set of co-ordinate axes in its plane is equal to the product of inertia of the section with respect to the centroidal axes parallel to the co-ordinate axes plus the product of the area and the co-ordinates of the centroid of the section with respect to the given set of co-ordinate axes.

Example 8.3 Fig. 8.14 shows an unequal channel section, determine its product of inertia I_{xy} and $I_{\bar{x}\bar{y}}$.

Solution Let us break up the section into 3 rectangular strips, I, II and III as shown and write the co-ordinates of their centroids with respect to the given set of axis *yox*.

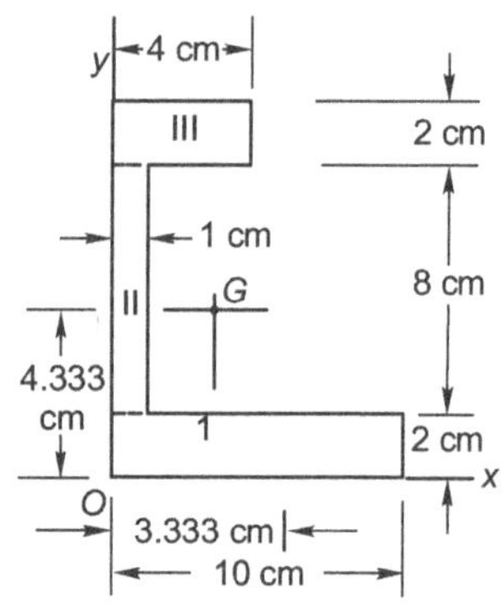

Fig. 8.14

Strip	Area	$\bar{x}$	$\bar{y}$	$A\bar{x}\bar{y}$
I.	20 cm^2	5 cm	1 cm	100 cm^4
II.	8 cm^2	0.5 cm	6 cm	24 cm^4
III.	8 cm^2	2 cm	11 cm	176 cm^4

Remember that the *product of inertia of these rectangular strips about their principal axes passing through the respective centroids is zero, because rectangular strips have two axes of symmetry.*

$(I_{xy})_{I} = 0 + 100$ cm^4 (Using the parallel axis theorem for product of inertia)

$(I_{xy})_{II} = 0 + 24$ cm^4

$(I_{xy})_{III} = 0 + 176$ cm^4

$I_{xy} = 300$ cm^4

To determine, $I_{\bar{x}\bar{y}}$, let us first determine the position of the centroid of the section

$$\bar{x} = \frac{20\times5+8\times0.5+8\times2}{20+8+8} = 3.333 \text{ cm}$$

$$\bar{y} = \frac{20\times1+8\times6+8\times11}{20+8+8} = 4.333 \text{ cm}$$

Area of the cross-section,

$$A = 20+8+8 = 36 \text{ cm}^2$$

$$I_{\bar{x}\bar{y}} = I_{xy} - A\bar{x}\bar{y} = 300 - 36\times3.333\times4.333 = -220 \text{ cm}^4.$$

Exercise 8.4 A *Z*-section of dimensions is shown in the Fig. 8.15. Determine its product of inertia about *yox*-axes. Also calculate product of inertia about centroidal axes.

[**Hint:** Divide the section into 3 areas as shown].

[**Ans:** 289 cm^4; $\bar{x} = 3.5$ cm, $\bar{y} = 5$ cm, $I_{\bar{x}\bar{y}} = -96$ cm^4].

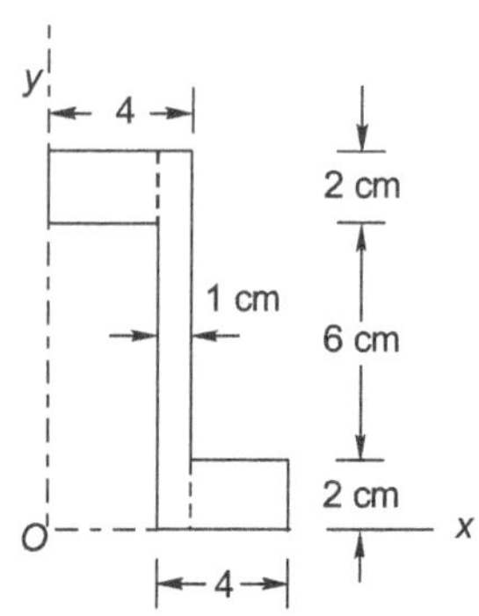

Fig. 8.15

Remember

- Moment of inertia of a section about *xx*-axis, $I_{xx} = \int y^2 dA$.
- Moment of inertia of a section about *yy*-axis, $I_{yy} = \int x^2 dA$.
- Moment of inertia of a triangle about base $= \dfrac{bh^3}{12}$.
- MI of a triangle about centroidal axis parallel to the base $= \dfrac{bh^3}{36}$ where b = base, h = altitude.
- MI of a rectangle about base $= \dfrac{bh^3}{3}$.
- MI of a rectangle about a centroidal axis parallel to the base $= \dfrac{bh^3}{12}$.
- MI of a parallelogram about base $= \dfrac{bh^3}{3}$ where h = altitude of parallelogram, b = base.
- Polar moment of inertia of a circular section $= \dfrac{\pi R^4}{2}$.
- MI of circular section about a diametral axis, $I_{xx} = I_{yy} = \dfrac{\pi R^4}{4}$.
- Polar moment of inertia of any section $I_{00} = I_{xx} + I_{yy}$.
- Perpendicular axis theorem $I_{00} = I_{xx} + I_{yy} = I_{x'x'} + I_{y'y'}$.
- Parallel axes theorem $I_{x'x'} = I_{xx} + Ah^2$.
- Product of inertia, $I_{xy} = \int\int xy\,dx\,dy$.
- About principal axes, product of inertia $\int xy\,dA = 0$.
- For any section having at least one axis of symmetry $\int xy\,dA = 0$, about axes containing the axis of symmetry.

MULTIPLE CHOICE QUESTIONS

8.1 **Fig. 8.16** shows a equilateral triangle with side *a*. *xx* and *yy* are centroidal axes. If I_{xx} = 100 cm^4, what is I_{yy}?

(*a*) 87.6 cm^4 (*b*) 100 cm^4
(*c*) 115.5 cm^4 (*d*) None of these

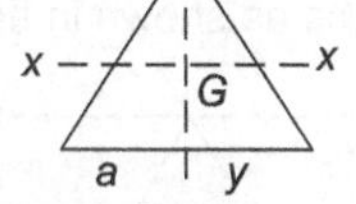

Fig. 8.16

8.2 **Fig. 8.17** shows a rectangle with sides *b* and *d* = 2*b*. If I_{xx} = 1000 cm^4, what is *b*?

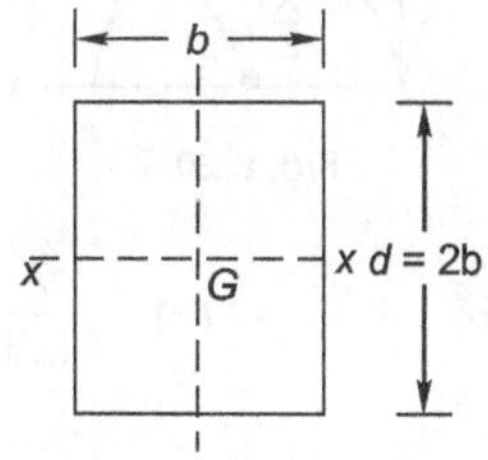

Fig. 8.17

(*a*) 6.22 cm (*b*) 5.6 cm
(*c*) 5.08 cm (*d*) None of these

8.3 **Fig. 8.18** shows a triangle, with base *ab* and centroidal axes *xx*. If I_{xx} = 150 cm^4, what is I_{ab}?

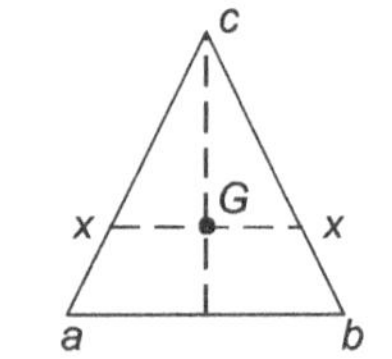

Fig. 8.18

(*a*) 50 cm^4 (*b*) 550 cm^4
(*c*) 450 cm^4 (*d*) None of these

8.4 **Figure 8.19** shows a parallelogram of sides 5 cm and 6 cm with included angle of 60°, what is its moment of inertia about base *AB*?

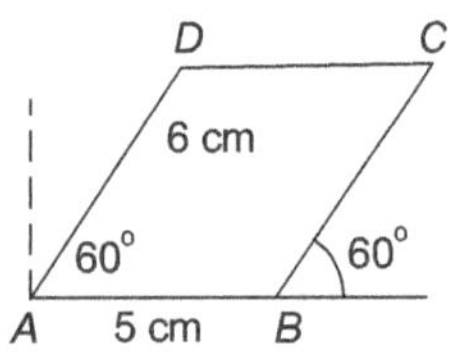

Fig. 8.19

(*a*) 360 cm^4 (*b*) 233.08 cm^4
(*c*) 180 cm^4 (*d*) None of these

8.5 A hollow circular section with outer diameter 10 cm and inner diameter 6 cm. What is its moment of inertia about a diametral axis?
(*a*) 1709 cm^4 (*b*) 854.5 cm^4
(*c*) 427.26 cm^4 (*d*) None of these

8.6 What is moment of inertia about *xx*-axis of the shaded area shown in **Fig. 8.20**?

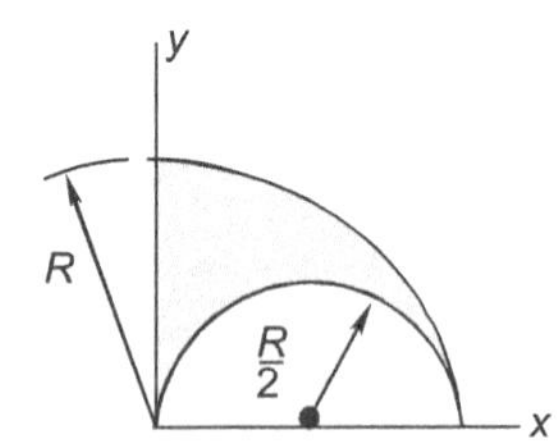

Fig. 8.20

(*a*) $\frac{3}{64}\pi R^4$ (*b*) $\frac{7}{128}\pi R^4$

(*c*) $\frac{9}{128}\pi R^4$ (*d*) None of these

8.7 An equilateral triangle of side *a* has a circular hole touching its 3 sides. *CG* of hole and triangle is same what is I_{xx} of shaded area **(Fig. 8.21)**
(*a*) 0.108a^4 (*b*) 0.106a^4
(*c*) 0.1025a^4 (*d*) None of these

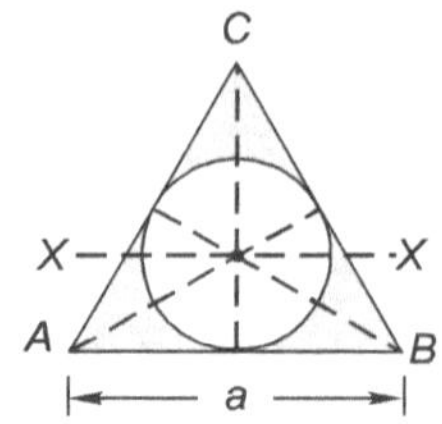

Fig. 8.21

8.8 A triangle *ABC*, with base *b* and attitude *h* is shown in the **Fig. 8.22.** What is the ratio of I_{EF}/I_{AB}?

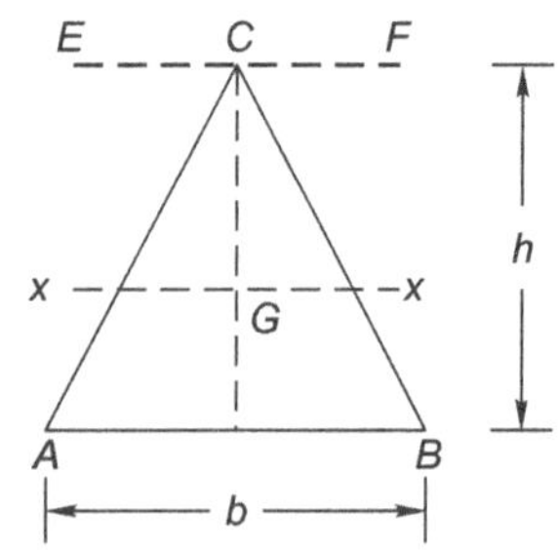

Fig. 8.22

(*a*) 3 (*b*) 2.0
(*c*) 1.5 (*d*) 0.33

8.9 xx, x_1x_1, and x_2x_2 are parallel axes of which xx is the centroidal axis. If moment of inertia of the figure about x_1x_1 axis is 10 m^4, what is the moment of inertia of the **figure 8.23** about x_2x_2 axis

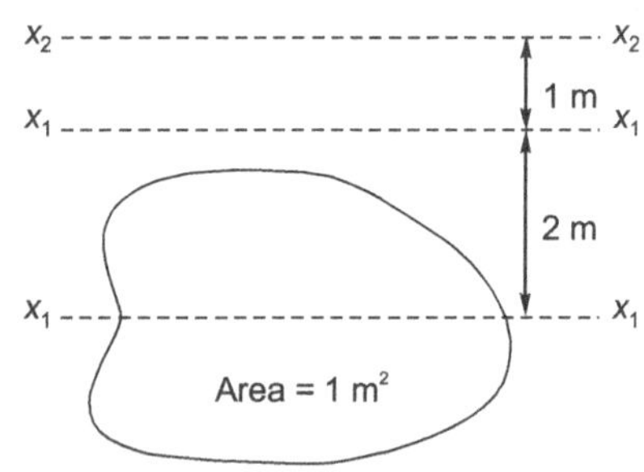

Fig. 8.23

(*a*) 10 m^4 (*b*) 11 m^4
(*c*) 14 m^4 (*d*) 15 m^4

[CSE, Prelim, CE : 2005]

8.10 What is the moment of inertia of the triangle with respect to xx axis as shown in **figure 8.24**

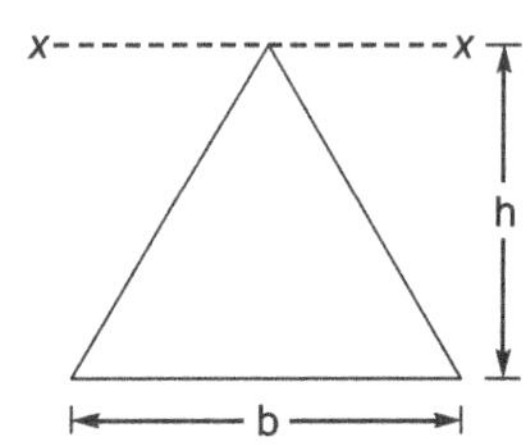

Fig. 8.24

(a) $\frac{bh^3}{12}$ (b) $\frac{bh^3}{6}$

(c) $\frac{bh^3}{3}$ (d) $\frac{bh^3}{4}$

[CSE, Prelim, CE : 2009]

Answers

8.1 (b)	8.2 (a)	8.3 (c)	8.4 (b)	8.5 (c)
8.6 (b)	8.7 (c)	8.8 (a)	8.9 (d)	8.10 (d)

EXPLANATIONS

8.1 (b)

For equilateral triangle, $I_{xx} = I_{yy}$ along centroidal axes.

8.2 (a)

$$I_{xx} = \frac{bd^3}{12} = \frac{b \times 8b^3}{12} = \frac{2}{3}b^4 = 1000$$

$b = 6.22$ cm.

8.3 (c)

$$I_{xx} = \frac{bh^3}{36}, I_{ab} = \frac{bh^3}{12} = 3I_{xx}.$$

8.4 (b)

$$I_{AB} = \frac{bh^3}{3} = \frac{5(0.866 \times 6)^3}{3} = 233.08 \text{ cm}^4.$$

8.5 (c)

$$I = \frac{\pi(D^4 - d^4)}{64} = \frac{\pi}{64}(10^4 - 6^4) = 427.26 \text{ cm}^4.$$

8.6 (b)

$$I_{xx} = \frac{\pi R^4}{16} - \frac{\pi}{8}\left(\frac{R}{2}\right)^4 = \pi R^4\left(\frac{1}{16} - \frac{1}{128}\right)$$

$$= \pi R^4 \times \frac{7}{128}.$$

8.7 (c)

$$\text{Radius of circle} = \frac{0.866}{3}a = 0.288666a$$

$$I_{xx} = 0.108a^4 - \frac{\pi}{4}(.288666a)^4 = .1025a^4$$

8.8 (a)

$$I_{AB} = \frac{bh^3}{12}, I_{EF} = \frac{bh^3}{36} + \frac{bh}{2}\left(\frac{4h^2}{9}\right) = \frac{bh^3}{36} + \frac{2bh^3}{9} = \frac{bh^3}{4}.$$

8.9 (d)

$I_{xx} = I_{xx} + 1 \times 4 = 10$

$I_{xx} = 6 \ m^4$

$P_{yy} = 6 + 1 \times 3^2 = 15 \text{ m}^4$

8.10 (d)

$$I_{xx} = \frac{bh^3}{36} + \frac{bh}{2}\left(\frac{2b}{3}\right)^2$$

$$= \frac{bh^3}{36} + \frac{2bh^3}{9} = \frac{bh^3}{4}$$

■■■■

09

CHAPTER

Virtual Work

9.1 Introduction

In the preceding chapters we have studied about the equilibrium of a body or a system of bodies by considering $\Sigma F_x = 0$, $\Sigma F_y = 0$, and $\Sigma M = 0$ at any point of the body or system of bodies. But there are special problems of a system of connected bodies, mechanics or mechanisms which can be *easily solved by using the principle of virtual work*. While using this concept of virtual work, the equations of virtual work for virtual displacement consistent with the constraints are developed. These equations do involve reactions or connected forces and if these reactions and connected forces are of no interest, then this method is very useful for determining unknown forces. We may solve as many number of active forces as there are independent equations made, using the principle of *virtual displacement.* The number of equations depend upon the *number of degrees of freedom of a system.* This method provides deeper insight into the behaviour of mechanical systems and stability of a system in equilibrium. Virtual work method is a tool to solve many mechanics problems which are otherwise difficult to solve by using equilibrium equations.

9.2 Virtual Work

Work done by a force is a scalar quantity and it is dot product of a force vector and a displacement vector at a point in a body as shown in **Fig. 9.1.**

Work done, $U = F\,.\,S$

$$= (F_x i + F_y j + F_z .k).(S_x i + S_y j + S_z .k)$$

$$= |F||S| \cos\alpha$$

= A scalar quantity as discussed earlier.

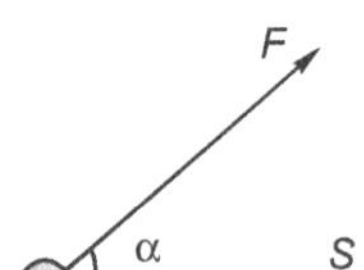

Fig. 9.1

In the principle of virtual work, a virtual displacement $\delta s = |\delta r|$ is given at a point of a body in equilibrium. This displacement is not real but only imaginary and all the forces acting on the body are in equilibrium *i.e.*, there is no resultant force in any direction, as a result this virtual work will be zero. Say a point A on a body in equilibrium is given virtual displacement δr and the point moves to A'. Then δr is the virtual displacement. Initial position vector of point A is $OA = \bar{r}$. **(Fig. 9.2)**

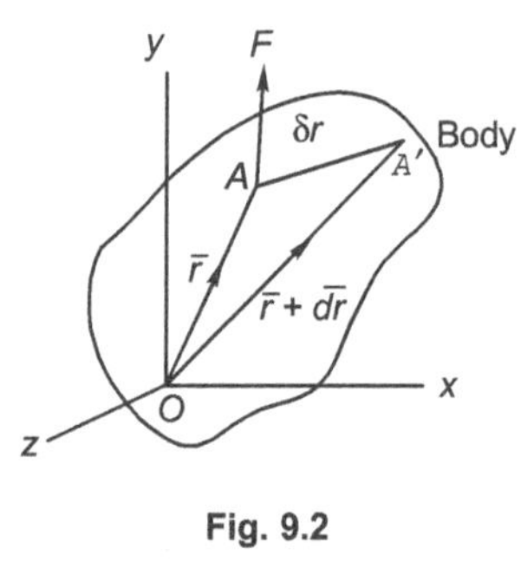

Fig. 9.2

Final position vector of point A', $OA' = \bar{r} + \overline{dr}$

Force F is acting on the body at point A, say

$$F = F_1 + F_2 + F_3 \ldots F_n$$

$$= \Sigma F_x i + \Sigma F_y j + \Sigma F_z .k$$

$$\delta r = \delta x.i + \delta y.j + \delta z.k$$

where magnitude of

$$F = \sqrt{(\Sigma F_x)^2 + (\Sigma F_y)^2 + (\Sigma F_z)^2}$$

Magnitude of $\delta r = \sqrt{(\delta x)^2 + (\delta y)^2 + (\delta z)^2}$

Virtual work, $U = F.\delta r$

$$= (\Sigma F_x i + \Sigma F_y j + \Sigma F_z .k).\delta r = 0$$

because $\Sigma F_x = 0,\ \Sigma F_y = 0,\ \Sigma F_z = 0$ since the body is in equilibrium.

Total work done on the body is zero due to virtual displacement. Therefore work done by some force may be positive and the work done by some other force may be negative. *If the force and virtual displacement are in the same direction then virtual work done is positive and if force and virtual displacement are in opposite directions then the virtual work done will be negative.* Principle of virtual work can be stated as follows:

If a system of connected bodies or mechanisms is in equilibrium under various external forces and this system is given virtual displacement from the position of equilibrium then total work done by the external forces during arbitrary displacement is zero.

This principle of virtual work was first used by Swiss mathematician J. Bernoulli in eighteenth century.

While computing the virtual work, we should remember that there are many forces encountered in statics which *do no work* as these forces are applied through fixed points *i.e.*, $dr = 0$ examples of such forces are:

(*i*) Reaction at a frictionless pin; body rotating about the pin.
(*ii*) Reaction at a frictionless contacting surface when two bodies are in relative motion.
(*iii*) Reaction at a roller moving along its track.
(*iv*) Weight of a body when its *CG* moves horizontally.
(*v*) Friction force acting on a wheel rolling without slipping as the point of contact is instantaneously it rest.

9.3 Work Done by a Couple

Say at two points of a body at a distance *r*, forces *F* and $-F$ are acting forming a couple as shown in **Fig. 9.3**. Point *A* moves from *A* to A' and point *B* moves to B' and then to B'' as shown. Point *A* moves through displacement δr_1 and to keep the distance between *A* and *B* to be the same point *B* also moves through displacement δr_1. Total work done by forces *F* and $-F$ at *A* and *B*.

$$\delta U = F.\delta r_1 + (-F).\delta r_1$$
$$= 0.$$

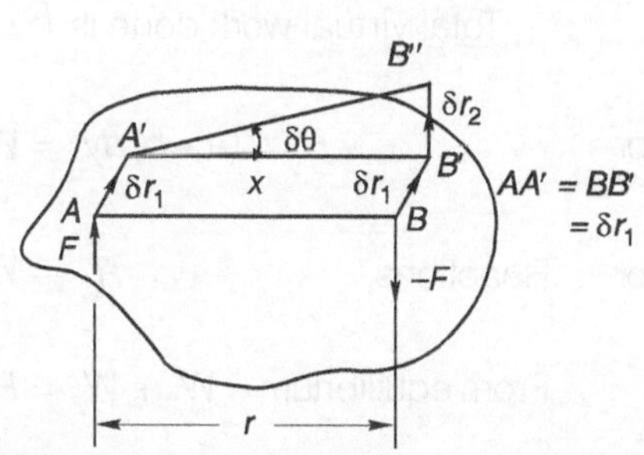

Fig. 9.3

Now point A' remains stationary and point B' moves to B'' due to the couple applied.

Work done, $\Delta U' = F.\delta r_2 = F.r.\delta\theta$ **(Fig. 9.3)**

Because, both points *A* and *B* lie on a rigid body, distance

$$A'B'' = A'B' = AB = r$$

or $\delta r_2 = r.\delta\theta$

But $F.r = $ Couple applied

Couple applied is clockwise but movement $\delta\theta$ is anticlockwise therefore this work will be a negative work. If both couple and displacement $\delta\theta$ are in the same sense (both clockwise or both in anticlockwise directions) then work done by couple will be positive.

Exercise 9.1 A crank and connecting rod arrange-ment is shown in **Fig. 9.4.** A moment M is applied on the crank and force P on the cross head. Determine the relation-ship between M and P.

[**Hint:** $M.\delta\theta + P.\delta x = 0$; use $R\sin\theta = L\sin\phi$].

[**Ans:** $\frac{M}{P} = R\sin\theta\left(1+\frac{R\cos\theta}{\sqrt{L^2 - R^2\sin^2\theta}}\right)$].

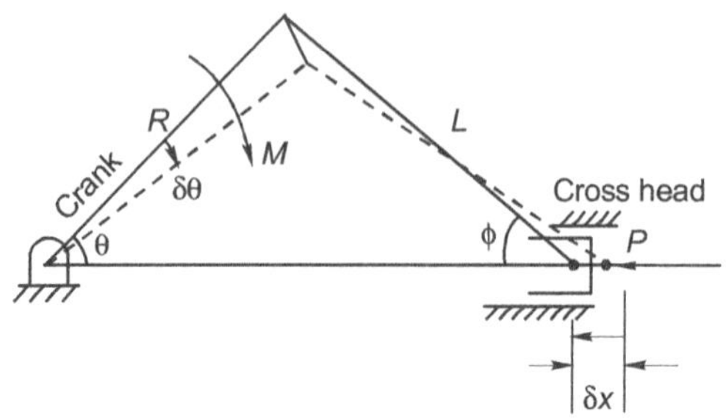

Fig. 9.4

9.4 Principle of Virtual Work Applied to Beams

Principle of virtual work can be applied to determine the reactions in the case of beams subjected to transverse loads. In such applications, keeping one end of the beam stationary, other end of the beam is given virtual displacement. This principle is more conveniently used in cases where two or more beams are connected at ends through hinges. Consider a beam $ABCD$ of length $a+b+c$ as shown in **Fig. 9.5**. The beam is hinged at end A and roller supported at C. There will be reactions at A and C due to vertical loads W_1 at B and W_2 at D. Let us give virtual displacement of $dy\uparrow$ at the point D, considering the beam hinged at A and due to virtual displacement the beam is lifted up from the support C, Fig. 9.6. Then virtual displacements at different points can be worked out as

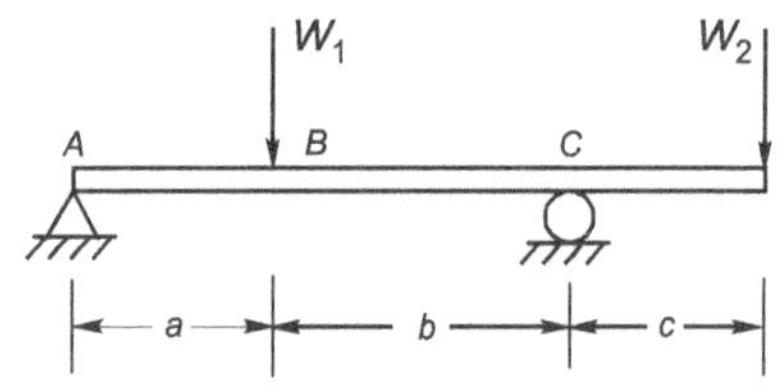

Fig. 9.5

$$dy_A = 0$$

$$dy_B = \frac{a}{a+b+c} \times dy$$

$$dy_C = \frac{a+b}{a+b+c} \times dy$$

$$dy_D = \frac{a+b+c}{a+b+c} \times dy = dy.$$

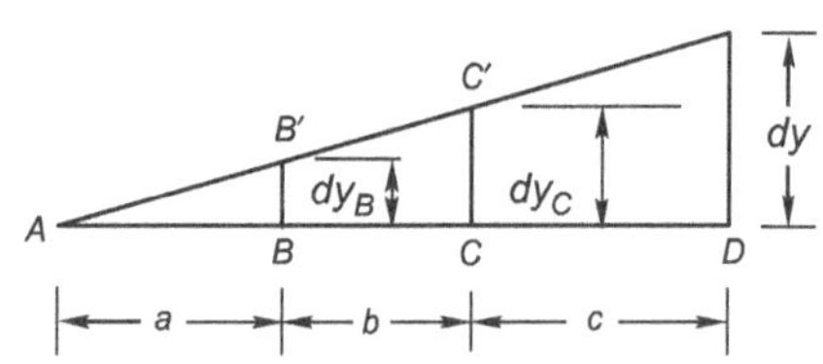

Fig. 9.6

Virtual displacements are upwards, but transverse loads are downwards, therefore virtual work done by loads is $-W_1.dy_B - W_2.dy_D$.

Support reactions are upwards, therefore virtual work done by reactions is $R_A \times dy_A + R_C \times dy_C$.

Total virtual work done is $R_A \times 0 + R_C \times \left(\frac{a+b}{a+b+c}\right) dy - W_1 \times \frac{a}{a+b+c} \cdot dy - W_2 \times dy = 0$

or $$R_C \times (a+b)dy = W_1(a)\, dy + W_2(a+b+c)dy$$

or Reactions, $$R_C = W_1 \times \frac{a}{a+b} + W_2 \times \frac{a+b+c}{a+b}$$

From equilibrium $W_1 + W_2 = R_A + R_C$

or Reaction, $$R_A = W_1 + W_2 - R_C.$$

Example 9.1 Two beams AB and BC of length 8 m and 10 m respectively are hinged at B. These are supported on roller at D and hinged at ends A and C. A roller support is provided at D, 6 m from A as shown in **Fig. 9.7**. Using the principle of virtual work, determine the force transmitted by the hinge B and the reaction at the support D, when a load W of 500 N acts at a point 7 m from D as shown in the **Fig. 9.7.**

Solution Let us give virtual displacement *dy* to the point *B* from the horizontal position *AC* of the beam as shown in **Fig. 9.8**.

Virtual displacements of points *D* and *E* are

$$dy_D = \frac{6}{8}dy = 0.75dy$$

$$dy_E = \frac{7}{10}dy = 0.7dy$$

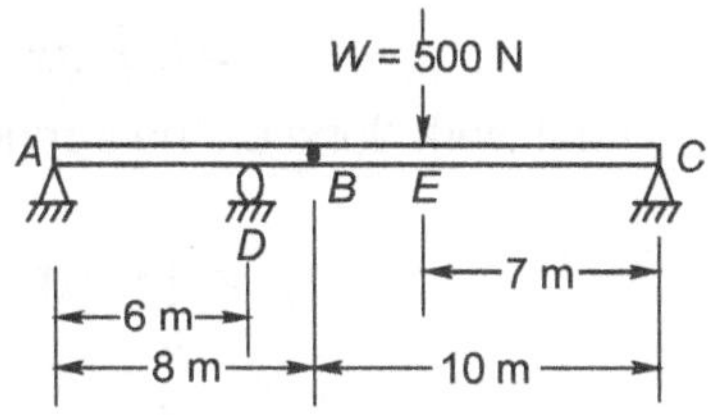

Fig. 9.7

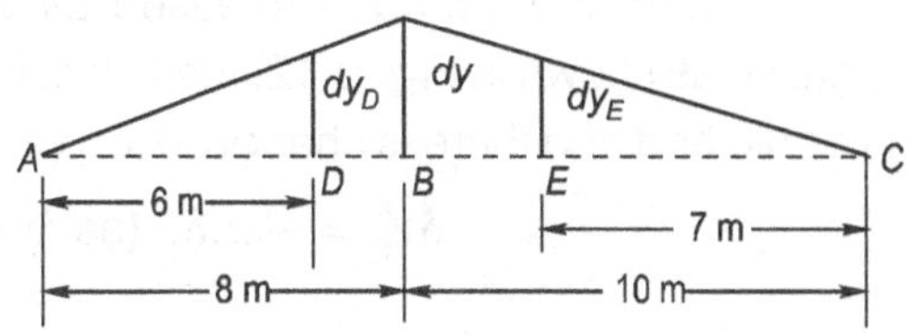

Fig. 9.8

Reaction at *D* is upwards and load at *E* is downwards. Virtual displacements at ends *A* and *C* are zero. Using the principle of virtual work

$$R_A \times 0 + R_D.dy_D - W.dy_E + R_C \times 0 = 0$$

$$R_D \times 0.75dy - 500 \times 0.7dy = 0$$

Reaction at D_1, $R_D = 350/0.75 = 466.66$ N.

Force transmitted by hinge *B*: Two determine this force F_B at hinge let us consider beam *ADB* only

$$R_D \times \delta y_D + F_B \times dy = 0 \text{ (using the principle of virtual work)}$$

$$466.66 \times 0.75dy = -F_B.dy$$

or $F_B = -350$ N (downwards)

To verify this let us take beam *BEC*. Using the principle of virtual work for *BEC*

$$F_B \times dy - 500 \times dy_E = 0$$

$$F_B.dy = 500 \times 0.7dy$$

$$F_B = 350 \text{ N} \uparrow$$

Free body diagrams for beams *ADB* and *BEC* can be drawn as shown in **Fig. 9.9**.

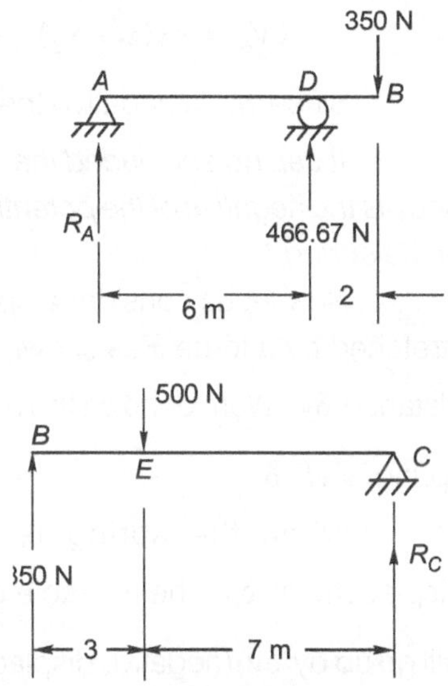

Fig. 9.9

Obviously the reactions will be

$$R_A = 116.67 \text{ N}\uparrow \text{ and } R_C = 150 \text{ N}\uparrow.$$

Exercise 9.2 Two beams *AB* and *CD* are supported on rollers at *E* and *C* as shown in **Fig. 9.10**. The beam *AB* is hinged at *A* and beam *CD* is hinged at *D*. Determine the reactions at the rollers using the method of virtual work.

[**Hint:** Give deflection *y* to support *E* and then analyse).

[**Ans:** R_E = 600 N, R_C = 571.43 N].

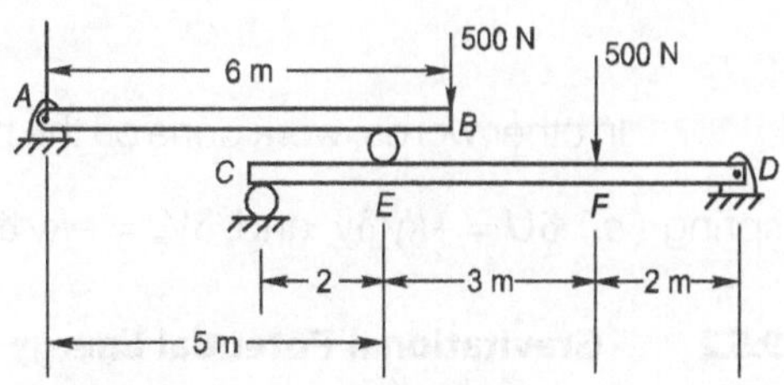

Fig. 9.10

9.5 Potential Energy and Stability

There are many mechanical systems which include springs as elastic members and the concept of total potential energy of the system *i.e.*, sum of elastic and gravitational potential energies of the system is helpful in the study of stability of equilibrium of such systems.

9.5.1 Elastic Potential Energy

Figure 9.11 shows a spring of stiffness *k* compressed by a force *F* say through a displacement δx. Since the displacement is in the direction of the force, work done on the spring, $\delta U = F.\delta x$ is positive

$= k.x.dx$

Total work done on the spring,

$$U_e = \int_0^x kx\,dx$$

$$U_e = \frac{1}{2}kx^2, \text{ stored as elastic potential energy}$$

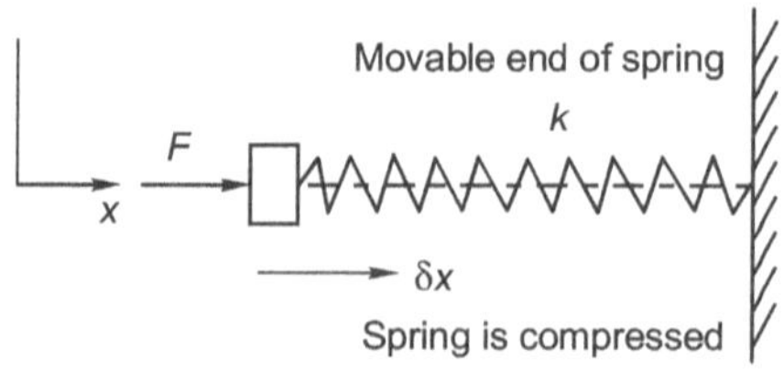

Fig. 9.11

When the spring is released by displacement δx, which is negative as shown in **Fig. 9.12**, then work is done by the spring on the movable end and there is change in its potential energy by an amount

$$\delta V_e = -kx.\delta x \text{ (as } \delta x \text{ is negative)}$$

(Final minus initial energy)

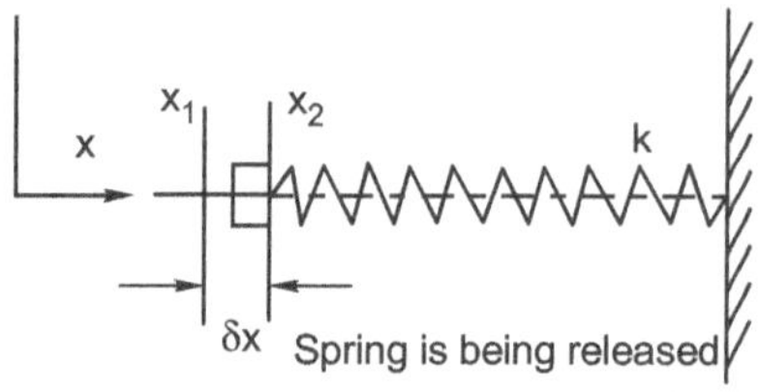

Fig. 9.12

During a decrease in the compression of the spring as it is relaxed from $x = x_2$ to $x = x_1$, change in potential energy of the spring

$$\delta V_e = kx(x_1 - x_2) = -kx.\delta x$$

Since δx is negative therefore δV_e is also negative.

It can be concluded that work done on the body is the negative of the potential energy change of the spring.

Now let us consider a spring of stiffness k stretched by a force F as shown in **Fig. 9.13** by a distance δy. Work done on the movable end of the spring $= +F.\delta y$.

When the spring is released by a displacement δy, the movable end of the spring will go up by δy (negative displacement as per the co-ordinate system shown in **Fig. 9.14**.

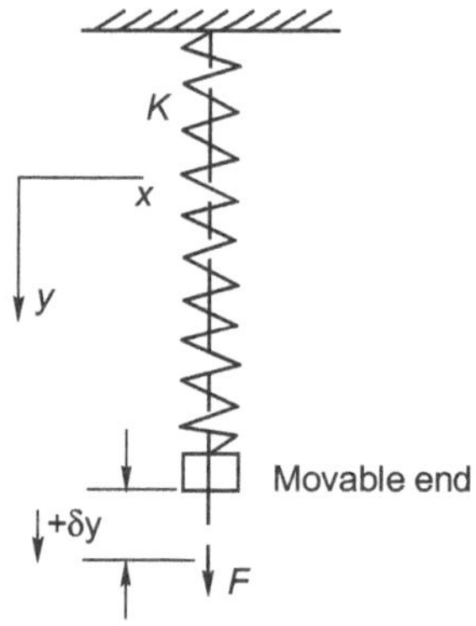

(Spring is being Stratched)

Fig. 9.13

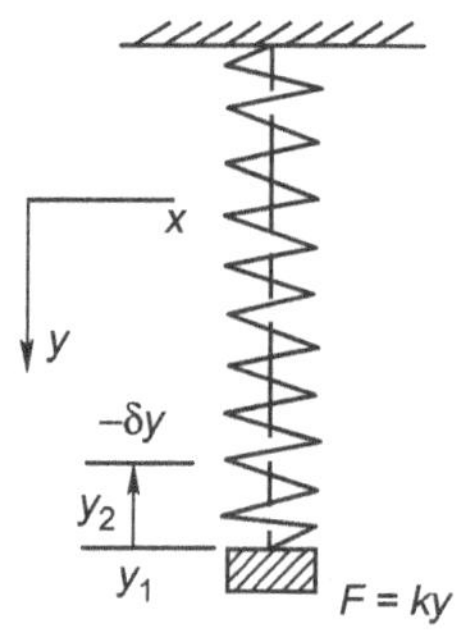

(Spring is being released)

Fig. 9.14

Change in potential energy $\quad \delta V_e = ky(y_2 - y_1)$

$$= -ky\,\delta y$$

In other words, work done on the body *i.e.*, $F\delta y = ky\delta y$ is negative of the potential energy change of the spring *i.e.*, $\delta U = +ky\delta y$ and $\delta V_e = -ky\delta y$.

9.5.2 Gravitational Potential Energy

Work done by the weight of the body is the same as the work done on the body by any other external active force. It is easier to understand that work done on a body is negative of the change of potential energy, through a gravitational force or weight. **Fig. 9.15** shows a body of mass m and weight $W = mg$ travelling through a small downwards displacement (from the datum plane), δh. Work done on the body, $\delta U = +W\delta h = mg\delta h$.

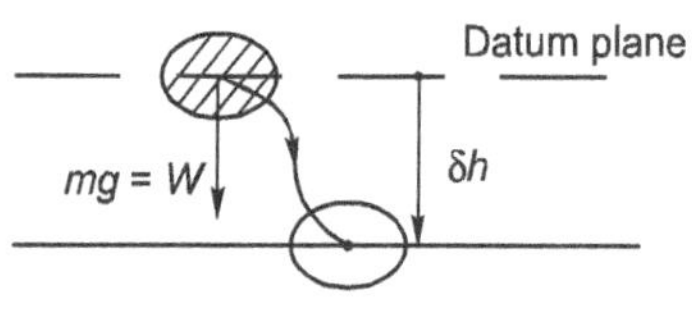

Fig. 9.15

Say at the datum plane, potential energy = 0 when the body has been displaced downward through δh there will be loss of potential energy *i.e.*, change in potential energy is $-mg\delta h$

or $\delta V_g = -mg\delta h$ (change in potential energy due to gravity)

If we consider that the body of mass m or weight $W = mg$ moves upwards through a displacement δh then there will be increase in potential energy due to gravity, $\delta V_g = +mg.\delta h$, while the work done.

$$\delta U = -W.\delta h = -mg.\delta h$$

(because W is downwards and displacement is upwards as shown in **Fig. 9.16**).

So it can be concluded that change in potential energy due to gravity also is negative of the work done on the body.

Moreover the gravitational potential energy is independent of the path followed in going to the new position. The path (for going up or for coming down) can be straight or curved as shown in **Figs. 9.15** and **9.16**).

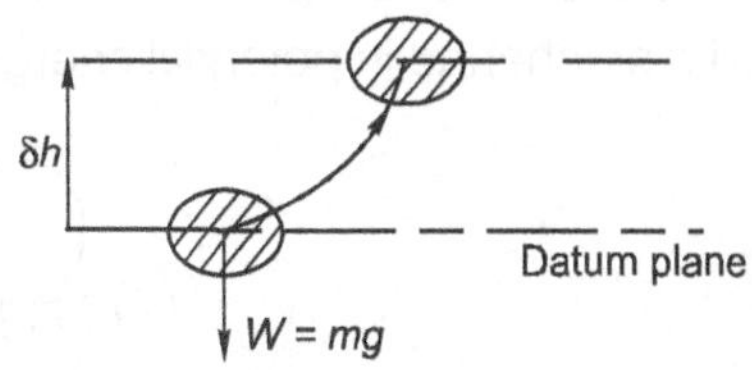

Fig. 9.16

So $\delta U = -\delta V_e$ and $\delta U = -\delta V_g$

Total work done on the body,

$$\delta U' = -(\delta V_e + \delta V_g)$$
$$= -\delta V$$

where $V = V_e + V_g$, total potential energy of the system.

Say a body or a particle of mass m is being guided in a fixed path with the help of force F as shown in **Fig. 9.17**. Spring force on the body is kx where x is the displacement in spring producing restoring force in spring equal to kx. The body is in equilibrium under the action of force F (an external active force), spring restoring force kx and weight of body mg. *The normal reaction on the particle from the smooth fixed path is omitted because it does no work.* F is the external active force which does work on the body through a virtual displacement δx. The work δV_g due the weight mg and work δV_e due to spring force kx give the change in potential energy of the system.

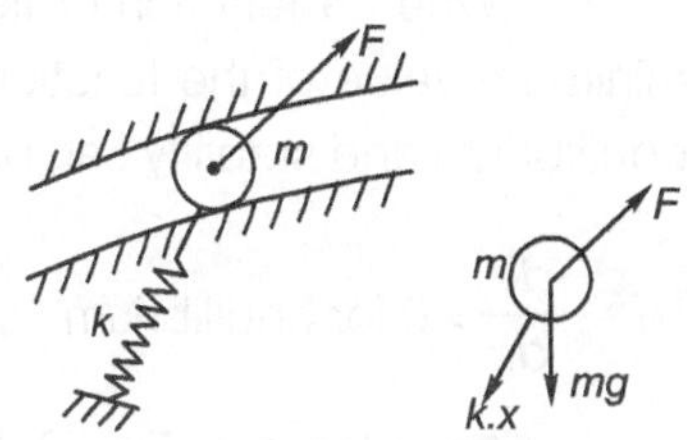

Fig. 9.17

Therefore for such mechanical systems which include springs as elastic members, principle of virtual work can be summarized as follows.

Virtual work done by all external active forces on a mechanical system in equilibrium = – (Total change in elastic and gravitational potential energy) for any or all virtual displacements consistent with the constraints.

9.5.3 Stability of Equilibrium

During a virtual displacement or virtual displacements, no work is done on a mechanical system *i.e.*, $\delta U = 0$, total work done due to external forces is zero. Therefore $\delta V = \delta\,(V_e + V_g)$ total change in potential energy due to elastic force in member and due to gravitational forces) will also be zero

or $\delta V = 0$ for equilibrium

In other words for equilibrium configuration of a mechanical system, total potential energy of the system is stationary. If the system has single degree of freedom, derivatives of V are continuous functions of a single variable say x *i.e.*, $\delta V = 0$, or mathematically

$$\frac{dV}{dx} = 0 \qquad ...(1)$$

If a system has several degrees of freedom, the partial derivative of total potential energy V, with respect to each co-ordinate must be zero for equilibrium.

There are three conditions under which Equation (1) is applicable, as follows:

(*i*) Total potential energy is minimum, a stable equilibrium is obtained

(*ii*) Total potential energy is maximum, an unstable equilibrium.

(*iii*) Total potential energy is constant, a neutral equilibrium is observed.

These three conditions are shown in **Fig. 9.18** through a roller on curved path and straight path.

Note that a small displacement away from the stable equilibrium position **Fig. 9.18 (*a*)** results in an increase in potential energy *V* and a tendency to return to the position of lower energy. While, a small displacement away from the unstable position as shown in **Fig. 9.18 (*b*)** results in a decrease in potential energy *V* and a tendency for the body to move away from the equilibrium position to the position of still lower energy. For the neutral equilibrium position, a small displacement in one way or the other results in a new equilibrium position, with no change in potential energy [**Fig. 9.18 (*c*)**].

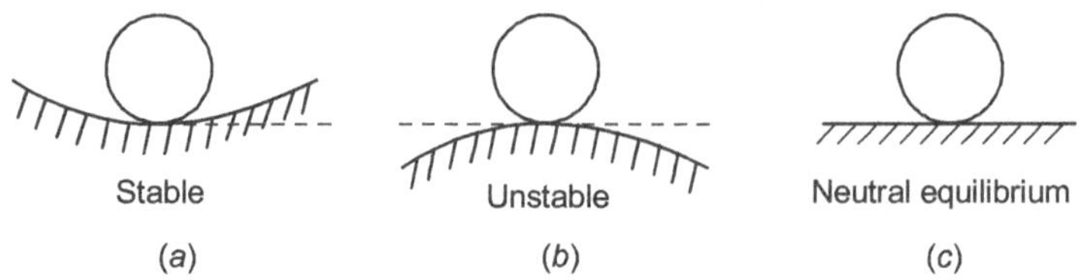

Fig. 9.18

When a function or its derivatives are continuous, the second derivative is (*i*) positive at a point of minimum value of the function (*ii*) negative at a point of maximum value of the function. So mathematically equilibrium and stability can be expressed as

$$\frac{dV}{dx} = 0 \text{ for equilibrium}, \quad \frac{d^2V}{dx^2} > 0 \text{ for stable equilibrium}, \quad \frac{d^2V}{dx^2} < 0 \text{ for unstable equilibrium}$$

Exercise 9.3 **Fig. 9.19** shows the cross-section of a uniform 50 kg ventilator door hinged at edge *O*. The door is controlled by a spring loaded cable which passes over a small pulley at *A*. The spring has a stiffness of 150 N per metre of stretch. Spring is undeformed when $\theta = 0°$. Determine the magnitude of angle θ for equilibrium.

[**Hint:** $x = AB = \text{Extension in spring} = 2 \times 1.2 \sin\frac{\theta}{2}$

$$V_e = \frac{1}{2}kx^2, \quad V_g = -0.6mg \times \sin\theta$$

$$V = V_e + V_g]$$

[**Ans:** $\theta = 53.7°$].

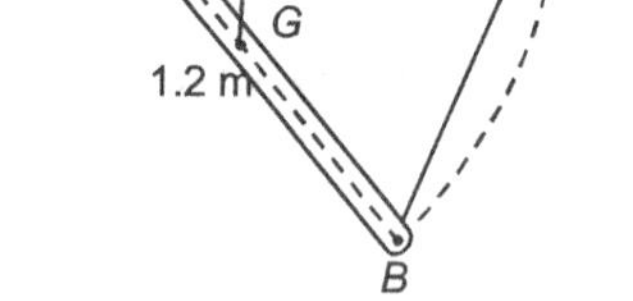

Fig. 9.19

9.6 Multiple Degrees of Freedom

Let us say that a system has multiple degrees of freedom $\theta, \phi, \alpha, \beta$ etc. Then principle of virtual work for such a system can be written as

$$\delta U = \frac{\partial U}{\partial \theta} \cdot \delta\theta + \frac{\partial U}{\partial \phi} \cdot \delta\phi + \frac{\partial U}{\partial \alpha} \cdot \delta\alpha + \frac{\partial U}{\partial \beta} \cdot \delta\beta + \ldots$$

Consider a double pendulum shown in **Fig. 9.20**, with two degrees of freedom θ and ϕ, which can vary independently. Two links *AB* and *BC* are pinned smoothly together and held in position by a horizontal force *P* as shown in **Fig. 9.20**. The weight of link *AB* is 2*W* while weight of link *BC* is *W*.

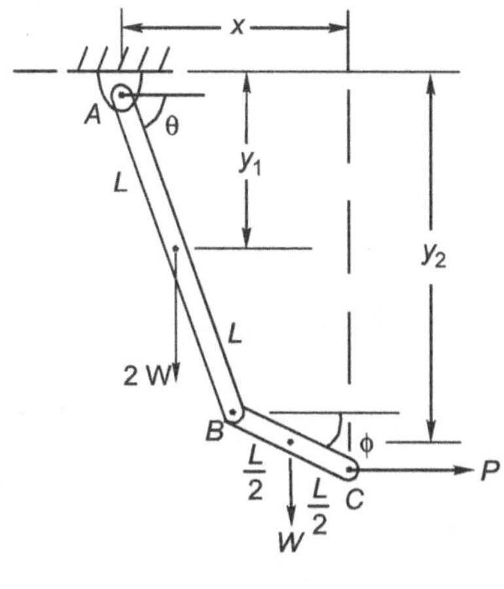

Fig. 9.20

Let us determine equilibrium position of this double pendulum.

The positions of active forces, 2 W, W and P are

$$y_1 = L\sin\theta$$

$$y_2 = 2L\sin\theta + \frac{L}{2}\sin\phi$$

$$x = 2L\cos\theta + L\cos\phi$$

Keeping θ constant and changing angle ϕ gives expression (for link BC)

$$\delta U = W.\delta y_2 + P.\delta x = 0$$

where $\delta y_2 = \frac{L}{2}\cos\phi\,\delta\phi$, as θ is constant

$\delta x = -L\sin\phi\ \delta\phi$ as θ is consant

So $W \times \frac{L}{2}\cos\phi\,\delta\phi - L\sin\phi\ \delta\phi.P = 0$

or $$\tan\phi = \frac{W}{2P} \quad ...(1)$$

Now keeping the angle ϕ constant and changing angle θ will give following equation

$$\delta U = 2W.\delta y_1 + W\delta y_2 + P.\delta x = 0$$

where $\delta y_1 = L\cos\theta\,\delta\theta$

$\delta y_2 = 2L\cos\theta.\delta\theta$

$\delta x = -2L\sin\theta.\delta\theta$

So $$\delta U = 2W\,[L\cos\theta\,\delta\theta] + W\,[2L\cos\theta\,\delta\theta] + P\,[-2L\sin\theta\,\delta\theta] = 0$$

or $2W\cos\theta + 2W\cos\theta - 2P\sin\theta = 0$

or $$\tan\theta = \frac{2W}{P}.$$

Example 9.2 A uniform ladder of length 7 m rests against a vertical wall with which it makes an angle of 45 degrees. The coefficient of friction between the ladder and the wall is 1/3, and between the ladder and the horizontal floor is 1/2. If a man weighing half as that of the ladder ascends it, how high will he be, when the ladder just starts to slip?

Solution:

Reaction at $A = R_A$

Reaction at $B = R_B$

Ladders starts to slip towards left at A, downward at B.

Friction force at $B = \frac{R_B}{3}(\uparrow)$

Friction force at $A = \frac{R_A}{2}$

Equilibrium

$$\frac{R_A}{2} = R_B$$

or $$R_A = 2R_B \quad ...(i)$$

$$W + \frac{W}{2} = R_A + \frac{R_B}{3}$$

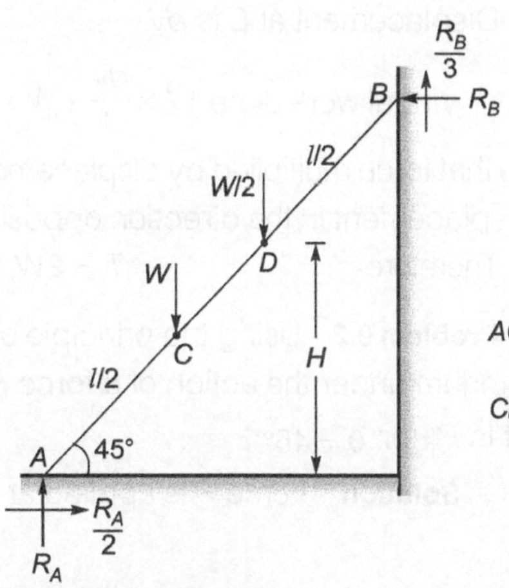

Fig. 9.21

$$1.5\text{ W} = R_A + \frac{R_A}{6} = \frac{7R_A}{6}$$

$$R_A = \frac{9}{7}W \quad \text{...(ii)}$$

$$R_B = \frac{R_A}{2} = \frac{9}{14}W \quad \text{...(iii)}$$

Moment about A

$$W \times 3.5\cos 45° + \frac{W}{2} \times H = W \times 3.5\cos 45° + \frac{W}{2} \times H$$

$$3.5W + \frac{WH}{2 \times 0.707} = \frac{4R_B}{3} \times 7 = \frac{4}{3} \times \frac{9}{14} W \times 7 = 6W$$

$$H = \frac{2.5W \times 1.414}{W} = 2.5\sqrt{2}\text{ m} = 3.535\text{ m}$$

$$AD = 5\text{ m (along the ladder)}$$

PROBLEMS

Problem 9.1 A square frame *ABCD* with a diagonal member *AC* is shown in **Fig. 9.22**. Weight of diagonal member *AC* is negligible while the weight of each bar *AB*, *BC*, *CD* and *DA* is *W* each. The square frame is suspended from joint *A*. Determine tension in diagonal member *AC* due to self weight of the frame.

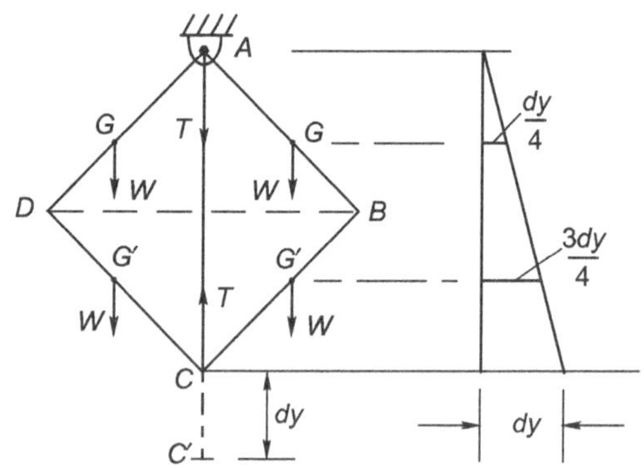

Fig. 9.22

Solution Let us give a virtual displacement $CC' = dy \downarrow$ at the point *C*. Due to symmetry we can say that:

(*i*) *CG* of members *AD* and *AB* moves down by $\frac{dy}{4}$.

(*ii*) *CG* of members *DC* and *BC* moves down by $\frac{3dy}{4}$.

(*iii*) Displacement at *A* is zero.

(*iv*) Displacement at *C* is *dy*.

Virtual work done $W \times \frac{dy}{4} + W \times \frac{dy}{4} + W \times \frac{3dy}{4} + W \times \frac{3dy}{4} - T \times 0 - T \times dy = 0.$

(Note that force multiplied by displacement in the direction of force gives positive virtual work and force multiplied by displacement in the direction opposite to the direction of force gives negative work).

Therefore $T = 2W$, tension in member *AC*.

Problem 9.2 Using the principle of virtual work, determine the magnitude of forces P_1 to hold the frame in equilibrium under the action of a force *P* applied on the frame as shown in **Fig. 9.23**, length of each link is 2*a*. What is P_1 for $\theta = 45°$?

Solution Force *P* is applied at joint *F*. Let us take *xy* coordinate with origin at *O* as shown in **Fig. 9.23**. Then

$x = 3a\cos\theta$ for the force *P*

$y = a\sin\theta$ for forces P_1 each

Then $\delta x = -3a\sin\theta\,\delta\theta$

$\delta y = +a\cos\theta.\delta\theta$

Using the principle of virtual work

$$P.\delta x + 2.P_1\delta y = 0$$

$$-P\times 3a\sin\theta.\delta\theta + 2P_1 a\cos\theta.\delta\theta = 0$$

or $$2P_1 = 3P\tan\theta$$

$$P_1 = 1.5P\tan\theta$$

If $$\theta = 45°,\ P_1 = 1.5P.$$

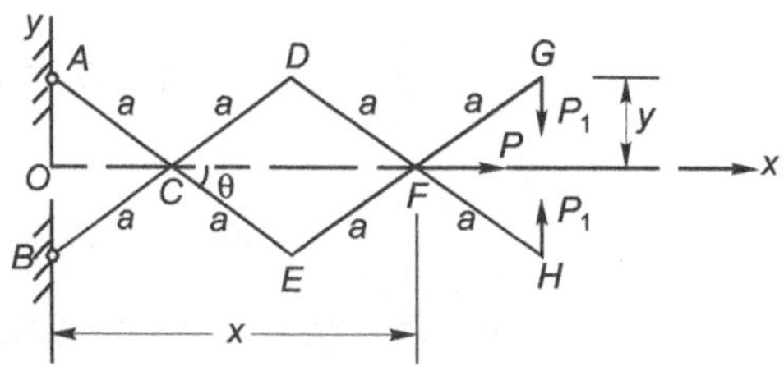

Fig. 9.23

Remember

- Method of virtual work can be used to determine unknown forces and moments in a system, without determining the reactions. As many number of active forces are determined as many independent equations are made using the principle of virtual displacements.
- Work done by a force is a scalar quantity and it is the dot product of a force vector and a displacement vector.
- Forces applied through fixed points do no work, such as (*i*) reaction at frictionless pin (*ii*) reaction at frictionless contacting surface (*iii*) reaction at roller support (*iv*) weight of the body if its *CG* moves horizontally (*v*) friction force acting on a wheel rolling without slipping.
- If the virtual displacement is along the direction of force, then positive virtual work is done, but if the virtual displacement is in the reverse direction to the direction of force applied then negative virtual work is done.
- If the moment applied on a body is clockwise and virtual angular displacement is anticlockwise then negative work is done. Similarly if the moment applied on a body is clockwise and virtual angular displacement is also in clockwise direction then positive virtual work is done.
- To determine reactions at supports in the case of beams carrying transverse loads, a virtual displacement is given to the beam at the support along the direction of reactions, then reactions are determined.
- Elastic potential energy of a spring,

$$V_e = \frac{1}{2}kx^2$$

where k = Stiffness of spring

x = Compression/extension in spring.

- A body is attached to a spring
  Work done on body = – (potential energy change of the spring).
- Gravitational potential energy, $V_g = mgh$ (h is the height of the CG of the body of mass m from datum).
- Total potential energy, $V = V_e + V_g$.
- Virtual work done by all external active forces on a mechanical system in equilibrium =–(total change in elastic and gravitational potential energy for any or all virtual displacements consistent with the constraints).
- $\frac{\partial V}{\partial x} = 0$ for equilibrium.

- $\frac{\partial^2 V}{\partial x^2} > 0$ for stable equilibrium.

  $\frac{\partial^2 V}{\partial x^2} < 0$ for unstable equilibrium.
- If a system has multiple degrees of freedom say θ, α, ϕ and U is the total potential energy then as per principle of virtual work.

  $$\partial U = \frac{\partial U}{\partial \theta}\cdot \delta\theta + \frac{\partial U}{\partial \alpha}\cdot \delta\alpha + \frac{\partial U}{\partial \phi}\cdot \delta\phi = 0$$

  For equilibrium, each individual term has to be zero independently.

PRACTICE PROBLEMS

9.1 Using the principle of virtual work, find the magnitude of horizontal force P applied at B of the bar is in equilibrium as shown in **Fig. 9.24.**

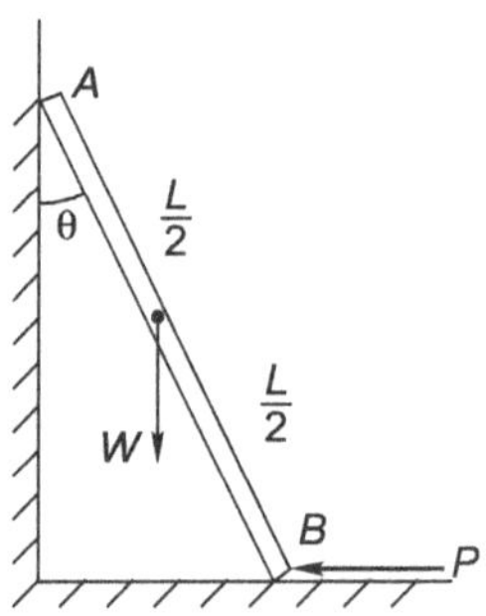

Fig. 9.24

[**Hint:** $x = L\sin\theta$, $y = \frac{L}{2}\cos\theta$].

[**Ans:** $\tan\theta = \frac{2P}{W}$].

9.2 A hexagon frame is made up of six bars of equal weight and equal length as shown in **Fig. 9.25**. Rod AF is fixed in a horizontal plane. A rod GH is fixed at the mid-points of rods AF and CD. If weight of each rod is W, then by using the principle of virtual work show that tension in member GH is equal to $3W$.

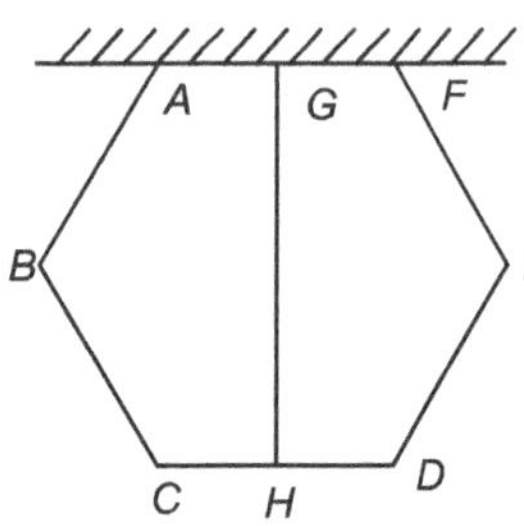

Fig. 9.25

[CSE, Prelims, CE : 2004]

9.3 Determine the horizontal reaction Q for the frame consisting of two rhombuses of sides 750 mm and 500 mm each as shown in **Fig. 9.26**. The frame carries two loads P as shown. If $\theta = 45°$, then what is the reaction Q?

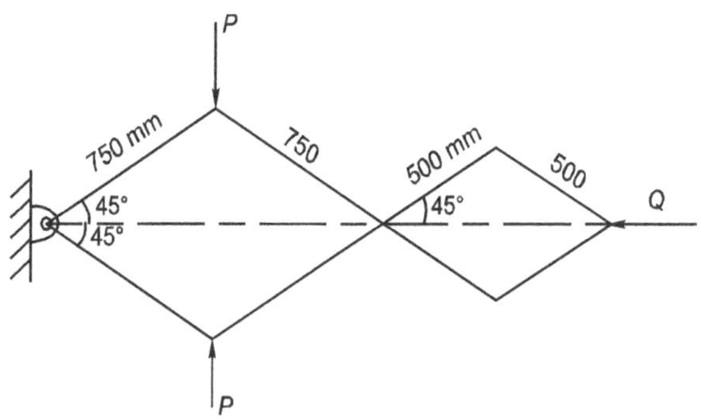

Fig. 9.26

[**Ans:** $Q = 0.6P$].

9.4 A beam ABC, 2 m long is held in equilibrium by the application of a force P as shown in **Fig. 9.27**. Using the principle of virtual work, find the magnitude of the force P when a weight of 4 kN is hung from the beam ABC at its mid-point.
[**Ans:** $P = 3.11$ kN].

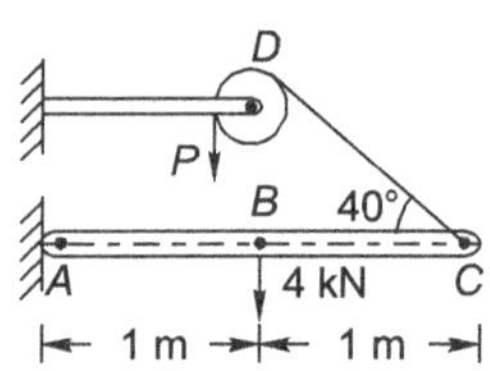

Fig. 9.27

9.5 **Figure 9.28**shows two hinged uniform bars AB and BC *of* having mass m and length L each, which is supported and loaded as shown. For a given value of P determine angle θ for equilibrium.

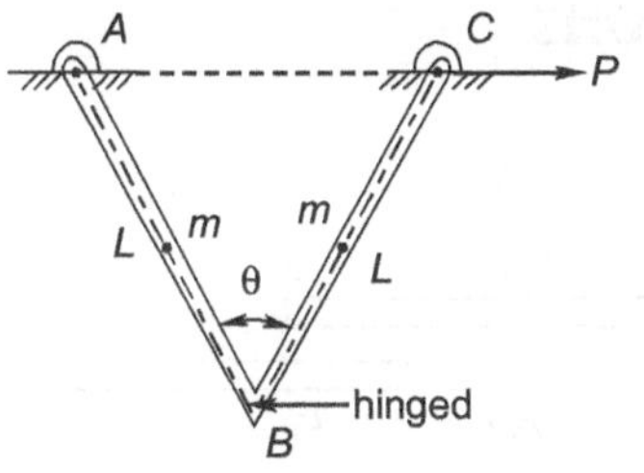

Fig. 9.28

[**Ans:** $\theta = 2\tan^{-1}\dfrac{2P}{mg}$].

9.6 Five bars *AB*, *BC*, *CD*, *DA* and *BD* each of equal length *L* and equal cross sectional area '*a*' are pin jointed so as to form a plane frame *ABCD* with a diagonal member *BD*. The frame is suspended from the joint *A* and a weight *W* is attached at the lower joint *C* as shown in **Fig. 9.29**. Neglecting self weight of the bars, determine magnitude of thrust in bar *BD*, using the principle of virtual work.

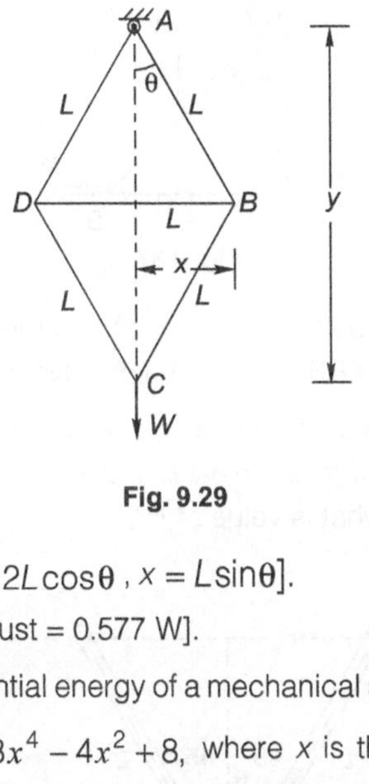

Fig. 9.29

[**Hint:** $y = 2L\cos\theta$, $x = L\sin\theta$].

[**Ans:** Thrust = 0.577 W].

9.7 The potential energy of a mechanical system is given by $V = 8x^4 - 4x^2 + 8$, where *x* is the position co-ordinate defining the configuration of a single degree of freedom system. Determine the values of *x* and stability condition of each.

[**Hint:** For equilibrium $\dfrac{dV}{dx} = 0$, for stability $\dfrac{d^2V}{dx^2} > 0$].

[**Ans:** $x = 0, \pm\dfrac{1}{2}$ for equilibrium, $x = 0$, unstable, $x = \pm\dfrac{1}{2}$, stable].

9.8 A simple truss consisting of 3 equilateral triangles is shown in **Fig. 9.30**. Using the principle of virtual work, determine the force in the top member *BC* of the truss.

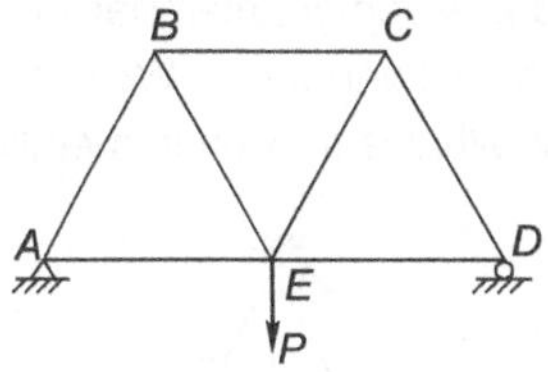

Fig. 9.30

[**Ans:** 0.577*P*].

9.9 Four bars of equal length *L* each are joined together at their ends in the form of a rhombus as shown in **Fig. 9.31**. Using the principle of virtual work, find the relation between forces F_1 and F_2 for equilibrium of the system in any configuration as defined by angle θ. Neglect friction at joints and weight of the bars.

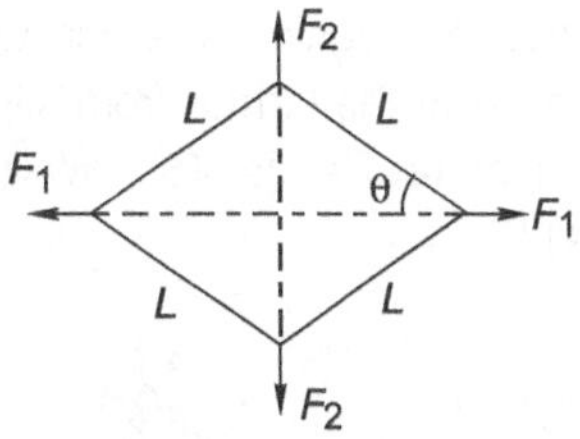

Fig. 9.31

[**Ans:** $F_2 = F_1\tan\theta$].

9.10 A uniform ladder of length *L* and weight 150 N, rests on a smooth floor at *B* and against a smooth vertical wall at *A* as shown in **Fig. 9.32**. A horizontal rope *CD* prevents the ladder from slipping. Rope is connected at the centre of the ladder. Using principle of virtual work, determine tension in the rope. Angle $\alpha = 30°$.

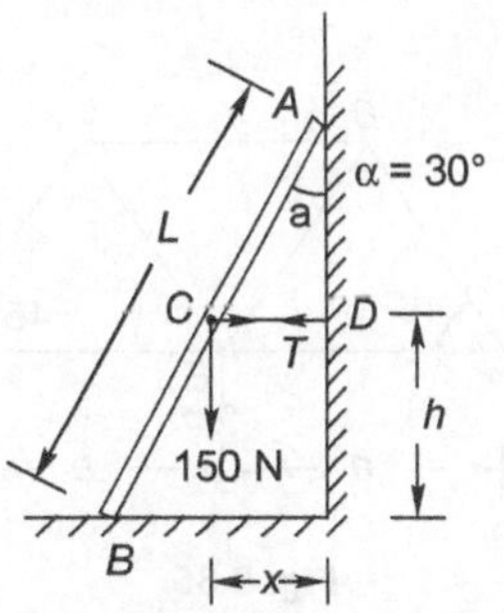

Fig. 9.32

[**Hint:** $x = \dfrac{L}{2}\sin\alpha$, $h = \dfrac{L}{2}\cos\alpha$].

[**Ans:** 86.55 N].

MULTIPLE CHOICE QUESTIONS

9.1 An equilateral triangular frame *ABC* of side *L* is shown in **Fig. 9.33**. A vertical member of *AD* of negligible weight is also connected. Weight of each side of frame is *W*. What is tension in member *AD*?

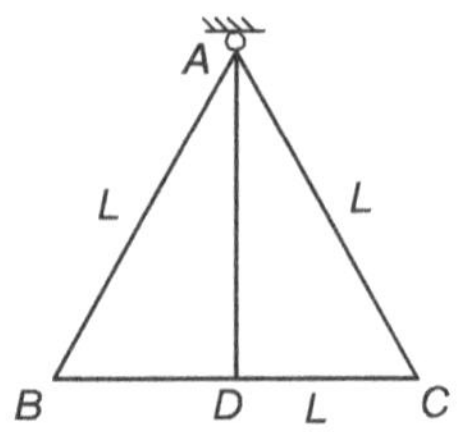

Fig. 9.33

(*a*) 3*W* (*b*) 2*W*
(*c*) 1.5*W* (*d*) None of these

9.2 A uniform ladder of length *L* and weight 100 N rest on a smooth floor and against a vertical wall. A horizontal rope *CD* prevent the ladder from slipping. Angle of inclination of ladder is 45°. What is tension *T* **(Fig. 9.34)**?

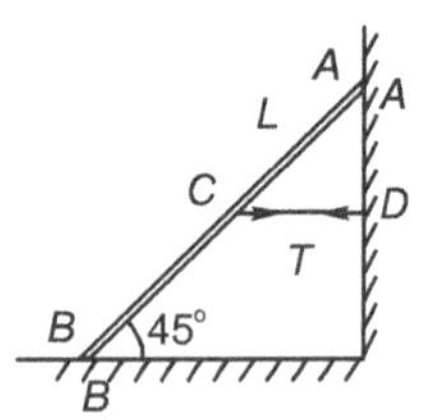

Fig. 9.34

(*a*) 70.7 N (*b*) 80.6 N
(*c*) 100 N (*d*) 141.4 N

9.3 A simple truss *ABCDE*, containing of 3 triangles is shown in **Fig. 9.35**. What is force in top member *BC*?

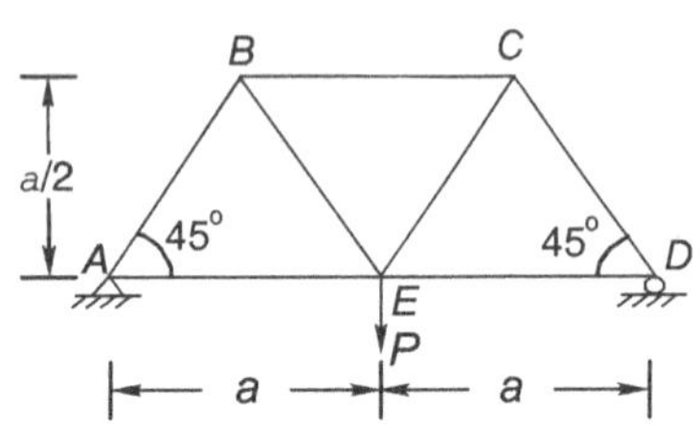

Fig. 9.35

(*a*) 0.5*P* (*b*) 0.707*P*
(*c*) *P* (*d*) 1.414*P*

9.4 Beams *AB* and *DF* carry load *W* as shown in **Fig. 9.36**. What is reaction at *F*?

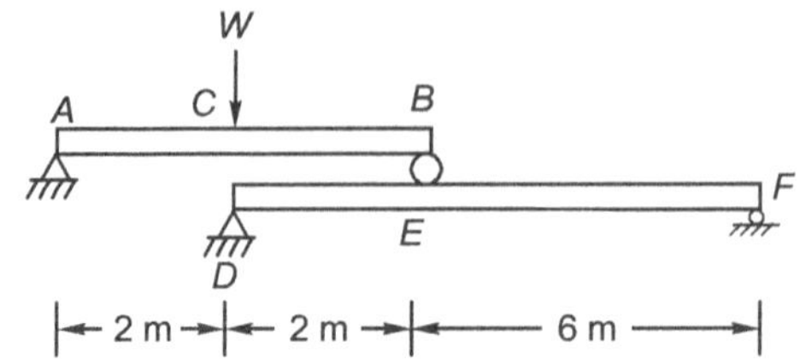

Fig. 9.36

(*a*) $\frac{W}{2}$ (*b*) $\frac{W}{4}$
(*c*) $\frac{W}{8}$ (*d*) $\frac{W}{16}$

9.5 A ladder of length *L*, weight *W* rests against a vertical wall and a smooth floor. A horizontal force *P* is applied on ladder to maintain its equilibrium. What is *P* in terms of *W* **(Fig. 9.37)**?

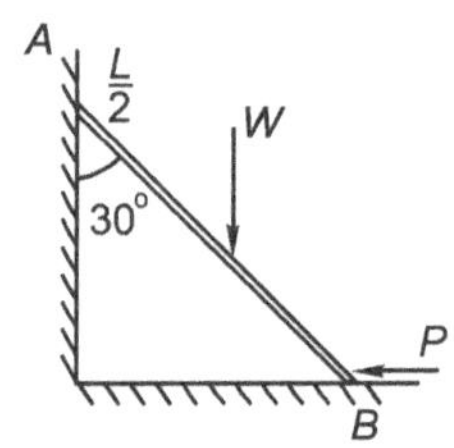

Fig. 9.37

(*a*) 0.29*W* (*b*) 0.5*W*
(*c*) 0.866*W* (*d*) None of these

9.6 **Figure 9.38** shows a two hinged uniform bars *AB* and *BC* of mass *m* each and length *L*. If θ = 90°, as shown, what is value of *P*?

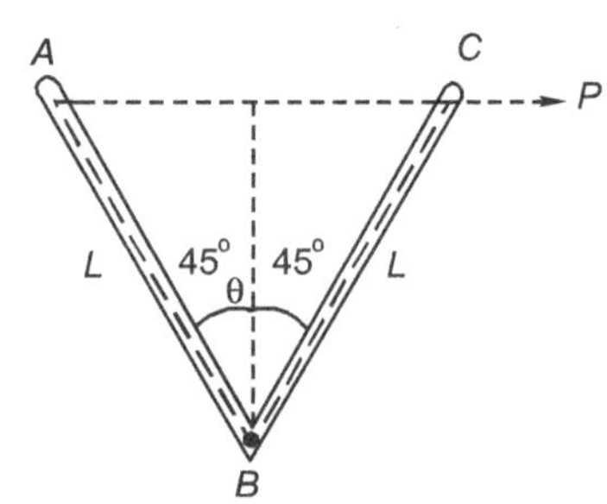

Fig. 9.38

(*a*) 1.0 mg (*b*) 0.5 mg
(*c*) 0.25 mg (*d*) None of these

9.7 A beam *AB*, 2 m long, uniform in cross-section is held in equilibrium by the application of a force *P* as shown in **Fig. 9.39**. The self weight of bar is 100 N. What is the magnitude of *P*?

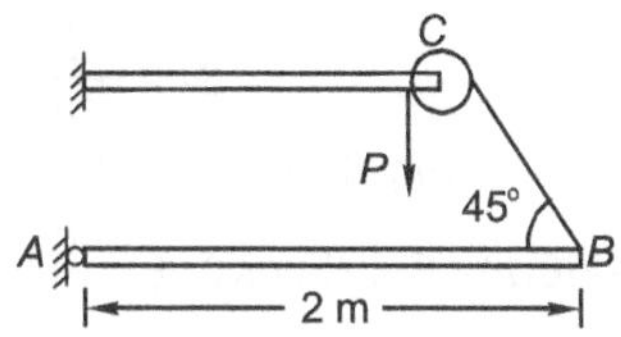

Fig. 9.39

(*a*) 141.4 N (*b*) 100 N
(*c*) 70.7 N (*d*) 50 N

9.8 Determine the horizontal reaction Q from the frame consisting of a rhombous of sides 0.5 m as shown. Frame carries two loads P as shown in **Fig. 9.40.**

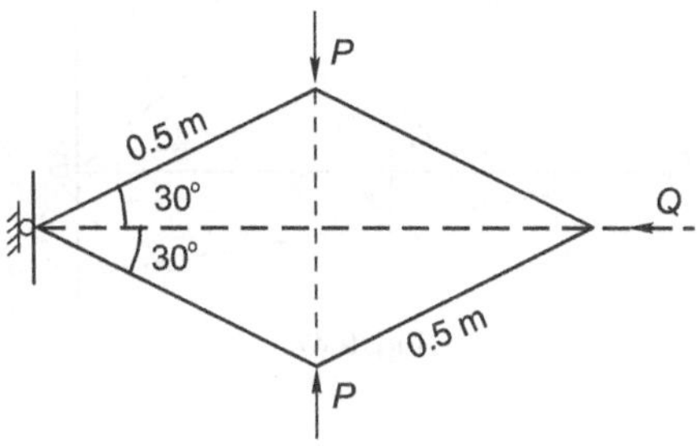

Fig. 9.40

(*a*) P (*b*) $1.5P$
(*c*) $1.377P$ (*d*) $1.732P$

9.9 The principle of virtual work states that the virtual work is zero for
(*a*) a body moving with constant linear velocity
(*b*) a body rotating with constant angular velocity
(*c*) a body in equlibrium
(*d*) a body moving with constant linear acceleration
[CSE, Prelim, CE : 2003]

9.10 A number of rods are hinged together to form three identical rhombous as shown in the **Fig. 9.41**. If a horizontal force of 1000 N is applied at A, what is the force P required at B for equilibrium?

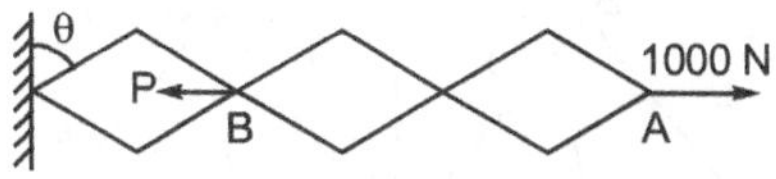

Fig. 9.41

(*a*) 3000 N (*b*) 1000 N
(*c*) 4000 N (*d*) 2000 N
[CSE, Prelim, CE : 2009]

9.11 If a system is in equilibrium and the position of the system depends upon many variables, the principle of virtual work states that the partial derivatives of its total potential energy with respect to each of the independent variable must be
(*a*) –1.0 (*b*) 0
(*c*) 1.0 (*d*) ∞
[GATE, 2006 : 2 Marks]

Answers

9.1 (b)	9.2 (c)	9.3 (c)	9.4 (c)	9.5 (a)
9.6 (b)	9.7 (c)	9.8 (d)	9.9 (c)	9.10 (a)
9.11 (b)				

EXPLANATIONS

9.1 (b)

Say CG of BC moves down by δy then CG of AB and AC will move down by $\delta y/2$ (**Fig. 9.42**).

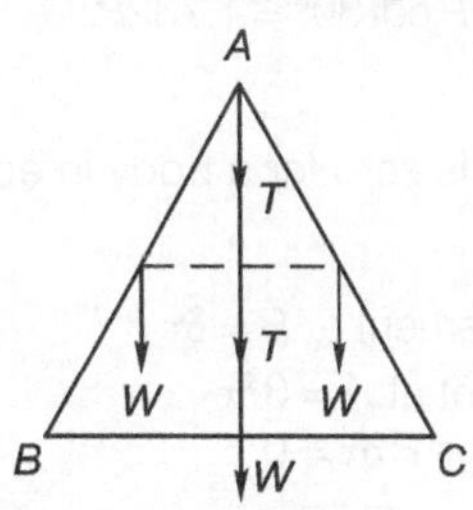

Fig. 9.42

Work done $= 2 \times W \times \frac{\partial y}{2} + W.\delta y$

$= 2W\delta y$

$2W.\delta y - T.\delta y = 0,\ T = 2W$

9.2 (c)

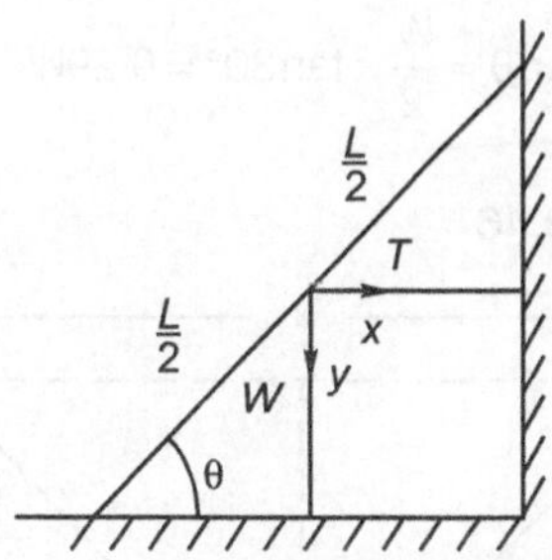

Fig. 9.43

(Fig. 9.43) $x = \frac{L}{2}\cos\theta$

$y = \frac{L}{2}\sin\theta$

$\delta x = -\frac{L}{2}\sin\theta.d\theta,\ \delta y = \frac{L}{2}\cos\theta.d\theta$

$$W.\delta y + T.\delta x = 0$$

$$W\cdot\frac{L}{2}\cos\theta.d\theta = \frac{L}{2}\sin\theta.d\theta.T$$

or $$T = W\cot\theta \quad \text{but } \theta = 45^\circ$$

$$T = W = 100 \text{ N}.$$

9.3 (c)

Say length of members (**Fig. 9.44**)

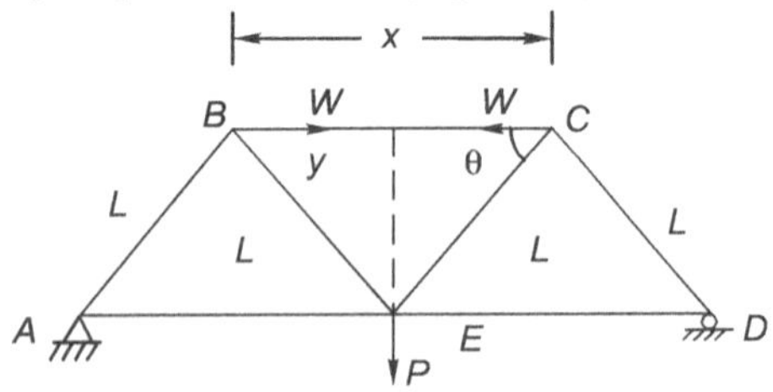

Fig. 9.44

$$AB = BE = EC = CD = L$$

$$y = L\sin\theta\,.\; x = L\cos\theta$$

$$dy = L.\cos\theta\,.\,d\theta, \quad dx$$

$$= -L\sin\theta\,.\,d\theta$$

Say force in top member is W

$$Pdy + W\delta y = 0$$

$$y = L\sin\theta,\; x = L\cos\theta$$

$$PL\cos\theta\, d\theta + W(-L\sin\theta)d\theta = 0$$

$$W = P\cot\theta = P, \text{as } \theta = 45^\circ$$

9.4 (c)

$$R_F = \frac{W}{2}\times\frac{2}{8} = \frac{W}{8}.$$

9.5 (a)

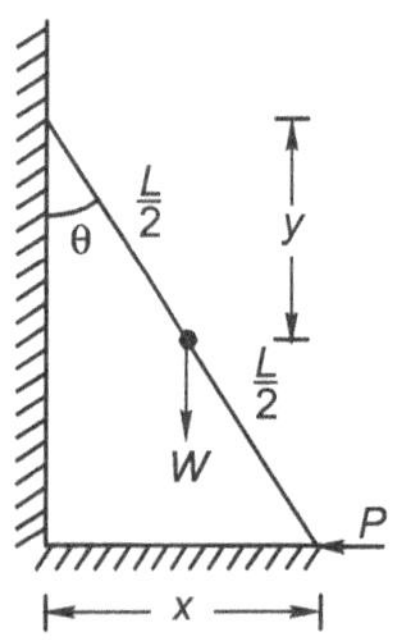

Fig. 9.45

$$y = \frac{L}{2}\cos\theta$$

$$x = L\sin\theta$$

$$\delta y = -\frac{L}{2}\sin\theta.d\theta$$

$$\delta x = L\cos\theta.d\theta$$

$$L\cos\theta.d\theta.P - \frac{L}{2}\sin\theta.d\theta.W = 0$$

$$P = \frac{W}{2}\times\tan\theta = \frac{W}{2}\times\tan 30^\circ = 0.29W.$$

9.6 (b)

See **Fig. 9.46**

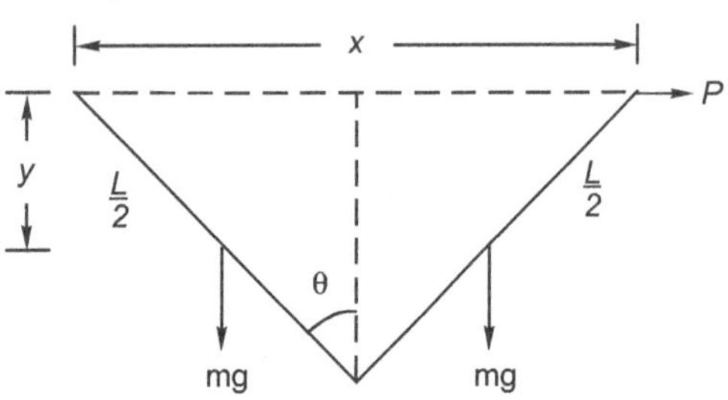

Fig. 9.46

$$x = 2L\sin\theta$$

$$y = \frac{L}{2}\cos\theta$$

$$\delta x = 2L\cos\theta.d\theta$$

$$\delta y = -\frac{L}{2}\sin\theta.d\theta$$

$$2mg.\delta y + P\delta x = 0$$

$$-2mg\frac{L}{2}\sin\theta\, d\theta + P\times 2L\cos\theta\, d\theta = 0$$

$$mg\sin\theta + 2P\cos\theta = 0$$

$$P = \frac{mg}{2}\times\tan\theta = \frac{mg}{2}, \text{ as } \theta = 45^\circ.$$

9.7 (c)

Self weight 100 N movement of G by y movement of P by $2\times 0.707y$ (**Fig. 9.47**)

$$100\,.\,y = 1.414y\,.\,P$$

$$P = 70.7 \text{ N}.$$

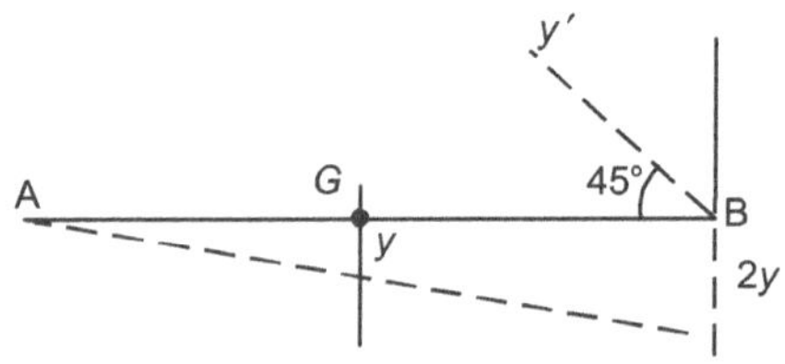

Fig. 9.47

9.8 (d)

See **Fig. 9.48**

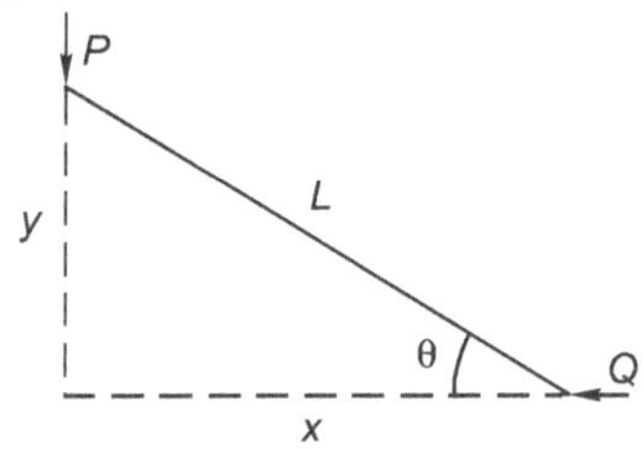

Fig. 9.48

$$x = L\cos\theta$$

$$y = L\sin\theta$$

$$\delta x = -L\sin\theta.d\theta$$

$$\delta y = +L\cos\theta.d\theta$$

$$+L\sin\theta.d\theta Q = L\cos\theta d\theta.P$$

$$\theta = P\cot\theta, \quad P\cot 30^\circ = 1.732P.$$

9.9 (c)

Virtual work is zero for a body in equlibrium.

9.10 (a)

Say displacement at $B = \delta x$

Displacement at $A = 3\delta x$

$1000\times(3\delta x) + P\,\delta x = 0$

$P = -3000 \text{ N} = \overleftarrow{3000}$

9.11 (b)

For equilibrium, partial derivation of its total potential energy with respect to each of the indisplacement variables must be zero.

■■■■

10

CHAPTER

Rectilinear Motion

10.1 Introduction

Straight line motion as motion of a train on a straight track, motion of a body under gravity in a straight vertical direction or motion of an automobile on a straight road are examples of rectilinear motion. In this type of motion particles of a body travel along parallel lines and the direction of the lines remains the same throughout the motion. There are two types of motion of translation *i.e.*, (*a*) rectilinear translation, (*b*) curvilinear translation as shown in **Figs. 10.1 (*a*)** and **(*b*)**. In rectilinear translation all the particles of a body travel along straight line paths and in *curvilinear translation*, all the particles of the body travel along *curved paths parallel to each other*. In the figure below, $v_A = v_B$, velocity of particles *A* and *B* is the same, while there is no relative displacement between particles *A* and *B* or relative velocity $v_{A/B} = v_{B/A} = 0$ *i.e.*, relative velocity of particle *A* with respect to particle *B* is zero, or relative velocity of particle *B* with respect to particle *A* is zero.

$$v_A = v_B$$

$$\frac{\partial v_A}{\partial t} = \frac{\partial v_B}{\partial t}, \text{ or acceleration, } a_A = a_B, \text{ same acceleration}$$

(*a*) (*b*)

Fig. 10.1

Acceleration of the particles also remains the same. *We can draw a conclusion for a body in rectilinear translation, i.e., all the points on the body travel with same velocity and same acceleration.*

If *S* is the displacement, *v* is the velocity and *a* is the acceleration, then with respect to time *t*

$$\frac{dS}{dt} = v_a = \text{average velocity}$$

$$\frac{\partial v_a}{\partial t} = \frac{\partial^2 S}{\partial t^2} = a_a, \text{ average acceleration.}$$

Fig. 10.2 (*a*) shows variation of displacement *S* with respect to time *t* for a particular rectilinear motion.

Average velocity at any instant, $v_{av} = \frac{\Delta S}{\Delta t}$, as shown.

Fig. 10.2 (*b*) shows variation of velocity v with respect to time t, for a particular rectilinear motion.

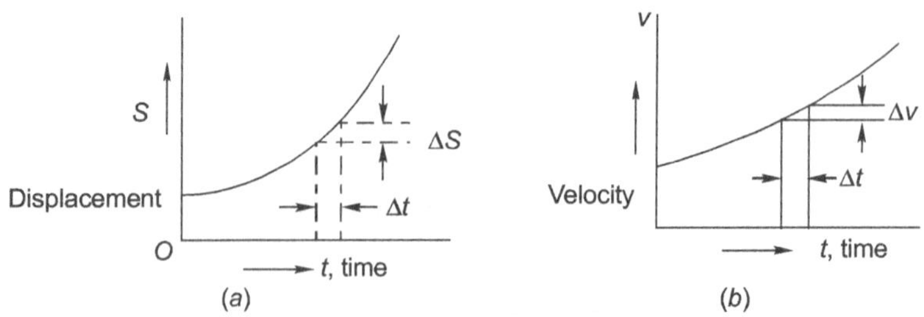

Fig. 10.2

Average acceleration of any instant, $a_{av} = \frac{\Delta v}{\Delta t}$, as shown

or in limits velocity $\quad v = \frac{dS}{dt}$

acceleration, $\quad a = \frac{dv}{dt} = \frac{d^2S}{dt^2}$

In this chapter we will discuss four different cases of rectilinear translation *i.e.*,

(*i*) Acceleration remains constant such as a rigid body falling under gravity with constant acceleration due to gravity g.

(*ii*) Acceleration as a function of time t, *i.e.*, $a = f(t)$.

(*iii*) Acceleration as a function of velocity, *i.e.*, $a = f(v)$.

(*iv*) Acceleration as a function of displacement, *i.e.*, $a = f(S)$.

10.2 Rectilinear Motion with Constant Acceleration

Say, acceleration = a, constant

u = initial velocity

t = time

So, final velocity $\quad v = u + at \qquad$ ...(1)

Displacement, $\quad S = ut + \frac{1}{2}at^2$, in t seconds $\qquad$ ...(2)

Moreover $\quad 2aS = v^2 - u^2$, there are three velocities.

where u is initial velocity, v is final velocity, a is acceleration and t is time.

Example 10.1 A particle moves along a straight line path such that its displacement is $S = 6t^2 + 8t + 5$ where S is in metres and t in seconds. Determine initial and final velocities, initial and final acceleration at time, $t = 0$ and $t = 10$ seconds.

Solution $\quad S = 6t^2 + 8t + 5$

$$\frac{dS}{dt} = 12t + 8$$

$t = 0$, velocity, $\quad u = 8$ m/s

$t = 10$ s, velocity, $\quad v = 12 \times 10 + 8 = 128$ m/s

Acceleration, $\quad a = \frac{dv}{dt} = \frac{d^2S}{dt^2} = 12 \text{ m/s}^2$

Constant acceleration, initial acceleration is 12 m/s^2 and final acceleration at $t = 10$ s, is also 12 m/s^2.

Exercise 10.1 The displacement of particle is given by $S = 2t^2 - 3t + 4$ m, where time t is in seconds. Determine velocity at time $t = 0$, $t = 3$ s and acceleration at $t = 0$, $t = 3$ s. After what time from start, velocity becomes zero?

[**Ans:** $u = -3$ m/s, $v = 9$ m/s; acceleration = 4 m/s^2 constant; $v = 0$ at $t = 0.75$ s].

10.3 Motion Under Gravity

Acceleration provided by the earth's centre to a body is constant and is equal to $g = 9.81$ m/s^2, which is constant. When a body is falling under gravity say with initial velocity u, the relationships between velocity and displacement at any instant are:

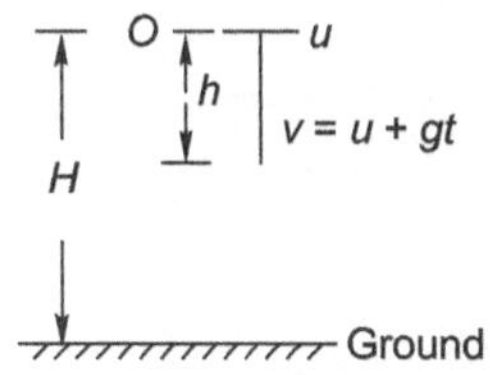

Fig. 10.3

$$v = u + gt \quad ...(1)$$

h = vertical distance covered

$$= ut + \frac{1}{2}gt^2, \quad ...(2)$$

$$v^2 - u^2 = 2gh \quad ...(3)$$

If $u = 0$ and height of the body from the ground is H, then from relation (3)

$$v^2 = 2gH$$

or

$$v = \sqrt{2gH}$$

Time taken by the body to touch the ground,

$$t = \frac{v}{g} = \frac{\sqrt{2gH}}{g} = \sqrt{\frac{2H}{g}}.$$

10.3.1 Motion against Gravity

If a body is projected upwards (against gravity, with initial velocity u then relationships will be as follows:

$$v = u - gt$$

$$h = ut - \frac{1}{2}gt^2$$

$$v^2 - u^2 = -2gh$$

Time when velocity becomes zero,

i.e., $v = 0$,

$$t = \frac{u}{g}$$

$$-u^2 = -2gH$$

$$H = \frac{u^2}{2g},$$

height through which the body goes up and then comes down under gravity.

Example 10.2 A ball is thrown upwards with a velocity of 30 m/s, from the elevation of an elevator shaft at a height of 20 m from the ground. At the same time, the elevator moves upwards with a constant velocity of 3 m/s from a height of 6 m from the ground. If $g = 9.81$ m/s^2, determine time when the ball will hit the elevator.

Solution Fig. 10.4 shows a ball being projected upwards at an initial velocity of 30 m/s from a height of 20 m from the ground and at the same time another elevator is moving up with a constant velocity of 3 m/s. At any instant t, height of the ball and elevator from the ground will be

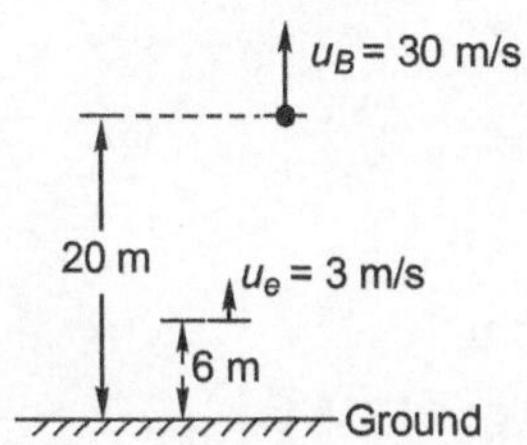

Fig. 10.4

$$h_B = 20 + u_B.t - \frac{1}{2}gt^2 \text{ (against gravity)}$$

$$u_B = 30 \text{ m/s}$$

$$h_B = 20 + 30t - \frac{1}{2} \times 9.81t^2$$

h_E = height of the elevator from the ground, which is moving with constant velocity of 3 m/s

$$h_E = 6 + 3 \,.\, t$$

(*a*) Time when the ball hits the elevator

$$h_B = h_E$$

$$20 + 30t - 4.905t^2 = 6 + 3t$$

$$14 + 27t - 4.905t^2 = 0$$

$$4.905t^2 - 27t - 14 = 0$$

$$t = \frac{27 \pm \sqrt{27^2 + 4 \times 14 \times 4.905}}{2 \times 4.905}$$

$$= \frac{27 \pm \sqrt{729 + 274.68}}{9.81} = \frac{27 + 31.68}{9.81} = 5.98 \text{ s}$$

Exercise 10.2 A stone is projected upwards with a velocity of 14.715 m/s from the top of a tower two seconds later a second stone is dropped from the same position and the two stones reach the ground together. Neglecting air resistance, find height of the tower, $g = 9.81$ m/s^2.

[**Hint:** $t_1 = \sqrt{\frac{2H}{g}}, t_2 = t_1 + 2; -H = 14.715\,(t_2) - \frac{1}{2} g \times (t_2)^2$].

[**Ans:** 19.62 m].

10.4 Acceleration as a Function of Time

Say acceleration,

$$a = f(t)$$

or

$$\frac{dv}{dt} = f(t)$$

$$\int_u^v dv = \int_0^t f(t)dt$$

or

$$v = u + \int_0^t f(t)dt \qquad ...(1)$$

Knowing the value of v from Equation (1)

$$v = \frac{dS}{dt}$$

Displacement,

$$dS = v\, dt$$

$$\int_{S_0}^{S} dS = \int_0^t v dt$$

$$S = S_0 + \int_0^t v.dt$$

where S_0 is the displacement at time $t = 0$.

Example 10.3 The acceleration of a particle is given by $a = 5t - 20$, where a is in m/s^2 and t in seconds. Determine velocity as a function of time. Initial velocity $u = 5$ m/s. When the velocity of the particle will be zero.

Solution Acceleration, $a = 5t - 20$ m/s^2

or
$$\frac{dv}{dt} = 5t - 20$$

$$\int_u^v dv = \int_0^t (5t - 20)dt$$

$$v = u + \frac{5t^2}{2} - 20t \text{, where } u \text{ is initial velocity}$$

Now initial velocity, $u = 5$ m/s

Velocity at any instant, $v = 5 + 2.5t^2 - 20t$ expression for velocity

time when velocity is zero,

$$v = 5 + 2.5t^2 - 20t = 0$$

or
$$t^2 - 8t + 2 = 0$$

$$t = \frac{8 \pm \sqrt{64 - 8}}{2} = \frac{8 \pm \sqrt{56}}{2}$$

$$= 0.25853,\ 7.74 \text{ second.}$$

Exercise 10.3 The motion of a particle is given by acceleration, $a = t^3 - 2t^2 + 5$, where a is in m/s^2 and t in seconds. The velocity is 3.8 m/s at $t = 1$ s. Calculate the velocity at $t = 3$ second.

[**Ans:** $v_3 = 10.4667$ m/s, $a_3 = 14$ m/s^2].

PROBLEMS

Problem 10.1 A body A starts from rest and travels along a straight path from origin O, with an acceleration of 3 m/s^2. Another body B starts from rest from O but after 4 seconds the body A has started. Body B starts with an acceleration of 4 m/s^2. How far from O, will the bodies A and B will be, when the body B overtakes body A (Fig. 10.5)?

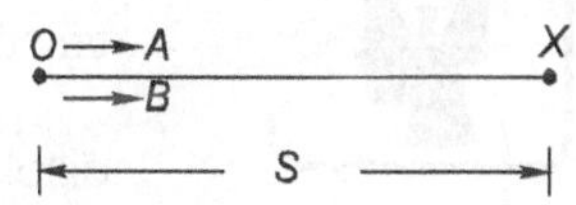

Fig. 10.5

Solution Say the time taken by body A to reach the point X, is t, then the time taken by body B will be $(t - 4)$ seconds to reach the same destination.

$$a_A = 3 \text{ m/s}^2,\ u_A = 0$$
$$a_B = 4 \text{ m/s}^2,\ u_B = 0$$

So
$$S = \frac{1}{2} a_A t^2 = \frac{3}{2} t^2 = 1.5t^2 \quad \text{...(1)}$$

$$S = \frac{1}{2}(a_B)(t-4)^2 = \frac{1}{2} \times 4\,(t-4)^2 = 2\,(t-4)^2 = 2t^2 - 16t + 32 \quad \text{...(2)}$$

From Equations (1) and (2),

$$1.5t^2 = 2t^2 - 16t + 32$$

or $0.5t^2 - 16t + 32 = 0$

or $t^2 - 32t + 64 = 0$

$$t = \frac{32 + \sqrt{32^2 - 4 \times 64}}{2} = \frac{32 \pm \sqrt{1024 - 256}}{2}$$

$$= \frac{32 \pm \sqrt{768}}{2} = \frac{32 - 27.713}{2} = 2.14 \text{ s} < 4 \text{ s, not admissible}$$

$$t = \frac{32+27.713}{2} = 29.8565 \text{ second}$$

Distance, $S = 1.5t^2 = 1.5 \times 29.856^2 = 1337.1$ m.

Problem 10.2 The velocity time diagram is a sine curve as shown in **Fig. 10.6**. Determine the distance travelled by the particle during the half cycle or time period $\frac{T}{2}$.

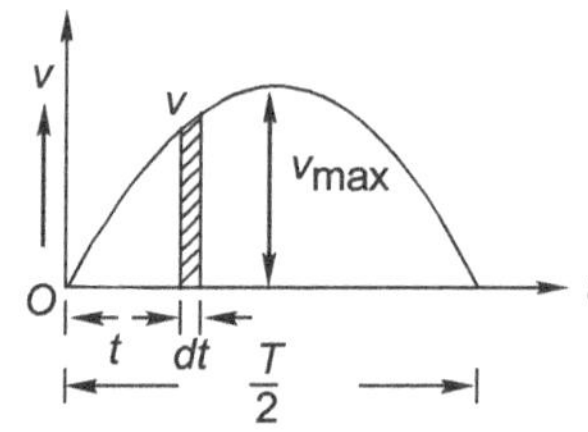

Fig. 10.6

Solution Velocity at any instant

$$v = v_{max} . \sin \frac{2\pi t}{T}$$

where T is time period of a sine curve v_{max} at $t = T/4$.

Consider the distance travelled through a small interval dt

$$dS = v\,dt = v_{max} . \sin \frac{2\pi t}{T} \cdot dt$$

or

$$\int ds = \int_0^{T/2} v_{max} \cdot \frac{\sin 2\pi t}{T} dt$$

$$S = v_{max} \frac{T}{2\pi} \left| -\cos \frac{2\pi t}{T} \right|_0^{T/2} = -v_{max} \cdot \frac{T}{2\pi} [\cos\pi - \cos 0°]$$

$$= -v_{max} \cdot \frac{T}{2\pi} [-1-1] = 2v_{max} \cdot \frac{T}{2\pi} = v_{max} \cdot \frac{T}{\pi}.$$

Remember

- In rectilinear translation, all the particles of a body travel along straight line paths.
- Uniform or constant acceleration

  velocity, $v = u + at$, where u is initial velocity, t is time

  displacement, $S = ut + \frac{1}{2} at^2$

  $v^2 - u^2 = 2aS$.
- Motion under gravity, acceleration $a = g$

  $v = u - gt$, motion against gravity

  $h = ut - \frac{1}{2} gt^2$

  $v^2 - u^2 = -2gh$.

PRACTICE PROBLEMS

10.1 A bus starts from rest from station A and accelerates at 0.5 m/s^2 until it reaches a speed of 10 m/s. It then proceeds at 10 m/s until the brakes are applied and it comes to rest at station B, 40 m beyond the point where the brakes were applied. Assuming uniform retardation determine the time required for the bus to travel from A to B, if distance between A and B is 350 m.

[**Hint:** $S_1 + S_2 + S_3 = 350$ m].

[**Ans:** 49 seconds].

10.2 The v-t diagram of a particle is given by a parabola of equation $v = at + bt^2$ as shown in **Fig. 10.7**. If $v_{max} = 4$ m/s and $T = 6$ s, determine distance travelled upto the time $t = \frac{T}{4}$.

[**Hint:** $a = 5.333$, $b = -1.7778$].

[**Ans:** 4 m].

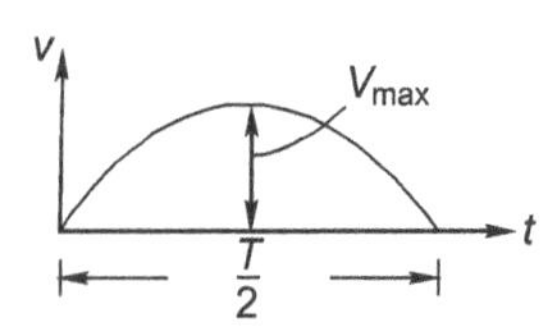

Fig. 10.7

10.3 For an aircraft, jet thrust remains constant or the acceleration remains constant during take off and can be taken equal to $g/3$. If the take off speed is 210 km/hour, calculate the distance and time from rest to take off the aircraft.
[Ans: 12.84 s, 520 m].

10.4 Determine the vertical velocity u with which a ball must be thrown upwards at A in order that the time for the flight to return to the bottom of the cliff is 3 seconds as shown in **Fig. 10.8**.

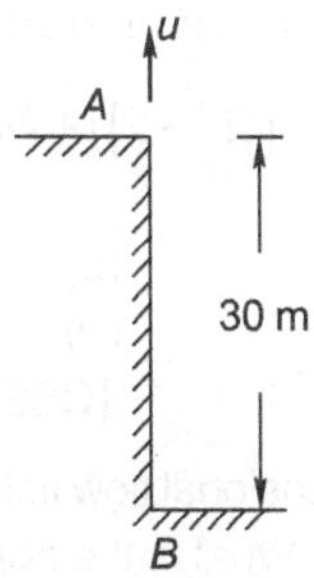

Fig. 10.8

[Ans: 4.715 m/s].

10.5 A particle has straight line motion as per equation $x = t^3 - 2t^2 - 4$ with x in metres and t in seconds. What is change in displacement when velocity changes from 5 m/s to 20 m/s?
[Hint: $v = \frac{dx}{dt}$; determine t_1 & t_2 for velocities v_1 and v_2].
[Ans: $\Delta s = 14.27$ m].

10.6 A bullet leaves the muzzle of a gun with a velocity of 800 m/s, after travelling through a distance of 80 cm from breech to muzzle with constant acceleration. Determine the time taken by the bullet to leave the gun.
[Hint: $v^2 - u^2 = 2as$]
[Ans: $a = 400{,}000$ m/s^2, $t = 0.002$ second].

10.7 A jet aircraft with a landing speed of 200 km/hr is brought to a speed of 36 km/hour in the available runaway of 700 metres. Calculate the average retardation of the aircraft during the braking period.
[Ans: 2.133 m/s^2].

10.8 A stone is dropped into a well and falls under gravity $g = 9.81$ m/s^2. It strikes the bottom of the well and the sound of impact of stone from the bottom of the well is heard 4 seconds after it is dropped. If the velocity of sound is 370 m/s, how deep the well is?

[Hint: $\sqrt{\frac{2H}{g}} + \frac{H}{370} = 4$ s].

[Ans: 71.17 m].

10.9 Determine the constant acceleration provided by a catapult of an aircraft carrier to produce a launch velocity of 300 km/hr in a distance of 105 m. Assume that the carrier is on anchor.
[Ans: 33.068 m/s^2].

MULTIPLE CHOICE QUESTIONS

10.1 A particle has straight line motion as per equation $x = t^3 - 2t^2 - 4$ with x in m and t in s. After 2 seconds from start, what is the acceleration?
(a) 16 m/s^2 (b) 12 m/s^2
(c) 8 m/s^2 (d) 4 m/s^2

10.2 A man in a balloon rising with constant velocity of 6 m/s propels a ball upwards with a velocity of 9 m/s, relative to the balloon. After what time interval will the ball return to the balloon $g = 9.8$ m/s^2?
(a) 0.5 s (b) 0.61 s
(c) 1.22 s (d) None of these

10.3 A jet aircraft with a landing speed of 56 m/s is brought to a speed of 10 m/s in runway distance of 750 m. What is the average retardation of aircraft?
(a) 2.10 m/s^2 (b) 2.02 m/s^2
(c) 1.98 m/s^2 (d) None of these

10.4 A bullet leaves the muzzle of a gun with a velocity of 750 m/s, after travelling through a distance of 0.75 m from breech to muzzle with constant acceleration. In how much time bullet leaves the gun?
(a) 0.04 s (b) 0.02 s
(c) 0.002 s (d) 0.004 s

10.5 A body starts from rest and travels along a straight path with acceleration of 2 m/s^2. Another body starts after 3 seconds but with acceleration of 4 m/s^2. At what time both will reach the same destination?
(a) 11.36 s (b) 7 s
(c) 10.24 s (d) None of these

10.6 The velocity time diagram is a sine curve shown in **Fig. 10.9**. How much distance is travelled by a particle in half cycle as shown?

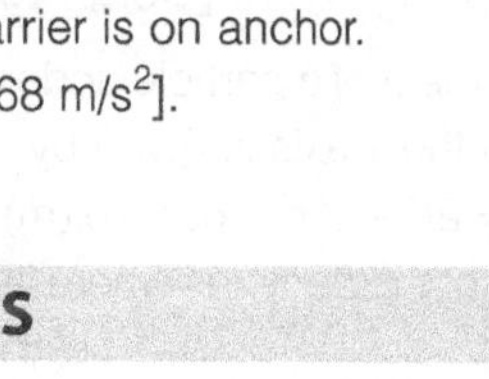

Fig. 10.9

(a) $V_{max}\frac{T}{2\pi}$ (b) $V_{max}\frac{2T}{\pi}$

(c) $V_{max}\frac{T}{\pi}$ (d) None of these

10.7 The velocity vector for a steady three dimensional flow field is described as

$$\vec{V} = yz^2\hat{i} + xy^2\hat{j} + (xy - 2xyz)\hat{k}$$

A point (1, 2, 3) what is the approximate value of the magnitude of the velocity?

(a) 21 (b) 18
(c) 10 (d) 4

[CSE, Prelim, CE : 2006]

10.8 A particle starting at time t = 0, moves a distance given by $x = t(1 - e^{-t})$ consider the following statements

1. Velocity of the particle when a long period of time has elapsed is one unit
2. Acceleration of the particle when a long period of time has elasped, is zero
3. Velocity of the particle increase on with time

Which of these statements are correct?

(a) 1 and 2 (b) 1 and 3
(c) 2 and 3 (d) 3 only

[CSE, Prelim, CE : 2002]

10.9 The displacement of a particle undergoing rectilinear motion along the x-axis is given by

$$x = 2t^3 - 21t^2 + 60t + 6 \text{ (m)}$$

Acceleration of the particles where its velocity is zero.

(a) 36 m/s^2 (b) ±18 m/s^2
(c) 9 m/s^2 (d) –9 m/s^2

[CSE, Prelim, CE : 2003]

10.10 In the following equations x represents position and t time. Which equation represents uniformly accelerated motion

(a) $x = t^3 - 6t^2 + 8t - 10$ (b) $x = t^2 - 4t + 9$
(c) $x = 10t - 12$ (d) $x = t^4 - 10t^3 + 6$

[CSE, Prelim, CE : 2004]

10.11 A stone dropped from the top of a tower is found to travel (5/9) of the height of the tower during the last second of the fall. What is the total time of fall?

(a) 2s (b) 3s
(c) 6s (d) 4s

[CSE, Prelim, CE : 2006]

10.12 A stone is released from an elevator, as it goes up with an acceleration a. If the acceleration due to gravity in g, what is the acceleration of the stone after release

(a) a, upward (b) g-a, upward
(c) g-a, downward (d) g, downward

[CSE, Prelim, CE : 2009]

10.13 The movement of a particle is defined by $\bar{r}(t) = (4t^2 + 5)t + (3t^2 + 2)j + 2k$, what is its position at t = 2s.

(a) (4, 3, 2) (b) (9, 5, 2)
(c) (21, 14, 2) (d) (5, 2, 2)

[CSE, Prelim, CE : 2010]

10.14 A steady 2-dimensional flow field has stream function $\phi = y^2 - x^2 + 3xy$. What is the approximate magnitude of velocity at point (1, 2)?

(a) 4 (b) 7
(c) 8 (d) 11

[CSE, Prelim, CE : 2006]

10.15 A car moving with uniform accelerations covers 450 m in a 5 second interval and covers 700 m in the next 5 second interval. Acceleration of car is

(a) 7 m/s^2 (b) 50 m/s^2
(c) 25 m/s^2 (d) 10 m/s^2

[GATE, 1998 : 1 Mark]

10.16 The time variation of the position of a particle in rectilinear motion is given by $x = 2t^3 + t^2 + 2t$. If v is the velocity and a the acceleration of the particle in consistent units, the motion started with

(a) V = 0, a = 0 (b) V = 0, a = 2
(c) V = 2, a = 0 (d) V = 2, a = 2

[GATE, 2005 : 1 Mark]

Answers

10.1 (c)	10.2 (b)	10.3 (b)	10.4 (c)	10.5 (c)
10.6 (c)	10.7 (a)	10.8 (a)	10.9 (b)	10.10 (b)
10.11 (b)	10.122 (c)	10.13 (c)	10.14 (c)	10.15 (d)
10.16 (d)				

EXPLANATIONS

10.1 (c)

$a = 6t$, $t = 2$ s, $a = 8$ m/s^2.

10.2 (b)

$$6t = 9t - \frac{1}{2}gt^2$$

$$6t = 9t - 4.9t^2$$

$$6 = 9 - 4.9t$$

$$t = \frac{3}{4.9} = 0.61 \text{ s.}$$

10.3 (b)

$$\frac{56^2-10^2}{2\times750}=a=\frac{3136-100}{1500}=2.024 \text{ m/s}^2.$$

10.4 (c)

$$\frac{750^2}{2\times0.75}=a=375000 \text{ m/s}^2$$

$$t=\frac{750}{375000}=0.002 \text{ s.}$$

10.5 (c)

$$t^2=2\times(t-3)^2$$

$$t^2=2t^2-12t+10$$

$$t^2-12t+18=0$$

$$t=\frac{12+\sqrt{144-72}}{2}=10.24 \text{ s.}$$

10.6 (c)

$$\text{Distance}=V_{max}\cdot\left(\frac{T}{2}\right)\left(\frac{2}{\pi}\right)=V_{max}\cdot\frac{T}{\pi}.$$

10.7 (a)

Point (1, 2, 3)

$$\vec{V}=2\times3^2\hat{i}+1\times2^2\hat{j}+(1\times2-2\times1\times2\times3)\hat{k}$$

$$=18i+4j+(-10)\hat{k}$$

$$|V|=\sqrt{18^2+4^2+(-10)^2}$$

$$=\sqrt{324+16+100}=\sqrt{440}\simeq21$$

10.8 (a)

$$x=1-e^{-t}-t(-t)e^{-t}$$

$$=1-e^{-t}+t^2e^{-t}=1+(t^2-1)e^{-t}$$

$t\to\infty$ (∞), infinity

$$V=1$$

$$\frac{d^2x}{dt^2}=t\times e^{-t}+2t\times e^{-t}+t^2\times e^{-t}(-t)$$

$$=3te^{-t}-t^3e^{-t}=e^{-t}[3t-t^3]$$

$a\to0$ at, $t=\infty$ (infinity)

10.9 (b)

$$x=2t^3-21t^2+60t+6$$

$$\frac{dx}{dt}=6t^2-42t+60=0$$

$$t^2-7t+10=0$$

$$(t-5)(t-2)=0$$

$$\simeq t=5, t=2 \text{ s}$$

$$\frac{d^2x}{dt}=12t-42$$

$t=5$, $a=+18$ m/s²

$t=2$, $a=-18$ m/s²

10.10 (b)

$$\frac{dx}{dt}=2t-4$$

$$\frac{d^2x}{dt}=2\text{, uniform acceleration}$$

10.11 (b)

$$H_t=\frac{1}{2}gt^2$$

$$H_{t-1}=\frac{1}{2}g(t-1)^2$$

$$H_t-H_{t-1}=\frac{5}{9}H=\frac{1}{2}g\left[t^2-t^2-1+2t\right]$$

$$=\frac{1}{2}\times g\times\frac{5}{9}t^2$$

$$5t^2=18t-9$$

$$5t^2-18t+9=0$$

$$t=3\text{ second}$$

10.12 (c)

g – a, downward acceleration of a < g.

10.13 (c)

$$t=25$$

Position, (21, 14, 2)

10.14 (c)

$$\phi=y^2-x^2+3xy$$

$$\frac{d\phi}{dt}=2y\frac{dy}{dx}-2x\frac{dx}{dt}+3x\frac{dy}{dt}+3t\frac{dx}{dt}$$

$$=(3y-2x)\frac{dx}{dt}+(2y+3x)\frac{dy}{dt}$$

$$=4\frac{dx}{dt}+7\frac{dy}{dt}$$

Point = (1, 2)

$$\text{Point}=\sqrt{16+49}\simeq8$$

10.15 (d)

$$450=54+\frac{1}{2}a\times25$$

$$1150=104+\frac{1}{2}a\times100$$

or

$$1800=204+\frac{1}{2}a\times100$$

$u=65$ m/s, $a=10$ m/s²

10.16 (d)

$$\frac{dx}{dt}=V=6t^2+2t+2$$

$$V=2, t=0$$

$$\frac{d^2x}{dt^2}=12t+2=a$$

$$a=2, t=0$$

■■■■

11

CHAPTER

Curvilinear Motion

11.1 Introduction

In the present chapter we will study about motion of a body *along a curved path*. The curved path can be in a horizontal plane or in a vertical plane or in a 3 dimensional co-ordinates system as shown in **Figs. 11.1 (*a*), (*b*)** and **(*c*)** in a three dimensional curved path.

Curved track as of a road or rail track (*a*) | Curved path in vertical plane as an uneven road (*b*) | Three dimensional curved path (*c*)

Fig. 11.1

The co-ordinates of points of path can be *x*, *y*, *z* or *r*, θ and *z* (polar co-ordinates) or *r*, θ, φ (spherical coordinates.

Common examples of a three dimensional curvilinear motion are (*a*) a point on a screw moving along the helical groove of a nut (*b*) rotating member in a mixer with periodic axial motion, (*c*) a robotic device rotating about a fixed vertical axis, while its arm extending and elevating simultaneously.

We will discuss the cases of cycloidal motion, curvilinear motion of an automobile along curved path in horizontal and vertical planes, normal and tangential velocities and accelerations in curvilinear motion.

11.2 Plane Curvilinear Motion

Let us consider a particle moving along a curved path *ab* in *xy* plane as shown in **Fig. 11.2**. At a particular instant at time *t*, particle is at position *P* and changes its position from *P* to P_1 in time Δt along curved path *ab* as shown. Position vector of point *P* is *r* and position vector of point P_1 is r_1. Change in position vector $\Delta r = r_1 - r$ (vectorially). Curved path from *P* to P_1 is Δs as shown in enlarged view. Please note that r_1 and *r* depend upon the co-ordinates chosen but Δr is independent of the co-ordinate system.

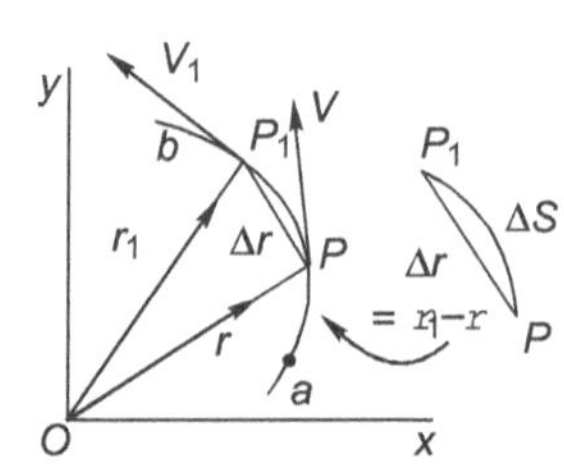

Plane curvilinear motion

Fig. 11.2

Average *velocity* of the particle from position P to P_1

$V_{av} = \frac{\Delta r}{\Delta t}$, direction along Δr, a vector quantity, speed of the particle from position P to P_1

$$v_{av} = \frac{\Delta s}{\Delta t} = \frac{\text{curved distance}}{\text{time interval}}, \text{ a scalar quantity.}$$

As the point P_1 comes closer to P, and time interval $\Delta t \to 0$, the magnitude of average velocity and average speed approaches one another.

V = limiting value of average velocity

$$= \lim_{\Delta t \to 0} \frac{\Delta r}{\Delta r} = \frac{dr}{dt} = \dot{r}, \text{ a derivative of vector quantity } r$$

i.e., derivative of position vector r at point P and Δr is the vector showing changes in magnitude and direction of position vector.

Magnitude of $V = |V| = \frac{ds}{dt} = \dot{S}$, a derivative of scalar quantity S, distance along curved path.

Average Acceleration: Say instantaneous velocity of a particle moving along curved path ab, at point P is V, tangential to the curve at point P. As the particle travels along curved path, goes to point P_1, say instantaneous velocity at point P_1 is V_1 tangential to the curved path at point P_1 as shown in **Fig. 11.3**. So velocity V at P has changed to velocity V_1 at P_1.

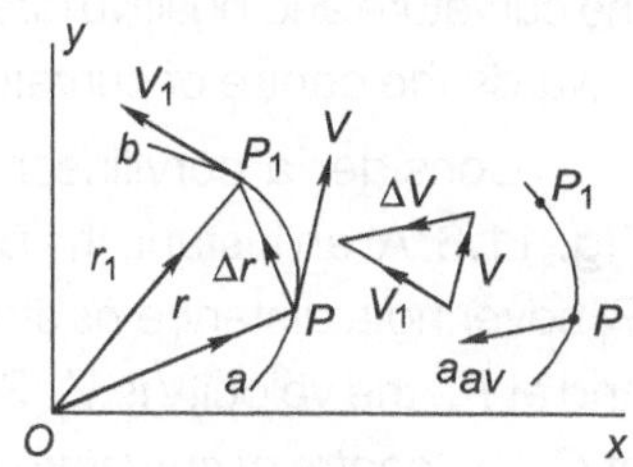

Fig. 11.3

Instantaneous velocity at P, $V_{dt \to 0} = \frac{dr}{dt}$

Instantaneous velocity at P_1, $V_{1\, dt \to 0} = \frac{dr_1}{dt}$

Change in velocity ΔV as shown in figure in time interval Δt,

Average acceleration of the particle, $a_{av} = \frac{\Delta V}{\Delta t}$, *is vector in the direction of* ΔV

Instantaneous acceleration can be defined as the limiting value of average acceleration as time interval Δt approaches zero.

Instantaneous acceleration, $a = \lim_{\Delta t \to 0} \frac{\Delta V}{\Delta t} = \frac{dV}{dt} = \dot{V}$

Instantaneous acceleration is a result of change in magnitude and change in direction of velocity. It can be noted that velocity at a particular point is tangential to the curved path at that point but instantaneous acceleration is neither tangential nor normal to the curved path. However we will study later on in the chapter that there are normal and tangential components of acceleration, where the normal component points towards the centre of curvature of the curved path.

11.3 Rectangular Coordinates

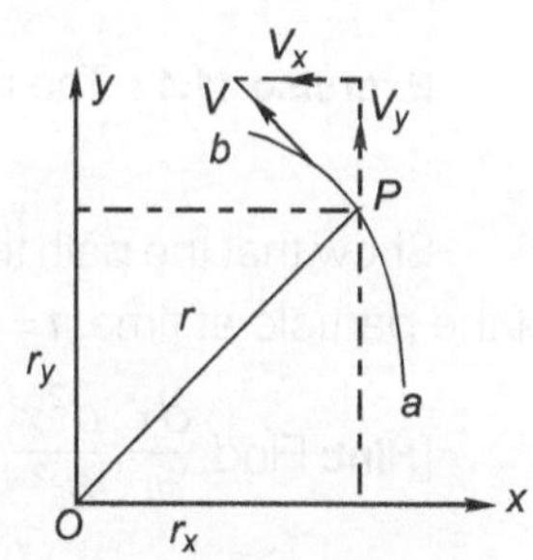

Fig. 11.4

Consider a particle travelling along a curved path ab as shown in **Fig. 11.4**. At any instant, time t, position vector of a point P as curved path is $r = f(t)$, a function of time. Position vector in terms of rectangular components is

$$r = r_x i + r_y j$$

where $r_x = f_1(t)$, function of time

$r_y = f_2(t)$, function of time

Velocity, $$V = \frac{dr_x}{dt} i + \frac{dr_y}{dt} j = \frac{df_1(t)}{dt} i + \frac{df_2(t)}{dt} j$$

$$dt \to 0$$

Similarly acceleration has two components a_x, a_y *i.e.*,

$$a = a_x i + a_y j$$

where $$a_x = \frac{d^2 r_x}{dt^2} = \frac{d^2 f_1(t)}{dt^2}, \quad a_y = \frac{d^2 r_y}{dt^2} = \frac{d^2 f_2(t)}{dt^2}.$$

11.4 Tangential and Normal Co-ordinates

In curvilinear motion, these are the most common direct co-ordinates, as these co-ordinates provide natural description of the curvilinear path of a body. **Fig. 11.5** shows a curved path *aP* P_1 P_2 *b*. At any instant, the tangential co-ordinates *t* is along tangent to the curve at the point and normal co-ordinate *n* is along the perpendicular to the tangent at that point, *n–t* co-ordinates are shown at different points. Note that *t*-co-ordinate is along the positive direction and *n*-coordinate *directs towards* the centre of the curvature and positive direction of normal co-ordinate is always towards the centre of curvature as *C*, C_1, C_2 etc. as shown.

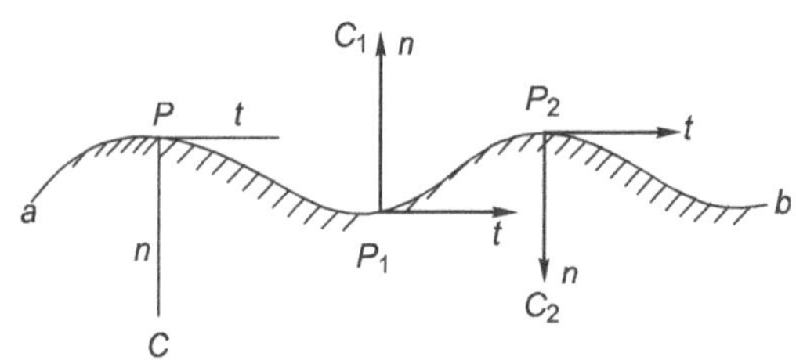

Fig. 11.5

Consider a curvilinear motion of a particle along path *ab* as shown in **Fig. 11.6**. At an instant, the particle is at *P* and after time δt, it approaches point P_1 covering a distance *ds* along the path as shown. Tangential velocity at *P* is *V* and at P_1, the velocity is V_1. Perpendiculars drawn to tangents at *P* and P_1 meet at *C*, *i.e.*, *centre of curvature*, distance *ds* subtending an angle $d\theta$ and radius of curvature is ρ as shown.

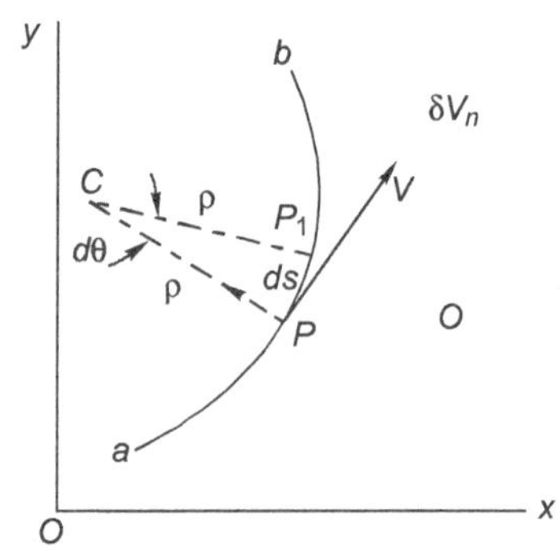

$$ds = \rho . d\theta$$

Magnitude of velocity, $$v = \frac{ds}{dt} = \frac{\rho d\theta}{dt}$$

Tangential acceleration, $$a_t = \frac{\delta V_t}{\delta t} = \dot{V} = \frac{d}{dt}(V) = \frac{d}{dt}\left(\frac{ds}{dt}\right)$$

$$= \frac{d^2 s}{dt^2} \quad ...(1)$$

Normal acceleration of a particle moving along a curve is given by

$$a_n = \frac{v^2}{\rho}$$

Exercise 11.1 The motion of a particle is described by $x - t$ and $y - t$, displacement time relations

$$x = 2(t+2)^2, \; y = 2(t+2)^{-2}$$

Show that the path travelled by the particle is a rectangular hyperbola. Determine velocity and acceleration of the particle at time, $t = 2$ seconds.

[**Hint:** Find $\frac{dx}{dt}, \frac{d^2x}{dt^2}, \frac{dy}{dt}, \frac{d^2y}{dt^2}$ at $t = 2$ s].

[**Ans:** $V_2 = 16i - \frac{1}{16} j$ m/s, $a_2 = 4i + \frac{3}{64} j$ m/s^2 $x.y = 4$, equation of rectangular parabola].

PROBLEMS

Problem 11.1 A car is moving in a curvilinear path at a constant speed as shown in **Fig. 11.7**. When the car passes through the dip at *A*, an acceleration 0.55 g acts on the mass centre *G* of the car, which is located at 0.5 m from the level of the road, determine the speed *V* of the car. The radius of curvature of the road is 96 m.

Solution Tangential speed is constant.

Tangential acceleration, $a_t = 0$

Radius of curvature of path = 96 mm

Location of *G* above road = 0.5 m

Radius of curvature upto *G* of car

$$= 96 - 0.5 = 95.5 \text{ m}$$

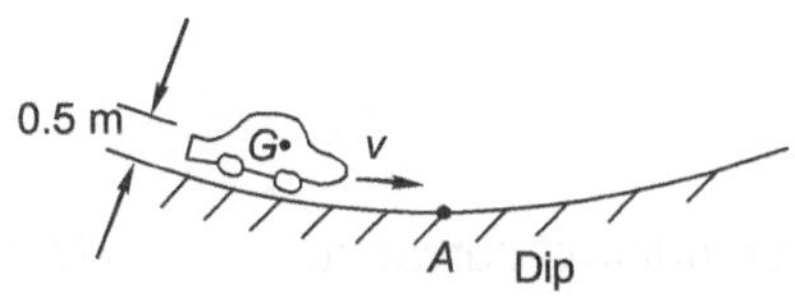

Fig. 11.7

Normal acceleration, $a_n = 0.55 \text{ g} = 0.55 \times 9.81 = 5.3955 \text{ m/s}^2$

$$= \frac{V^2}{\rho}$$

or $V^2 = \rho \times a_n = 95.5 \times 5.3955 = 515.27$

V = average speed = 22.7 m/s.

Problem 11.2 A car starting from rest moves along a curved road of radius $R = 300$ m. The car attains a speed of 72 kmph in 20 seconds. What are tangential and normal accelerations at (*a*) $t = 0$ s, (*b*) $t = 10$ s (**Fig. 11.8**)?

Solution Speed after 20 s, $v = \frac{72000}{3600} = 20$ m/s

Initial speed $v_0 = 0$

Time $t = 20$ s, $v = 20$ m/s

Tangential acceleration, $a_t = \frac{20}{20} = 1 \text{ m/s}^2$

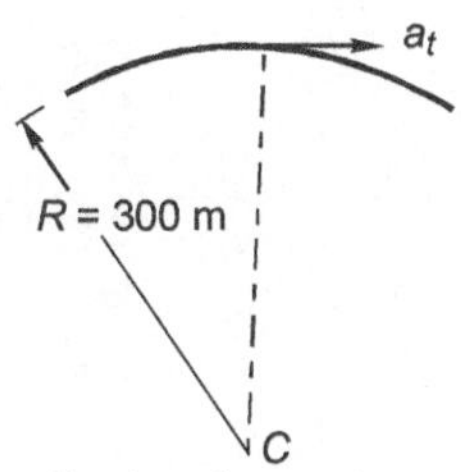

Fig. 11.8

(*a*) At time $t = 0$ s

$$v_0 = 0, \quad a_n = \frac{v_0^2}{R} = 0$$

$$a_t = 1 \text{ m/s}^2$$

(*b*) At time $t = 10$ s, $v = 10 \times 1 = 10$ m/s

$$a_t = 1 \text{ m/s}^2$$

$$a_n = \frac{v^2}{R} = \frac{10^2}{300} = 0.333 \text{ m/s}^2.$$

Problem 11.3 A car travels along a depression in a road, the equation of depression being $x^2 = 200y$, speed of the car is constant and is equal to 20 m/s. What is the acceleration of the car at the deepest point *C* in the depression, as shown in **Fig. 11.9**? What is the radius of curvature at the depression at the point *C*?

Solution $x^2 = 200y$

or $y = 0.005x^2$ ...(1)

or $$\frac{dy}{dx} = 0.005 \times 2x = 0.01x$$

$$\frac{d^2y}{dx^2} = 0.01$$

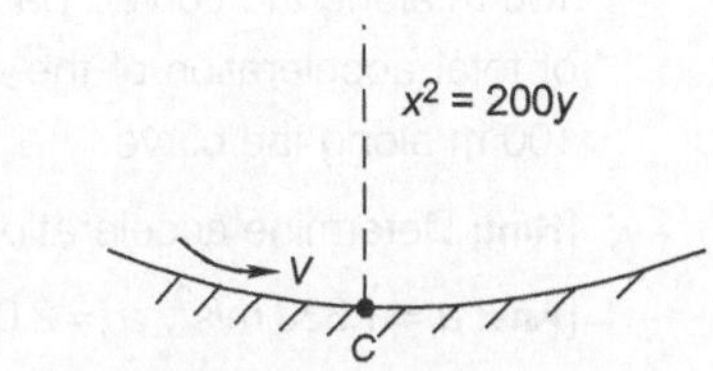

Fig. 11.9

at point C, $\frac{dy}{dx} = 0$ (slope is zero)

If R is radius of curvature, then

$$R = \frac{\left[1+\left(\frac{dy}{dx}\right)^2\right]^{1.5}}{\frac{d^2y}{dx^2}} = \frac{1}{\frac{d^2y}{dx^2}}$$

or radius of curvature, $R = \frac{1}{0.01} = 100$ m

Constant velocity, $v = 20$ m/s

Normal acceleration, $a_n = \frac{v^2}{R} = \frac{20^2}{100} = 4$ m/s^2.

Remember

- If $\Delta r = r' - r$, vectorial change in position vector

  V = average velocity $= \frac{\Delta r}{\Delta t} = \frac{ds}{dt} = \frac{\text{distance covered along curved path}}{\text{short interval}, dt}$.
- Instantaneous velocity, $V = \frac{dr}{dt}$.
- Instantaneous acceleration, $a = \lim_{dt \to 0} = \frac{dV}{dt} = \dot{V}$.
- Curve joining the ends of velocity vectors (drawn from any arbitrary point) is known as hodograph. At any instant, acceleration vector is tangential to the hodograph.
- Rectangular co-ordinates

  $r_x = f_1(t);\; r_y = f_2(t);\; r = r_x i + r_y j$

  $$V = \frac{dr_x}{dt} i + \frac{dr_y}{dt} j = \frac{df_1(t)}{dt} i + \frac{df_2(t)}{dt} j.$$
- Acceleration $a = a_x i + a_y j = \frac{d^2 f_1(t)}{dt^2} i + \frac{d^2 f_2(t)}{dt^2} j$.
- Tangential and normal co-ordinates

  $a = \sqrt{a_t^2 + a_n^2}$, where $a_n = \frac{v^2}{R} = \frac{(\text{Average velocity})^2}{\text{Radius of curvature}}$.

PRACTICE PROBLEMS

11.1 A racing car is moving along a curved path of radius 240 m. Speed of the car is increased at a uniform rate from 15 m/s to 25 m/s while travelling a distance of 150 m along the curved path. Determine magnitude of total acceleration of the car after it has travelled 100 m along the curve.

[**Hint:** Determine acceleration, velocity at $s = 100$ m]

[**Ans:** $a_t = 1.333$ m/s^2, $a_n = 2.048$ m/s^2, $a = 2.444$ m/s^2].

11.2 The position vector of a point moving in xy plane is given by, $r = \left[\frac{4}{3}t^3 - \frac{1}{2}t^2\right] i + \frac{t^4}{6} j$, where r is in metres and t is in seconds. Determine the angle between the velocity V of acceleration a at $t = 2$ s.

[**Hint:** Determine $\theta_1 = \tan^{-1}\frac{V_y}{V_x}$;

$\theta_2 = \tan^{-1}\frac{a_y}{a_x}$ at $t = 2$ s].

[**Ans:** $V_2 = 14i + 5.33j$ m/s $a_2 = 15i + 8j$ m/s^2, $\theta = \theta_2 - \theta_1 = 7.22°$]

11.3 Belt speed over a pulley changes uniformly from 4 m/s to 6 m/s over an interval of 2 seconds. Calculate the magnitude of accelerations at points P_1 and P_2 half way through the time interval **(Fig. 11.10)**.

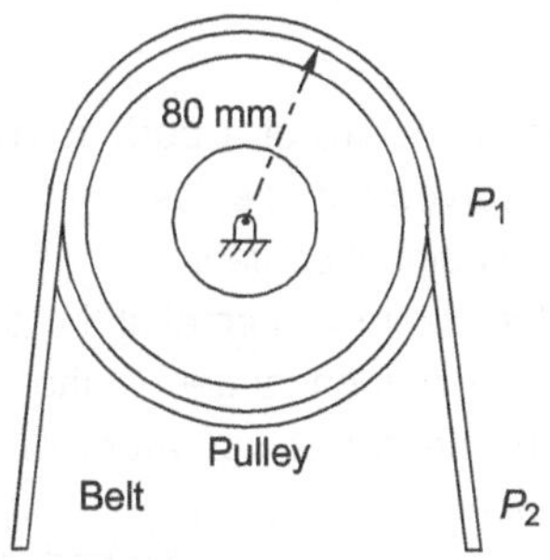

Fig. 11.10

[Hint: $a = \frac{6-4}{2} = 1\text{ m/s}^2 = a_t$

$V_{av} = \frac{4+6}{2} = 5\text{ m/s}^2$

$\rho = 0.080$ m.

[Ans: At P_1, $a_n = \frac{V_{av}^2}{\rho} = 312.5\text{ m/s}^2$,

$a_t = 1\text{ m/s}^2$, $a = 312.5016\text{ m/s}^2$

At P_2, $a_n = 0$, $a_t = 1\text{ m/s}^2$].

11.4 A car driver is travelling along a curved path of radius 250 m at a speed of 20 m/s. If he slows down the car by applying the brakes and slow down the speed of the car to 10 m/s in 10 seconds find: (*i*) normal and tangential acceleration at the time of application of brakes.

[Hint: $a_t = -1\text{ m/s}^2$, $v_0 = 20$ m/s].

[Ans: (*i*) $a_t = -1\text{ m/s}^2$, $a_n = 1.6\text{ m/s}^2$].

11.5 A magnetic tape runs over an idler pulley of radius 80 mm. At the contact point *A*, at time $t = 0$. Velocity is $V = 2.5$ m/s. If the total acceleration at the point of contact *A* is inclined at an angle of 6° to the tangent at time $t = 0$, determine the time in which the pulley will be brought to rest at uniform deceleration assuming no slip between pulley and tape **(Fig. 11.11)**.

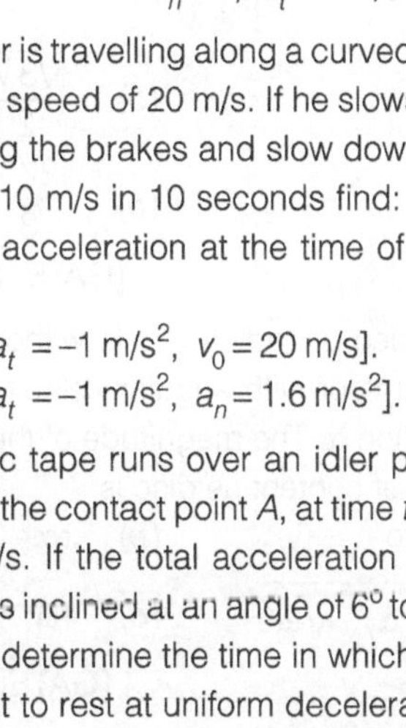

Fig. 11.11

[Hint: $a_n = \frac{v^2}{R} = a_t \tan 6°$].

11.6 A particle describes the path $y = 4.9x^2$ where *x* and *y* are in metres. The *x* component of the velocity is constant *i.e.*, $V_x = 2$ m/s constant. Assuming $x = 0$, $y = 0$ at $t = 0$, determine (*i*) velocity vector, (*ii*) acceleration vector as a function of time.

[Hint: $x = 2t$, $x^2 = 4t^2$, $y = 19.6t^2$].

[Ans: $V = 2i + 39.2t\,j$ m/s, $a = 39.2j\text{ m/s}^2$].

11.7 A car starts from rest on a curved path of radius 200 m and accelerates at constant tangential acceleration of 0.8 m/s^2. After what time resultant acceleration of the car becomes 1.0 m/s^2]?

[Hint: $a_n = \sqrt{1-0.8^2}\text{ m/s}^2$].

[Ans: $t = 13.693$ seconds].

11.8 A truck is moving on a curved path as shown in **Fig. 11.12**. *G* of the truck is located at a distance of 1.2 m from the level of the road. When the truck reaches the top of the hump at *A*, its *G* experiences an acceleration of 5.4 m/s^2. What is the radius of curvature of the road at *A*, if the average speed of the truck is 80 km/hour?

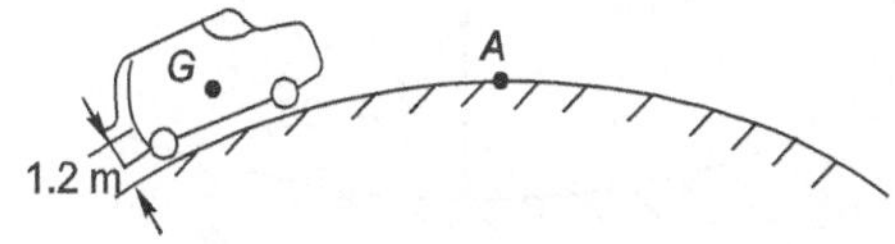

Fig. 11.12

[Hint: $a_n = \frac{v^2}{(R+1.2)}$].

[Ans: 90.25 m].

11.9 A car driver anticipates a dip and a hump in road applies brakes to the car. At position *A*, speed of car is 90 kmph which is reduced to 36 kmph at point *C*, covering a distance *AC* = 100 m as shown in **Fig. 11.13.** Determine normal accelerations at *A*, *B* and *C*. What is the total acceleration at *A*?

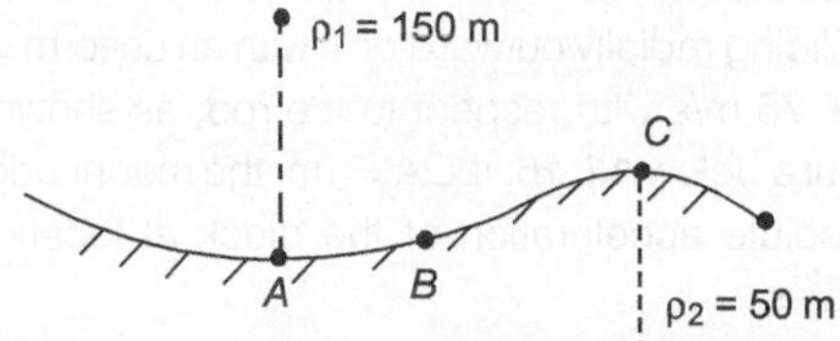

Fig. 11.13

Note: Point *B* is point of inflexion, $\rho = \infty$.

[Ans: $a_{n_A} = 4.1667\text{ m/s}^2$, $a_{n_B} = 0$, $a_{n_C} = 2\text{ m/s}^2$

total acceleration at *A*, $a = 4.925\text{ m/s}^2$].

MULTIPLE CHOICE QUESTIONS

11.1 A car is moving in a curvilinear path at a constant speed when the car passes through dip it is subjected to normal acceleration of 0.6*g*. If radius of curvature of path is 100 m. What is speed of car **(Fig. 11.14)**?

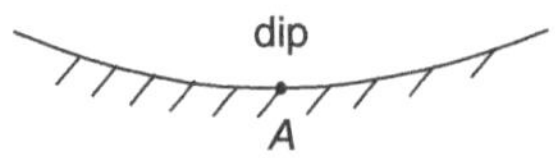

Fig. 11.14

(*a*) 24.25 m/s (*b*) 37.5 m/s
(*c*) 5.88 m/s (*d*) None of these.

11.2 A car starting from rest moves along a curved road of radius 150 m. Car attains a speed of 72 kmph in 20 seconds. What is normal acceleration at $t = 10$ s?

(*a*) 1 m/s^2 (*b*) 2.667 m/s^2
(*c*) 0.667 m/s^2 (*d*) None of these

11.3 A car travels along a depression in a road, the equation of depression is $x^2 = 100y$ speed of the car is constant and is equal to 10 m/s. What is acceleration of the car at the deepest point *A* **(Fig. 11.15)**?

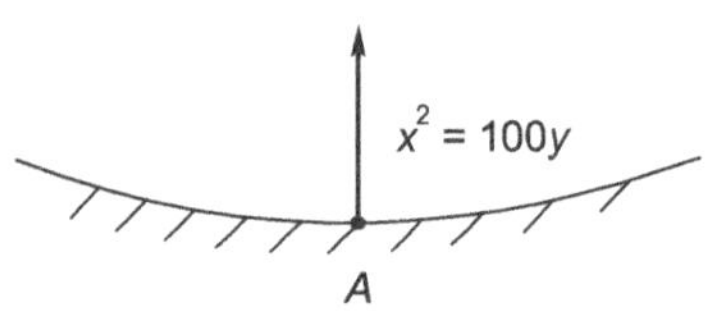

Fig. 11.15

(*a*) 4 m/s^2 (*b*) 2 m/s^2
(*c*) 1 m/s^2 (*d*) None of these

11.4 A car driver is travelling along a curved path of radius 250 m at a speed of 25 m/s. If he slows down the car by applying brakes and slow down the speed of the car to 10 m/s in 10 seconds. What are normal and tangential acceleration at the time of application of brakes?

(*a*) -1 m/s^2, 2 m/s^2 (*b*) $+1$ m/s^2, 2.5 m/s^2
(*c*) 2.5 m/s^2, -1.5 m/s^2 (*d*) 1.25 m/s^2, -1 m/s^2

11.5 A line *OB* is rotating with a constant angular velocity of 3 rad/s in counter clockwise direction and a block is sliding radially outward on it with an uniform velocity of 0.75 m/s with respect to the rod, as shown in the **figure** below **11.16**. If *OA* = 1m, the magnitude of the absolute acceleration of the block at location A in m/s^2 is

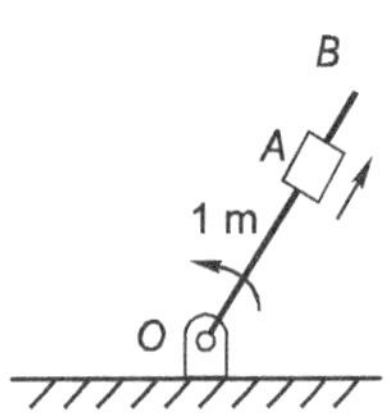

Fig. 11.16

(*a*) 3 (*b*) 4
(*c*) 5 (*d*) 6

[GATE-2013]

11.6 Instantaneous centre of a body rolling sliding on a stationary curved surface lies

(*a*) at the point of contact
(*b*) on the common normal at the point of contact
(*c*) on the common tangent at the point of contact
(*d*) at the centre of curvature of the stationary surface

[GATE, 1992 : 1 Mark]

11.7 A circular disc of a radius R rolls without slipping at a velocity V. The magnitude of the velocity of the point P is

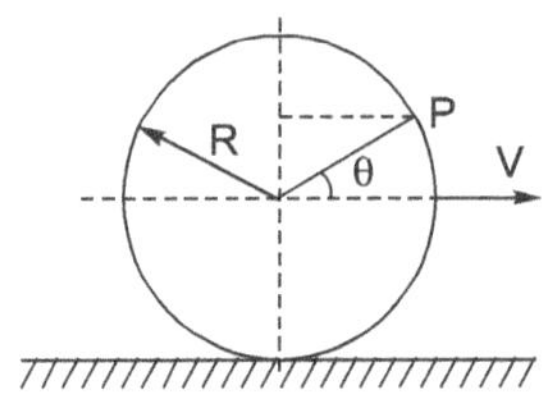

Fig. 11.17

(*a*) $\sqrt{3}V$ (*b*) $\frac{\sqrt{3}V}{2}$
(*c*) $\frac{V}{2}$ (*d*) $\frac{2V}{\sqrt{3}}$

[GATE, 2008 : 2 Marks]

11.8 A solid disc of radius r rolls without slipping an a horizontal floor with angular velocity ω and angular acceleration α. The magnitude of the acceleration at the point of content as disc is

(*a*) zero (*b*) $r\alpha$
(*c*) $\sqrt{(r\alpha)^2 + r\omega^2}$ (*d*) $r\omega$

[GATE, 2012 : 1 Mark]

11.9 A circular object of radius r rolls without slipping an a horizontal level floor with the center having the velocity V. The velocity at the point of content between the object and floor is

(*a*) zero
(*b*) V in the direction of motion
(*c*) V opposite to the direction of motion
(*d*) V vertically upward from the floor

[GATE, 2014: 1 Mark (set-1)]

11.10 The cylinder shown below rolls without slipping. In which direction does the friction force act? Toward which of the following points is the acceleration of the point of contact A on the cylinder directed?

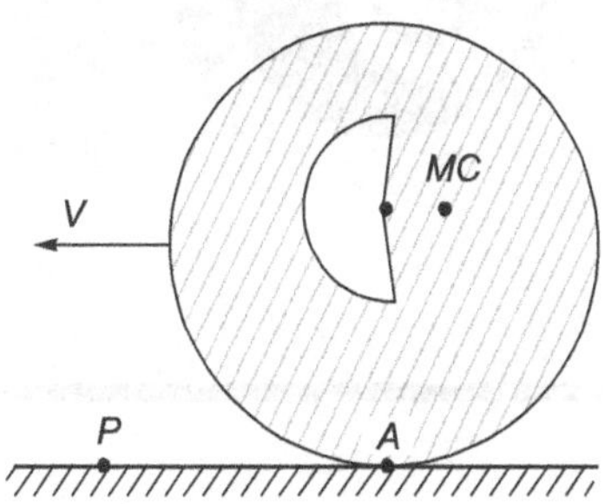

Fig. 11.18

(a) The mass centre
(b) The geometric centre
(c) The point P as marked
(d) None of the above

[GATE, 1993 : 1 Mark]

Answers

11.1 (a)	11.2 (c)	11.3 (b)	11.4 (c)	11.5 (c)
11.6 (a)	11.7 (a)	11.8 (a)	11.9 (a)	11.10 (b)

EXPLANATIONS

11.1 (a)

$$a_n = 0.6\times 9.8 \text{ m/s}^2 \frac{V^2}{R}$$

$V^2 = 0.6 \times 9.8 \times 100 = 588$, $V = 24.25$ m/s.

11.2 (c)

$$a_t = \frac{72000}{20\times 3600} = \frac{20 \text{ m/s}^2}{20}$$

$$a_t = 1 \text{ m/s}^2 \quad V = 10 \text{ m/s}^2$$

$$a_n = \frac{100}{150} = 0.667 \text{ m/s}^2.$$

11.3 (b)

$$y = \frac{x^2}{100}, \quad \frac{dy}{dx} = \frac{x}{50}, \quad R = 50 \text{ m}$$

$$\frac{d^2y}{dx^2} = \frac{1}{50} \quad a_n = \frac{V^2}{R} = \frac{100}{50} = 2 \text{ m/s}^2.$$

11.4 (c)

$$a_t = \frac{10-25}{10} = -1.5 \text{ m/s}^2$$

$$a_n = \frac{v^2}{R} = \frac{25^2}{250} = \frac{625}{250} = 2.5 \text{ m/s}^2.$$

11.5. (c)

Coriollis component of acceleration

$= 2\,v\omega = 2 \times 0.75 \times 2 = 3 \text{ m/s}^2$,

Normal acceleration $= \omega^2 \times OA = 2 \times 2 \times 1 = 4 \text{ m/s}^2$,

Absolute acceleration $= (3^2 + 4^2)^{0.5} = 5 \text{ m/s}^2$

11.6 (a)

At the point of contact

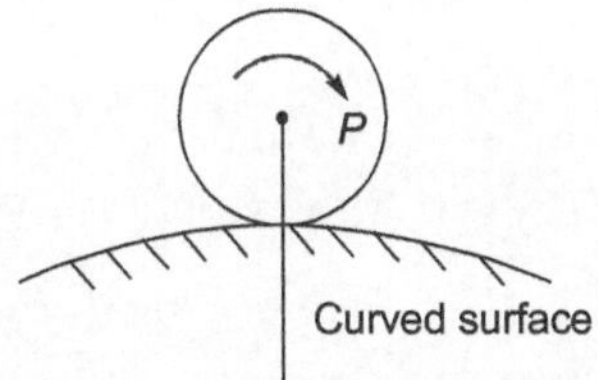

Fig. 11.19

11.7 (a)

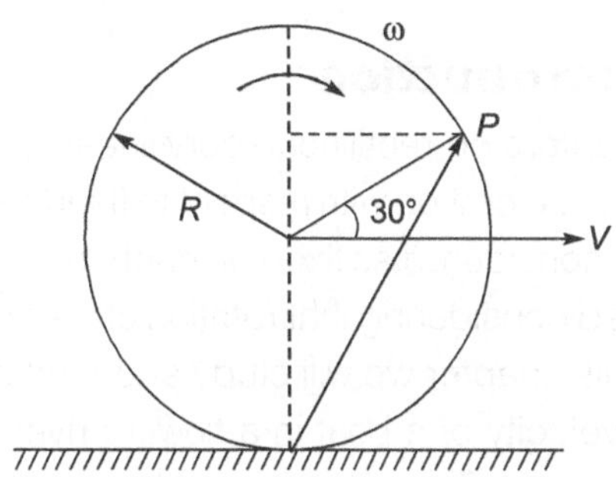

Fig. 11.20

$$OP = \sqrt{(1.5R)^2 + (0.866R^2)} = \sqrt{3}R$$

$$V_p = w\sqrt{3}R = \sqrt{3}V$$

11.8 (a)

At point of content acceleration due to rotation (α_n) – acceleration due to translation (α_v) = 0.

11.9 (a)

Point of contact is instantaneous centre of rotation, where velocity Is zero.

11.10 (b)

Force of friction acts towards point P acceleration at A acts towards GC (geometric centres)

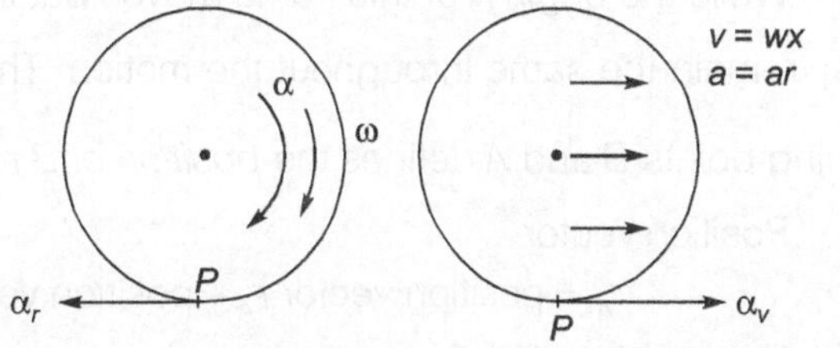

Fig. 11.21

■■■■

12

CHAPTER

Relative Velocity

12.1 Introduction

In the chapters on rectilinear, curvilinear and projectile motions we have determined absolute velocities and absolute acceleration with respect to fixed frame of reference *i.e.*, *x*, *y*, *z* fixed co-ordinate axis. This co-ordinate system is considered as fixed on earth, but the earth is rotating, therefore velocity and acceleration of an object are studied considering that rotation of earth is negligible and references on earth are fixed.

In this chapter we will study about relative velocity of an object with reference to a moving frame, as the study of velocity of a boat in a flowing river or in other words we study the velocity of the boat relative to the velocity of water. There are many other cases which are subject matter of this chapter as:

(*a*) Velocity of a ship with reference to another moving ship.

(*b*) Velocity of one train with reference to another moving train.

(*c*) Velocity of rain drop with reference to a another moving object.

For the analysis of such problems we consider two sets of frames of axes as: (*i*) frame of fixed co-ordinates, (*ii*) frame of moving co-ordinate axes.

12.2 Motion Relative to a Frame in Translation

Let us consider two frames (*i*) frame attached to the earth as *fixed frame*, (*ii*) a *moving frame*, not rigidly connected to fixed frame. Take two particles *A* and *B* in space, at position vectors r_A and r_B *with reference to xyz-axes on fixed frame.* Now consider a system of axes centred at *A* with *oz*-axis parallel to A'_z- axis, plane *xy* parallel to plane $\zeta\eta$ as shown in Fig. 12.1.

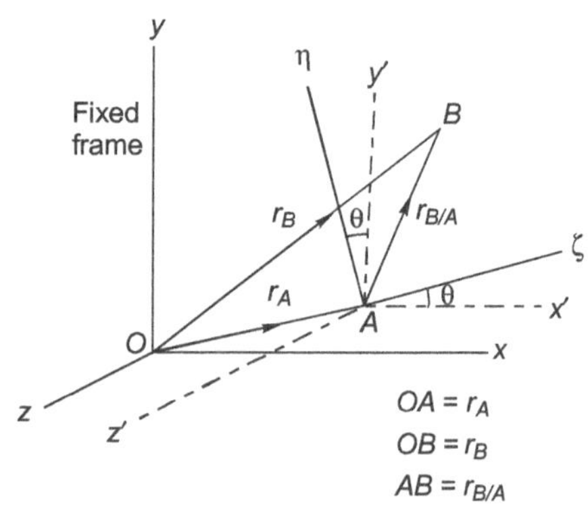

Fig. 12.1

While the *origin A* of this frame moves but the orientation of axes ζ, η remain the same throughout the motion. The position vector $r_{B/A}$ joining points *B* and *A* defines the *position of B relative* to the position of *A* in frame $\zeta\eta z'$.

Position vector

$$r_A = \text{position vector } r_A + \text{position vector } r_{B/A}$$

or $$r_B = r_A + r_{B/A} \qquad ...(1)$$

Differentiating Equation (1) with respect to time, *t*

$$\frac{dr_B}{dt} = \frac{dr_A}{dt} + \frac{dr_{B/A}}{dt} \qquad ...(2)$$

or $\dot{r}_B = \dot{r}_A + \dot{r}_{B/A}$

Velocity of B = velocity of A + relative velocity of B with respect to velocity of A

or $V_B = V_A + V_{B/A}$

or $V_{B/A} = V_B - V_A$ ...(3)

Relative velocity of B with respect to velocity of A

= Velocity of B – velocity of A

Similarly differentiating Equation (3) with respect to time, t

$$\dot{V}_{B/A} = \dot{V}_B - \dot{V}_A \quad ...(4)$$

Relative acceleration of a body B with respect to acceleration of body A

= Acceleration of body B – acceleration of body A

Note that absolute motion of body B is defined as the motion of body A combined with relative motion of body B with respect to the moving frame attached at A.

Remember the following:

(*i*) Relative velocity of body B with respect to body A

= Absolute velocity of body B – absolute velocity of body A.

(*ii*) Relative acceleration of body B with respect to body A

= Absolute acceleration of body B – absolute acceleration of body B.

(*iii*) Frame $\zeta\eta z'$, attached to body A is in translation *i.e.*, when it moves with A, it maintains its orientation defined by angle θ as in **Fig. 12.1**.

Example 12.1 A car A is moving towards East at a constant speed of 15 m/s. As the car A crosses the intersection of the road, another car B starts from rest with an acceleration of 5.4 km/hr/second. Car B is moving towards South from a starting point 60 m North of the point of intersection as shown in **Fig. 12.2**. Determine the position, velocity of the car B relative to car A; 3 seconds after the car A crosses the point of intersection.

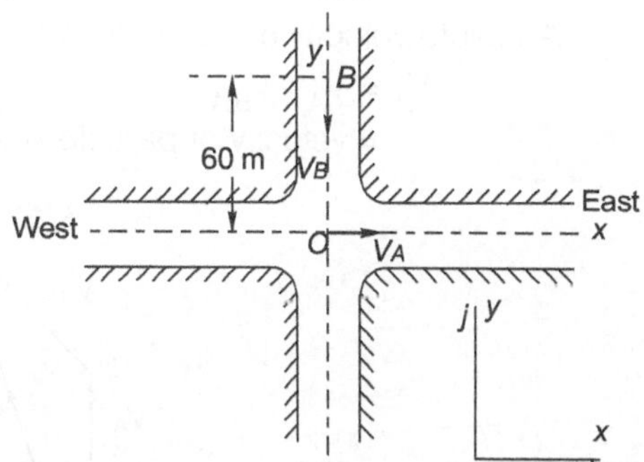

Fig. 12.2

Solution Car A

Velocity, $V_A = 15$ m/s

Acceleration, $a_A = 0$ (as given)

Car B

Initial velocity, $V_B = 0$

Acceleration, $a_B = 5.4$ km/hr/s

$$= \frac{5.4 \times 1000}{3600} = 1.5 \text{ m/s}^2 \downarrow$$

(in negative y-direction)

$= -1.5j$ m/s^2

After 3 seconds $V_A = 15$ m/s (constant)

x_A = distance covered

$= 15 \times 3 = 45$ m

$V_B = 0 \quad 3 \times 1.5 = 4.5$ m/s

$= -4.5j$ m/s

Velocity Diagram

Absolute velocity of B, $V_B = V_A + V_{B/A}$

$-4.5j = +15i + V_{B/A}$

or $V_{B/A} = -15i - 4.5j$ (as shown in **Fig. 12.3**)

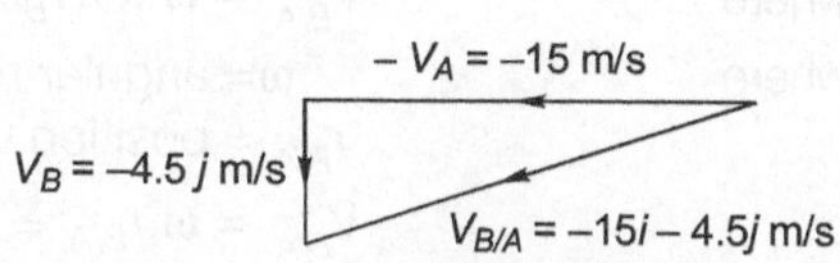

Velocity Diagram

Fig. 12.3

Exercise 12.1 Two ships *A* and *B* leave a port at the same time. Ship *A* is travelling north west at 30 kmph and ship *B* is travelling south west at 25 kmph, as shown in **Fig. 12.4**. Determine (*a*) the speed of *B* relative to the ship *A*, (*b*) at what time they will be 200 km apart?

[**Hint:** $V_A = -21.21i + 21.21j$ kmph,

$V_B = -17.675i - 17.675j$ kmph

$V_{B/A} = +3.535i - 38.885j$ kmph;

$\text{Time} = \dfrac{200}{|V_{B/A}|}$].

[**Ans:** $V_{B/A} = 3.535i - 38.885j$ kmph,

$|V_{B/A}| = 39.05$ kmph, 5.12 hour].

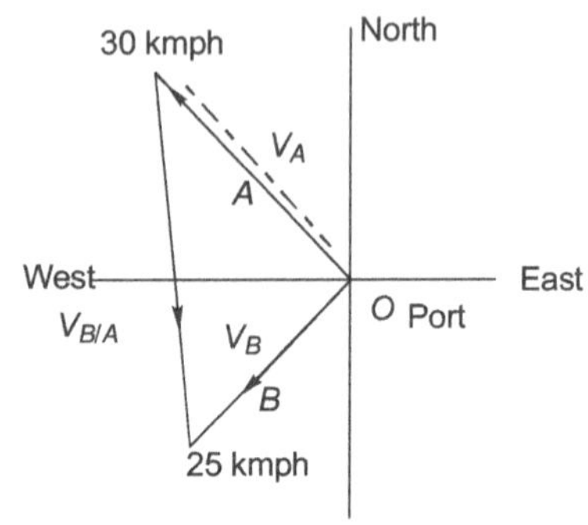

Fig. 12.4

12.3 Absolute and Relative Velocity in Plane Motion

Any plane motion of a body can be replaced by (*i*) translation, defined by the motion about any reference point *A* (as shown in **Fig. 12.5**) plus (*ii*) a simultaneous rotation about the same point *A* as shown.

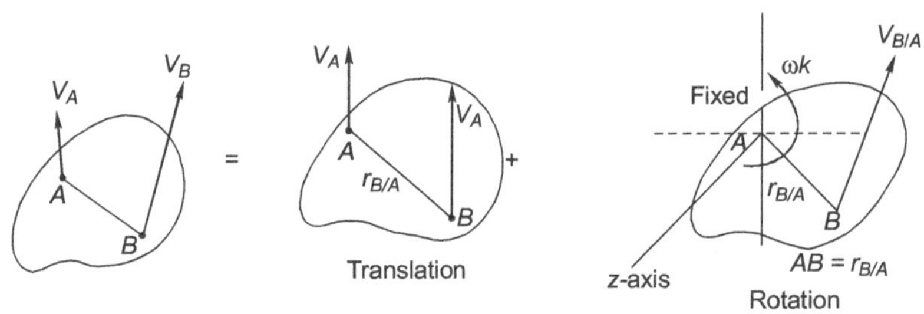

Fig. 12.5

Absolute velocity of the particle *B*,

$V_B = V_A + V_{B/A}$

= velocity of particle *A* + relative velocity of particle *B* with respect to particle *A* (**Fig. 12.6**)

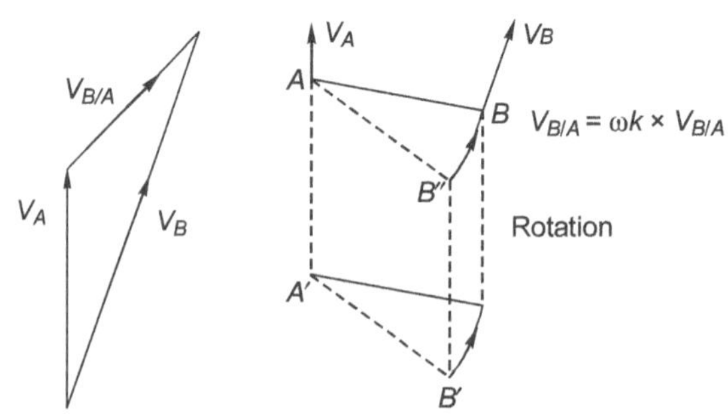

Fig. 12.6

where $V_{B/A} = \omega.k \times r_{B/A}$

where ω = angular rotation about *z*-axis passing through *A*, *k* is unit vector

$r_{B/A}$ = position vector of *B* relative to point *A*

or $V_{B/A} = \omega.r_{B/A}$ = angular velocity × distance from *A* to *B*

or angular velocity $\omega = \dfrac{V_{B/A}}{r_{B/A}}$.

PROBLEMS

Problem 12.1 Rain drops are falling vertically at a uniform speed of 18 m/s. Wind is blowing from west to east in a horizontal direction at a speed of 12 m/s. A man is running towards east at a speed of 8 m/s. What should be the inclination of umbrella with respect to the horizontal?

Solution First of all identify the velocities with respect to *x*-*y*-axes as shown

$V_R = -18j$ m/s, rain

$V_w = +12i$ m/s, wind

$V_m = +8i$ m/s

(Fig. 12.7), man

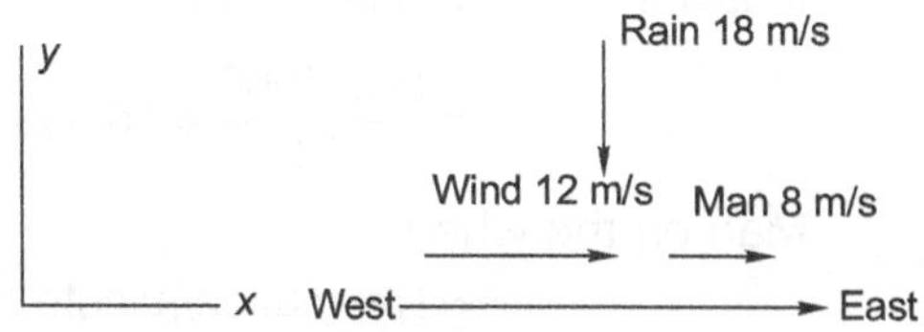

Fig. 12.7

Water drops are simultaneously subjected to wind velocity and vertical velocity

$$V'_R = V_R + V_w = -18j + 12i \text{ m/s}$$

Relative velocity of rain with respect to man

$$V'_{R/m} = V'_R - V_m = -18j + 12i - 8i$$

$$= -18j + 4i \text{ m/s}$$

$$|V'_{R/m}| = \sqrt{18^2 + 4^2} = 18.44 \text{ m/s}$$

$$\tan\theta = \frac{18}{4} = 4.5$$

$\theta = 77.5°$ (umbrella inclination) **(Fig. 12.8).**

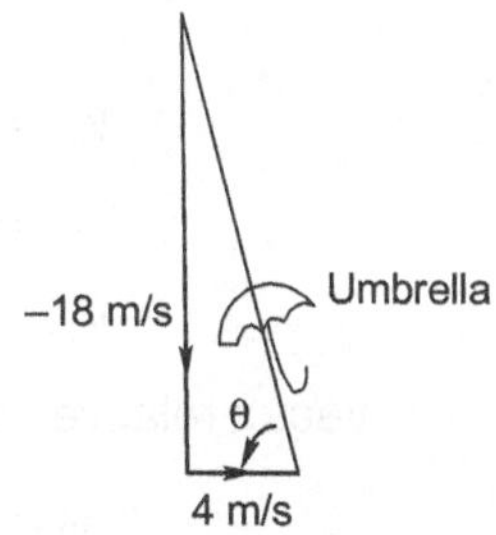

Fig. 12.8

Problem 12.2 The velocities of boats *A* and *C* are shown in **Fig. 12.9**. The relative velocity of boat *B* with respect to boat *A* is 2 m/s, ∠40°. What is the relative velocity of boat *B* with respect to boat *C*? What is the change in the position of *B* with respect to *C* in 15 seconds?

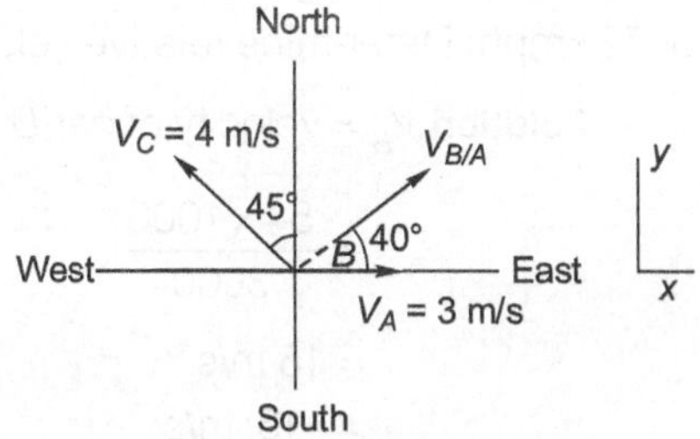

Fig. 12.9

Solution $V_C = -0.707 \times 4i + 0.707 \times 4j$

$$= -2.828i + 2.828j \text{ m/s}$$

$$V_A = +3i \text{ m/s}$$

$$V_{B/A} = 2 \times \cos 40° i + 2 \times \sin 40° j \text{ m/s}$$

$$= 2 \times 0.766i + 2 \times 0.643j \text{ m/s}$$

$$= 1.532i + 1.286j \text{ m/s}$$

$$V_{B/A} = V_B - V_A = 1.532i + 1.286j \text{ m/s}$$

$$V_B = 3i + 1.532i + 1.286j \text{ m/s}$$

$$= 4.532i + 1.286j \text{ m/s}$$

$$V_{B/C} = V_B - V_C = 4.532i + 1.286j + 2.828i - 2.828j$$

$$= 7.36i - 1.546j \text{ m/s}$$

$$|V_{B/C}| = \sqrt{7.36^2 + 1.546^2} = \sqrt{54.1696 + 2.3901}$$

= 7.52 m/s, relative velocity of *B* with respect to velocity of

Change in position = 15 × 7.52 in 15 seconds

= 112.81 m.

Problem 12.3 A ferris wheel of radius 8 m is rotating at 3 revolutions per minute. A car on a straight road is accelerating at 2 m/s^2. At a particular instant when the speed of the car is 54 kmph, determine (*a*) relative velocity of the car with respect to a man sitting on ferris wheel at an angle of 45° as shown in Fig. 12.10.

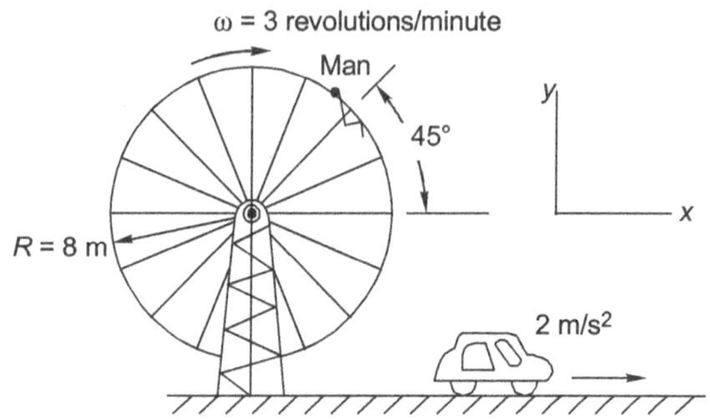

Fig. 12.10

Solution V_C = velocity of car

$$= \frac{54 \times 1000}{3600} = 15 \text{ m/s} = +15i \text{ m/s}$$

Man on the wheel

$$\omega = 3 \text{ revolution/minute}$$

$$= \frac{3 \times 2\pi}{60} = 0.314 \text{ radian/sec}$$

$$V_m = \text{velocity of man} = \omega R$$

$$= 0.314 \times 8 = 2.512 \text{ m/s (absolute)}$$

$$V_m = +1.78i - 1.78j \text{ m/s (Fig. 12.11)}$$

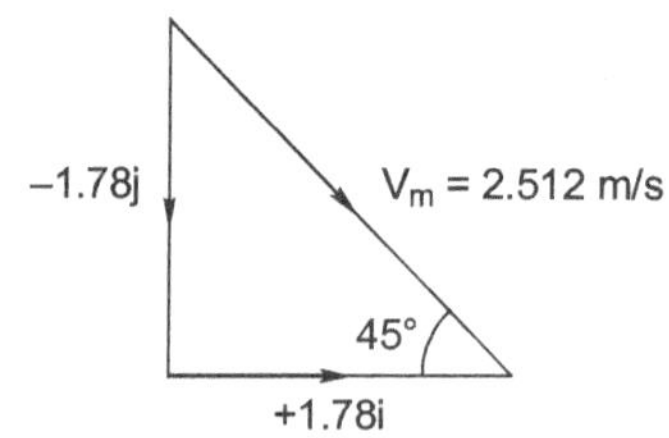

Fig. 12.11

Vector relative velocity (Fig. 12.11)

$$V_{c/m} = V_c - V_m = 15i - 1.78i + 1.78j$$

$$= 13.22i + 1.78j \text{ m/s}$$

$$V_{c/m}, \text{ magnitude} = \sqrt{13.22^2 + 1.78^2} = 13.34 \text{ m/s.}$$

Problem 12.4 Car *A* is moving along a curved road of radius *R* = 150 m, with a constant speed of 54 kmph. Car *B* is accelerating at 1.5 m/s^2 in the direction shown in Fig. 12.12, and at a particular instant reaches a speed of 72 kmph. Deter-mine relative velocity of *B* with respect to car *A*.

Solution V_B = velocity of car *B*

$$= \frac{54 \times 1000}{3600}$$

$$= 15 \text{ m/s}$$

$$= -15i \text{ m/s}$$

Acceleration of car *B*

$$a_B = -1.5i \text{ m/s}^2$$

$$V_A = \text{velocity of car } A$$

$$= \frac{72000}{3600} = 20 \text{ m/s}$$

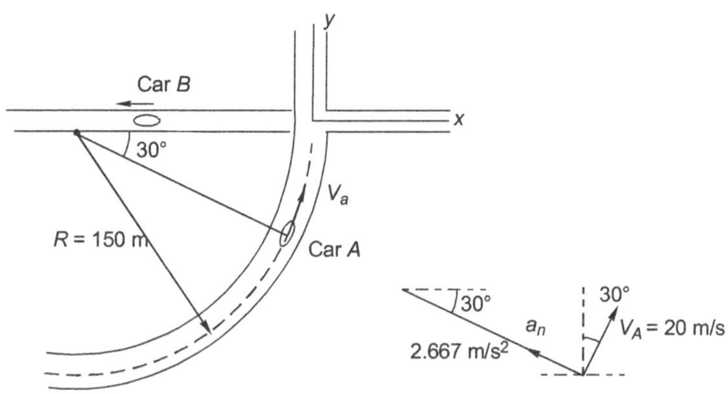

Fig. 12.12

$$= 20 \times \sin 30^\circ\, i + 20 \times \cos 30^\circ\, j \text{ m/s}$$

$$= 10i + 17.32j \text{ m/s}$$

$$V_{B/A} = \text{Relative velocity of car } B \text{ with respect to car } A$$

$$= V_B - V_A = -15i - 10i - 17.32j = -25i - 17.32j \text{ m/s}$$

$$|V_{B/A}| = \sqrt{25^2 + 17.32^2} = 30.41 \text{ m/s}$$

Problem 12.5 A rod AB is moving along vertical and horizontal planes as shown in **Fig. 12.13**. End B of the rod is moving at a velocity 3 m/s. Choose A as a reference point, express the given motion as equivalent to a translation with B and a simultaneous rotation about A. Length of $AB = 0.5$ m.

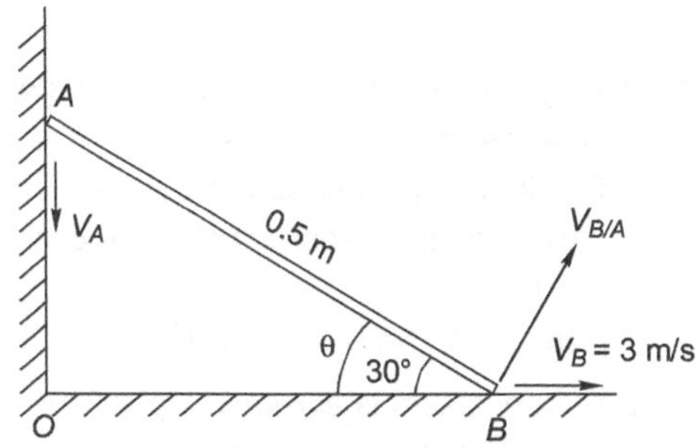

Fig. 12.13

Solution If we choose A as the reference, end A is moving downwards with a velocity V_A and end B is moving with velocity $V_B = 3$ m/s.

The motion of the rod AB is equivalent to sum of (*i*) motion of translation $A-A'$, $B-B'$ with velocity V_B plus (*ii*) motion of rotation with angular velocity ω, anticlock-wise. Take $ob = V_B$ to some suitable scale. From o, draw a line parallel to velocity V_A, from b draw a line parallel to velocity $V_{B/A}$, perpendicular to line AB at the particular instant. Complete the velocity triangle. From the velocity triangle oab **(Fig. 12.14)**.

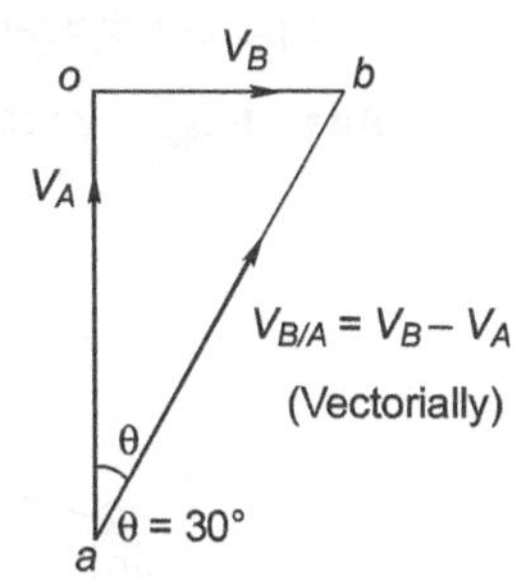

Fig. 12.14

$$V_A = \frac{V_B}{\tan\theta} = \frac{3}{\tan 30°} = \frac{3}{0.577} = 5.2 \text{ m/s}$$

Relative velocity of B with respect to A

$$V_{B/A} = \frac{V_B}{\sin\theta} = \frac{3}{0.5} = 6 \text{ m/s}$$

ω, angular velocity of the rod $= \dfrac{V_{B/A}}{\text{length } AB} = \dfrac{6}{0.5} = 12$ rad/s.

Remember

- No motion is absolute motion. Motion is measured with respect to co-ordinate axes which are fixed on earth, but the earth is moving.
- For most engineering problems, the effect of motion of earth is negligible.
- In many engineering problems motion of an object is analysed with respect to an observer who is moving *i.e.*, relative motion.
- Position vector r_B = position vector r_A + position vector $r_{B/A}$.
- $V_B = V_A + V_{B/A}$ relative velocity of B with respect to A,

  $V_{B/A} = V_B - V_A$ **(Fig. 12.15)**.

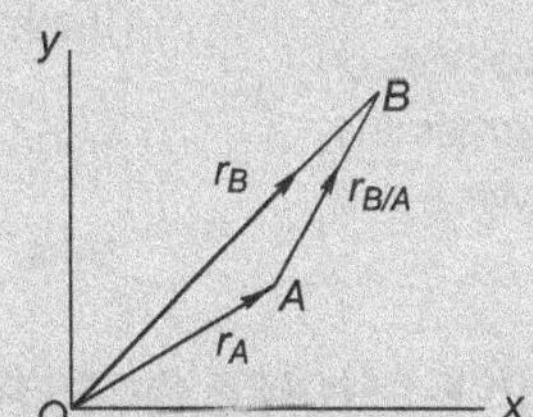

Fig. 12.15

- Acceleration, $a_B = a_A + a_{B/A}$.

  Relative acceleration of B with respect to A

  $a_{B/A} = a_B - a_A$

  $V_{B/A} = \omega \times r$ = angular velocity × distance between points B and A.
- Any plane motion of a body can be replaced by (*i*) motion of translation about any point A+ (*ii*) motion of rotation about the same point A

  $V_B = V_A + V_{B/A}$.

PRACTICE PROBLEMS

12.1 A cyclist travelling towards east at a speed of 9 km/hour feels that the wind is blowing directly from north. On doubling his speed, the cyclist feels the wind to blow from north-east. Find the direction and speed of the wind.

[**Ans:** *N–W*, 12.728 km/hour].

12.2 A mechanism is shown in Fig. 12.16. Bar *AB* has a constant angular velocity of 4 rad/s, counter clockwise. Determine the angular velocity of bars *BC* and *CD*.

[**Hint:** $V_B = \omega \times AB$, is perpendicular to *AB* at point *B*, V_C is perpendicular to *CD* at point *C*].

[**Ans:** $\omega_{BC} = 2$ rad/s, (*ccw*), $\omega_{CD} = 6$ rad/s (*ccw*)].

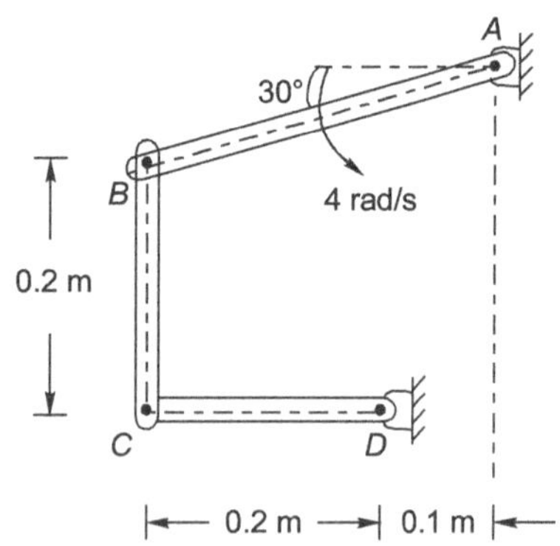

Fig. 12.16

12.3 A rod *AB*, 1 m long is moving along vertical and horizontal planes as shown in Fig. 12.17. End *A* of rod is moving downwards with a constant velocity of 2 m/s. Choose *B* as a reference point, express the given motion as equivalent to a translation with *A* and simultaneous rotation about *B*.

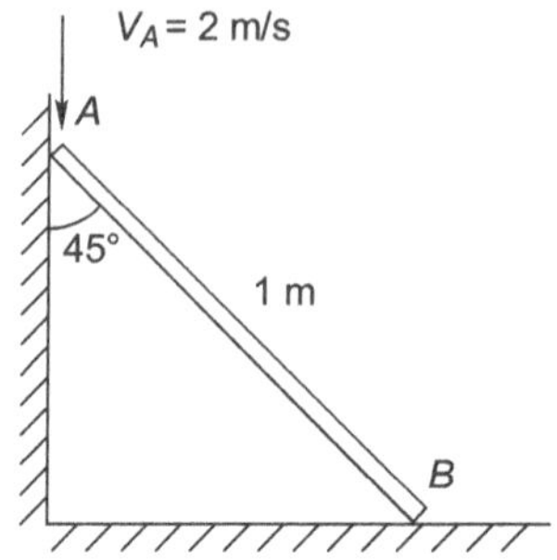

Fig. 12.17

[**Ans:** $V_B = \overrightarrow{2 \text{ m/s}}$, $V_{B/A} = 2.828$ m/s,
Angular velocity of rod= 2.828 rad/s (*ccw*)].

12.4 A railway coach having ordinary cross seats is moving at 3 m/s along the platform. A person runs at 4 m/s on the platform in such a direction that he enters the railway coach parallel to the cross seats. What is the direction of velocity of man? With what velocity he enters the coach?

[**Hint:** $V_{m/c} = 4\sin\theta i + 4\cos\theta j - 3j = 4\sin\theta i$].

[**Ans:** $\theta = 41.4°$ with the platform direction, with 2.645 m/s velocity man enters the coach].

12.5 A scooterist is travelling on a straight road with a constant velocity of 12 m/s and a train is running on a track parallel to the road passes the scooterist in 16 seconds, when it runs in the same direction and in 6 seconds when it runs in the opposite direction of scooterist. If the speed of the train is the same in both the cases, determine the length of the train and its speed.

[**Hint:** Say L = length of the train; V_m, V_t velocity of man and train $\frac{L}{V_m + V_t} = 6$, $\frac{L}{V_t - V_m} = 16$].

[**Ans:** 230.4 m; 26.4 m/s].

MULTIPLE CHOICE QUESTIONS

12.1 A car *A* moves with velocity 10 m/s towards east. Another car *B* moves towards south with velocity of 10 m/s. What is the relative velocity of *A* with respect to the velocity of *B* (**Fig. 12.18**)?

(*a*) 14.14 m/s NE
(*b*) 14.14 m/s SE
(*c*) 20 m/s NE
(*d*) 7.07 m/s SE

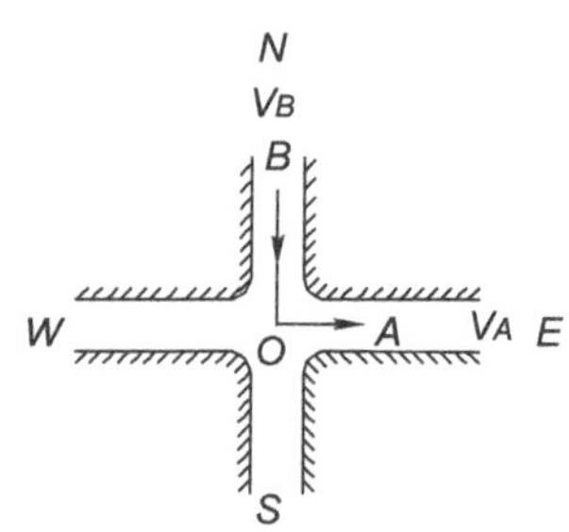

Fig. 12.18

12.2 Two ships *A* and *B* start from *O* with velocities of 100 kmph NE and 100 kmph NW. What is the distance between the two after 2 hours (**Fig. 12.19**)?

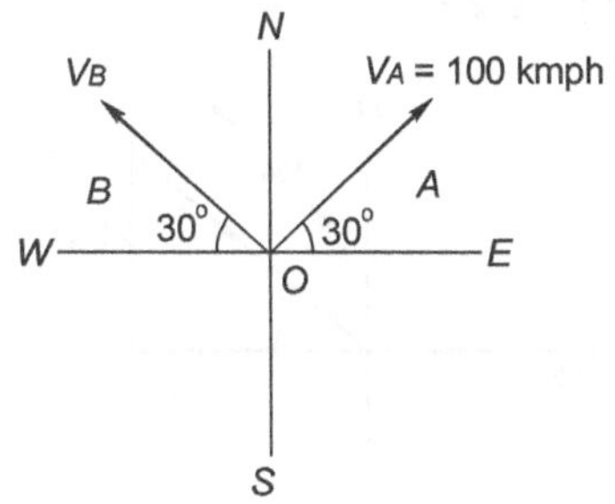

Fig. 12.19

(*a*) 200 km (*b*) 346.4 km
(*c*) 400 km (*d*) 461.9 km

12.3 A rod *AB*, 1 m long placed against a vertical wall and ground floor, as shown in Fig. 12.20. Velocity of end *B* of rod is 2 m/s at the particular instant as shown in **Fig. 12.20**. What is angular velocity of rod at this instant?

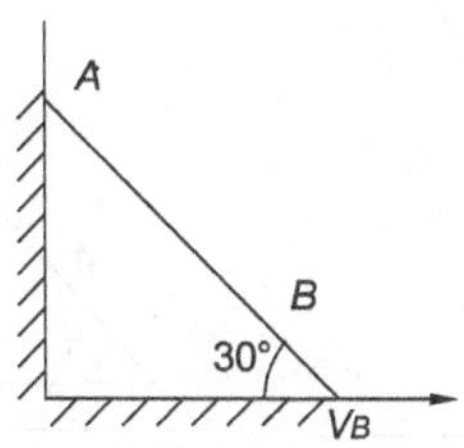

Fig. 12.20

(*a*) 1.732 rad/s (*b*) 2 rad/sec
(*c*) 2.31 rad/s (*d*) 4 rad/s

12.4 A mechanism is shown in **Fig. 12.21**. Bar *AB* has a constant angular velocity of 5 rad/sec counter-clockwise. What is the angular velocity of bar *CD*?

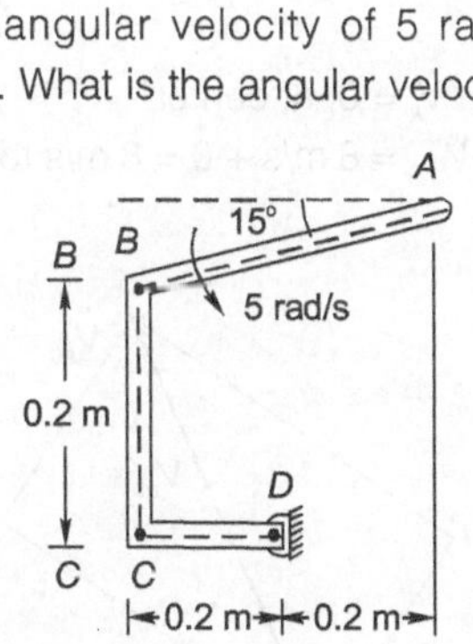

Fig. 12.21

(*a*) 10 rad/s (*b*) 10.35 rad/s
(*c*) 8.66 rad/sec (*d*) None of these

12.5 A cyclist travelling towards east at a speed of 9 km/hour feels that the wind is blowing directly from north. What is the speed of wind?

(*a*) 12.73 kmph NE (*b*) 9 kmph NW
(*c*) 6.303 kmph NE (*d*) None of these

12.6 A railway coach having ordinary cross seats is moving at 3 m/s along the platform. A person runs at 4 m/s on the platform in such a direction that he enters the railway coach parallel to cross seats. What is the direction of velocity of man?

(*a*) 41.8° (*b*) 45°
(*c*) 48.6° (*d*) None of these

12.7 As shown in figure, a person A is standing at the centre of a rotating platform facing person B who is riding a bicycle, heading East. The relevant speeds and distances are shown in given figure person, a bicycle, heading East. At the instant under consideration, what is the apparent velocity of B as seen by A? (**Fig. 12.22**)

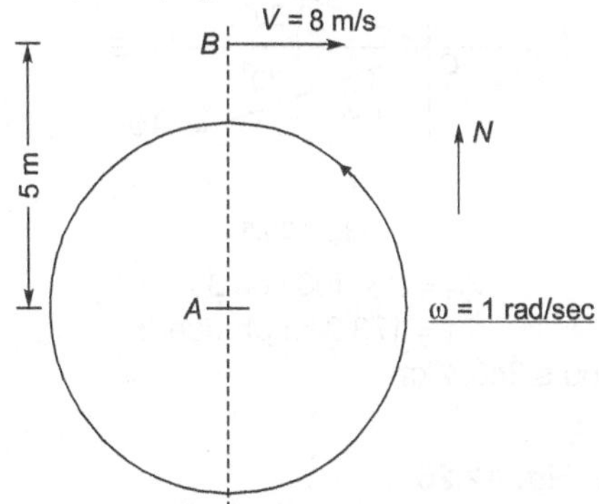

Fig. 12.22

(a) 3 m/s heading East
(b) 3 m/s heading West
(c) 8 m/s heading East
(d) 13 m/s heading East

[GATE, 1999 : 2 Marks]

12.8 A shell is fired from a cannon. At the instant the shell is just about to leave the barrel, its velocity relative to the barrel is 3 m/s, while the barrel is swinging upwards with a constant angular velocity of 2 rad/s. The magnitude of the absolute velocity of the shell is

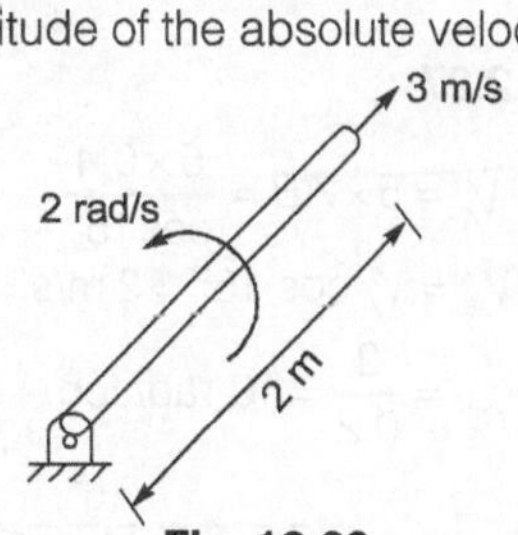

Fig. 12.23

(a) 3 m/s (b) 4 m/s
(c) 5 m/s (d) 7 m/s

[GATE, 2005 : 2 Marks]

Answers

12.1 (a)	12.2 (b)	12.3 (d)	12.4 (a)	12.5 (a)
12.6 (a)	12.7 (c)	12.8 (c)		

EXPLANATIONS

12.1 (a)

See Fig. 12.24.

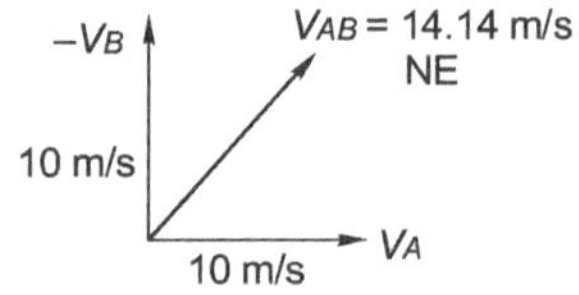

Fig. 12.24

12.2 (b)

See Fig. 12.25

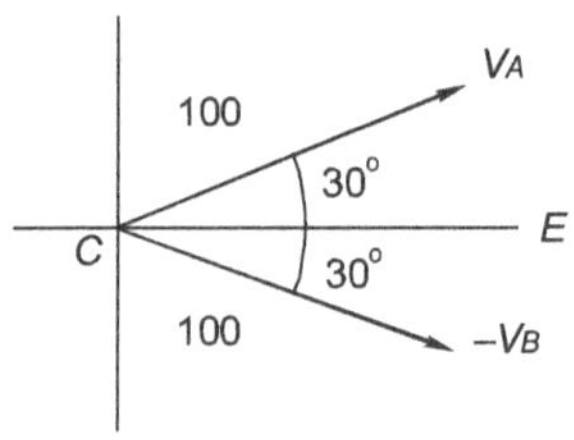

Fig. 12.25

$$V_{AB} = 2 \times 100 \cos 30^\circ$$
$$= 173.2 \text{ kmph due } E$$

2 hours 346.4 km.

12.3 (d)

See Fig. 12.26

$V_B = 2$ m/s; $V_{BA} = 4$ m/s

$\omega_{BA} = 4/1 = 4$ rad/sec.

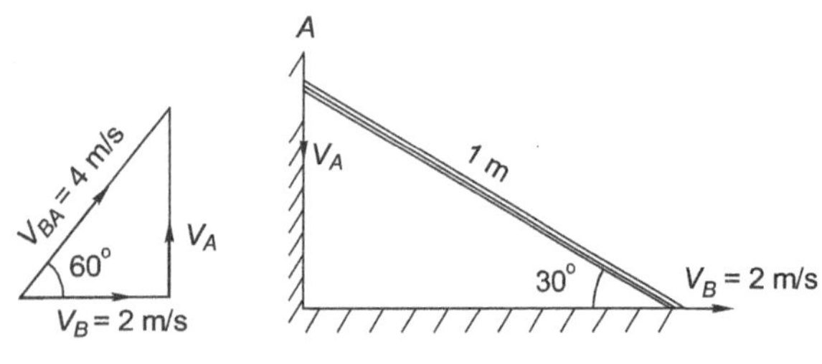

Fig. 12.26

12.4 (a)

See Fig. 12.27

$$V_A = 5 \times AB = \frac{5 \times 0.4}{\cos 15^\circ}$$
$$V_A' = V_A \cos 15^\circ = 2 \text{ m/s} = V_C$$
$$V_{CD} = \frac{2}{0.2} = 10 \text{ rad/sec.}$$

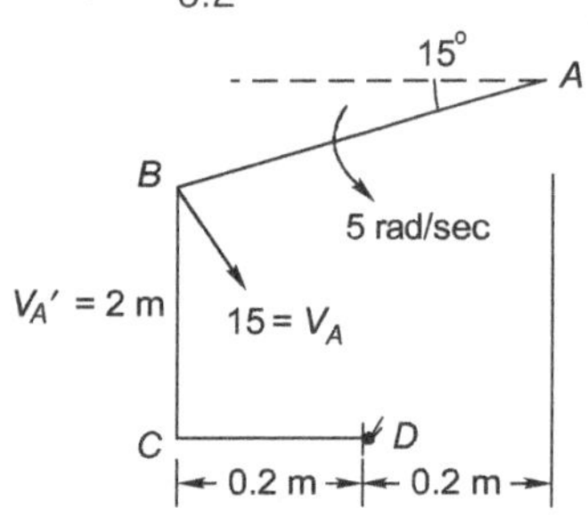

Fig. 12.27

12.5 (a)

See Fig. 12.28

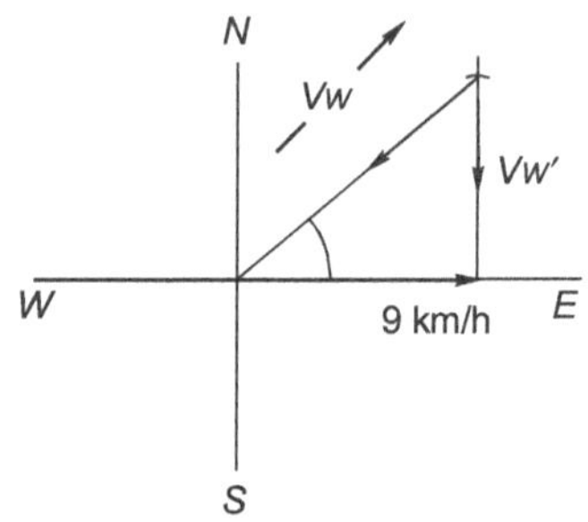

Fig. 12.28

$$V_c - V_w = V_w'$$
$$V_w = \frac{9}{0.707} = 12.73 \text{ km/hr (NE).}$$

12.6 (a)

See Fig. 12.29

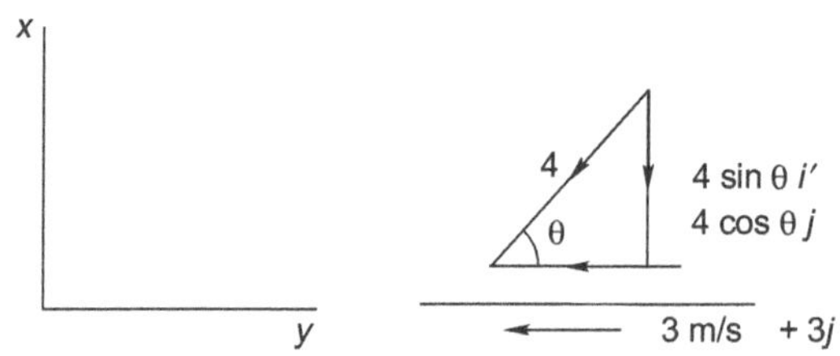

Fig. 12.29

$$4 \cos \theta\, j - 3j + 4 \sin \theta\, i = 4 \sin \theta\, i$$
$$\cos \theta = 0.75, \quad \theta = 41.8^\circ.$$

12.7 (c)

$V_A = 0$, at center

$V_{B/A} = 8 \text{ m/s} + 0 = 8$ m/s towards east

12.8 (c)

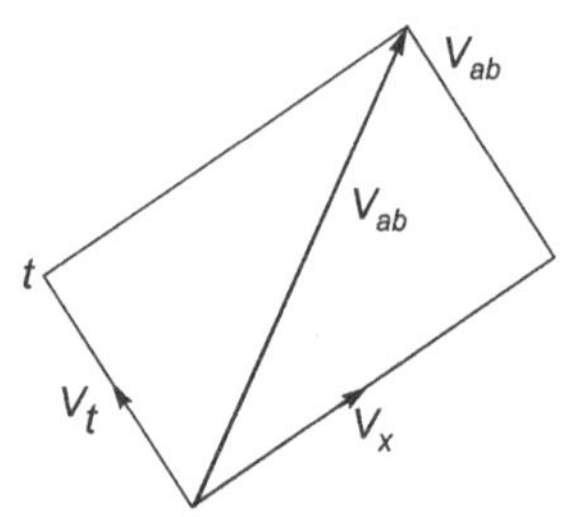

Fig. 12.30

$$V_{radial} = 3 \text{ m/s}$$
$$V_{tangent} = 2 \times 2 = 4 \text{ m/s}$$
$$V_{abs} = \sqrt{3^2 + 4^2} = 5 \text{ m/s}$$

■■■■

13

CHAPTER

Motion of Rotation

13.1 Introduction

Invention of a wheel is the biggest invention of civilization where the wheel rotates about an axis. All the particles on wheel rotate in concentric circles of different radii, along parallel circular planes. **Fig. 13.1** shows a wheel rotating about axis OO', showing particles of wheel rotating in parallel circular planes.

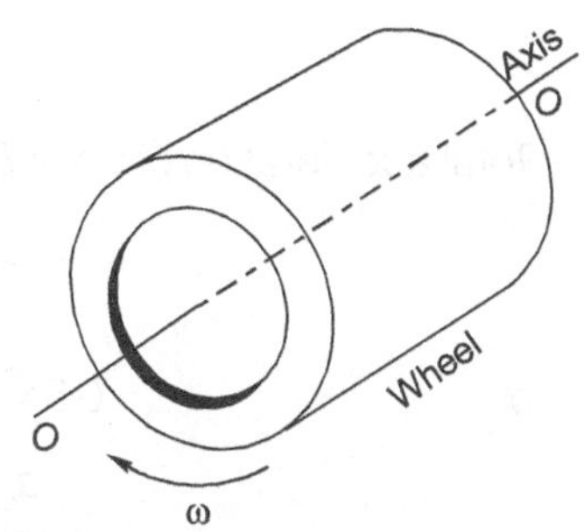

Fig. 13.1

There is another special type of motion *i.e.*, *curvilinear translation*, in which particles of a body move in parallel circular planes but of *same radius* as shown in **Fig. 13.2.**

A plate *ABCD* is held by two links at ends *A* and *B*. During the curvilinear translation any *straight line drawn on the plate* will remain in the same direction. *Line EF remains straight and occupies position* $E'F'$ *during the curvilinear translation.* It can be observed that in such a motion, all the particles of the body move along parallel circles (circle AA' is parallel to circle CC', circle BB' is parallel to circle AA'), but of same radius. However their centres of rotation are different.

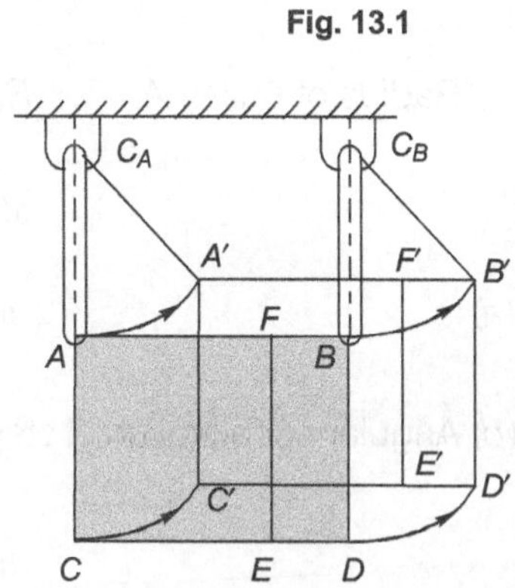

Fig. 13.2

Considering the same rectangular plate having motion of rotation. Plate is rigidly connected to a link *OE*. Link is supported by a pin at *O*. Plate moves in anticlockwise direction by an angular displacement θ as shown. All the particles of the plate and the link have rotated by an angle θ about the same axis of rotation passing through *O*. Axis is perpendicular to the plane of the plate as shown in **Fig. 13.3**. Particles located at *A*, *B*, *C* and *D* are moving in concentric circles of radii *OA*, *OB*, *OC* and *OD*. Therefore in motion of rotation all the particles of the body move in concentric circles of different radii about the same *fixed axis.* Particles located at axis of rotation have zero velocity and zero acceleration.

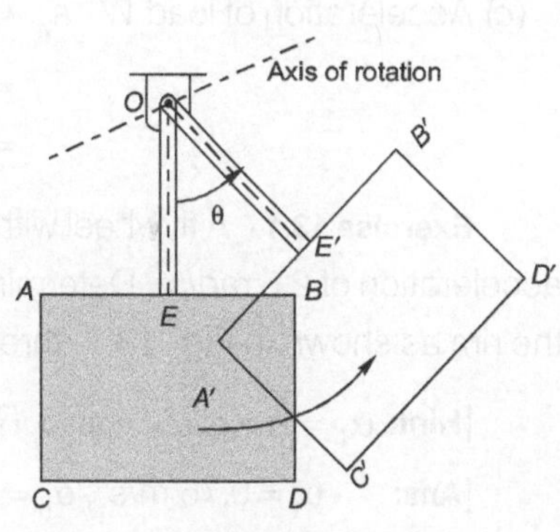

Fig. 13.3

13.2 Rotation about a Fixed Axis

Let us consider a disc of radius R rotating about fixed axis z-z passing through its center O. At a particular instant say

$$\omega = \text{angular velocity in rad/sec}$$

$$\alpha = \text{angular acceleration in rad/sec}^2$$

Linear velocity of point A as periphery

$$V_A = \omega R$$

Linear velocity of point B at radius

$$V_B = \omega r$$

∴ Tangential accelerations as A,

$$a_t = \alpha R$$

Normal accelerations at A

$$a_n = \omega^2 R \text{ (towards the center)}$$

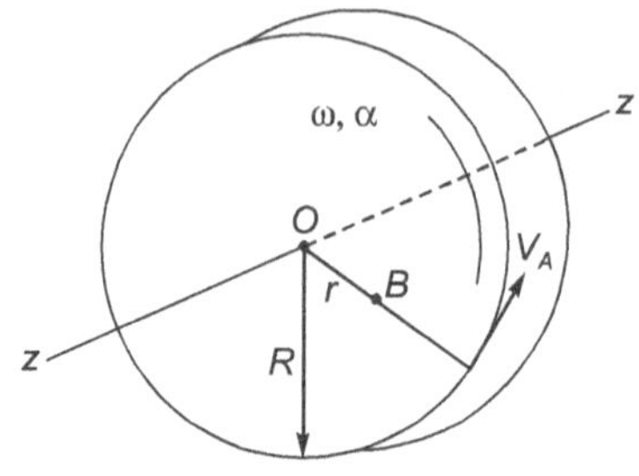

Fig. 13.4

Example 13.1 A load W is connected to a stepped pulley by one of the inextensible cables as shown in **Fig. 13.5**. Motion of pulley A is controlled by a cable running at constant acceleration of 0.15 m/s^2 as shown in the figure, such that pulley rotates in clockwise direction. Total acceleration of the point D on pulley A is 0.54 m/s^2. If R_1 = 6 cm and R_2 = 10 cm, determine: (*a*) angular velocity of stepped pulley, (*b*) angular acceleration of stepped pulley, (*c*) acceleration of load W.

Solution Constant acceleration of cable

$$= 0.15 \text{ m/s}^2$$

$$= \text{Tangential acceleration, } a_t$$

Total acceleration of point D = 0.54 m/s^2 = a

$$a = \sqrt{a_t^2 + a_n^2} \text{ (Fig. 13.6)}$$

or

$$0.54^2 = 0.15^2 + a_n^2$$

$$a_n = \text{normal acceleration}$$

$$= \sqrt{0.2916 - 0.0225} = \sqrt{0.2691} = 0.5188 \text{ m/s}^2$$

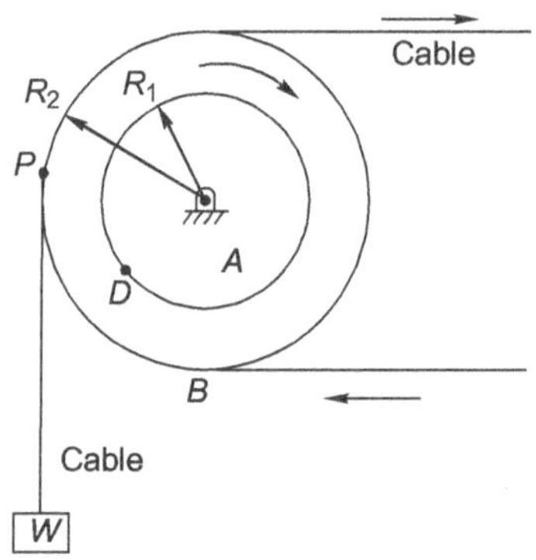

Fig. 13.5

Radius of pulley A, $R_1 = 0.06$ m

$$\omega^2 = \frac{0.51875}{0.06} = 8.64583$$

(*a*) $\omega = 2.94$ rad/s, angular velocity at a particular instant

= angular velocity of stepped pulley

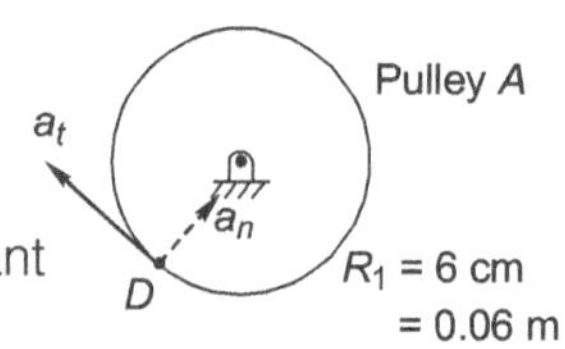

Fig. 13.6

(*b*) Angular acceleration of stepped pulley,

$$\alpha = \frac{a_t}{R_1} = \frac{0.15}{0.06} = 2.5 \text{ rad/s}^2$$

(*c*) Acceleration of load W, $a_W = \alpha \times R_2$

$$= 2.5 \times 0.1 = 0.25 \text{ m/s}^2$$

= tangential acceleration of a point P on outer cable or on pulley B.

Exercise 13.1 A flywheel with diameter 600 mm starts from rest with constant angular acceleration of 2.5 rad/s^2. Determine the tangential and normal acceleration of a point P on the rim as shown in **Fig. 13.7**, three seconds after the motion begins.

[**Hint:** $\alpha_t = R \times \alpha, \alpha_n = \omega_3^2 \times R$].

[**Ans:** $\alpha_t = 0.75$ m/s^2, $\alpha_n = 16.875$ m/s^2].

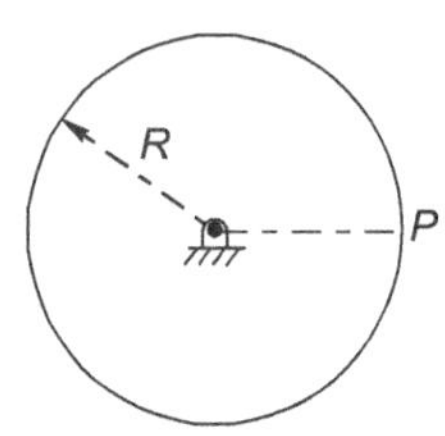

Fig. 13.7

13.3 Various Types of Motion of Rotation

In rectilinear motion we have studied various types of motion on (1) uniform, (2) variable, where acceleration can be dependent on displacement, velocity and time. Similarly there are various types of motion of rotation about a fixed axis and are classified as (*i*) uniform motion with no acceleration, (*ii*) uniformly accelerated or decelerated motion, (*iii*) motion with variable acceleration.

(*i*) **Uniform motion of rotation**

Angular displacement $\theta = \theta_0 + \omega.t$

where θ_0 = initial angular displacement at time $t = 0$

ω = uniform angular velocity in rad/s

θ = angular displacement at time t in radians.

(*ii*) **Uniformly accelerated/decelerated motion**

Angular velocity at any instant,

$$\omega = \omega_0 + \alpha.t$$

where ω_0 = initial angular velocity in radian/s

α = constant angular acceleration in rad/s^2

t = time

Angular displacement, $\theta = \theta_0 + \omega_0 t + \frac{1}{2}\alpha.t^2$

$$\omega^2 - \omega_0^2 = 2\alpha(\theta - \theta_0)$$

Similarly if α is uniform deceleration, expression can be written as

$$\omega = \omega_0 - \alpha.t$$

$$\theta = \theta_0 + \omega_0.t - \frac{1}{2}\alpha.t^2$$

$$\omega^2 - \omega_0^2 = -2\alpha(\theta - \theta_0).$$

(*iii*) **Variable Acceleration:** There are various cases of variable acceleration where (*a*) acceleration varies with time (*b*) acceleration varies with angular velocity, (*c*) acceleration varies with angular displacement, which will be discussed with the help of examples.

Example 13.2 A shaft is rotating with constant angular acceleration. It was observed that 8 seconds after the observation has started, shaft was rotating at 6 rad/s, while 12 seconds after it was observed that shaft has made angular displacement of 80 radians. Determine (*a*) angular acceleration, (*b*) total time taken by the shaft to come to rest, (*c*) angular displacement till the shaft comes to rest.

Solution Say initial velocity = ω_0

Velocity after 8 seconds, $\omega_8 = \omega_0 + 8\alpha = 6$ rad/s

or $\omega_0 = \omega_8 - 8\alpha$, where α is the angular acceleration

$$\omega_0 = (6 - 8\alpha)$$

Moreover $(\omega_{12}^2 - \omega_0^2) = 2.\alpha(\theta - \theta_0)$

But $\theta - \theta_0 = 80$ radians

$$\omega_{12}^2 - \omega_0^2 = 2.\alpha(80) = 160\alpha \quad ...(1)$$

$$\omega_{12} = \omega_0 + 12\alpha, \text{ putting the value of } \omega_0$$

$$= 6 - 8\alpha + 12\alpha = 6 + 4\alpha$$

So $\omega_{12} = 6 + 4\alpha,\ \omega_0 = 6 - 8\alpha$

Putting these values in Equation (1)

$$(6+4\alpha)^2 - (6-8\alpha)^2 = 160\alpha$$

$$36 + 48\alpha + 16\alpha^2 - (36 - 96\alpha + 64\alpha^2) = 160\alpha$$

$$144\alpha - 48\alpha^2 = 160\alpha$$

or

$$-48\alpha^2 = 16\alpha$$

$$\alpha = -0.333 \text{ rad/s}^2$$

$$\omega_0 = 6 - 8\alpha = 6 - 8(-0.333) = 8.664 \text{ rad/s}$$

$$\alpha = -0.333 \text{ rad/s}^2$$

$$\theta - \theta_0 = 100 \text{ revolutions} = 200\pi \text{ radians}$$

$$\omega_f = \text{final velocity} = 0$$

$$\omega_0 = 8.664 \text{ rad/sec}$$

$$\omega_f = \omega_0 - \alpha.t$$

$$0 = 8.664 - 0.333\ .\ t$$

Time, $t = 26$ seconds, *i.e.*, in 26 seconds shaft will come to rest

Angular displacement till the shaft comes to rest

$$= \omega_0 t - \frac{1}{2}\alpha.t^2$$

$$= 8.664 \times 26 - \frac{1}{2} \times 0.333 \times 26^2$$

$$= 225.264 - 112.666 = 112.66 \text{ radians}$$

$$\theta = \omega_{av}.t$$

$$= \frac{1}{2} \times 8.664 \times 26 = 112.64 \text{ radians}$$

Exercise 13.2 A shaft is uniformly accelerated from 8 revolutions per second to 12 revolutions per second in 5 seconds. The shaft continues to accelerate at this rate for the next 10 seconds also. There after the shaft rotates with a uniform angular speed. Find the total time to complete 500 revolutions.

[**Hint:** Calculate ω_0, ω_5; $\alpha = \frac{\omega_5 - \omega_0}{5}$, $\omega_{max} = \omega_5 + 10\alpha$

$$\theta = \theta_1 + \theta_2 = \text{during acceleration} + \text{during uniform motion})$$

$$= 500 \times 2\pi \text{ radians}].$$

[**Ans:** $15 + 14.5 = 29.5$ seconds].

13.4 Instantaneous Centre of Rotation

Any plane motion can be considered as an equivalent to a combined motion of rotation and translation and at a *particular instant*, the angular velocity of the particles of the body is the same as if the body is rotating about a *particular axis*, which is perpendicular to the plane of the body. This axis is called the *Instantaneous axis of rotation* (and it may change its position with time) and the point where this axis intersects the plane of the body is called *Instantaneous centre of rotation.*

Consider four cases as shown in **Fig. 13.8**. In **Fig. 13.8(*a*)**, a body is shown with velocities V_A and V_B of its two particles *B* and *A* respectively, such that $V_B = V_A$, parallel and equal in magnitude. This shows that the body is having motion of pure translation, in which all the particles of the body move along straight lines parallel to each other. In **Fig. 13.8(*b*)** V_B is parallel to V_A but $V_B \neq V_A$, showing thereby that body is moving with rotation. Join the extreme ends of velocity vectors and extend this line till it intersects the line *BA* (joining the two points *B* and *A* on the body). Two lines intersect at *I*, *i.e.*, instantaneous centre of rotation, in which

$$V_B = \omega \times BI, \text{ (where } BI \text{ is radial distance)}$$

$$V_A = \omega \times AI, \text{ (where } AI \text{ is radial distance)}$$

The body is rotating with angular velocity ω. If the velocity V_A is known then

$$\omega = \frac{V_A}{AI} \text{ and } V_B = \omega \times BI$$

Figure 13.8(*c*) shows a body with velocity V_A at *A* and velocity V_B at *B*. These velocity vectors are not parallel. Draw a line from point *B*, perpendicular to velocity vector V_B and another line from point *A*, perpendicular to the velocity vector V_A, these two drawn lines intersect at point *I i.e.*, instantaneous centre of rotation. Say velocity V_A is known then angular velocity of body,

$$\omega = \frac{V_A}{IA}$$

Velocity of the point *B*, $V_B = \omega . IB = \frac{V_A}{I_A} \times IB$

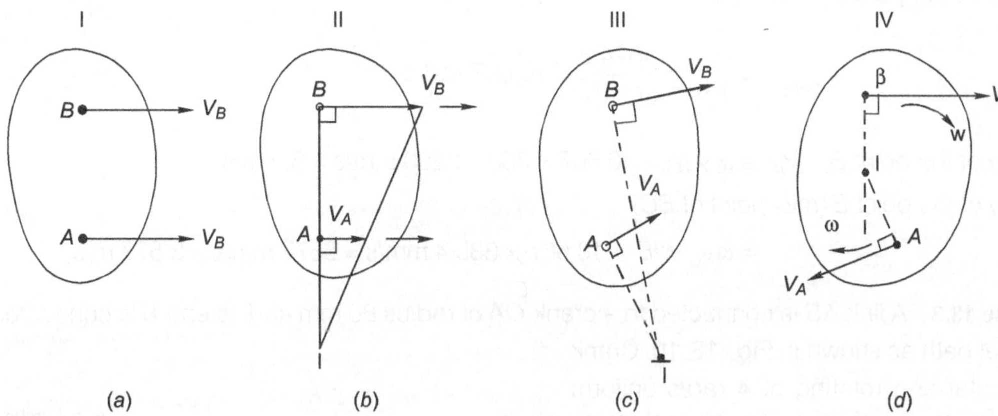

Fig. 13.8

In cases (*b*) and (*c*), instantaneous centre of rotation lies outside the body, however in some cases, this centre may lie within the body as shown in **Fig. 13.9 (*d*)**. The perpendicular to velocity vectors at *B* and *A* intersect at *I*, which lies on the body.

If velocity V_A *of point* A is known, then angular velocity at the particular instant,

$$\omega = V_A / IA$$

Velocity of the point *B*, $V_B = \omega \times IB = V_A \times \frac{IB}{IA}$

The position of the instantaneous centre of rotation does not remain fixed but goes on changing as the body moves. The instantaneous centres describe a path which is known as *centrode. While the velocity of the instantaneous centre is zero, but this centre does not have zero acceleration.*

Example 13.3 Arm *DC* of a four bar linkage has a clockwise angular velocity of 16 rad/s in the position shown in the **Fig. 13.9**, when $\theta = 45°$. Determine velocity of the point *B*, velocity of point *E* (mid-way between *B* and *C*) and angular velocity of link *BC* at the instant shown.

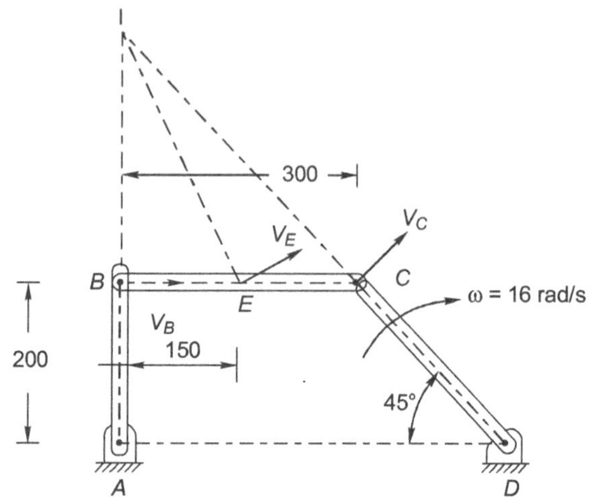

Fig. 13.9

Solution Crank rotates at 16 rad/s in clock-wise direction.

Crank radius, $CD = 200\sqrt{2} = 282.8$ mm

Velocity of point *C*, $V_C = \omega \times CD = 16 \times 282.8$
$= 4524.8$ mm/s
$= 4.525$ m/s

V_C is perpendicular to line *CD* at point *C*.

The instantaneous centre of rotation will lie on a line perpendicular to velocity vector V_C, or on extended line *DC*. Then at this instant, link *AB* will tend to rotate about axis *A* and V_B will be perpendicular to line *AB* at the point *B*. The instantaneous centre of rotation will lie on a line perpendicular to velocity V_B or on extended line *AB*. Extended lines *AB* and *DC* intersect at *I*, which is the instantaneous centre of rotation for connecting rod *BC*.

$$\angle IDA = 45^{\circ}$$

Distance $IB = 300$ mm as $BC = 300$ mm

$$IC = 300\sqrt{2} \text{ mm}, \; IE = \sqrt{300^2 + 150^2} = 335.4 \text{ mm}$$

Angular velocity of connecting rod,

$$\omega_{BC} = \frac{V_C}{IC} = \frac{4524.8}{300\sqrt{2}} = 10.667 \text{ rad/s}$$

Velocity of the point *B*, $V_B = \omega \times IB = 10.667 \times 300 = 3200$ mm/s $= 3.2$ m/s

Velocity of the point *E* (mid-point of *BC*)

$$= \omega_{BC} \times IE = 10.667 \times 335.4 \text{ mm/s} = 3577 \text{ mm/s} = 3.577 \text{ m/s}.$$

Exercise 13.3 A link *AB* is connected to a crank *OA* of radius 80 mm and its end *B* is connected to move in a horizontal path as shown in **Fig. 13.10**. Crank *OA* is horizontal and rotating at 4 rad/s uniform angular velocity. Determine velocity of end *B* of link. What is the velocity of the centre *C* of link *AB*?

[**Hint:** Determine $\angle\theta$;

$\theta = 48.6^{\circ}$,

$BA' = 160\cos\theta = 105.81$ mm,

Distance $IC = 80$ mm].

[**Ans:** $V_A = 320$ mm/s,

$$\omega_{AB} = \frac{320}{105.81} = 3.024 \text{ rad/s}$$

$V_C = 3.024 \times 80 = 241.9$ mm/s,

V_C is perpendicular to *IC* at point *C*].

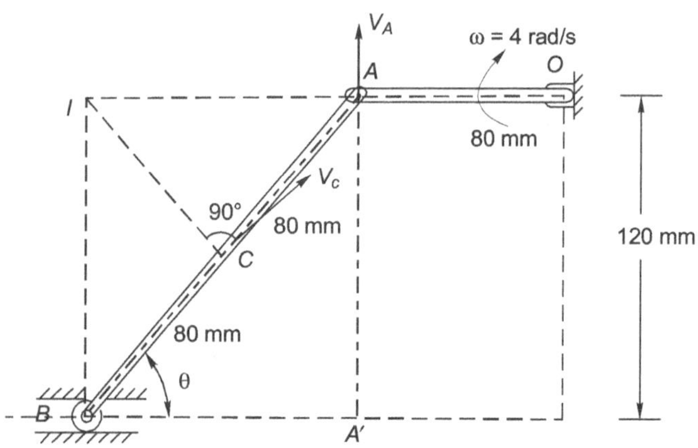

Fig. 13.10

PROBLEMS

Problem 13.1 A swing bridge turns through 90° in 100 seconds, the bridge is uniformly accelerated from rest for the first 30 seconds. Subsequently it turns with a uniform angular velocity for the next 50 seconds. Then the motion of the bridge is uniformly retarded for the last 20 seconds. Determine (*a*) angular acceleration, (*b*) maximum angular velocity and (*c*) angular retardation of the bridge swing.

Solution Fig. 13.11 shows the variation of angular velocity with respect to time as given in the problem.

Say maximum angular velocity $= \omega$ rad/s

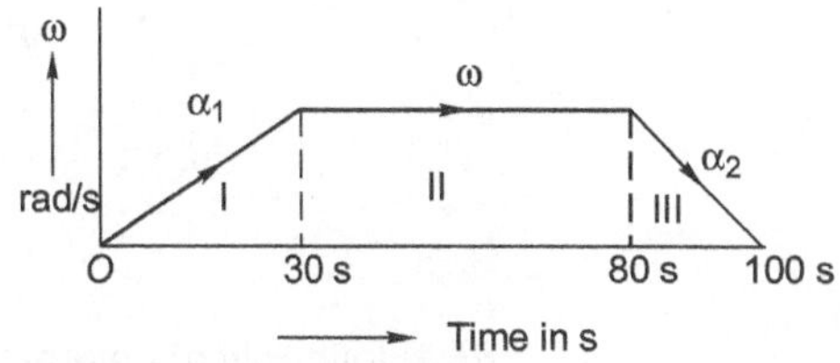

Fig. 13.11

Average velocity during acceleration $= \dfrac{\omega}{2}$

Total angle travelled $= \dfrac{\pi}{2}$ radian

So, $$\frac{\omega}{2}(30)+\omega(50)+\frac{\omega}{2}\times 20 = \frac{\pi}{2}$$

or $$75\omega = \frac{\pi}{2}$$

$$\omega = \frac{\pi}{150} = 2.09\times10^{-2} \text{ rad/s}$$

During Acceleration for 30 seconds (Initial velocity is zero)

Say α_1 = angular acceleration

$$\omega = 0+30\times\alpha_1$$

or $$\alpha_1 = \frac{\omega}{30} = \frac{2.09\times10^{-2}}{30} = 6.98\times10^{-4} \text{ rad/sec}^2$$

During Deceleration for 20 seconds (final velocity is zero)

α_2 = angular deceleration

$$0 = \omega-\alpha_2\times 20$$

$$\alpha_2 = \frac{2.09\times10^{-2}}{20} = 10.45\times10^{-4} \text{ rad/s}^2.$$

Problem 13.2 A circular disc suspended by a slender angular rod as shown in **Fig. 13.12**, perform torsional oscillations. Its angle of rotation measured from equilibrium position is given by $\theta = \theta_0 \cos\omega t$. Determine angular velocity and angular acceleration of the disc.

The maximum angular velocity of the disc is 3π rad/sec and disc makes 2 oscillations per second.

Solution $\theta = \theta_0 \cos\omega t$

Angular velocity $\dot{\theta} = -\theta_0\omega\sin\omega t$ ...(1)

Time period, $T = \dfrac{1}{2}$ second, as the disc makes 2 oscillations per second

$$= \frac{2\pi}{\omega} = \frac{1}{2}$$

Therefore, $\omega = 4\pi$ rad/sec

Maximum angular velocity, $\dot{\theta} = \theta_0 . \omega = \theta_0 \times 4\pi = 3\pi$ rad/sec (as given)

Therefore, $\theta_0 = 0.75$

Angular velocity, $\dot{\theta} = -0.75\times4\pi\sin4\pi t = -3\pi\sin4\pi t$ ...(2)

Angular acceleration $\ddot{\theta} = -\theta_0\omega^2\cos\omega t$ ...(3)

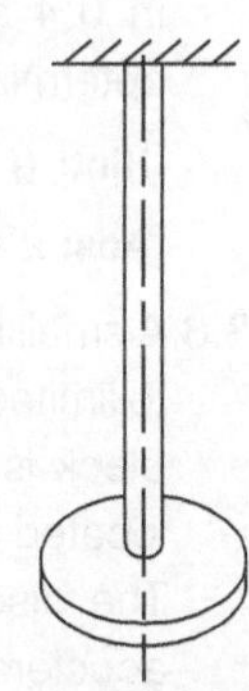

Fig. 13.12

Remember

- In motion of *rotation*, particles of a body move along parallel circular planes. These circular planes make concentric circles about a fixed axis.
- During curvilinear translation, particles of a body move along parallel circular planes of same radius but with different axes of rotation.
- A body is rotating with angular acceleration α and angular velocity (ω) at a particular instant, reduces if a particle as a body is R.

then αR = tangential acceleration of the particle = a_t

ωR = tangential linear velocity of the particle = v

a_n = normal acceleration of particle towards center of rotation

$$= \frac{v^2}{R} = \frac{\omega^2 R^2}{R} = \omega^2 R$$

- Uniformly accelerated motion

$$\omega = \omega_0 + \alpha t, \qquad \text{where } \alpha = \text{angular acceleration}$$

$$\theta = \theta_0 + \omega_0 t + \frac{1}{2}\alpha\tau^2$$

$$\omega^2 = \omega_0^2 + 2\alpha(\theta - \theta_0)$$

- At any instant, the angular velocity of the particles of a body in plane motion, is the same as if the body were rotated about a certain axis perpendicular to the plane of the body. This axis is called the *instantaneous axis of rotation.* The point where the axis intersects the plane of the body is said to be the *instantaneous centre of rotation.*

PRACTICE PROBLEMS

13.1 A small grinding wheel is run by an electric motor with a rated speed of 3000 rpm. When the power is switched on, the motor reaches its rated speed in 4 seconds and when power is switched off the unit comes to rest in 1 minute. Assuming uniform acceleration and then uniform retardation determine total number of revolutions made by the wheel in reaching its rated speed and in coming to rest.
[**Ans:** θ = 10053.12 radians; 1600 revolutions].

13.2 During the starting phase of a computer storage disc, which started from rest and completed 3 revolutions in 0.4 second, assuming uniform acceleration determine angular acceleration of the disc.

[**Hint:** $\theta = 6\pi$ radian, t = 0.4 s].

[**Ans:** 235.6 rad/s^2].

13.3 A small block B on the rotating disc starts sliding due to limited static friction, if the total acceleration on the block is 3.5 m/s^2, as shown in **Fig. 13.13**. Block is located at a radius of 0.3 m from the axis of rotation. The disc starts from rest at t = 0 and is uniformly accelerated at 4 rad/s^2. Determine the time t and angular velocity of the plate when the block starts sliding.

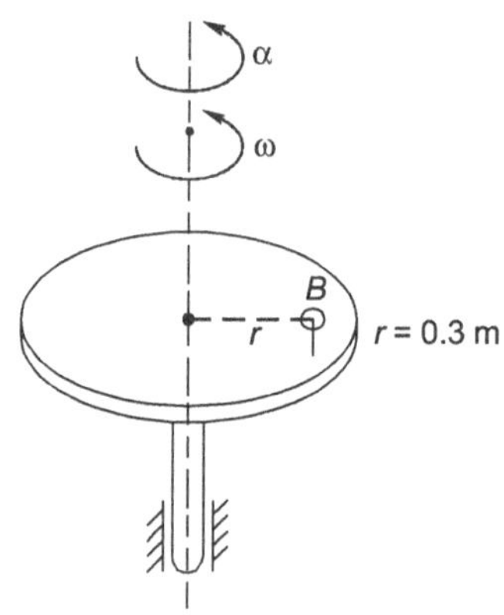

Fig. 13.13

[**Hint:** $a_t = \alpha . r$, $a_n = \omega^2 . r$].

[**Ans:** t = 0.828 s, ω = 3.312 rad/s].

13.4 A mixing drum of 150 mm outside radius rests on two castors, each of 20 mm radius as shown in **Fig. 13.14**. In a span of t seconds, the drum rotates by 15 revolutions, when its speed increased from 30 rpm to 50 rpm. Assuming no slip between drum and castors, determine (*a*) time interval, (*b*) angular acceleration of castors.

[**Hint:** $\omega_2^2 - \omega_1^2 = 2\alpha\theta$, $(\omega_2 - \omega_1) = \alpha t$].

[**Ans:** 22.52 s; 0.698 rad/s^2].

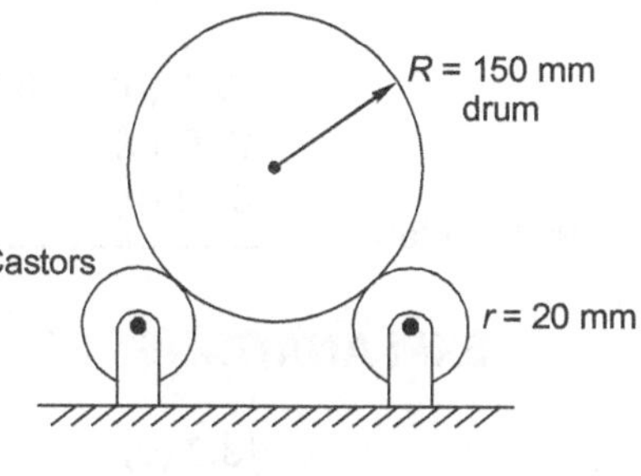

Fig. 13.14

MULTIPLE CHOICE QUESTIONS

13.1 A shaft rotating at 100 rpm about a fixed axis, changes its speed to 900 rpm is 20 seconds. What is average angular acceleration in rad/s^2

(*a*) 4.71 (*b*) 4.19
(*c*) 2.1 (*d*) None of these

13.2 A rotor decreases uniformly from 1800 rpm to rest in 40 seconds. How many revolutions are made before the disc comes to rest?

(*a*) 3600 (*b*) 900
(*c*) 600 (*d*) None of these

13.3 A belt moves without slipping over two idler pulleys *A* and *B* of radius 60 mm and 100 mm respectively (**Fig. 13.15**) at a particular instant, velocity of point *C* is 0.48 m/s. What is normal acceleration over pulley *A*?

(*a*) 10 m/s^2 (*b*) 6 m/s^2
(*c*) 3.84 m/s^2 (*d*) None of these

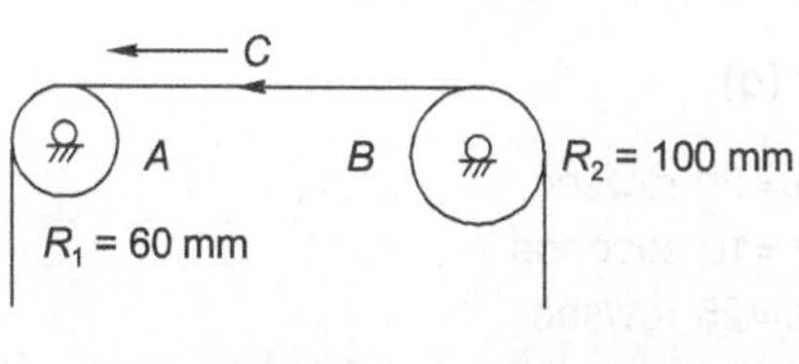

Fig. 13.15

13.4 During the starting phase of a computer storage disc, which started from rest and completes 3 revolutions in 0.5 second, what is angular acceleration of disc in rad/s^2

(*a*) 37.7 (*b*) 75.4
(*c*) 150.7 (*d*) None of these

13.5 A circular disc suspended by a slender angular rod performs torsional oscillations. Angular rotation is given by $\theta = \theta_0 \cos \omega t$. If disc makes 2 oscillations per second, what is angular velocity ω in rad/s

(*a*) 4π (*b*) 3π
(*c*) 2π (*d*) π

13.6 How many revolution a shaft has to rotate to achieve a rated speed of 30 Hertz in 4 seconds?

(*a*) 15 (*b*) 30
(*c*) 60 (*d*) 120

[CSE, Prelim, CE : 2005]

13.7 The rotation of a rigid body is governed by

$\phi = 2.5 \sin\left(\frac{\pi t}{4}\right)$ what is angular velocity at t = 2s?

(*a*) 18 rad/sec (*b*) $\frac{\pi}{4}$ rad/sec
(*c*) $\frac{\pi}{8}$ rad/sec (*d*) Zero

[CSE, Prelim, CE : 2007]

13.8 The acceleration of a particle starting from rest varies with time according to the relation, $a = -S\omega^2 \sin\omega T$. What is displacement of the particle at time T

(*a*) S sinωt (*b*) Sω cosωt
(*c*) Sω sin ωt (*d*) $-\frac{1}{2}\left(S\omega^2 \sin\omega t\right)t^2$

[CSE, Prelim, CE : 2006]

13.9 A grinding wheel rotates at 3000 rpm. When power supply is cut off, the wheel stops completely in 10 seconds. What is the number of revolutions made by the what before coming to rest

(*a*) 3000 (*b*) 1000
(*c*) 500 (*d*) 250

[CSE, Prelim, CE : 2007]

13.10 A wheel starting from rest is accelerated at the rate of 10 rad/sec^2 for an interval of 10 s. It is then made to stop in an interval of 10 s by applying brakes. What is the total rotation of the wheel?

(*a*) 500 radian (*b*) 750 radian
(*c*) 1000 radian (*d*) 1200 radian

[CSE, Prelim, CE : 2008]

13.11 A stone of mass m at the end of a string of length *l* is whirled in a vertical circle at a constant speed. The tension in the string will be maximum when the stone is

(a) at the top of the circle
(b) half way down from the top
(c) quarter way down from the top
(d) at the bottom of the circle

[GATE, 1994: 1 Mark]

Answers

13.1 (b)	13.2 (c)	13.3 (c)	13.4 (c)	13.5 (a)
13.6 (c)	13.7 (d)	13.8 (a)	13.9 (d)	13.10 (c)
13.11 (d)				

EXPLANATIONS

13.1 (b)

$$\omega_0 = \frac{2\times 100\times\pi}{60} = 10.47 \text{ rad/s}$$

$$\omega_f = \frac{2\times 900\pi}{60} = 94.25 \text{ rad/s}$$

$$t = 20 \text{ s}$$

$$\alpha = \frac{94.25-10.47}{20} = 4.19 \text{ rad/s}^2.$$

13.2 (c)

rps= revolution per second = 1800/60 = 30
time= 40 seconds
Number of revolution,

$$N = \frac{\text{rps}+0}{2}\times t = \frac{30+0}{2}\times 40 = 600.$$

13.3 (c)

$$\omega = \frac{0.48}{0.06} = 8 \text{ rad/sec}$$

$a_n = \omega^2 . r = 64\times 0.06 = 3.84 \text{ m/s}^2$.

13.4 (c)

$$\theta = 6\pi = \omega_0\, t + \frac{1}{2}\alpha . t^2$$

$$\omega_0 = 0$$

$$6\pi = \frac{1}{2}\times\alpha\times 0.5^2,\quad \alpha = \frac{6\pi}{0.125} = 150.7 \text{ rad/s.}$$

13.5 (a)

$$T = \frac{1}{2}s = \frac{2\pi}{\omega},\ \omega = 4\pi \text{ rad/s.}$$

13.6 (c)

$\omega = 30$ rev/sec
$t = 4$ seconds
$\omega_0 = 15$ rev/sec
$\theta = \omega_0 . t = 60$ revolution

13.7 (d)

$$\phi = 2.5\sin\left(\frac{\pi t}{4}\right)$$

$$\frac{d\phi}{dt} = \omega = 2.5\times\frac{\pi}{4}\left(\frac{\pi t}{4}\right)$$

$$t = 2 \text{ s},\ \cos\left(\frac{\pi\times 2}{4}\right) = 0$$

$$\omega = 0$$

13.8 (a)

$$a = -S\omega^2\sin\omega t = \frac{d\omega}{at}$$

$$\omega = +\frac{S\omega^2}{\omega}\cos\omega t + C_1$$

$\omega_0 = 0$ at $t = 0$, $C_1 = 0$

$$\omega = \frac{S\omega^2}{\omega}\cos\omega t = \frac{d\theta}{dt}$$

$\theta = +S\sin\omega t + C_2$
$\theta = 0$, at $t = 0$, $C_2 = 0$, $\theta = S.\sin\omega t$

13.9 (d)

$\omega = 50$ rev/sec
$t = 10$ seconds
$\omega_{av} = 25$ rev/sec
$\theta = \omega_{av}\times t = 250$ revolution

13.10 (c)

$\alpha = 10$ rad/sec^2
$\omega_{10} = 100$ rad/sec
$\omega_f = 0$, $\omega_{av} = 50$ rad/sec
$t = 10 + 10 = 20$ seconds
$\theta = 50\times 20 = 1000$ radian

13.11 (d)

At the bottom of the circle
Tension = $mg + m\omega^2 L$

■■■

14

CHAPTER

Kinetics of a Particle

14.1 Introduction

In Kinetics, we study about the magnitude and direction of forces involved in the motion of a particle or a body. Newton's Second Law of motion states that a force is necessary for the change of state of rest or of uniform motion of a body and in many engineering problems, Newton's Second Law is directly applied for the study of acceleration of a body. There are various types of accelerations such as linear acceleration, tangential acceleration, normal acceleration, angular acceleration of a body in motion of translation and or in motion of rotation.

The bodies possess definite shape and size. However when the *motion of all the particles of a body is defined by same parameter then the body can be simplified as a particle*. Such as a coach of a train, when all the particles of the coach possess same velocity and same acceleration, then it can be classified simply as a *particle*. Similarly the motion of a rocket or satellite can be analysed considering them to be particles of given mass. While analysing the effect of a force during any motion, the particle usually selected as the one located at the mass centre of the body and the resultant of all the forces acting on the translating body passes through its mass centre.

14.2 Newton's Second Law of Motion

Several forces F_1, F_2, F_3, ..., F_n are acting on a body as shown in **Fig. 14.1**. If the resultant of these forces is not zero, then body will experience acceleration in the direction of the resultant force

$$F_R = \sum_{i=1}^{n} F_i \neq 0$$

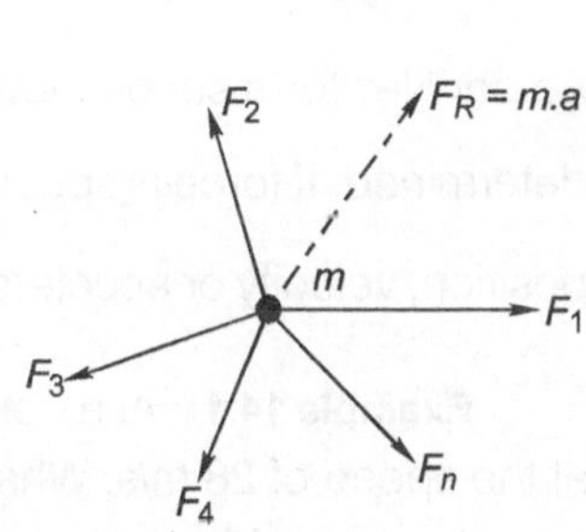

Fig. 14.1

and the body/or the particle will move in the direction of resultant force such that $F_R = m \,.\, a_R$, where a_R is acceleration in the direction of F_R.

In a three dimensional case, resultant force can be expressed as

$$\sum F = F_R = \left(\sum F_x\right) i + \left(\sum F_y\right) j + \left(\sum F_z\right) k$$

$$= F_{xR} i + F_{yR} j + F_{zR} k$$

Similarly acceleration $\quad a_R = a_x i + a_y j + a_z k$

where a_x, a_y, a_z are components of acceleration along 3 co-ordinates axes.

Now let us consider that a particle is subjected to forces F_1, F_2, F_3 and particle is subjected to accelerations a_1, a_2 and a_3 respectively, these acceleration components are proportional to forces applied *i.e.*,

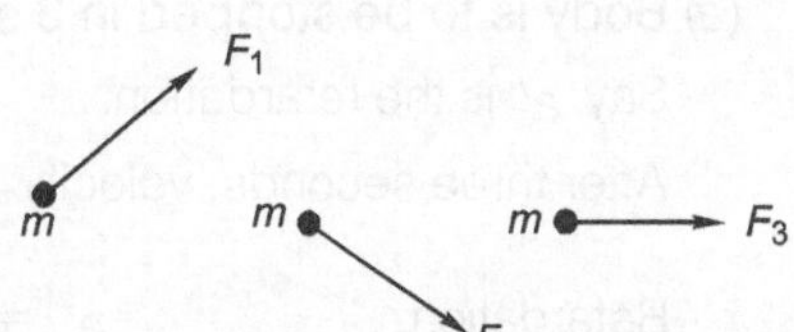

Fig. 14.2

i.e., $a_1 \propto F_1,\ a_2 \propto F_2;\ a_3 \propto F_3$

or $$\frac{F_1}{a_1} = \frac{F_2}{a_2} = \frac{F_3}{a_3} = a \text{ constant } = m \text{ (mass)}$$

So m is a measure of some property of a particle that does not change and remains constant. *It is a measure of inertia of a particle.* More is the mass, more is the inertia of the particle to change its state of rest or of uniform motion.

Equation of motion subjected to external forces $\sum F = m \cdot a$

$$\sum (F_x i + F_y j + F_z k) = m\,(a_x i + a_y j + a_z k).$$

14.2.1 Reference System

During the analyses of motion of a body or particle, system of reference cannot be arbitrary. These co-ordinate axes must have constant orientation with respect to space. Therefore the origin of the co-ordinate system should be attached to the sun, which is the mass center of solar system. Such a system of axes is known as *Newtonian frame of reference.* A system of axes attached to the earth does not constitute a Newtonian frame of reference, because the earth rotates with respect to the space and is accelerated with respect to the sun. However no appreciable error is caused in the solution of common engineering problems if acceleration of a body is determined with respect to the axes attached to the earth.

14.2.2 System of SI Units

In SI units, a Newton is a force acting on a body of mass 1 kg producing an acceleration of 1 m/s^2 in the body as shown in **Fig. 14.3**.

In a *vertical direction*, a body is attracted by the earth towards its center with acceleration g = 9.81 m/s^2 (on the surface of the earth). Force on body of mass 1 kg is attracted by a force of 9.81 Newton. This *force of attraction* by earth is known as *weight of the body*.

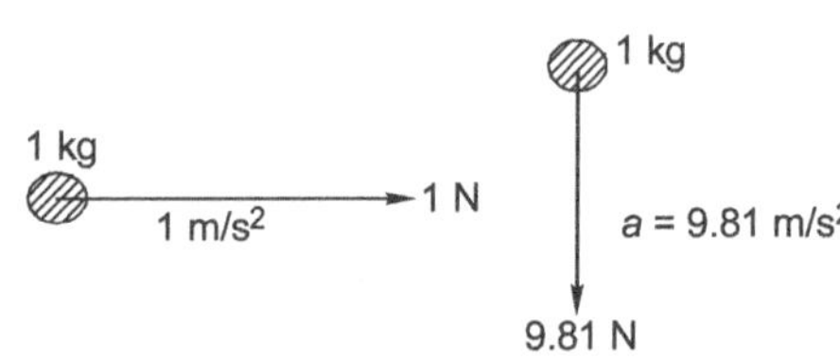

Fig. 14.3

14.3 Types of Problems

As per Newton's second law, $\sum F = ma$, if acceleration is specified then force required by a body of mass m is determined. If force is specified then acceleration can be determined. However if the force is a function of time, position, velocity or acceleration, then equation $\sum F = ma$, becomes a *differential equation.*

Example 14.1 At a certain instant a body of mass 18 kg falling freely under gravity was found to be falling at the speed of 28 m/s. What force will stop the body in (*a*) 3 seconds, (*b*) 30 metres (**Fig. 14.4**)?

Solution Mass, m = 18 kg

Speed V = 28 m/s

g = 9.81 m/s^2

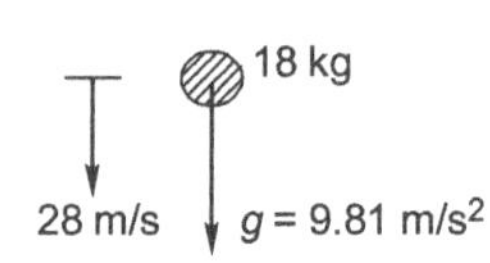

Fig. 14.4

Body is falling under gravity with initial speed of 28 m/s.

(*a*) **Body is to be stopped in 3 seconds:**

Say a' is the retardation.

After three seconds, velocity = 28 + 3 × 9.81 = 57.43 m/s

Retardation, $$a' = \frac{57.43}{3} = 19.143 \text{ m/s}^2$$

Force required, $F = ma' = 18 \times 19.143 = 344.57$ N

(b) Body is to be stopped in 30 metres

Say retardation $= a'$ m/s^2

$$28^2 = 2a' \times S = 2 \times a' \times 30$$

$$a' = 13.067 \text{ m/s}^2$$

But the body is falling under gravity, with acceleration g.

Retardation which must be provided by the force,

$$a'' = a' + g$$

$$a'' = 13.067 + 9.81 = 22.877 \text{ m/s}^2$$

Retarding force required, $P = ma'' = 18 \times 22.877 = 411.786$ N.

Exercise 14.1 A man of mass 68 kg dives vertically downwards into a swimming pool from a tower of height 21 m. He was found to go down in water by 2.5 m and then started rising. What is the average resistance of water? Neglect air resistance.

[**Hint:** During diving in water, due to upward thrust of water, retardation downward can be neglected]

[**Ans:** 5604.4 N].

Example 14.2 A force P is applied to a cart as shown in **Fig. 14.5**, which starts from rest. Determine the velocity and displacement of the cart at $t = 6$ seconds. The force time history is shown in **Fig. 14.6**. Neglect friction.

Solution Say force

$$P = kt^2$$

where k is a constant and t is in seconds

At $t = 6$ s, $P = 54$ N

$$54 = k \times 6^2$$

or constant, $k = \dfrac{54}{36} = 1.5$

So $P = 1.5t^2$ ...(1)

$= 15 \times a$, where 15 kg is mass and a is acceleration in m/s^2

$$1.5t^2 = 15a$$

or $a = 0.1t^2$

$$\frac{dV}{dt} = 0.1t^2$$

$$dV = 0.1t^2\, dt \quad \text{...(2)}$$

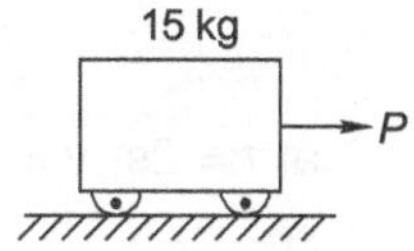

Fig. 14.5

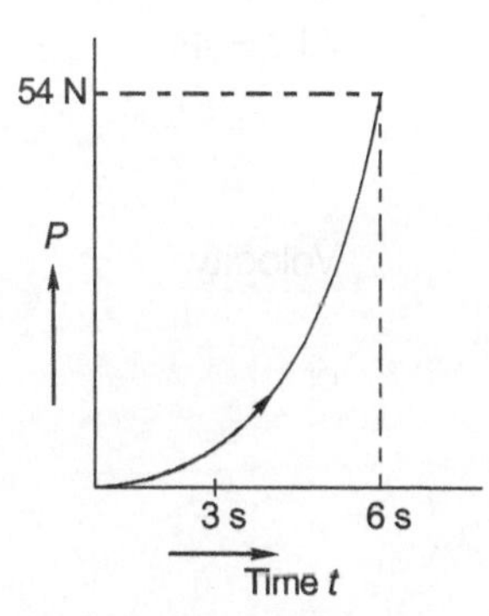

Fig. 14.6

Integrating both sides of Equation (2)

$$V = \left|\frac{0.1t^3}{3}\right|_0^6$$

$$V = \frac{0.1 \times 6^3}{3} = 7.2 \text{ m/s at } t = 6 \text{ s}$$

Again $V = \dfrac{0.1t^3}{3} = \dfrac{t^3}{30}$

or $\dfrac{dS}{dt} = \dfrac{t^3}{30}$

$$dS = \frac{t^3}{30}dt \quad \text{...(3)}$$

Integrating Equation (3)

$$S = \left|\frac{t^4}{120}\right|_0^6, \text{ as graph is only upto 6 seconds}$$

$$S = \frac{6^4}{120} = 10.8 \text{ m}$$

Example 14.3 A particle moves on a vertical line with an acceleration, $a = 3\sqrt{v}$, at $t = 2$ sec, its displacement is 8 m and velocity is 6 m/s. Determine its displacement, velocity and acceleration at time t = 4s, if a is acceleration, v is velocity.

Solution:

$$a = 3\sqrt{v}$$

or
$$\frac{dv}{dt} = 3\sqrt{v}$$

or
$$\frac{dv}{\sqrt{v}} = 3\,dt$$

$$\int\frac{dv}{\sqrt{v}} = 3\,dt$$

$$2\sqrt{v} = 3t + C_1 \quad \text{where } C_1 \text{ is constant of integration}$$

at t = 2s, v = 6 m/s,

so
$$2\sqrt{6} = 3\times 2 + C_1$$

or constant, $C_1 = 4.899 - 6 = -1.101$

So
$$2\sqrt{v} = 3t - 1.101 \quad \text{...(ii)}$$

At t = 4s

$$2\sqrt{v_4} = 4\times 3 - 1.101 = 10.899$$

Velocity,
$$v_4 = \left(\frac{10.899}{2}\right)^2 = 29.697 \text{ m/s}$$

or
$$\sqrt{v} = 1.5\,t - 0.5505$$

$$v = 2.25\,t^2 + 0.303 - 1.6515\,t$$

$$\frac{ds}{dt} = 2.25\,t^2 - 1.6575 + 0.303$$

$$S = \frac{2.25t^3}{3} - \frac{1.6515t^2}{2} + 0.303t + C_2$$

where, C_2 is another constant of integration.

At t = 2s, s = 8m

So,
$$S = 0.75\times 2^3 - \frac{1.6515}{2}\times 4 + 0.303\times 2 + C_2$$

$$S = 6 - 3.303 + 0.606 + c_2$$

$$S = 3.303 + c_2$$

$$c_2 = +4.697$$

or
$$S = 0.75t^3 - \frac{1.6515}{2}t^2 + 0.303t + 4.697$$

$$t = 4\text{s}$$

$$S_4 = 0.75 \times 64 - \frac{1.0515}{2} \times 16 + 0.303 \times 4 + 4.697$$

$$= 48 - 13.212 + 1.212 + 4.697$$
$$= 40.697 \text{ m}$$

Exercise 14.2 A force P is applied to a trolley as shown in **Fig. 14.7**, which *starts from rest*. Determine the velocity and displacement of the trolley at $t = 4$ s. The force history is shown. Neglect friction. Mass of trolley is 15 kg.

[**Hint:** $P = kt$].

[**Ans:** $V = 8$ m/s, $S = 10.667$ m].

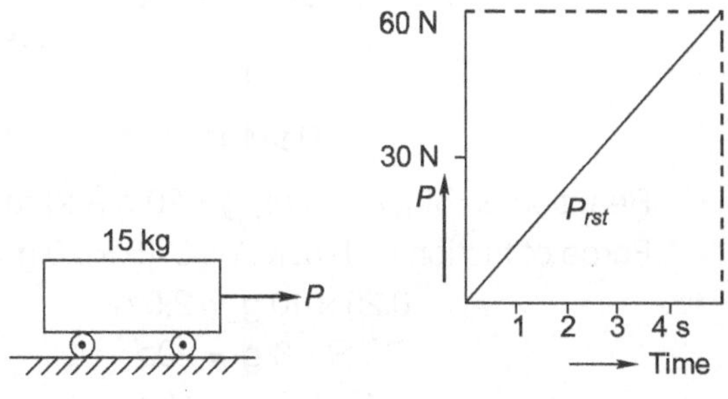

Fig. 14.7

14.4 Motion of Two Bodies in Contact

When two bodies in contact are simultaneously subjected to an external force, then the *contact pressure* between the two surfaces is of importance for analysing the motion of two bodies.

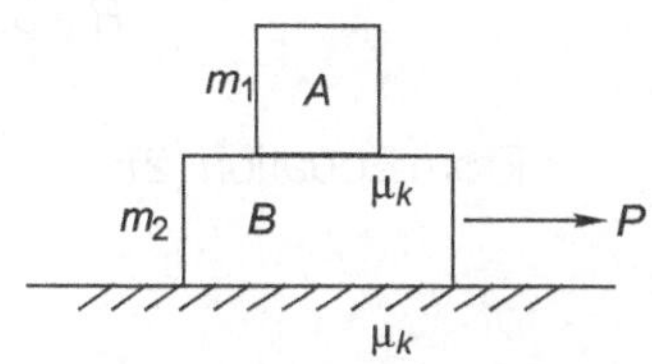

Fig. 14.8

Fig. 14.8 shows two blocks A and B in contact and a force P. Say the coefficient of friction between two surfaces is μ_k. The free body diagram of two masses is shown in **Fig. 14.9**. For block A of mass m_1.

Resultant force on mass m_1

$$F_R = \mu_k m_1 g \text{ as shown.}$$

For block B of mass m_2.

Resultant force will be

$$F_R = P - \mu_k m_1 g - \mu_k (m_1 + m_2) g$$
$$= m_2 a, \text{ where } a \text{ is acceleration of block } B.$$

Note that *friction force* $\mu_k m_1 g$ will tend to move the block A *towards right*.

Mass B will tend to move with acceleration as per the resultant force, F_R on the block B.

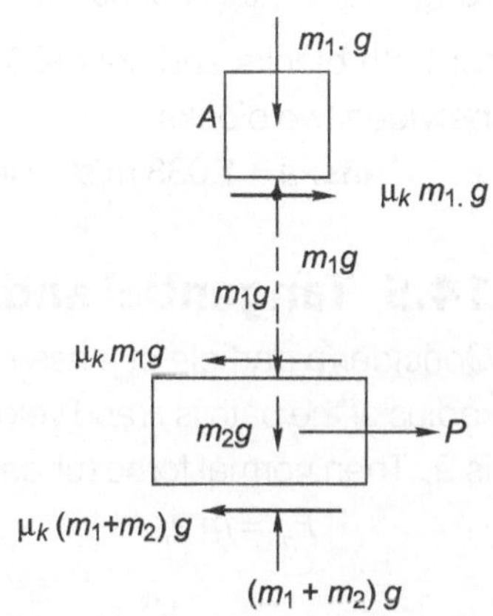

Fig. 14.9

Take another case when two blocks are placed side by side as shown in **Fig. 14.10**. The resultant force in this case will be

$$P - m_1 g . \mu_1 k - m_2 g \mu_2 k = (m_1 + m_2)\, a, \text{ where } a \text{ is acceleration.}$$

To maintain contact between two masses it is necessary that $\mu_{k1} < \mu_{k2}$.

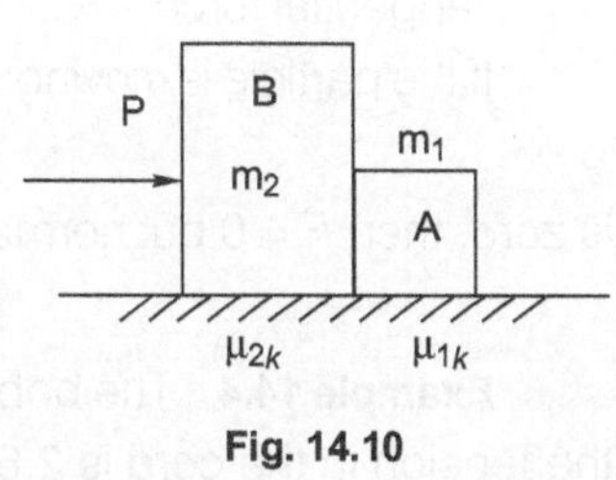

Fig. 14.10

Example 14.3 Two blocks A and B of masses 20 kg and 10 kg are on a rough horizontal floor. A force P pushes the masses at acceleration of 4 m/s^2. What is the magnitude of P and what is the reaction between the two blocks? Coefficient of kinetic friction between block A and floor is 0.15 and that between the block B and floor is 0.25. (**Fig 14.11**).

Solution Fig. 14.12 shows the free body diagrams of two blocks.

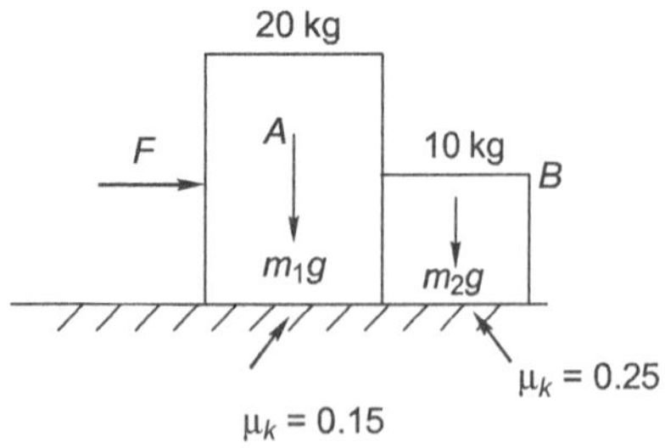

Fig.14.11

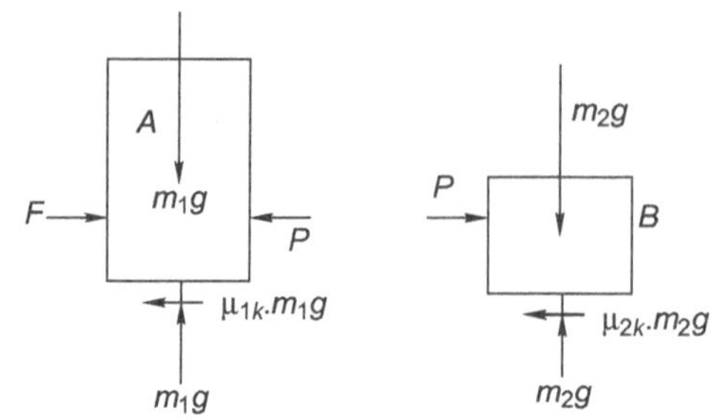

Free body diagram

Fig. 14.12

Reaction on block A is $m_1\ g = 20$ g and reaction on block B is $m_2\ g = 10$ g.
Force of friction on block A is 0.15×20 g = 3 g and force of friction on block B is

$$0.25 \times 10\text{ g} = 2.5\text{ g}$$

$$P - R - 3\text{ g} = 20a \qquad ...(1)$$

$$R - 2.5\text{ g} = 10a \qquad ...(2)$$

Adding the two equations

$$R - 5.5\text{ g} = 30a = 30 \times 4 = 120\text{ N}$$

$$F = 120 + 5.5 \times 9.81 = 120 + 53.96\text{ N} = 173.96\text{ Newton}$$

From Equation (2) R = reaction between blocks

$$= 10a + 2.5\text{ g} = 10 \times 4 + 2.5 \times 9.81$$

$$= 40 + 24.525 = 64.525\text{ Newton.}$$

Exercise 14.3 Two blocks A and B of masses 30 kg and 20 kg are resting on a rough horizontal floor as shown in Fig. 14.13. A force F = 150 N is applied on A. If μ_k for both blocks and floor is 0.2. Determine (1) acceleration of masses, (2) reaction between two blocks.

[**Ans:** a = 1.038 m/s^2, Reaction between blocks = 60 N].

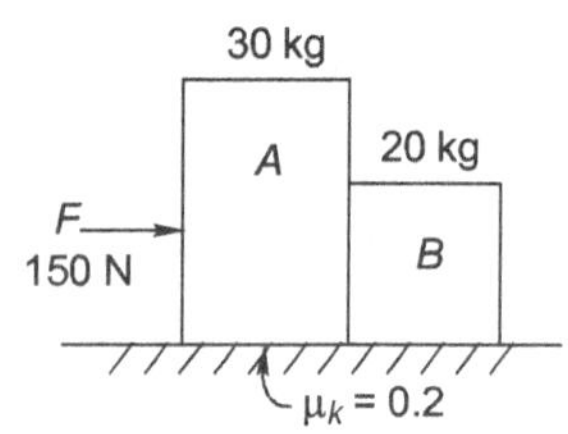

Fig. 14.13

14.5 Tangential and Normal Forces in Curvilinear Motion

Consider a particle of mass m traversing a curvilinear path abc as shown in Fig. 14.14. At a particular instant, radius of the path is R and velocity of the particle is V and tangential acceleration is a_t. Then normal force (or centripetal force) acting on the particle will be

$$F_n = m\,a_n$$

$$= m\frac{V^2}{R} \text{ (directed towards the centre of curvature, } C\text{)}$$

Tangential force $F_t = m\,.\,a_t$

If the particle is moving with constant velocity V and tangential acceleration is zero, then $F_t = 0$ but normal force or centripetal force, $F_n = \dfrac{mV^2}{R}$.

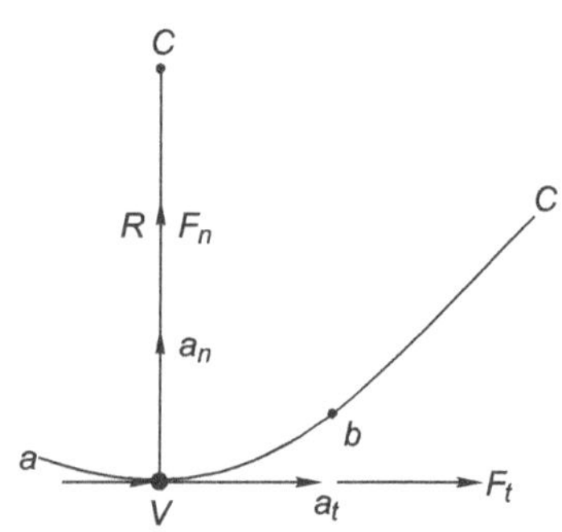

Fig. 14.14

Example 14.4 The bob of 1.0 m pendulum describes an arc of a circle. If the tension in the cord is 2.6 times the weight of the bob, and position of pendulum is as shown in Fig. 14.15, determine velocity and acceleration of the bob in the position.

Solution Fig. 14.16 shows the action of forces $T = 2.6$ mg and mg (weight of bob) in the position shown.

Tangential force, $F_t = m\,g \sin 30^\circ$

$$= 0.5m\,g \quad ...(1)$$

Normal force towards the centre O

$$F_n = T - m\,g \cos 30^\circ$$
$$= 2.6m\,g - 0.866m\,g$$
$$= 1.734m\,g \quad ...(2)$$

Tangential acceleration, a_t

$$m\,.\,a_t = 0.5m\,g$$
$$a_t = 0.5\,g = 0.5 \times 9.81 = 4.905 \text{ m/s}^2$$

Normal acceleration, a_n

$$m\,.\,a_n = 1.734m\,g$$
$$a_n = 1.734 \times g = 1.734 \times 9.81 = 17.01 \text{ m/s}^2$$

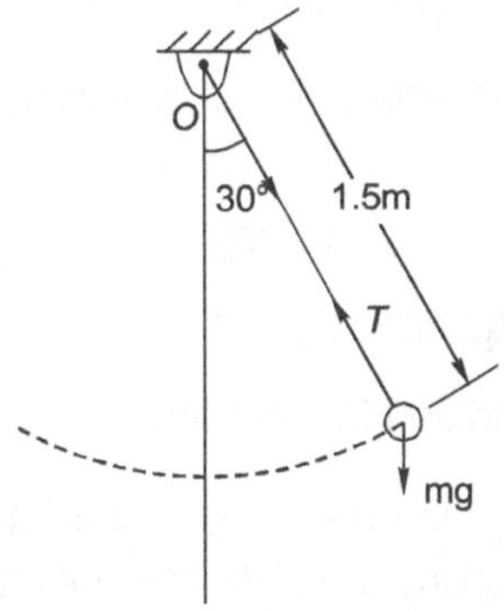

Fig. 14.15

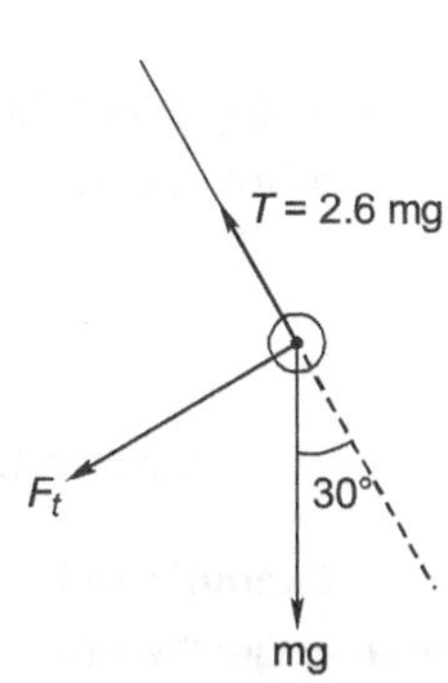

Fig. 14.16

Now normal acceleration, $a_n = \dfrac{V^2}{R}$, where $R = 1.0$ m, length of cord of pendulum

$$V^2 = a_n \times R = 17.01 \times 1.0 = 17.01$$

Velocity, $V = 4.124$ m/s.

Exercise 14.4 A ball of 2 kg mass is revolving in a horizontal circle as shown in Fig. 14.17. If the length of the cord is 1 m and maximum allowable tension in cord is 40 N, determine (*a*) maximum allowable speed, (*b*) corresponding value of angle θ.

[**Hint:** $\dfrac{mg}{T} = \cos\theta$, $F_n = T\sin\theta$].

[**Ans:** $\theta = 60.62°$, $V = 3.9$ m/s].

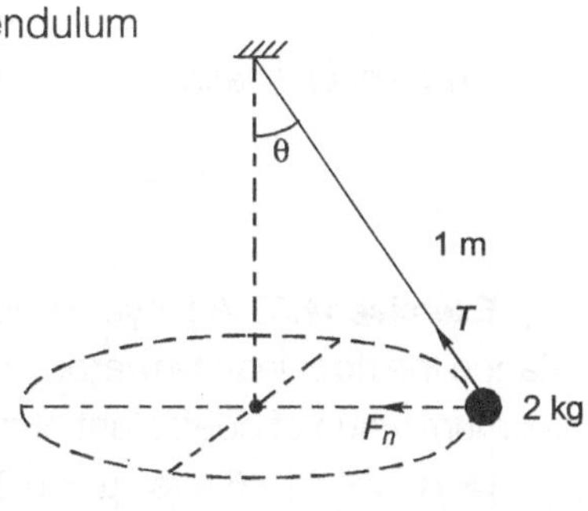

Fig. 14.17

14.6 Motion of a Bicycle Rider Along a Circular Track

Consider a bicycle rider, with a bicycle going around a circular track of radius R.

Say m = mass of the rider and bicycle

Gravitation force $= mg$

Say CG of bicycle and rider lies at G at a height of h from the ground (Fig. 14.18).

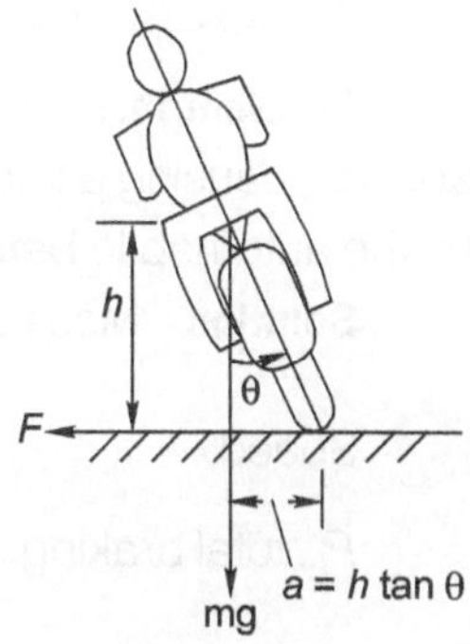

Fig. 14.18

Centrifugal force on bicycle and rider $= \dfrac{mV^2}{R} = CPF$, centripetal force

The moment of CPF tends to turn the rider outwards and rider leans inwards. Moment Ra (ccw) develops about G to balance the moment Fh (cw) (Fig. 14.19).

Vertical reaction $R = mg$

$$mga = Fh = \frac{mV^2}{R} \times h$$

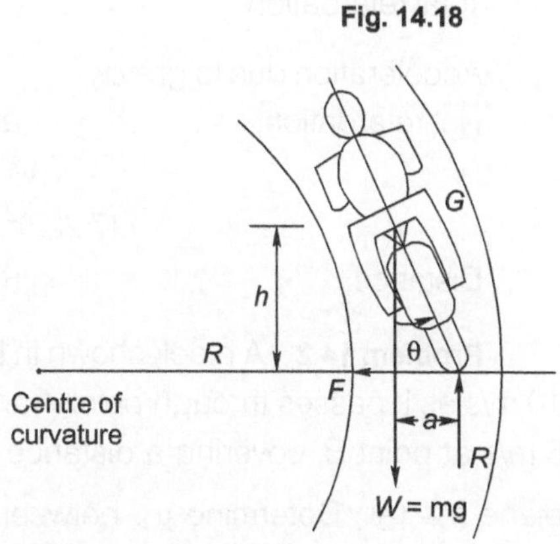

Fig. 14.19

$$g\frac{a}{h} = \frac{V^2}{R} \quad \text{or} \quad \frac{a}{h} = \frac{V^2}{gR} \quad \text{or} \quad \tan\theta = \frac{V^2}{gR}$$

when F is greater than the limiting friction, skidding occurs.

In this case

$$F > \mu mg$$

or

$$mg\tan\theta > \mu mg$$

$\tan\theta > \mu$, condition for skidding.

Example 14.5 A bicycle rider goes around a circular track of radius 100 m, with a velocity of 15 m/s. To what angle the rider will lean inwards? What is the maximum value of coefficient of friction between tyre of bicycle and road to avoid skidding?

Solution Radius of tracks, $R = 100$ mm

Velocity, $V = 15$ m/s

$g = 9.81$ m/s^2

Angle of inclination $\tan\theta = \frac{V^2}{Rg} = \frac{15^2}{100 \times 9.81} = 0.23$

$\theta = 12.95^\circ$

$\mu < 0.23$ to avoid skidding.

Exercise 14.5 A bicycle rider goes around a circular track of radius 50 m with a velocity V m/s. To maintain balance the rider leans inwards by an angle of 15°. What is the velocity of the rider? To avoid skidding, what is the maximum value of coefficient of friction between road and tire of bicycle?

[**Ans:** $V = 11.46$ m/s; $\mu < 0.268$].

PROBLEMS

Problem 14.1 A 1000 kg automobile is driven down a 5° incline at a speed of 80 kmph, when the brakes are applied, causing a total braking force of 5.8 kN to be applied to the automobile. Determine the distance travelled by the automobile before coming to rest.

Solution Mass of automobile, $m = 1000$ kg

Speed, $V = \frac{80{,}000}{3{,}600} = 22.222$ m/s

F_b, total braking force $= 5.8$ kN $= 5{,}800$ N

Total retardation $= \frac{F_b}{m} = \frac{5{,}800}{1{,}000} = 5.8$ m/s^2

Acceleration due to gravity $= g\sin 5^\circ = 9.81 \times \sin 5^\circ = 9.81 \times 0.0871 = 0.8545$ m/s^2

Net retardation, $a = 5.8 - 0.8545 = 4.9455$ m/s^2

$$V^2 = 2a \times S$$

$$(17.222)^2 = 2 \times 4.9455 \times S$$

Distance, $S = 29.986$ m.

Problem 14.2 A block shown in **Fig. 14.20** has a velocity of 10 m/s as it passes through point A and its velocity is reduced to 5 m/s at point B, covering a distance of 50 m along the inclined plane $\theta = 15°$. Determine μ_k between the block and the inclined plane.

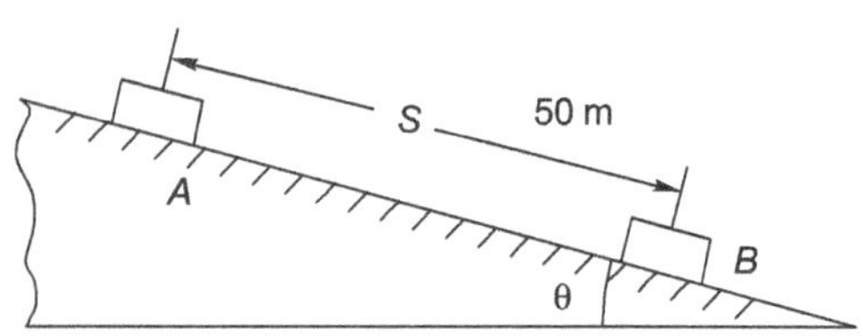

Fig. 14.20

Solution

$$\text{Say mass of block} = m$$

$$\text{Component of } mg \text{ along plane} = mg\sin\theta$$

$$\text{Reaction of inclined plane} = mg\cos\theta$$

$$\text{Force of friction} = \mu_k \, mg\cos\theta$$

Since the velocity is reduced

$$\mu_k mg\cos\theta > mg\sin\theta$$

$$\text{Initial velocity} = 10 \text{ m/s}$$

$$\text{Final velocity} = 5 \text{ m/s}$$

Distance covered, $S = 50$ m

$$5^2 - 10^2 = -2\alpha.S = -2\alpha.50$$

$$-75 = -2\alpha.50 = -100\alpha$$

Retardation $= 0.75 \text{ m/s}^2$

Net force $\mu_k mg\cos\theta - mg\sin\theta = m \times a$

or $\mu_k g\cos\theta - g\sin\theta = 0.75 \text{ m/s}^2$

Putting the values of $g\cos\theta, g\sin\theta$

$$\mu_k \times 9.81\cos 15° - 9.81 \times \sin 15° = 0.75$$

$$\mu_k \times 9.81 \times 0.9659 - 9.81 \times 0.2588 = 0.75$$

$$\mu_k = \frac{3.289}{9.81 \times 0.9659} = 0.347.$$

Remember

- Magnitude and direction of forces required for a particular motion is studied in *kinetics*.
- As per Newton's second law of motion, a force is necessary to change the state of rest or of uniform motion of a body.
- Bodies are of definite size and shape and are not of the size of a point, yet a body can be considered as a particle if the motion (velocity, acceleration and displacement) of all the particles of the body are the same respectively.
- $\sum F = ma$, resultant force = mass × acceleration.
- No appreciable error in the solution of most engineering problems is caused if acceleration a is determined with respect to the axes attached to the earth (as earth rotates with respect to outer space).
- 1 kgf = 9.81 Newton.
- A Newton is a force acting on a body of mass 1 kg and producing an acceleration of 1 m/s^2.
- In any problem, try to identify the net force *i.e.*,
  Net force = applied force – resisting force = mass × acceleration.
- In many problems, force of gravity or a component of force of gravity is the applied.
- Resisting force is generally due to kinetic friction *i.e.*, $\mu_k \times$ normal reaction.

PRACTICE PROBLEMS

14.1 A man is raising himself and the platform on which he stands with a uniform acceleration of 4 m/s^2 by means of a rope and a pulley arrangement shown in **Fig. 14.21**. The man has a mass of 80 kg and platform has a mass of 50 kg. Assuming pulley and rope as massless and frictionless, determine tension in the rope *ABC*. What is the force of contact exerted on the man by the platform.

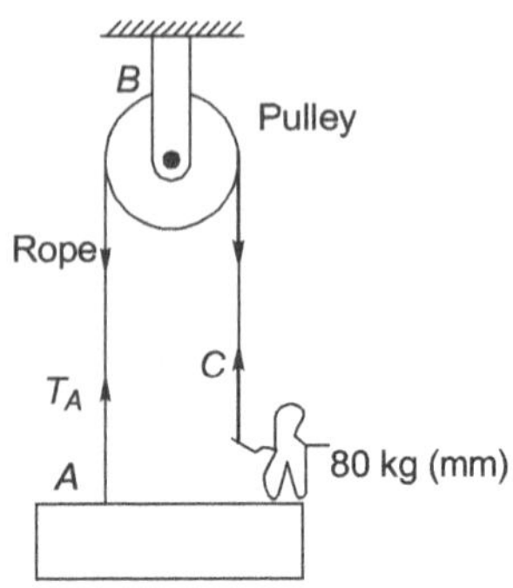

Fig. 14.38

[**Hint:** Tension in rope at *A*, *B*, *C*

$$T_B = T_A + T_C$$
$$(T_A + T_C) - (80 + 50)\,g = (80 + 50)\,a$$
$$a = 4 \text{ m/s}^2$$
$$T_A + T_C = 897.65 \text{ N}$$

Force exerted by man F_m,

$$897.65 + F_m = 80g + 80a].$$

[**Ans:** $F_C = 207.15$ N].

14.2 The acceleration of a 40 kg carriage *A* is controlled by tension *T* exerted in control cable that passes around two circular pegs fixed to the carriage. Determine the value of *T* required to shunt the downward acceleration of carriage to 2 m/s^2, as shown in **Fig. 14.22**. The coefficient of friction between control cable and peg is 0.20.

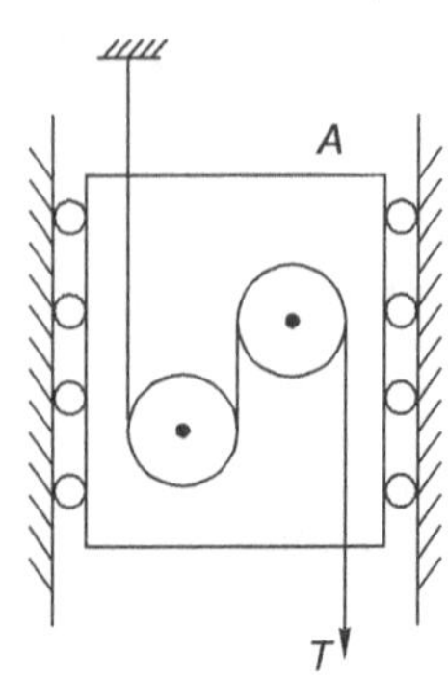

Fig. 14.22

[**Hint:** Use $\dfrac{T'}{T} = e^{\mu\theta}$, $\theta = 2\pi$].

[**Ans:** $T = 124.3$ Newton].

14.3 Two masses of 16 kg and 20 kg are connected with each other by an inextensible thread and accelerated by the force of gravity of a mass of 36 kg. The masses of threads and pulleys are negligible. Determine acceleration of the system. What are thread tensions?

[**Hint:** $36\text{ g} = 72a$, $a = \dfrac{g}{2} = 4.965 \text{ m/s}^2$

$$36\text{ g} = T_1 = 36a;\quad T_2 = 16a = 16 \times 4.905$$
$$T_1 = 36\,(g - a) = 18\text{ g}].$$

[**Ans:** $a = 4.905$ m/s^2, $T_1 = 176.58$ N; $T_2 = 78.48$ N].

14.4 A train weighing 2000 kN without locomotive starts to move with constant acceleration along a straight horizontal track and in the first 1 minute acquires a speed of 60 kmph. Determine tension in the draw bar between the locomotive and train if the total resistance to motion due to friction and air resistance is constant and equal to 0.6% of the weight of the car.

[**Hint:** Tension = *ma* + resistance].

[**Ans:** 68.63 kN].

14.5 A small block starts from rest at point *A* as shown in **Fig. 14.23** and slides down the inclined plane *AB*, 20 m long. What distance the block will travel along the horizontal track *BC* before coming to rest. The coefficient of friction between track and block is 0.30. It is assumed that the block will start moving along *BC* with same velocity which it acquired while sliding down. Coefficient of friction between plane *AB* and track is 0.3.

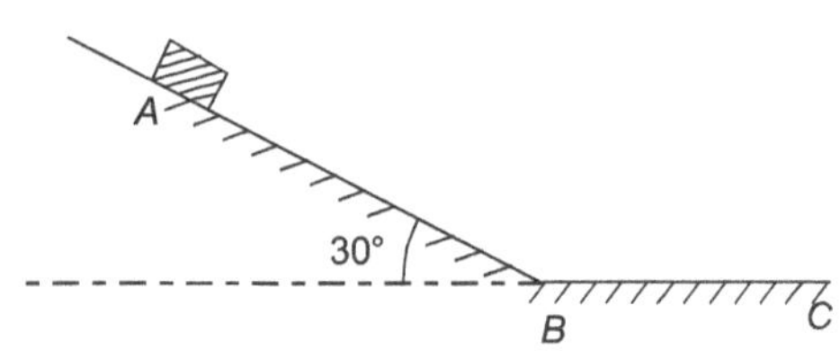

Fig. 14.23

[**Hint:** Along *AB* $\quad ma = mg\sin\theta - mg\cos\theta \times \mu$.

Along *BC*, $ma' = -\mu\, mg$].

[**Ans:** 16 m].

MULTIPLE CHOICE QUESTIONS

14.1 A police investigation of tire marks shows that a car travelling along a straight-road skidded for a total distance of 50 m, after the brakes were applied. Coefficient of friction between tyres and pavement is estimated to be 0.6. What was the probable speed of car when brakes were applied?
(*a*) 87.3 kmph (*b*) 61.75 kmph
(*c*) 356.6 kmph (*d*) None of these

14.2 A 10 kg block rests on a horizontal plane as shown in **Fig. 14.24**. Find magnitude of force *P* to give the block an acceleration of 2 m/s^2. Coefficient of kinetic friction between block and plane is $\mu_R = 0.3$

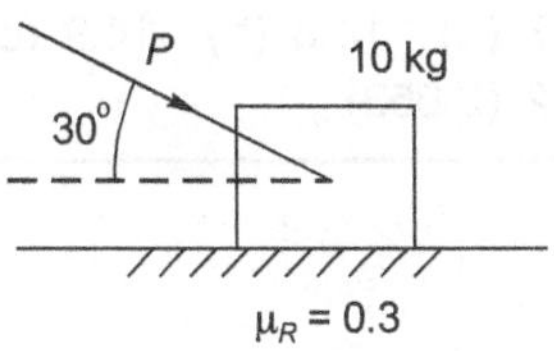

Fig. 14.24

(*a*) 48.65 N (*b*) 59 N
(*c*) 69 N (*d*) None of these

14.3 A box is dropped on a conveyor belt moving at 2 m/s. If the box is initially at rest and coefficient of friction between box and belt is 0.25, how far it will take for the slipping to stop?
(*a*) 0.203 m (*b*) 0.815 m
(*c*) 1.63 m (*d*) None of these

14.4 A body weighs 600 N on earth, what is its weight on moon, $g_{moon} = 1.4$ m/s^2?
(*a*) 85.63 N (*b*) 600 N
(*c*) 4200 N (*d*) None of these

14.5 A multiple coaches train has 600 tonnes of mass. Its resistance to motion is 1% of the train weight. The electric motors can provide a tracking force of 200 kN. In how much time train will have a velocity of 36 kmph?
(*a*) 23 s (*b*) 30 s
(*c*) 42.53 s (*d*) None of these

14.6 An automobile of mass 1000 kg traverses a 400 m curve at a constant speed of 36 kmph. Assuming no banking of curve, what must be the force of the tyres on the road to maintain equilibrium
(*a*) 1000 N (*b*) 500 N
(*c*) 250 N (*d*) 125 N

14.7 A body of mass 100 kg moves under the action of a force $P = 10t - t^2$ N where *t* is in seconds. When the body will reverse its speed?
(*a*) 25 s (*b*) 15 s
(*c*) 10 s (*d*) 7.5 s

14.8 What is the greatest speed at which a car can negotiate a corner of radius 50 m on a horizontal track without skidding, coefficient of friction between tyres and track is 0.4?
(*a*) 22 m/s (*b*) 14 m/s
(*c*) 11 m/s (*d*) 3 m/s.

14.9 A body of mass 2 kg is pulled-up by a force *P* = 20 N as shown in **Fig. 14.25**. What is acceleration of the body?

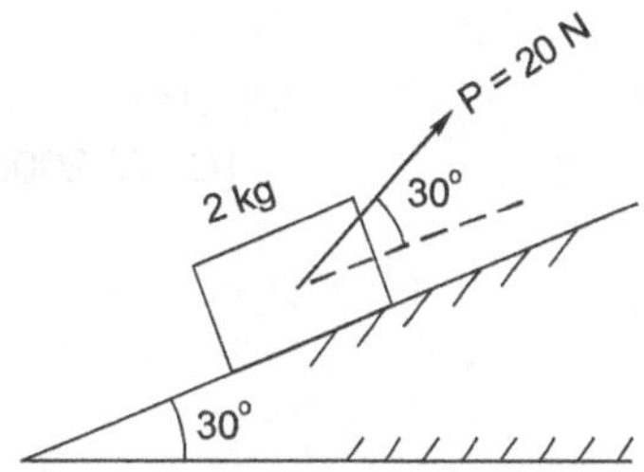

Fig. 14.25

(*a*) 5 m/s^2 (*b*) 4.33 m/s^2
(*c*) 3.755 m/s^2 (*d*) None of these

14.10 An elevator (lift) consists of the elevator cage and a counter weight, of mass *m* each. The cage and the counterweight are connected by chain that passes over a pulley. The pulley is coupled to a motor. It is desired that the everator should have a maximum stopping time of t seconds from a peak speed v. If the inertias of the pulley and the chain are neglected, the minimum power that the motor must have is

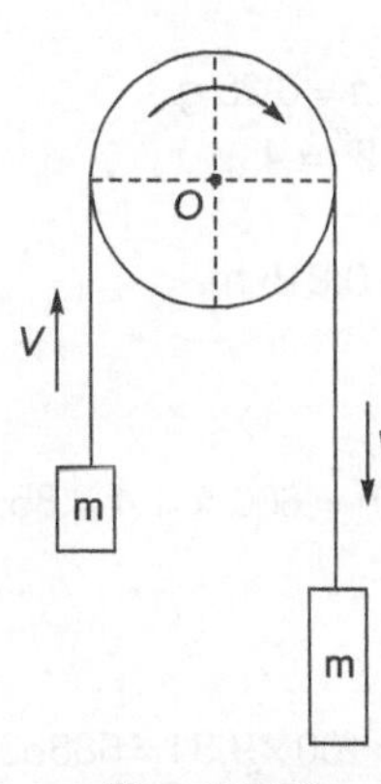

Fig. 14.26

(a) $\frac{1}{2}mv^2$ (b) $\frac{mv^2}{2t}$
(c) $\frac{mv^2}{t}$ (d) $\frac{2mv^2}{t}$

[GATE 2005 : 2 Marks]

14.11 A uniform rigid rod of mass M and length L is hinged at one end as shown in the adjacent figure. A force P is applied at a distance of 2L/3 from the hinge so that the rod swings to the right. The reaction at the hinge is

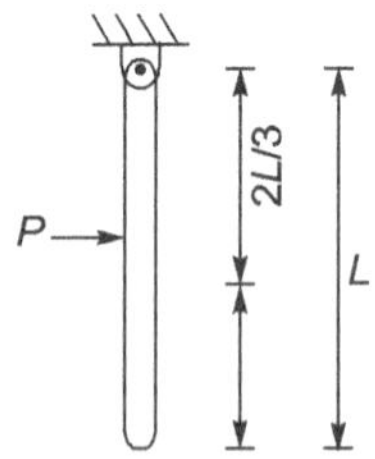

Fig. 14.27

(a) –P (b) 0
(c) P/3 (d) 2P/3

[GATE 2009 : 2 Marks]

14.12 A uniform slender rod (8 m length and 3 kg mass) rotates in a vertical plane about a horizontal axis 1 m from its end as shown in the figure. The magnitude of the angular acceleration (in rad/s^2) of the rod at the position shown is ______

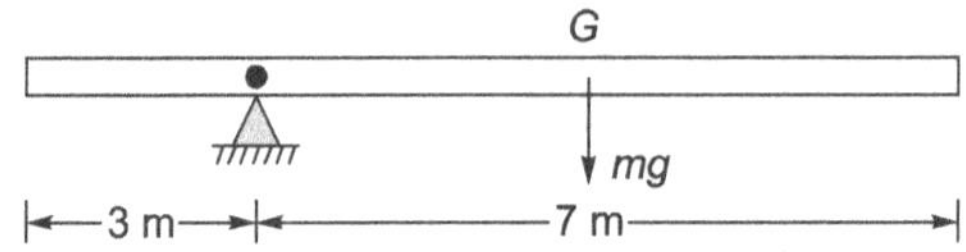

Fig. 14.28

[GATE 2014 : 2 Marks (set-4)]

Answers

14.1 (a)	14.2 (c)	14.3 (b)	14.4 (a)	14.5 (c)
14.6 (c)	14.7 (b)	14.8 (b)	14.9 (c)	14.10 (c)
14.11 (b)	14.12 (2.053)			

EXPLANATIONS

14.1 (a)

$a = 0.6g = 0.6 \times 9.81 = 5.886 \text{ m/s}^2$

$2as = V^2,\ V = \sqrt{2 \times 5.886 \times 50} = 24.26 \text{ m/s}.$

$= 87.3$ kmph

14.2 (c)

$P\cos 30 - (98.1 + P\sin 30^\circ) \times 0.3$

$= 10 \times 2$

$0.866P - 29.43 - 0.15P = 20$

$0.716P = 49.43$

$P = 69$ N

14.3 (b)

$0.25\ mg = ma,\ a = 0.25\ g$

$2\,(0.25\ g) \times s = 2^2 = 4$

$$s = \frac{4}{0.5 \times 9.81} = 0.815 \text{ m}.$$

14.4 (a)

Weight on moon = 600 × 1.4 = 85.63 N.

14.5 (c)

Resistance

$= 1/100 \times 600 \times 100 \times 9.81 = 58860$ N

$F_{net} = 200{,}000 - 58860 = \mathbf{141140\ N}$

$$a = \frac{141140}{600{,}000} = 0.235 \text{ m/s}^2$$

$v = 36000/3600 = 10$ m/s

$t = 10/.235 = 42.53$ s.

14.6 (c)

$$a_n = \frac{V^2}{R} = \frac{10^2}{400} = 0.25 \text{ m/s}^2$$

$F_n = 1000 \times 0.25 = 250$ N.

14.7 (b)

$P = 10t - t^2$ N

$100 \times a = 10t - t^2$

$a = 0.1t - 0.01t^2$

$$\frac{dV}{dt} = 0.1t - 0.01t^2$$

$dV = (0.1t - 0.01t^2)\,dt$

$$V = \frac{0.1t^2}{2} - \frac{0.01t^3}{3} + C_1$$

$V = 0,\ C_1 = 0$ velocity will reverse

$$\frac{0.1t^2}{2} = \frac{0.01t^2}{3}, \text{ so } t = 15 \text{ s}.$$

14.8 (b)

$$a_n = 0.4g = 0.4 \times 9.81 \text{ m/s}^2 = \frac{V^2}{R}$$

$V = \sqrt{0.4 \times 9.81 \times 50} = \mathbf{14\ m/s}.$

14.9 (c)

$F_{net} = 20\cos 30^\circ - 2 \times g \times \sin 30^\circ$

$= 1.732 - 2 \times 9.81 \times 0.5 = 7.51$

$$a = \frac{7.51}{2} = 3.755 \text{ m/s}^2.$$

14.10 (c)

Initial velocity = V

Final velocity = 0

$$KE = \frac{1}{2}mV^2 + \frac{1}{2}mV^2 = mV^2$$

$$\text{Power} = \frac{mV^2}{t}$$

14.11 (b)

$$\text{Torque} = T \times \frac{2L}{3} = I\alpha = \frac{mL^2}{3} \times \alpha$$

Angular acceleration,

$$\alpha = \frac{2P}{mL}$$

Linear acceleration,

$$a = \left(\frac{2P}{ML}\right) \times \frac{1}{2} = \frac{P}{M} \text{ (at three CG)}$$

So, $P - R = M.a = \frac{M \times P}{M} = P$

Reaction, R = 0

14.12 (2.053 rad/s²)

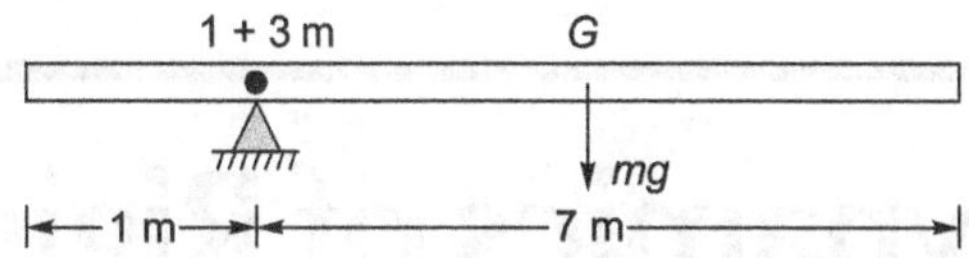

m = 3 kg

Torque about 0 = 3 × 3 × g

= 9 × 9.81 = 88.29 Nm

$$I_0 = \frac{mL^2}{12} + m \times 3^2 = \frac{3 \times 8^2}{12} + 9 \times 3$$

$$= 16 + 27 = 43 \text{ kg-m}^2$$

$$\alpha = \frac{88.29}{43} = 2.053 \text{ rad/s}^2$$

■■■■

15

CHAPTER

Dynamics of Rigid Bodies in Translation

15.1 Introduction

In the last chapter, we considered a body as a particle and studied the effect of applied force on the motion of the body as a particle, because all the particles on the body move with same velocity, same acceleration and same displacement. But in an actual body, its shape and size are of definite magnitude and *acceleration of mass centre* of the body may be different from acceleration of any other point of the body. So there is definite relationship, between size, shape, mass of the body and the motion produced.

In this chapter we consider only the motion of translation and there is *no motion of rotation.*

Therefore, $\sum M = 0$, summation of moments about any point on the body is zero

Summation of forces, $\sum F = F_R$, resultant force

$= m \times \bar{a} =$ mass × acceleration

where $\bar{a}$ is the *resultant acceleration of the mass centre of the body.*

In the case of coplanar forces

$\sum F_x = F_{Rx}$, summation of forces in *x*-direction

$= m \times \bar{a}_x$, where $\bar{a}_x$ is the component of acceleration in *x*-direction

$\sum F_y = F_{Ry}$, summation of forces in *y*-direction

$= m \times \bar{a}_y$, where $\bar{a}_y$ is the component of acceleration in *y*-direction

Figure 15.1 shows a body of mass *m* subjected to forces $F_1, F_2, F_3, \ldots, F_n$, such that

$F_R =$ resultant of forces $F_1, F_2, F_3, \ldots, F_n$

$= \sum (F_1 + F_2 + F_3 \ldots F_n)$ vectorially

$= m\bar{a}$, where $\bar{a}$ is the acceleration of the mass centre *G*.

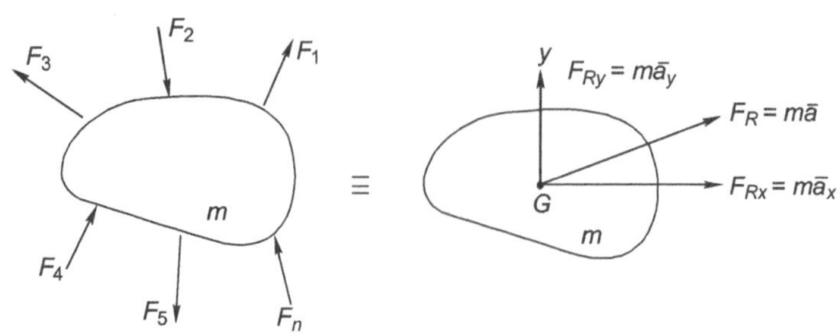

Fig. 15.1

In a two dimensional case, $\bar{a} = \bar{a}_x i + \bar{a}_y j$

Similarly $F_{Rx} i + F_{Ry} j = F_R.$

15.2 D' Alembert's Principle

After the determination of acceleration of the mass centre, in order to find support reactions and other unknown forces, an inertia force $m\bar{a}$ is applied on the body in the negative direction. So that the body is brought to *dynamic equilibrium* and equations of equilibrium can be used to determine *unknown reaction and forces*.

In this chapter we will take up a few cases of rigid bodies in which D' Alembert's principle will be applied, (*a*) inertia forces is applied on the body in negative direction of F_R, (*b*) unknown forces are determined.

Consider a sphere of mass *m*, a force *P* is applied on the sphere so that it moves towards right with acceleration *a*. The coefficient of friction between sphere and ground is μ.

Normal reaction, $R = mg$

$$P = m\bar{a} + \mu mg \text{ (Fig. 15.2)} \quad ...(1)$$

$m\bar{a}$ and μmg in direction opposite to that of *P*.

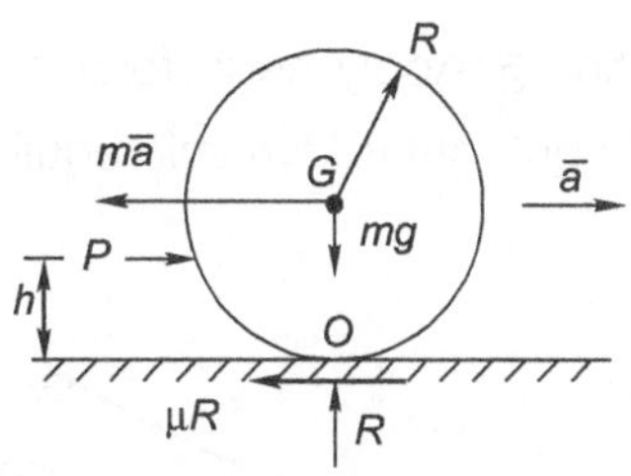

Fig. 15.2

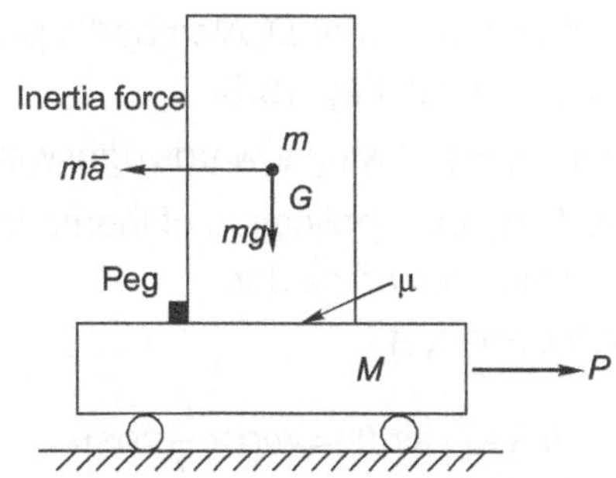

Fig. 15.3

Now apply inertia force $F_i = -m\bar{a}$ at *G*, mass centre of sphere of mass *m*. Taking moments about point of contact *O*, point of application *P* can be determined

$$m\bar{a} \times R = P \times h$$

or

$$h = \frac{m\bar{a}R}{P} \quad ...(2)$$

A block is placed on a cart and due to application of external force cart of mass *M* and block of mass *m* are moving with acceleration $\bar{a}$ towards right as shown in **Fig. 15.3**. Coefficient of friction between block and upper surface of cart is μ.

$$P = (M + m)\,\bar{a}$$

Apply the inertia force $m\bar{a}$ at the mass centre *G* of the block.

Force of friction $= \mu mg$

If $m\bar{a} > \mu mg$, *the upper block will tend to slide towards left.* To prevent sliding of the upper block towards left, a peg may be provided on the cart. But then depending upon $m\bar{a}$ and *mg* force acting on the *block, the block will try to tip at the peg.*

A small block of mass *M* moves with an acceleration $\bar{a}$ along the horizontal surface under the action of applied force *P*. During the motion, the bob of mass *M* attached to the block remains inclined at an angle θ as shown in the **Fig. 15.4**. Apply inertia force $m\bar{a}$ at *G* of the bob, as shown. Taking moments about *G* of block

$$mg \times L\sin\theta = m\bar{a} \times L\cos\theta$$

or $$\tan\theta = \frac{\bar{a}}{g}$$

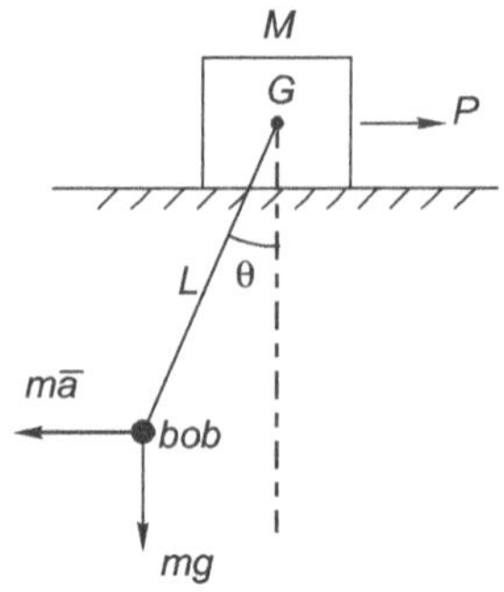

Fig. 15.4

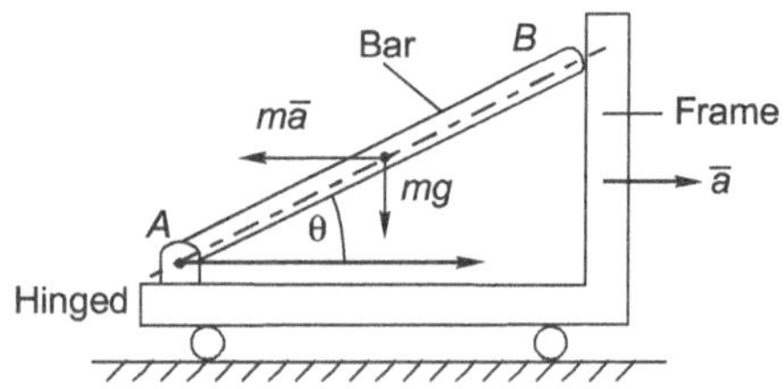

Fig. 15.5

To explain further, the use of *D' Alembert's* principle, consider one more example of a bar hinged to a moving frame as shown in the **Fig. 15.5**.

Frame and bar *AB* are moving towards right with an acceleration $\bar{a}$. Apply inertia force at *G* of the bar. Say length of the bar is *L*. By the application of inertia force, the bar is brought to dynamic equilibrium. **Fig. 15.6** shows the free body diagram of the bar.

Taking moments about *A*

$$m\bar{a} \times \frac{L}{2}\sin\theta = mg \times \frac{L}{2}\cos\theta$$

If $$m\bar{a}\frac{L}{2}\sin\theta > mg\frac{L}{2}\cos\theta$$

The bar *AB* will be lifted and contact at point *B* with the frame is *lost*.

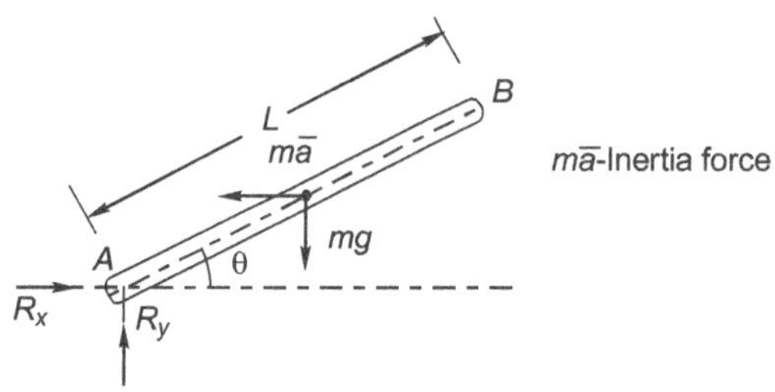

Free body diagram of bar *AB*

Fig. 15.6

Taking another case of a mass going up in an elevator at an acceleration $\bar{a}$

Say acceleration, $\bar{a} = 4\ \text{m/s}^2$ upward

Inertia force on man $F_1 = ma \downarrow$

Mass of the man $= 60$ kg

Gravitational force plus inertia force

$$= mg + m\bar{a}$$
$$= 60\,(9.81 + 4)$$
$$= 13.81 \times 60 = 828.6\ \text{N}$$

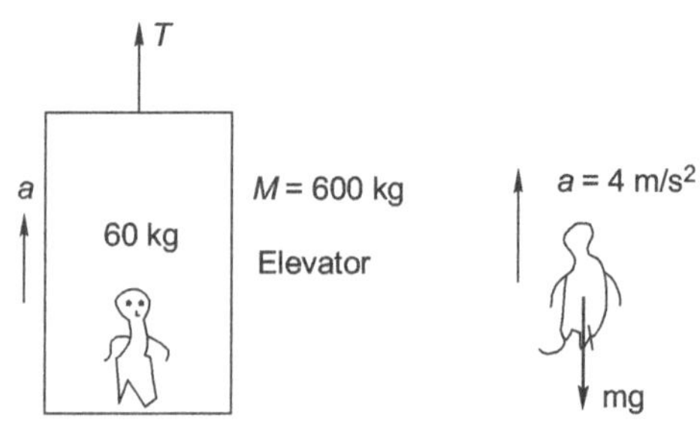

Fig. 15.7

There is apparent gain in the weight of the man. Similarly if the elevator is coming down under acceleration, there will be apparent loss in the weight of the man.

Let us take examples of different cases.

Example 15.1 A block of mass 80 kg rests on a cart as shown in **Fig. 15.8**. The coefficient of friction between block *A* and cart is 0.35. Determine the acceleration of the cart where (*a*) when the block slides off the cart, (*b*) when the block *tends to tip* about an edge.

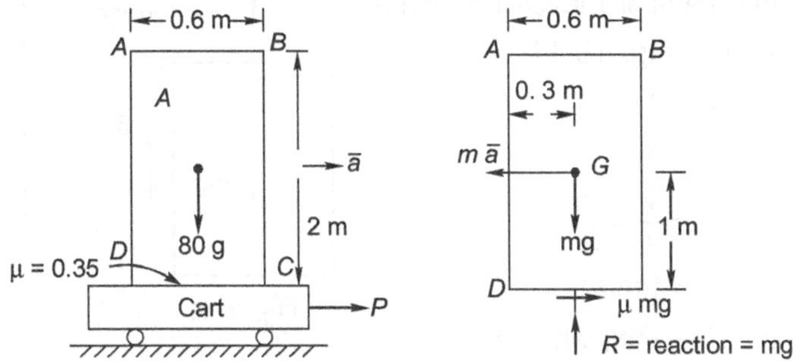

Fig. 15.8

Solution (*a*) When the block tends to slide off the cart.

Due to friction between block and cart, the block moves alongwith the cart at an acceleration $\bar{a}$ m/s^2. Inertia force $m\bar{a}$ at G of block will act in opposite direction. Block will slide off the cart if

$$m\bar{a} > \mu mg$$

or

$$\bar{a} > \mu g$$

$$\bar{a} > 0.35 \text{ g}$$

(*b*) When the block tends to tip about an edge.

For the block to tip about an edge D as shown in figure

$$m\bar{a} \times 1 \text{ m} > mg \times 0.3$$

$$\bar{a} > 0.3 \text{ g}$$

As the dimension of the block are such that the block *will first tend to tip* about an edge than to slide off the cart.

Exercise 15.1 A cabinet of mass 40 kg is mounted on castors as shown in Fig. 15.9. The cabinet can move freely on floor ($\mu = 0$). If a force $P = 300$ N is applied at a height H, calculate (*a*) acceleration of the cabinet, (*b*) value of H for which castor will not tip at edge A.

[**Hint:** $P = m\bar{a}$ for H consider tipping at edge A putting inertia force $m\bar{a}$ at G].

[**Ans:** $\bar{a} = 7.5$ m/s^2, $M_A = 0$, $H = 583.76$ mm].

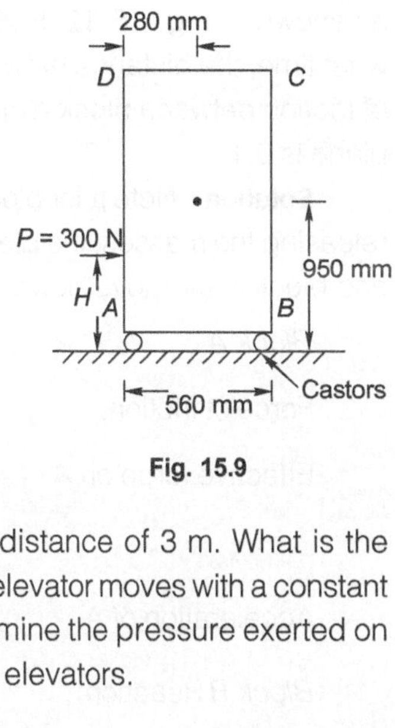

Fig. 15.9

Example 15.2 An elevator of gross weight 4.8 kN starts moving upwards with a constant acceleration and acquires a velocity of 3 m/s after travelling a distance of 3 m. What is the magnitude of cable tension during this accelerated motion? (*b*) While stopping, the elevator moves with a constant retardation from constant velocity of 3 m/s and comes to rest in 3 seconds. Determine the pressure exerted on the floor of the elevator by the feet of man weighing 65 kg, during stopping of the elevators.

Solution (*a*) Initial velocity= 0 m/s

Final velocity, $V = 3$ m/s

Distance, $S = 3$ m

$$V^2 = 2\bar{a} \times S$$

Acceleration, $\bar{a} = \dfrac{3^2}{2 \times 3} = 1.5 \text{ m/s}^2$

Cable Tension $T = M.g + M\bar{a}$, where M is mass of elevator Mg is weight of elevator

$$T = 4800 + \frac{4800}{9.81} \times 1.5, \text{ (Fig. 15.10)}$$

Cable Tension, $T = 4800 + 733.95 = 5533.95$ N

(*b*) During stopping the elevator moves up with a deceleration a' m/s^2, as shown in **Fig. 15.11**.

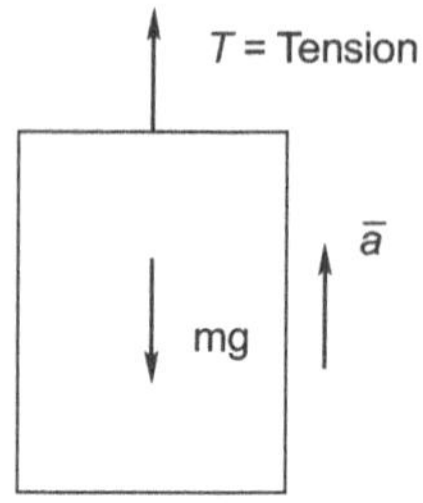

Fig. 15.10

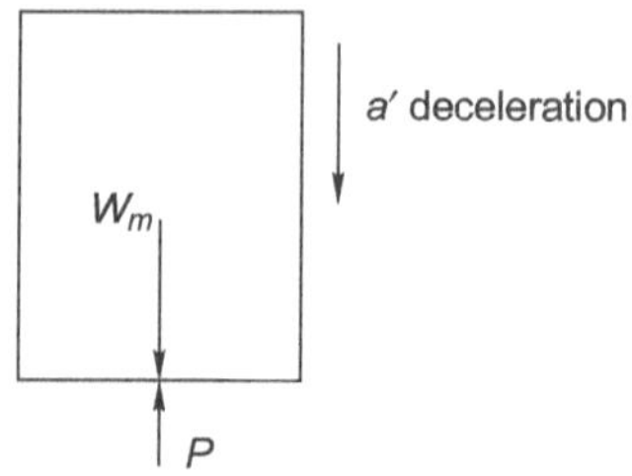

Fig. 15.11

Initial velocity, $V = 3$ m/s

Time, $t = 3$ s

Deceleration, $a' = \frac{V}{t} = \frac{3}{3}$

$= 1$ m/s^2

$mg - P = ma'$, where m is mass of man and P is pressure exerted by man

$$P = mg - ma' = 65 \times 9.81 - 65 \times 1 = \mathbf{572.65\ N}$$

(Note that apparent weight of man is reduced).

Exercise 15.2 A lift of mass 1200 kg starts from rest and moves upward with constant acceleration and covering a distance of 12 m while attaining a velocity of 3 m/s. Find the pull in the cable hoisting the lift. When the lift stops, its speed is reduced from 4 m/s to 1.5 m/s in 5 seconds. What will be the pull in the cable during stopping?

[**Ans:** 12.222 kN, 11.172 kN].

Example 15.3 Two blocks *A* and *B* of 6 kg and 10 kg masses respectively are placed on an inclined plane with angle of inclination of 30° as shown in **Fig. 15.12**. Both the masses are released from rest. After what time, the distance between the two masses will be 6 m? Coefficient of friction between block *A* and planes 0.2 and that between block *B* and plane is 0.1.

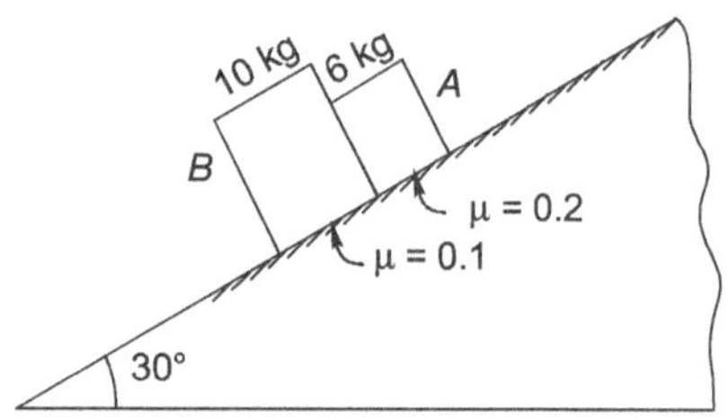

Fig. 15.12

Solution Note μ for block *B* is less than μ for block *A*, therefore after releasing the masses, the block *B* will travel faster with more acceleration and block *A* will move slower with less acceleration.

Block A $R_A = 6g\cos\theta = 6g\cos 30°$

Force of friction, $F_A = 0.2 \times 6 \times g\cos 30° = 10.1945$ N

Effective force on *A* $= 6g\sin\theta - F_A$

$= 6 \times 9.81 \times \sin 30° - 10.1945 = 19.2355$ N

Acceleration of *A*, $a_A = \frac{19.2355}{6} = 3.206$ m/s^2

Block B Reaction, $R_B = 10g\cos 30° = 10 \times 9.81 \times \cos 30° = 84.9546$ N

Force of friction $F_B = 0.1 \times 84.9546 = 8.495$ N

Effective force on block *B* $= mg\sin\theta - F_B$

$= 10 \times 9.81 \times \sin 30° - 8.495 = 40.555$ N

Acceleration of block *B*, $a_B = \frac{40.555}{10} = 4.0555$ m/s^2

as shown in **Fig. 15.13**.

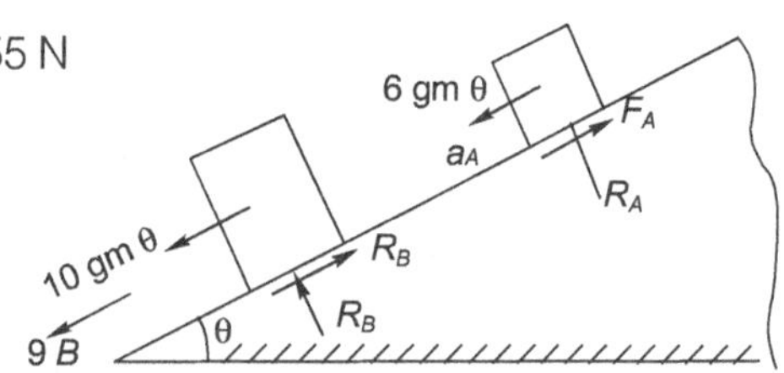

Fig. 15.13

Say after time *t*, distance between block is 6 m

$$S_A = \frac{1}{2}a_A t^2; \qquad S_B = \frac{1}{2}a_B \times t^2$$

$$S_B - S_A = \frac{t^2}{2}(4.0555 - 3.206) = 0.42475t^2 = 6 \text{ m}$$

Time, $t = 3.76$ second.

Exercise 15.3 Two blocks A and B of masses 10 kg and 5 kg respectively are held on an inclined plane with 30° angle of inclination. Distance between the blocks along the plane is 5 m when they are released as shown in **Fig. 15.14**. Coefficient of friction between blocks A and plane is 0.15 and that between block B and plane is 0.25. After what time from release, the block A will collide with block B and at what velocity?

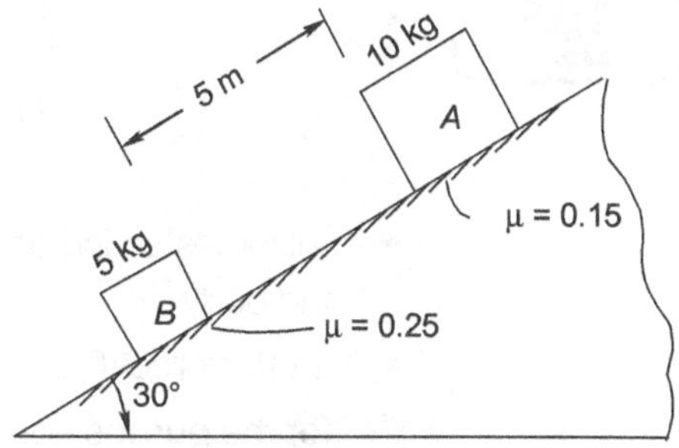

Fig. 15.14

[**Hint:** $\mu_A < \mu_B$, $V_A > V_B$]

[**Ans:** 3.43 s, 12.45 m/s].

Problem 15.1 A uniform slender bar AB of length L rests on car seat as shown in **Fig. 15.15**. The bar makes an angle of 35° with the vertical. Determine the retardation a for which the bar will begin to tip forward. Assume that friction at B is sufficient to prevent slipping.

Solution Car is moving towards right, so deceleration of car will be towards left as shown. Inertia force $m\bar{a}$ can be applied at G of slender bar AB. Weight of the bar mg will act vertically downwards. Taking moments about end B of the bar

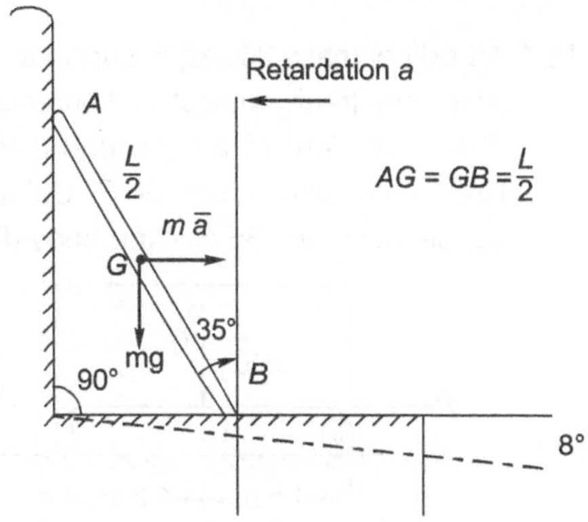

Fig. 15.15

$$m\bar{a} \times \frac{L}{2}\cos 35° = mg \times \frac{L}{2}\sin 35°$$

$$\bar{a} = g\tan 35° = 9.81 \times 0.70 = 6.87 \text{ m/s}^2.$$

Problem 15.2 System shown in the **Fig. 15.16** is initially at rest. Neglecting friction determine (a) Force F required if velocity of collar B becomes 5 m/s in 2 seconds after start, (b) Tension in the cable.

Solution Velocity of collar,

$$V = 5 \text{ m/s}$$

Time, $t = 2$ s

Acceleration of roller,

$$a = \frac{5}{2} = 2.5 \text{ m/s}^2$$

For collar $F - 2T = 5a$ ...(1)

where T is cable tension.

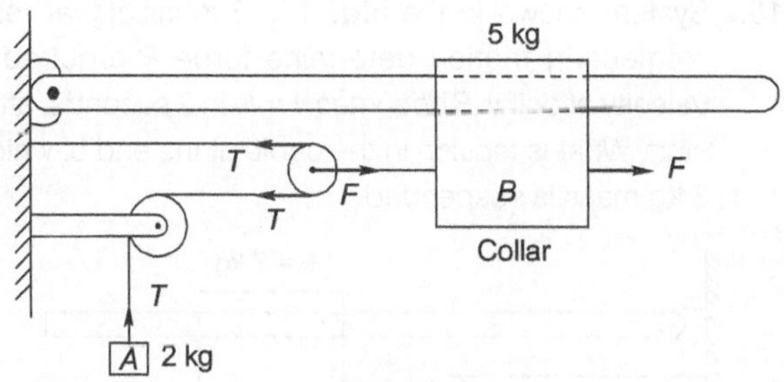

Fig. 15.16

Note that if acceleration of collar B is 2.5 m/s², then acceleration of mass A will be 2.5 × 2 = 5 m/s² because if collar moves by distance x in horizontal direction, then during the same time, mass A goes up by $2x$ distance.

Mass A $T - 2g = 2 \times a'$

where a' = acceleration of mass A = 5 m/s²

$$T = 2g + 2a' = 2 \times 9.81 + 2 \times 5 = 29.62 \text{ N}$$

Collar: From equation (1)

$$F = 2T + 5a = 2 \times 29.62 + 5 \times 2.5 = 59.74 + 12.5 = 71.74 \text{ Newton.}$$

Remember

- Resultant force, $F_R = \sum F = m\bar{a}$, where m is mass of body and $\bar{a}$ is acceleration at G of the body.
- $$\sum F = F_{Rx} i + F_{Ry} j$$
  $$F_{Rx} = m\bar{a}_x,\; F_{Ry} = m\bar{a}_y$$
  $$\bar{a} = \bar{a}_x i + \bar{a}_y j.$$
- All forces acting on a body are equivalent to a single force F_R acting at the mass centre G of the body.
- Inertia force, $F_i = m\bar{a}$ is applied at the mass centre to bring the body to dynamic equilibrium for the purpose of study.
- Inertia force method is used to determine support reactions.

PRACTICE PROBLEMS

15.1 A body of mass 200 kg is supported by wheels at A which roll freely without friction and by a skid at B. The coefficient of friction under the skid is 0.45. Determine the value of P. So as to cause an acceleration of 0.35 g of the body (**Fig. 15.17**).

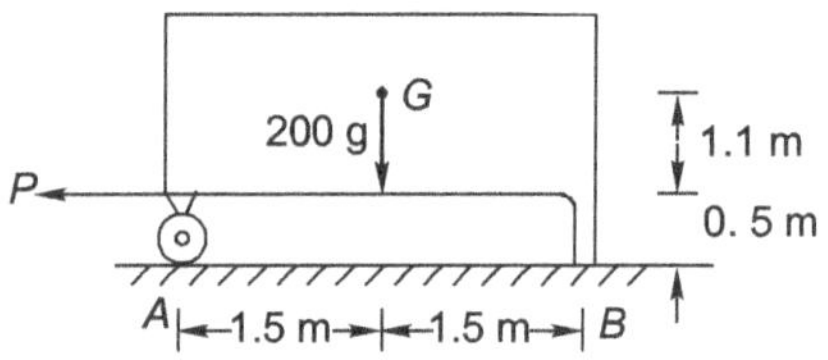

Fig. 15.17

[**Hint:** Apply inertia force $m\bar{a}$ at G, calculate reaction at B, $P - \mu R_B$ = effective force $= m\bar{a}$]

[**Ans:** 1202.75 N].

15.2 System shown in the **Fig. 15.18** is initially at rest. Neglecting friction determine force P required if velocity of collar B becomes 4 m/s in 2 seconds, after start. What is tension in the cable at the end of which 3 kg mass is suspended.

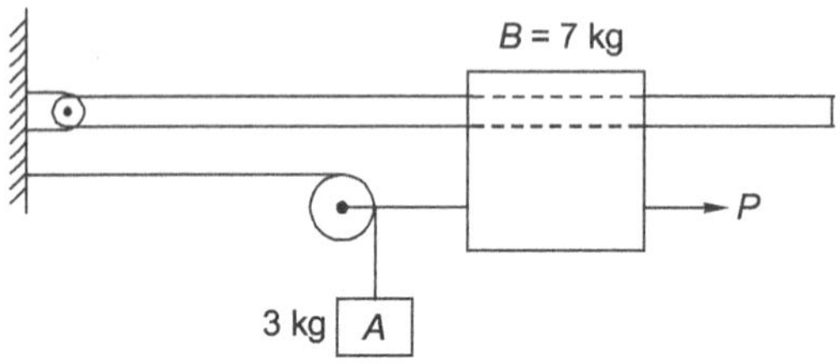

Fig. 15.18

[**Hint:** Acceleration of block A and block B is the same].

[**Ans:** P = 49.43 N; Tension = 35.43 N].

15.3 A uniform rectangular block (0.5 m × 1.0 m and 40 kg mass) is placed on a rough inclined plane, $\mu = 0.2$ as shown in **Fig. 15.19**. Determine the maximum and minimum values of d. So that the block does not overturn as it slides up the plane of inclination of 30°.

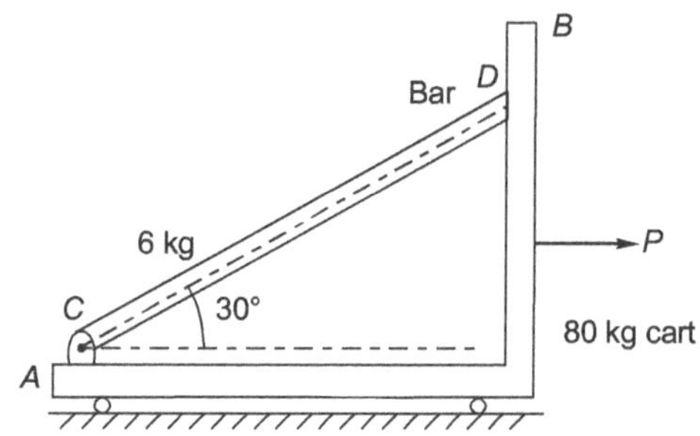

Fig. 15.19

[**Hint:** $40\bar{a} = 600\text{ N} - mg\sin\theta - \mu\, mg\cos$

Use inertia force $40\bar{a}$, take moments, $M_A = 0$, then, $M_B = 0$ $\bar{a} = 8.396\text{ m/s}^2$].

[**Ans:** d = 0.307 m to 0.585 m].

15.4 A homogeneous bar CD has a mass of 6 kg and is attached by a frictionless pin at C and rests against a smooth vertical part of the cart. The cart has a mass of 80 kg and is pulled to the right by a horizontal force P as shown in **Fig. 15.20**. At what value of P will the bar exert no force on cart at D? Draw a free body diagram of the bar CD.

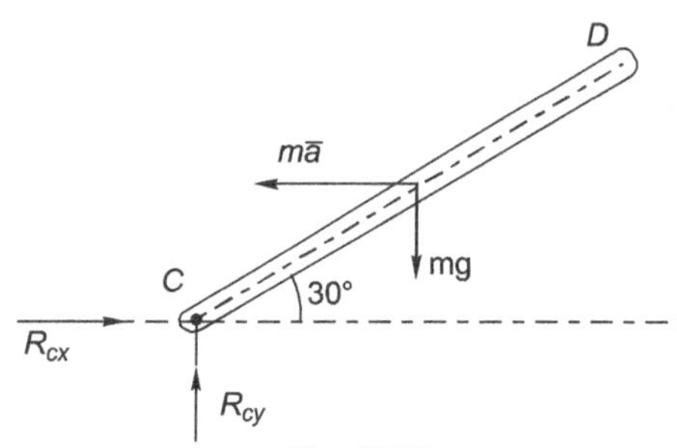

Fig. 15.20

[**Ans:** 1461.2 N (**Fig. 15.20**)].

MULTIPLE CHOICE QUESTIONS

15.1 A uniform bar *AB* of mass *m* is maintained inside the smooth surface of a cylindrical drum translating and acceleration at 0.6 g. Find angle θ to maintain bar in same position during motion (**Fig. 15.21**).

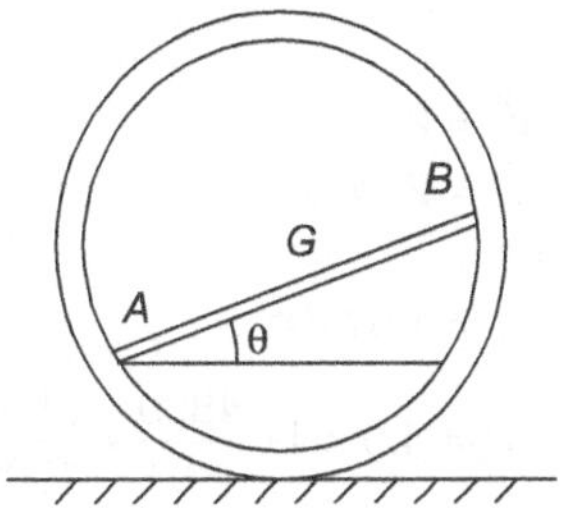

Fig. 15.21

(*a*) 59° (*b*) 55°
(*c*) 45° (*d*) 43.5°

15.2 A cabinet of mass 50 kg is mounted on castors, but castors are locked and slide along the floor with coefficient of friction between castors and floor equal to 0.3. If *P* = 350 N and cabinet is not to tip about *A* or *B* castors, what is acceleration of cabinet (**Fig. 15.22**)?

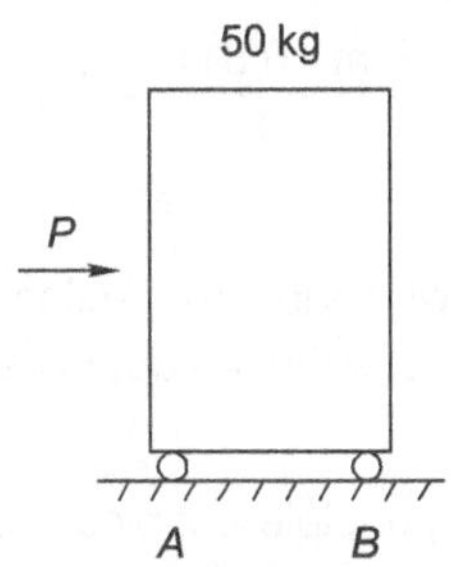

Fig. 15.22

(*a*) 7 m/s^2 (*b*) 4.06 m/s^2
(*c*) 3.5 m/s^2 (*d*) None of these

15.3 A mine cage of weight 12 kN starts from rest and moves downward with a constant acceleration travelling a distance of *S* = 30 m in 15 seconds. What is tension in cable during this time?

(*a*) 12.326 kN (*b*) 12 kN
(*c*) 11.674 kN (*d*) None of these

15.4 An elevator weighing 200 kg is moving upward with a uniform velocity of 4 m/s. Power is put off and friction opposing motion is 100 N. In how much distance elevator will stop

(*a*) 32 m (*b*) 0.86 m
(*c*) 0.76 m (*d*) None of these

15.5 The driver of an automobile travelling along a straight road suddenly applies brakes and automobile slides for a distance of 9.81 m in 2 seconds before coming to rest. Assuming constant acceleration, find coefficient of friction between tyres and the road

(*a*) 0.7 (*b*) 0.5
(*c*) 0.3 (*d*) None of these

15.6 An elevator of gross weight 4.8 kN starts moving upwards with a constant acceleration and acquires a velocity of 3 m/s after travelling through a distance of 3 m. What is cable tension during this accelerated motion?

(*a*) 6.268 kN (*b*) 4.8 kN
(*c*) 5.534 kN (*d*) None of these

15.7 A balloon is rising-up with an acceleration of 3 m/s^2. What fraction of weight of balloon must be emptied out in the form of loose sound in order to make acceleration 6 m/s^2. Take g = 9.8 m/s^2?

(*a*) 0.19 (*b*) 0.38
(*c*) 0.33 (*d*) 0.16

15.8 A vehicle is moving up an machine when the driver applies its brakes and the vehicle retards at 0.5 m/s^2. Then according to D' Aterbett's principle

(*a*) inertia force will be directed in the direction of motion
(*b*) inertia force will be directed opposite to the direction of motion
(*c*) magnitude of inertia force will depend on the angle of inclination
(*d*) magnitude of inertia force will depend on coefficient of friction between the wheels and the road surface

[CSE, Prelim, CE : 2002]

15.9 D Aterbert's principle

(*a*) is based on the law of conservation of energy
(*b*) is based on the law of conservation of angular momentum
(*c*) provides no special advantages over Newton's law
(*d*) enables a dynamic problem to be treated akin to a problem of statics

[CSE, Prelim, CE : 2003]

15.10 Where is D'Aterbert's principle applied

(*a*) in the design of long columns
(*b*) to determine stresses induced in thick cylinders
(*c*) to reduce a problem of kinetics to an equivalent problem of statics
(*d*) in the analysis of many geotechincal problem

[CSE, Prelim, CE : 2006]

Answers

15.1 (a)	15.2 (b)	15.3 (c)	15.4 (c)	15.5 (b)
15.6 (c)	15.7 (a)	15.8 (a)	15.9 (d)	15.10 (c)

EXPLANATIONS

15.1 (a)

$$mg\frac{L}{2}\cos\theta = m\bar{a}\frac{L}{2}\sin\theta$$

$$g\cot\theta = \bar{a} = 0.6\text{ g}$$

$$\cot\theta = 0.6 = 59^\circ.$$

15.2 (b)

$$P - \mu mg = ma$$

$$350 - 0.3 \times 50 \times 9.81 = 50a$$

$$a = \frac{202.85}{50} = 4.057 \text{ m/s}^2.$$

15.3 (c)

$$S = \frac{1}{2}at^2,\ 30 = \frac{1}{2}a \times 15^2,\ a = 0.2666 \text{ m/s}^2$$

$$T = 12000 - \frac{12000}{9.81} \times 0.266 = 12000 - 326.2 \text{ N}$$

$$= 11674 \text{ N} = 11.674 \text{ kN}.$$

15.4 (c)

Resistance= $200g + 100$ N = 2062 N

$$a = \frac{2062}{200} = 10.31 \text{ m/s}^2$$

$$\frac{V^2}{2a} = S = \frac{4^2}{2 \times 10.31} = 0.76 \text{ m}.$$

15.5 (b)

$$S = 9.81 \text{ m/s} \quad t = 2 \text{ s}$$

$$V_{av} = \frac{9.81}{2} = 4.905 \text{ m/s}$$

$$V = 2V_{av} = 9.81 \text{ m/s}$$

$$V^2 = 2as \quad 9.81^2 = 2 \times a \times 9.81$$

acceleration = 4.905 m/s = μg

$$\mu = 0.5.$$

15.6 (c)

$$V^2 = 2a\,.\,s$$

$$a = 2a \times 3$$

acceleration,

$$a = 1.5 \text{ m/s}^2$$

$$T = 4.8 \text{ kN} + \frac{4800}{9.81} \times \frac{1.5 \times 1}{1000}$$

$$= 4.8 + 0.734 = 5.534 \text{ kN}.$$

15.7 (a)

$$T = m \times 3 + 9.81 \times m = m' \times 6 + 9.81m'$$

$$12.81 \text{ m} = m'(6 + 9.81)$$

$$\frac{m'}{m} = \frac{12.81}{15.81} = 0.81$$

$$\mathbf{m' = 0.81\ m}$$

$$\frac{m - m'}{m} = \frac{m - 0.81\text{ m}}{m} = 0.19.$$

15.8 (a)

Inertia force will be in the direction of motion because retarding force act in the opposite direction.

15.9 (d)

D'Alerberti's principle enables a dynamics problem to be treated akin to a probelm of statics.

15.10 (c)

To reduce a problem of kinetics to an equivalent problem of statics.

■■■■

16

CHAPTER

Mass Moment of Inertia

16.1 Introduction

It is important to learn about mass moment of inertia of a rotating body. The damage which can be done by a wheel of a train is much larger than the damage which can be done by the wheel of a car. Similarly because of heavy inertia of masses of a train, much more work is done by a train than by a car. Moreover, in a rotating machine as an IC engine, the speed is controlled by inertia of a flywheel *i.e.*, the fluctuations about the mean torque are controlled.

To have clear understanding about moment of inertia of a mass, let us determine *angular momentum of a body* rotating about an axis. Consider that a body of mass m is rotating about an axis aa' with angular speed ω as shown in the **Fig. 16.1**. Consider an elementary mass dm_1, at radius r_1 from axis of rotation aa' having linear velocity $V_1 = \omega r_1$. Then linear momentum of mass dm_1

$$= dm_1 \times V_1$$

but linear velocity, $V_1 = \omega r_1$.

Therefore linear momentum of mass dm_1

$$= dm_1 \times r_1 \times \omega$$

$$= \omega r_1\, dm_1$$

Moment of linear momentum about axis of rotation

$$= r_1 \times \omega r_1\, dm_1$$

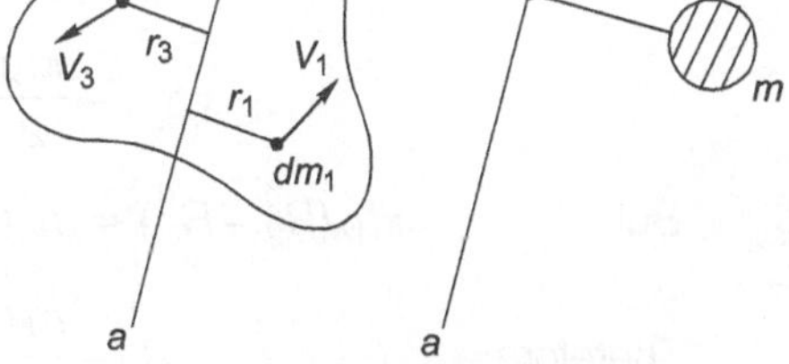

Fig. 16.1

Moment of linear momentum about an axis is termed as angular momentum about that axis of rotation, therefore angular momentum of mass dm_1 about axis $aa' = \omega r_1^2\, dm_1$. Similarly angular momentum of elementary masses dm_2 and dm_3 will be $\omega r_2^2\, dm_2$, $\omega r_3^2\, dm_3$ respectively. The body can be divided into elements of masses dm_1, dm_2, dm_3, ... etc. If a torque is applied to the body about the axis aa', the mass assumed initially at rest will start rotating about axis aa' and the time required by the body to reach a given speed ω, is proportional to mass dm of an element and to the square of the distance r from axis of rotation. Total angular momentum

$$= \omega\,(r_1^2\, dm_1 + r_2^2\, dm_2 + r_3^2\, dm_3 + \ldots)$$

$$= \omega . I$$

The product r^2. dm provides a measure of inertia of the small element i.e., the resistance offered by an element when we try to set the body in motion. Total resistance offered by the body is measured by the sum of

$r_1^2\,dm_1 + r_2^2\,dm_2 + r_3^2\,dm_3 \ldots r_n^2 dm_n$

Increasing the number of elements, we may say that the moment of inertia is equal, in the limit to the integral

$$I = \int r^2 dm = mk^2$$

where k = radius of gyration of the body with respect to axis aa'

or radius of gyration, $$k = \sqrt{\frac{I}{m}}$$

The radius of gyration k represents the distance at which the entire mass of the body may be located if its moment of inertia with respect to axis of rotation is to remain unchanged. The radius of gyration is expressed in metres and mass in kilogram (kg). Therefore moment of inertia of a mass is expressed in kg . m^2.

Example 16.1 Determine the mass moment of inertia of a thin ring of radii R_1, R_2 and axial thickness t about z–z-axis as shown in **Fig. 16.2**. The density of ring is ρ.

Solution Mass of the ring,

$$m = \rho t \pi\,(R_2^2 - R_1^2)$$

Consider an elementary ring of radial thickness dr at radius r, then

$$dm = \text{mass of elementary ring} = 2\pi r\,dr . t . \rho$$

$$= 2\pi t\rho . r\,dr$$

$$r^2\,dm = 2\pi t\rho r^3\,dr$$

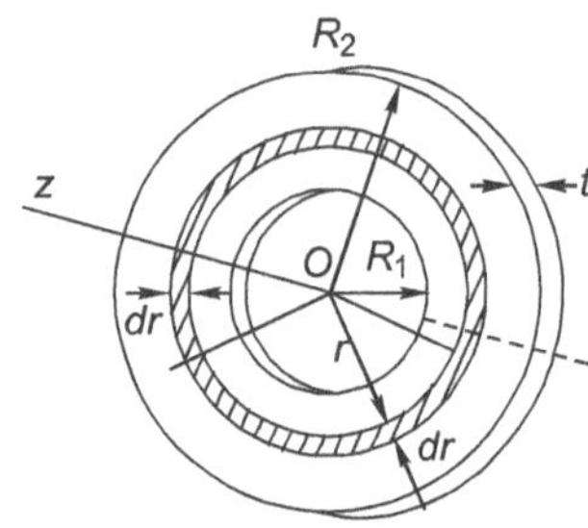

Fig. 16.2

Mass moment of inertia, $$I = \int r^2 dm$$

$$= \int_{R_1}^{R_2} 2\pi t\rho r^3 dr = 2\pi t\rho \left|\frac{R_2^4}{4} - \frac{R_1^4}{4}\right|$$

$$= \frac{\pi t\rho}{2} \times (R_2^4 - R_1^4) = \frac{\pi t\rho}{2}(R_2^2 + R_1^2)(R_2^2 - R_1^2)$$

But $\pi t\rho\,(R_2^2 - R_1^2) = m$, mass of ring

Therefore $$I = \frac{m\,(R_2^2 + R_1^2)}{2}$$, mass moment of inertia.

Exercise 16.1 A steel ring 10 cm thick is of inner radius 20 cm and outer radius 30 cm. If density of steel 7800 kg/m^3, determine mass moment of inertia of steel ring.

[**Ans:** 7.964 kg . m^2].

Exercise 16.2 Determine the mass moment of inertia of a steel rod 15 mm in diameter and 1 m long about an axis through one end and perpendicular to the rod. Density of steel is 7830 kg/m^3.

[Hints : $mL^2/3$]

[**Ans:** 0.4612 kg m^2].

16.2 Parallel-Axes Theorem

Consider a system of *Oxyz* co-ordinate axes through a body of mass *m*. *Say G is the centre of gravity of the body at co-ordinates* $(\bar{x}, \bar{y}, \bar{z})$ with respect to *xyz* co-ordinate. Consider an elementary mass *dm* with co-ordinates

$x'\ y'\ z'$ with respect to $Gx'y'z'$ co-ordinate axes passing through the centre of gravity of the body and parallel to the *Oxyz* co-ordinate axes. Co-ordinates of *dm* with respect to axes *Oxyz* are (as shown in Fig. 16.3)

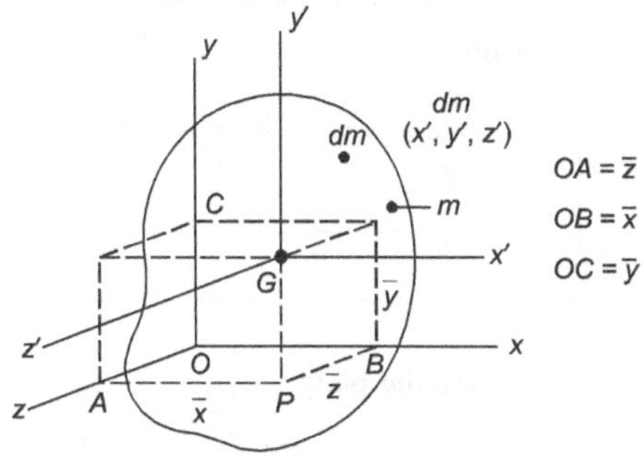

Fig. 16.3

$$x = x' + \bar{x}$$
$$y = y' + \bar{y}$$
$$z = z' + \bar{z}$$

Mass moment of inertia of the body about *x*-axis

$$I_{xx} = \int (y^2 + z)^2 dm = \int [(y' + \bar{y})^2 + (z' + \bar{z})^2]\, dm$$

$$= \int y'^2 dm + \int 2y'y'\, dm + \int \bar{y}^2 dm + \int z'^2 dm + \int 2zz' dm + \int \bar{z}^2 dm$$

$$= \int (y'^2 + z'^2)\, dm + 2\bar{y} \int y' dm + 2\bar{z} \int z' dm + \int \bar{y}^2 dm + \int \bar{z}^2 dm$$

Since $Gx'y'z'$ are centroidal co-ordinate axes, therefore

$$\int y' dm = \int z' dm = 0 \text{ (first moment of mass about centroidal axes is zero)}$$

Therefore, $I_x = \bar{I}'_x + m\bar{y}^2 + m\bar{z}^2$, since $\bar{y}$ and $\bar{z}$ are constants

$$I_{xx} = \bar{I}_{x'x'} + m\bar{y}^2 + m\bar{z}^2$$

$$I_{xx} = \bar{I}_{x'x'} + m(\bar{y}^2 + \bar{z}^2), \text{ where } \bar{I}_x = \int (y'^2 + z'^2)\, dm$$

where $\bar{I}_{x'x'}$ = Moment of inertia of the body about centroidal axis $x'x'$

Similarly $I_{yy} = \bar{I}_{y'y'} + m(\bar{x}^2 + \bar{z}^2)$ and $I_{zz} = \bar{I}_{z'z'} + m(\bar{x}^2 + \bar{y}^2)$.

Body	Mass	Mass moment of inertia
Rectangular Block 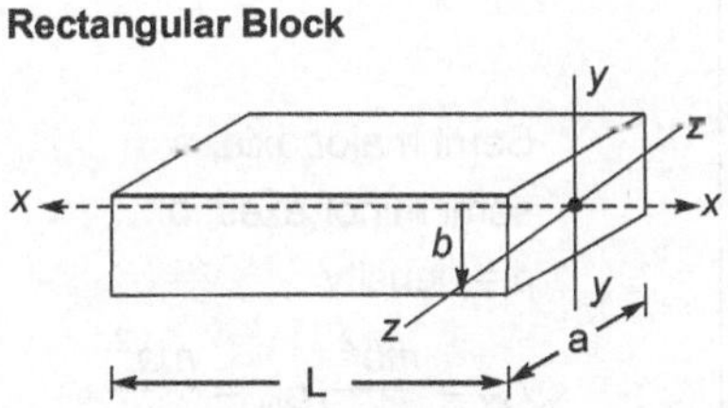	ρ = density $m = \rho abL$	$I_{xx} = m\left[\frac{a^2}{12} + \frac{b^2}{12}\right]$ $I_{yy} = \frac{ma^2}{12}$ $I_{zz} = \frac{mb^2}{12}$
Cylinder 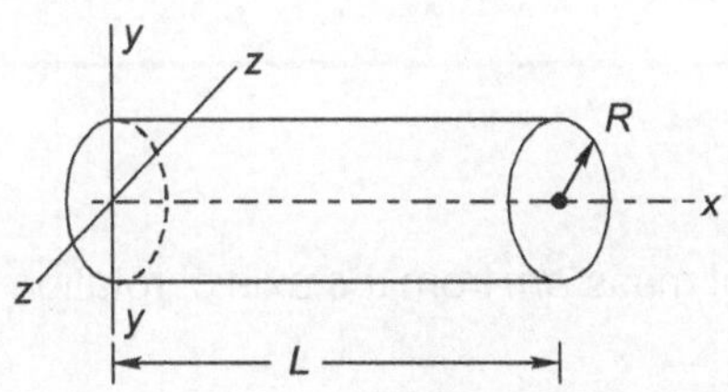	$m = \rho\pi R^2 L$	$I_{xx} = \frac{mR^2}{2}$ $I_{yy} = I_{zz} = m\left(\frac{R^2}{4} + \frac{L^2}{3}\right)$

Body	Mass	Moment of inertia
Rod 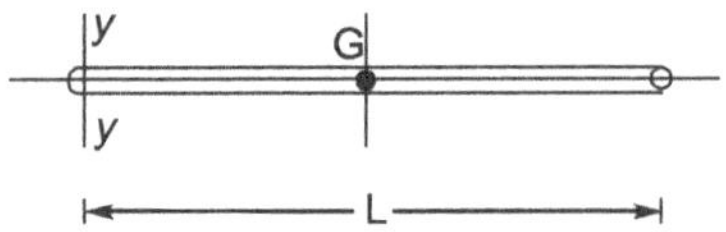	m = mass of bar	$I_{yy} = \dfrac{mL^2}{3}$ $I_G = \dfrac{mL^2}{12}$
Triangular plate 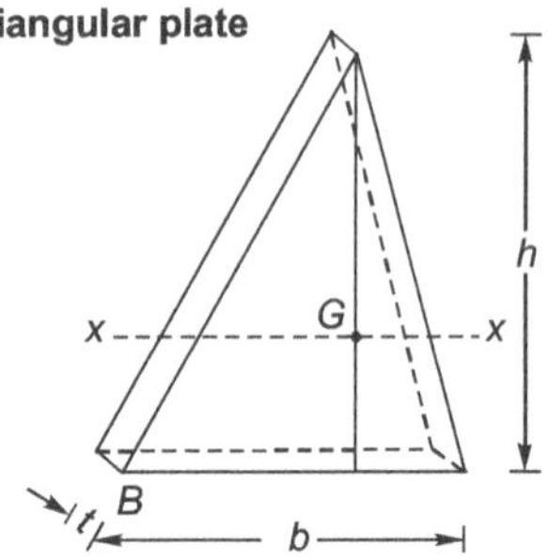	$m = \rho\tau\dfrac{bh}{2}$	$I_{xx} = \rho t\left(\dfrac{bh^3}{36}\right) = m\left(\dfrac{h^2}{18}\right)$
Solid sphere 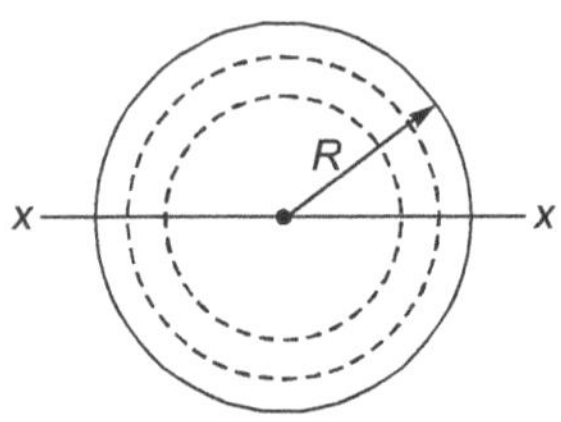	$m = \dfrac{4\rho\pi R^3}{3}$	$I_{xx} = \dfrac{2mR^2}{5}$
Hollow sphere 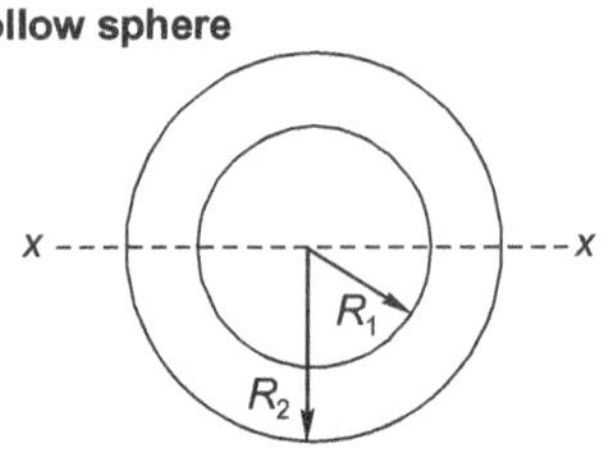	$m = \rho\dfrac{4}{3}\pi\left(R_2^3 - R_1^3\right)$	$I_{xx} = \dfrac{2}{5}m\left[\dfrac{R_2^5 - R_1^5}{R_2^3 - R_1^3}\right]$
Eleptical plate 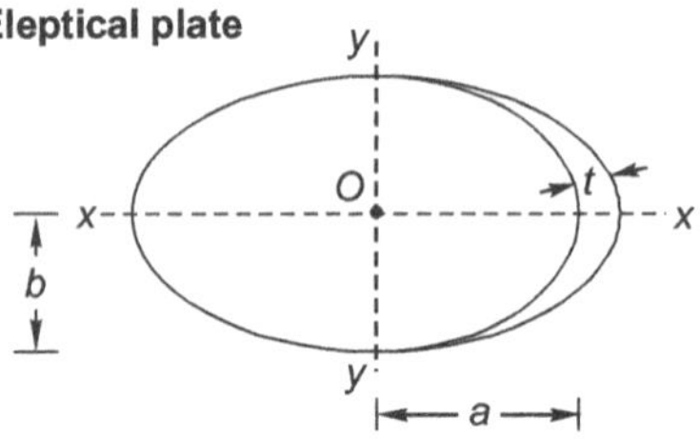	$m = \rho\pi abt$	Semi major axis, a semi minor axes, b ρ = density $I_{xx} = \dfrac{mb^2}{4}, I_{yy} = \dfrac{ma^2}{4}$

Remember

- Mass moment of inertia,

$$I_m = \int r^2 dm = mk^2$$

where r = perpendicular distance of mass dm from the axis of rotation
m = mass of the body
k = radius of gyration.

- The radius of gyration represents the distance at which the entire mass of the body is assumed to be concentrated if its moment of inertia with respect to axis of rotation is to remain unchanged.
- $I_{xx} = \int (y^2 + z^2)\, dm,\ I_{yy} = \int (x^2 + z^2)\, dm,\ I_{zz} = \int (x^2 + y^2)\, dm.$
- Parallel axis theorem $I_{xx} = \bar{I}_{x'x'} + m(\bar{y}^2 + \bar{z}^2)$

  where $\bar{I}_{x'x'}$ = Moment of inertia of the body about centroidal axis $x'x'$

  $\bar{y}, \bar{z}$, distance of centroid of body from axis under consideration *i.e.*, *x*, *y*, *z*-axes.
- Moment of inertia of thin plates

  $I_{xx_{mass}} = \rho . t . I_{xx_{area}}$

  where ρ = Mass density, t = thickness of plate.

MULTIPLE CHOICE QUESTIONS

16.1 An ellipse of semi major axis '*a*' and semi minor axis '*b*' is cut from a flat plate of thickness *t*. Mass of elliptical plate is *m*. What is mass moment of inertia of this plate about an axis passing through centroid of elliptical section and perpendicular to surface of plate?

(*a*) $\frac{m}{4}(a^2 + t^2)$ (*b*) $\frac{m}{4}(b^2 + t^2)$

(*c*) $\frac{m(a^2 + b^2)}{4}$ (*d*) None of these

16.2 An equilateral triangular plate of side '*a*' is cut from a plate of thickness *t*. Mass of plate is *m*. What is mass moment of inertia of the plate about an axis passing through centroid of triangular plate and perpendicular to surface of plate?

(*a*) $ma^2/12$ (*b*) $ma^2/18$

(*c*) $ma^2/36$ (*d*) None of these

16.3 A rhombus plate of side *a* mass *m* is made of a thin plate. What is mass moment of inertia of plate about *x*-axis (**Fig. 16.4**)?

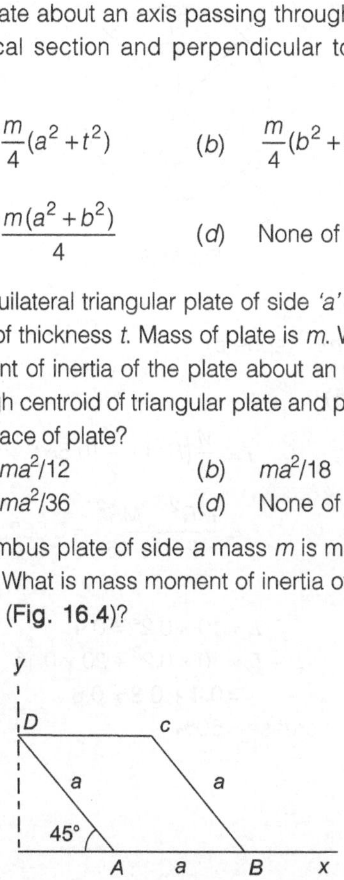

Fig. 16.4

(*a*) $ma^2/12$ (*b*) $ma^2/6$

(*c*) $ma^2/4$ (*d*) $ma^2/3$.

16.4 A hollow disc of inner radius *R* and outer radius R_2 is cut from a flat plate of thickness *t*. Mass of disc is *m*. What is its mass moment of inertia about an axis passing through its centre and perpendicular to surface of disc?

(*a*) $\frac{m(R_2^2 + R_1^2)}{4}$ (*b*) $\frac{m(R_2^2 + R_1^2)}{2}$

(*c*) $\frac{m(R_2^2 - R_1^2)}{2}$ (*d*) None of these

16.5 What is mass moment of inertia of a sphere of mass *m* and radius *R* about a diametral axis?

(*a*) $0.25mR^2$

(*b*) $0.3125mR^2$

(*c*) $0.4mR^2$

(*d*) None of these

16.6 Consider a flywheel whose mass *M* is distributed almost equally between a heavy, ring-like rim of radius *R* and a concentric disk-like feature of radius *R*/2. Other parts of the flywheel, such as spokes, etc, have negligible mass. The best approximation for α, if the moment of inertia of the flywheel about its axis of rotation is expressed as αMR^2, is ______.

[GATE, Prelim, CE : 2002]

16.7 A rigid body shown in the **fig. 16.5(a)** has a mass of 10 kg. It rotates with a uniform angular velocity 'ω'. A balancing mass of 20 kg is attached as shown in **fig. 16.5(b)**. The percentage increase in mass moment of inertia as a result of this addition is

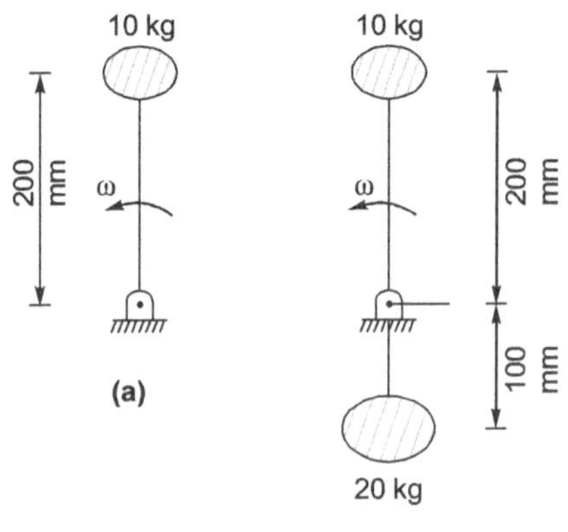

Fig. 16.5 (b)

(a) 25% (b) 50%
(c) 100% (d) 200%

[GATE 2004 : 2 Marks]

Answers

16.1 (c)	16.2 (a)	16.3 (b)	16.4 (b)	16.5 (c)
16.6 (0.5625)		16.7 (b)		

EXPLANATIONS

16.1 (c)

$$I_{z_2} = \frac{m}{4}(a^2+b^2). \text{ (Fig. 16.6).}$$

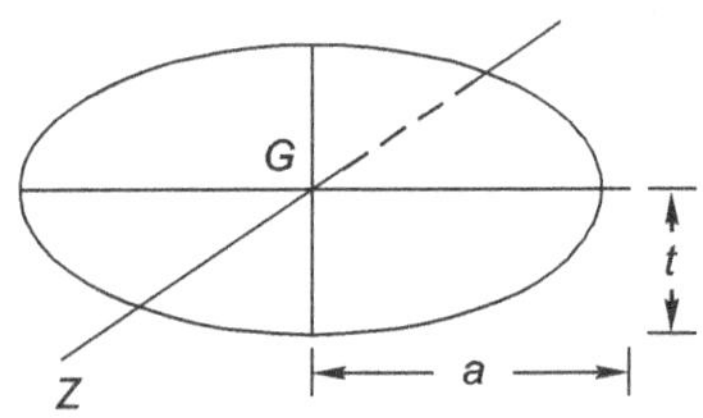

Fig. 16.6

16.2 (a)

$$I_{z_2} = \frac{ma^2}{12}. \text{ (Fig. 16.7).}$$

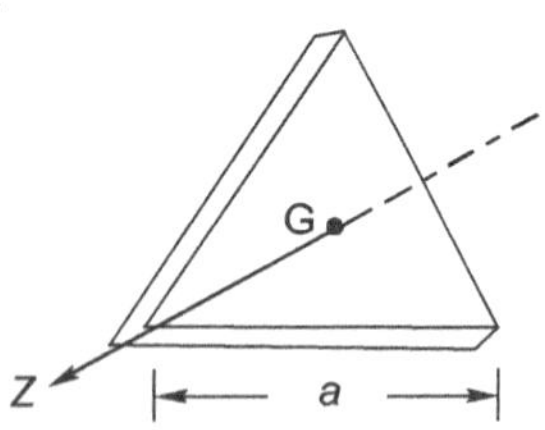

Fig. 16.7

16.3 (b)

$$I_x = \frac{ma'^2}{3} \text{ (Fig. 16.8)}$$

$$= \frac{m}{3}(a\times 0.707)^2 = \frac{0.5ma^2}{3} = \frac{ma^2}{6}.$$

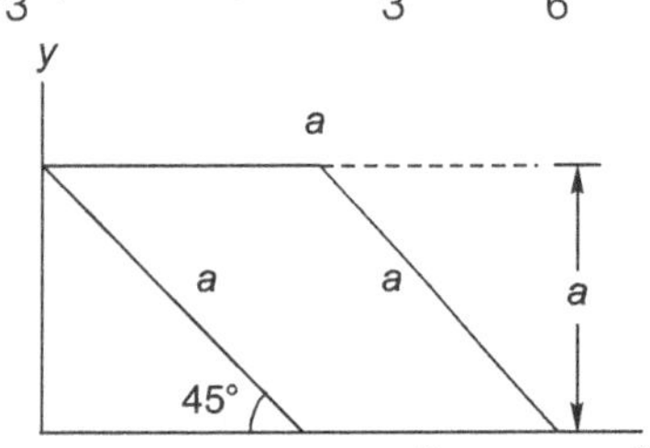

Fig. 16.8

16.4 (b)

$$I_{zz} = \frac{m(R_2^2+R_1^2)}{2} \text{ (Fig. 16.9)}$$

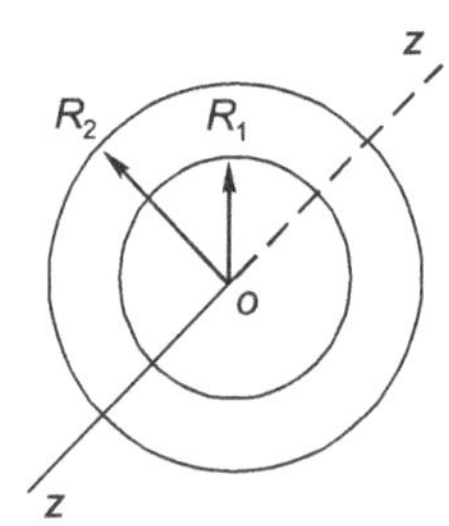

Fig. 16.9

16.5 (c)

$I = 0.4\ mR^2$ (sphere).

16.6 (0.5625)

$$I = \frac{M}{2}\left(R^2\right)+\frac{M}{2}(0.5R)^2\times\frac{1}{2}$$

$$= \frac{MR^2}{2}+\frac{MR^2}{16} = 0.5625$$

16.7 (b)

$$I_1 = 10\times 0.2^2 = 0.4$$

$$I_1 + I_2 = 10\times 0.2^2 + 20\times 0.1^2$$

$$= 0.4 + 0.2 = 0.6$$

% increase = 50%

■■■■

17

CHAPTER

Dynamics of Rigid Bodies in Plane Motion

17.1 Introduction

In the case of dynamics of a particle, the motion of a body as a whole is considered and forces are considered at its mass centre. But a body has definite size, shape and mass. The motion of its mass centre may not be the same as the motion of particles located away from the mass centre.

Any system of external forces acting on a body can be reduced to a single resultant force acting at the mass centre and a moment about the mass centre. If the deformations produced in the body due to external forces acting on it, are negligible and are not taken into account the body is said to be a *rigid body*. Majority of problems in engineering dynamics fall in this category of dynamics of rigid bodies in plane motion. In this chapter we do not analyse the deformations in the body and the body is assumed to be rigid. **Fig. 17.1** shows a rigid body subjected to external forces F_1, F_2, F_3, F_n. This system of external forces can be reduced to a *single force and a moment system at any point* or in the present case say at the mass centre G of the body, such that

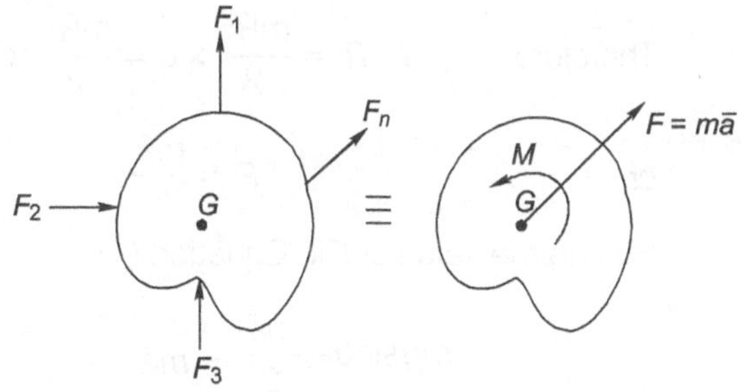

Fig. 17.1

$$\sum F_i = F = m\bar{a}, \quad \text{where } \bar{a} \text{ is}$$

the acceleration vector of the mass centre

$$\sum M_i = M = I_G . \alpha, \quad \text{where } \alpha = \text{angular acceleration}$$

where I_G is the mass moment of inertia of the body about its mass centre and α is the angular acceleration of the body about the mass centre at a particular instant.

In *x-y* co-ordinate system, scalar equations of motion are

$$\sum F_x = m\bar{a}_x, \text{ summation of forces in } x \text{ direction}$$

$$\sum F_y = m\bar{a}_y, \text{ summation of forces in } y \text{ direction}$$

$$\sum \bar{M} = \bar{I} . \alpha, \text{ summation of moments of forces about mass centre } G.$$

where $\bar{a}_x$ and $\bar{a}_y$ are the components of mass centre acceleration $\bar{a}$,

$\bar{I}$ = moment of inertia of the rigid body about an axis through G, its mass centre = I_G

α = angular acceleration of body.

Example 17.1 A cylinder of mass m is rolling down an inclined plane of angle of inclination θ as shown in the **Fig. 17.2**. The coefficient of friction between cylinder and plane is μ. Determine (*a*) linear acceleration of mass centre G of cylinder if the cylinder is to roll without slipping, (*b*) linear acceleration of mass centre if the frictional force is insufficient for rolling without slipping. Take μ as the coefficient of kinetic friction between cylinder and plane.

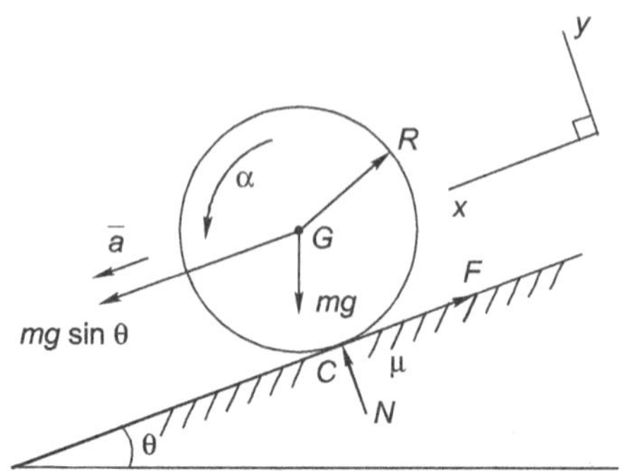

Fig. 17.2

Solution (*a*) *Rolling without slipping.* Let us take xy co-ordinates along and perpendicular to the inclined plane as shown.

$$\sum F_x = mg\sin\theta - F = m\bar{a} \quad ...(1)$$

where $\bar{a}$ is linear acceleration of mass centre and F is the frictional force on cylinder

$$\sum F_y = mg\cos\theta - N = 0, \text{ where } N \text{ is the normal reaction}$$

$$\sum M_G = \bar{I}.\alpha \text{ where } I \text{ is moment of inertia of cylinder and } \alpha \text{ is the angular acceleration}$$

$$M_G = F \times R$$

$$\bar{I} = \frac{mR^2}{2}$$

$$\alpha = \frac{\bar{a}}{R} \text{ (for rolling without slipping), where } \bar{a} = \text{mass centre acceleration}$$

Therefore $$F.R = \frac{mR^2}{2} \times \alpha = \frac{mR^2}{2} \times \frac{\bar{a}}{R} = \frac{mR\bar{a}}{2}$$

or $$F = \frac{m\bar{a}}{2} \quad ...(2)$$

Putting the value of F in Equation (1)

$$mg\sin\theta - \frac{m\bar{a}}{2} = m\bar{a}$$

or linear acceleration of mass centre of cylinder,

$$\bar{a} = \frac{2}{3}g\sin\theta$$

Angular acceleration $$\alpha = \frac{\bar{a}}{R} = \frac{2}{3} \times \frac{g}{R} \times \sin\theta \quad ...(3)$$

(*b*) Friction force $$F = \mu N = \mu mg\cos\theta$$

Say $$\mu mg\cos\theta < \frac{m\bar{a}}{2}, \text{ calculated in part } (a).$$

Then rolling will occur with some slip and linear acceleration

$$a \neq \alpha.R$$

Rolling and slip will occur simultaneously for the cylinder.

Now $$\sum F_x = mg\sin\theta - F = mg\sin\theta - \mu mg\cos\theta$$

$$= m\bar{a}$$

or linear acceleration of mass centre,

$$\bar{a} = g\sin\theta - \mu g\cos\theta \quad ...(4)$$

$$\sum M_G = \bar{I}\alpha = FR$$

$$\frac{mR^2}{2} \times \alpha = \mu mg\cos\theta \times R$$

$$\alpha = \frac{2\mu g\cos\theta}{R} \quad \text{...(5)}$$

Please note that $\alpha R \neq \bar{a}$, because cylinder rolls down with some occasional slips.

Exercise 17.1 A homogeneous cylinder of mass m = 40 kg and radius R = 0.1 m is placed on inclined plane as shown in **Fig. 17.3**. If the maximum coefficient of friction between cylinder and plane is 0.12, determine whether the cylinder will roll without slipping or cylinder will roll with slip.

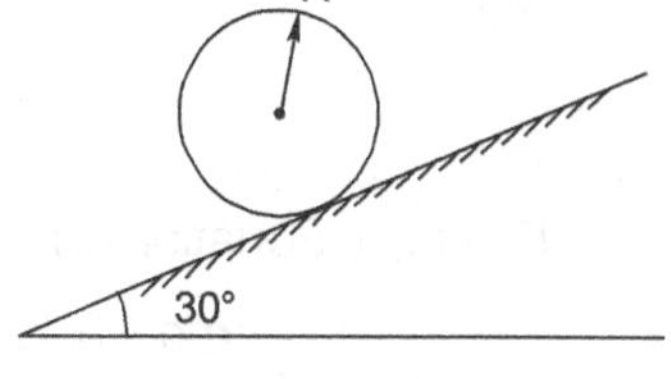

Fig. 17.3

[**Hint:** $\bar{a} = \frac{2}{3} g\sin\theta$, if $\mu mg\cos\theta < \frac{m\bar{a}}{2}$ rolling will occur with slip].

[**Ans:** Cylinder will roll with slipping].

PROBLEMS

Problem 17.1 A thin uniform bar of mass 9 kg and length 2 m is held at an angle of 30° to the horizontal as shown in **Fig. 17.4**, by means of two vertical inextensible strings. Suddenly the string at right hand end breaks, leaving the bar to swing about axis A. Determine the tension in the string at end A and angular acceleration of the bar immediately after the string breaks.

Solution Say $\bar{a}_y$ = acceleration of bar at mass centre G in y direction

and T = tension in string at end A

Then $$mg - T = m\bar{a}_y \quad \text{...(1)}$$

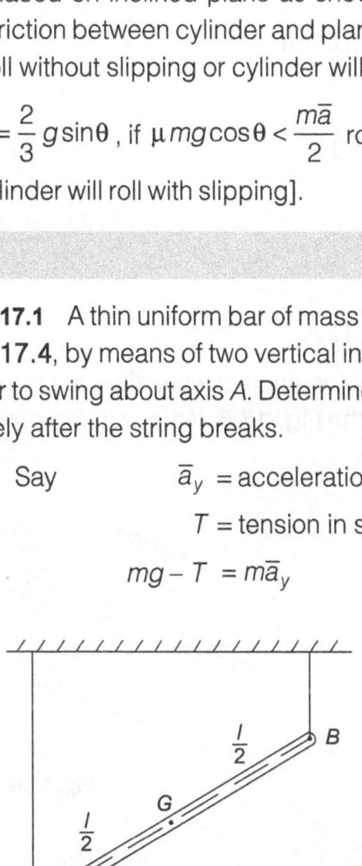

Fig. 17.4

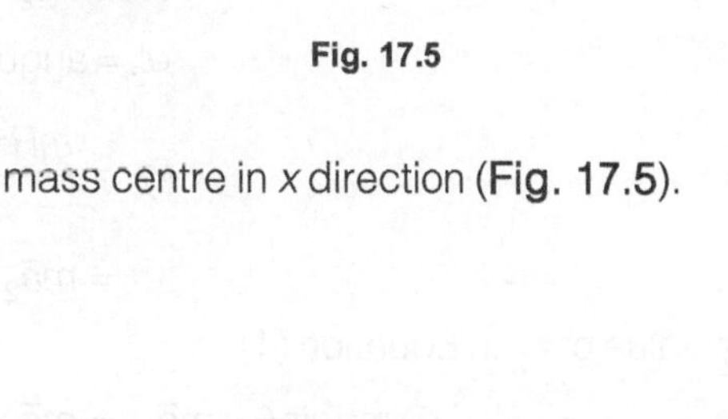

Fig. 17.5

No force in x direction, so $\bar{a}_x = 0$, acceleration of mass centre in x direction (**Fig. 17.5**).

Moment about G, $$M_G = T \times \frac{l}{2}\cos\theta$$

where l = 2 m, length of bar

θ = 30° as shown

Moment $$M_G = T \times 1 \times \cos 30° = 0.866T \text{ Nm}$$

$$= I_G . \alpha$$

I_G = moment of inertia of bar about axis of rotation through G

$$\frac{ml^2}{12} = \frac{9\times 2^2}{12} = 3.00 \text{ kg m}^2$$

α = angular acceleration of the bar

Therefore $I\alpha = 3\alpha = 0.866T$

or $T = 3.464\alpha$ ...(2)

Putting the Equation (1)

$mg - 3.464\alpha = m\bar{a}_y$, ...(3)

where $\bar{a}_y$ = acceleration in y direction

$$= \frac{l}{2}\cos\theta.\alpha = 1\times\cos 30°\times\alpha = 0.866\alpha$$

Putting in Equation (3)

$mg - 3.464\alpha = m\times 0.866\alpha$ where $= m = 9$ kg

$9\times 9.81 - 3.464\alpha = 9\times 0.866\alpha$

or angular acceleration, $\alpha = \dfrac{88.29}{11.258} = 7.842 \text{ rad/s}^2$

Tension in string, $T = 3.464\alpha = 3.464\times 7.842 = 27.17$ N.

Problem 17.2 A solid homogeneous cylinder and a hoop are in contact when they are released from the state of rest. Mass and radius of both are the same, as shown in the **Fig. 17.6**. If they roll without slipping, what will be the gap between them after 1 second?

Solution Say of mass of cylinder = m

Radius of cylinder = R

Moment of inertia of cylinder, $I_{G1} = \dfrac{mR^2}{2}$

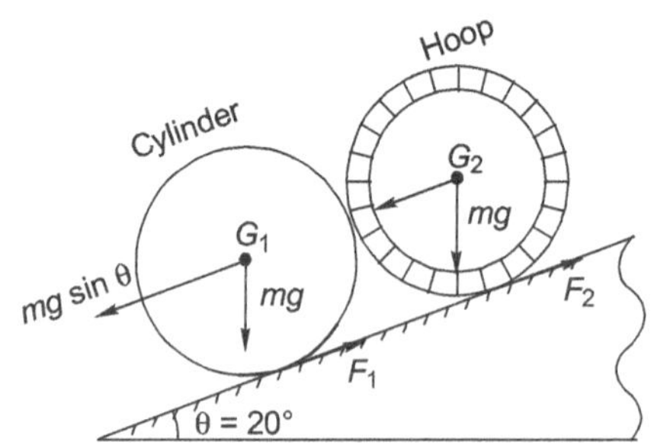

Fig. 17.6

Hoop

Then mass of hoop = m

Radius of hoop = R

Moment of inertia of hoop, $I_{G2} = mR^2$

Say acceleration of mass centre $= \bar{a}_2$

Then $mg\sin\theta - F_2 = m\bar{a}_2$ ...(1)

$F_2\times R = I_{G2}\times\alpha_2$

where α_2 = angular acceleration of hoop

or $F_2 = \dfrac{mR^2\times\alpha_2}{R} = mR\alpha_2$

$= m\bar{a}_2$, as $\bar{a}_2 = \alpha_2 R_2$

Putting the value of F_2 in Equation (1)

$mg\sin\theta - m\bar{a}_2 = m\bar{a}_2$

or acceleration, $\bar{a}_2 = \dfrac{g\sin\theta}{2}$ ...(2)

Cylinder $mg\sin\theta - F_1 = m\bar{a}_1$ ...(3)

where a_1 is the acceleration of mass centre of cylinder and F_1 is the force of friction at contact point.

Moreover $\quad F_1 \times R = I_{G1} \times \alpha_1$, where α_1 is the angular acceleration of cylinder

$$= \frac{mR^2}{2} \times \alpha_1$$

or $\quad F_1 = \frac{mR\alpha_1}{2}$ but $R\alpha_1 = \bar{a}_1 = \frac{m\bar{a}_1}{2}$

Putting the value of F_1 in Equation (3)

$$mg\sin\theta - \frac{m\bar{a}_1}{2} = m\bar{a}_1$$

acceleration $\quad \bar{a}_1 = \frac{g\sin\theta}{1.5}$

Now mass centre acceleration of cylinder is more than the mass centre acceleration of hoop

$$\bar{a}_1 = \frac{9.81 \times \sin 20^\circ}{1.5} = 2.237 \text{ m/s}^2$$

$$\bar{a}_2 = \frac{9.81 \times \sin 20^\circ}{2} = 1.68 \text{ m/s}^2$$

Distance covered by hoop and cylinder is 1 second

$$S_1 = \frac{1}{2} \times 2.237 \times 1^2 = 1.1185 \text{ m}$$

$$S_2 = \frac{1}{2} \times 1.68 \times 1^2 = 0.84$$

Gap between the two after 1 second

$$S_1 - S_2 = 1.1185 - 0.84 = 0.2785 \text{ m.}$$

Problem 17.3 A uniform rod *AB* of mass *m* and length *l* as shown in **Fig. 17.7** is released from rest when angle $\theta = 75^\circ$. Assuming that the friction force between end *A* of the rod and the horizontal surface is sufficient after release of rod, determine (*i*) normal reaction and force of friction at end *A* of rod, (*ii*) minimum value of coefficient of friction μ for the motion of rod.

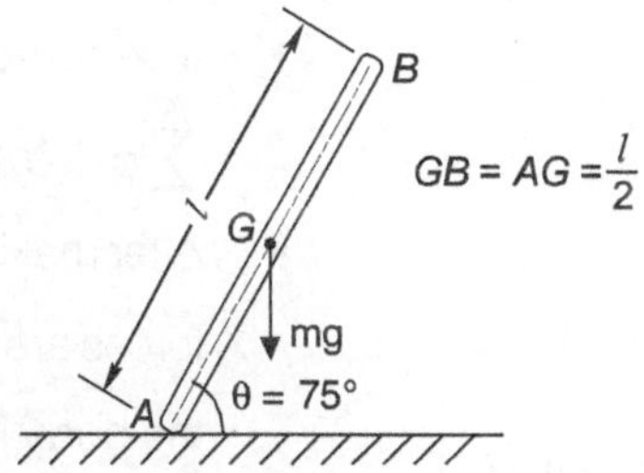

Fig. 17.7

Solution After the release of the rod from rest, rod will rotate about end *A* and therefore the point *A* becomes the instantaneous centre of rotation.

Moment of inertia of the rod about *A*

$$I = \frac{ml^2}{3} \quad \text{...(1)}$$

Moment of the force *mg* of bar about *A*

$$M = mg\frac{l}{2}\cos\theta = mg\frac{l}{2} \times \cos 75^\circ$$

$$= 0.1294mgl \quad \text{...(2)}$$

Angular acceleration of the rod about axis through end *A*

$$\alpha = \frac{M}{I} = 0.1294mgl \times \frac{3}{ml^2} = \frac{0.3882g}{l} \quad \text{...(3)}$$

Acceleration of *G* of rod $\quad = \alpha \cdot \frac{l}{2} = \frac{0.3882g}{l} \times \frac{l}{2} = 0.1941g$

$$\bar{a} = 0.1941g \quad \text{...(4)}$$

Inertia force at $\quad G = m\bar{a} = m \times 0.1941g$

Using D' Alembert's principle, inertia force, $m\bar{a}$ is applied in the opposite direction as shown in **Fig. 17.7**. So as to determine reactions at *A*.

Vertical reaction, $N = mg - m\bar{a}\cos\theta$

$= mg - m\bar{a}\cos 75° = mg - 0.2588m\bar{a}$

$= mg - 0.2588m \times 0.1941g$ putting the value of $\bar{a}$

$= 0.95mg$

The bar will tend to slip towards left, therefore force of friction *F* will be towards right as shown in **Fig. 17.8**.

$F = m\bar{a}\sin\theta$

$= m \times 0.1941g \times \sin 75°$

$= 0.1941 \times 0.9659mg = 0.1875mg$

Coefficient of friction, $\mu = \frac{F}{N} = \frac{0.1875mg}{0.95mg}$

$= 0.1974$ (minimum value required).

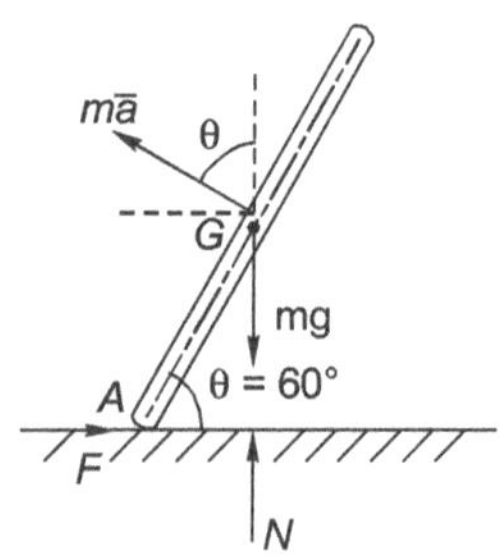

Fig. 17.8

Remember

- External forces acting on a body under general plane motion are equivalent to a force and or a couple acting at mass centre *G* of the body under motion of translation and rotation as shown in the **Fig. 17.9**.

  Moment, $M = I\alpha$

  Force, $F = m\bar{a}$.
- In *x-y* co-ordinates

  $\sum F_x = m\bar{a}_x,\ \sum F_y = m\bar{a}_y,\ \sum M = \bar{I}.\alpha.$
- After making the equations of accelerations $\bar{a}_x, \bar{a}_y, \alpha$, inertia forces are applied $m\bar{a}_x$, $m\bar{a}_y$ in the opposite direction to bring the body in dynamic equilibrium.
- Making the equation of equilibrium, unknown forces as reactions, frictional forces etc. are determined.

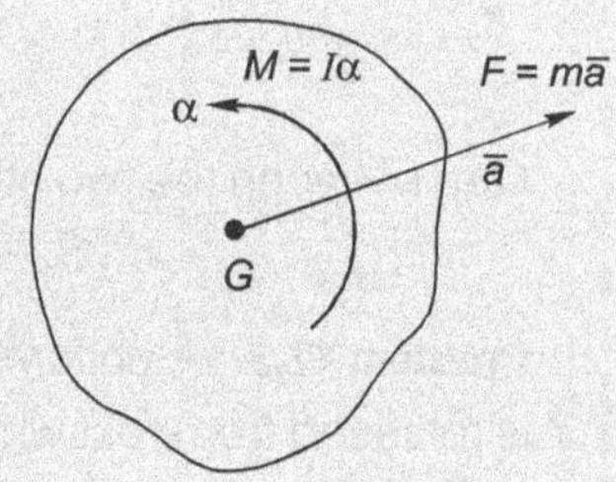

Fig. 17.9

PRACTICE PROBLEMS

17.1 A bar *AB* of mass *m* and length *l* is held in position as shown in **Fig. 17.10** by a horizontal string attached to the wall. If end *B* is placed on *smooth floor* and if the system is released in the position shown, determine (*i*) instantaneous acceleration of end *B* of the bar, (*ii*) tension in the string, (*iii*) normal reaction at *B*.

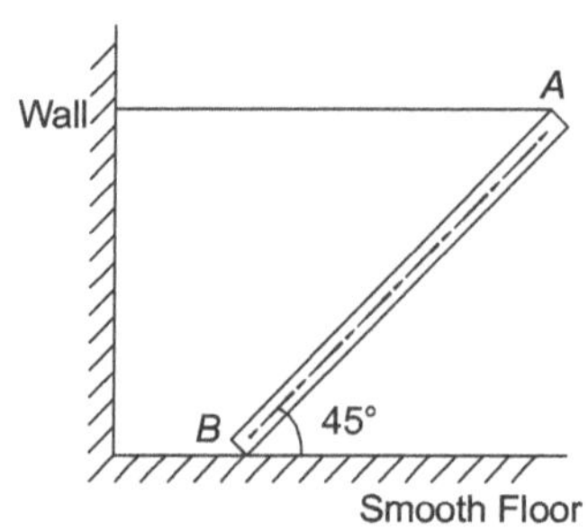

Fig. 17.10

[**Hint:** $T = ma_x$ at G, $N + ma_y = mg$].

[**Ans:** (*i*) $a_B = g$, (*ii*) Tension $= \frac{mg}{2}$,

(*iii*) Normal reaction $= \frac{mg}{2}$].

17.2 A solid homogeneous sphere and a hoop are in contact when they are released from the state of rest. Mass and radius of both are the same, as shown in **Fig. 17.11**. If they roll without slipping, what are their velocities after 3 seconds from start.

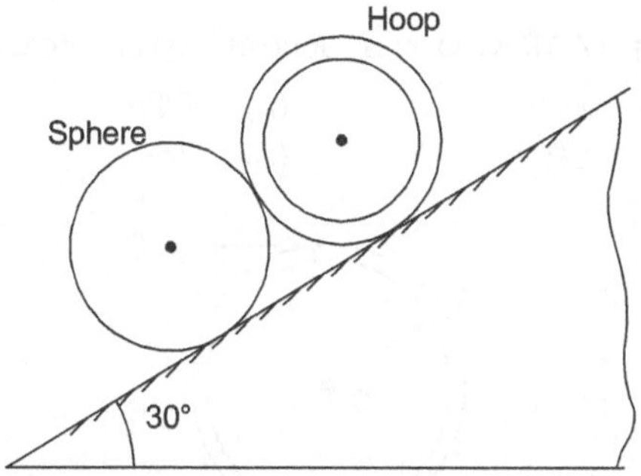

Fig. 17.11

[**Hint:** $I_S = 0.4mR^2$, $I_h = mR^2$, sphere will travel faster].
[**Ans:** $V_S = 10.51$ m/s, $V_h = 7.358$ m/s].

17.4 A uniform rod of mass m and length L as shown in **Fig. 17.12** is released from rest when angle $\theta = 60°$. Assuming that the friction force between end A of the rod and the horizontal surface is sufficient after release of rod, determine (*i*) normal reaction and force of friction at end A, (*ii*) minimum value of coefficient of friction μ for the motion of rod.

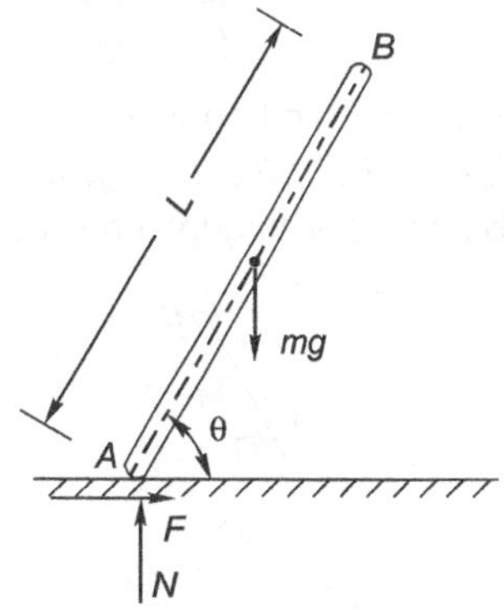

Fig. 17.12

[**Ans:** $N = \frac{13}{16}mg$, $F = \frac{2.6}{8}mg$, $\mu = \frac{F}{N} = 0.4$].

17.5 A thin walled cylinder is on a horizontal platform as shown in the **Fig. 17.13**. The horizontal platform has acceleration of a m/s^2. Determine the acceleration of the centre of the cylinder assuming rolling without slipping.

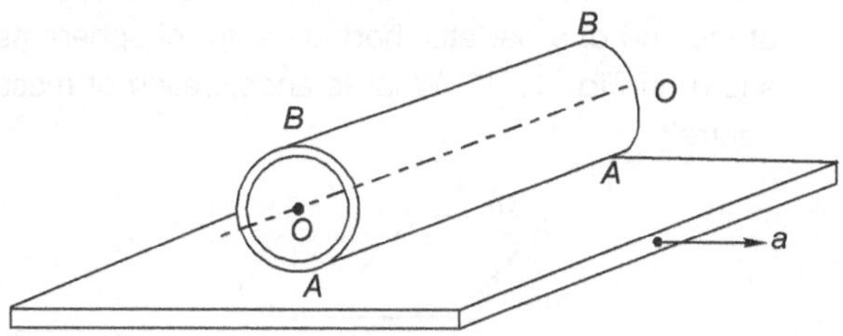

Fig. 17.13

[**Hint:** Cylinder is not rotating about axis OO. Edge AA has acceleration of a m/s^2. So cylinder is rolling about edge BB].
[**Ans:** Acceleration of the edge BB = 0
Acceleration of the edge AA = a,
Acceleration of the centre $OO = a/2$].

MULTIPLE CHOICE QUESTIONS

17.1 A homogeneous cylinder of mass m and radius R is placed on inclined plane with angle of inclination θ. What should be the maximum angle θ, so that cylinder rolls without slipping? The coefficient of friction between cylinder and inclined plane is 0.12 (**Fig. 17.14**)

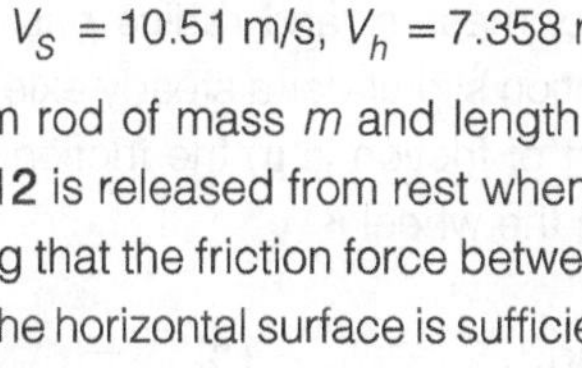

Fig. 17.14

(*a*) 15° (*b*) 20°
(*c*) 25° (*d*) 30°

17.2 A hoop of mass m and radius R is rolling down on inclined plane without slipping. Angle of inclination of plane is 30°. What is linear acceleration of center of hoop (**Fig. 17.15**)?

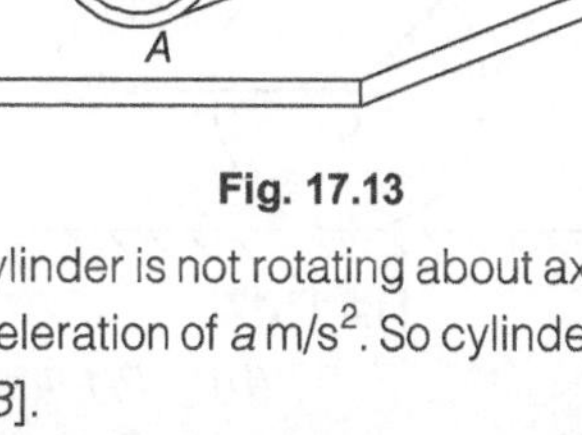

Fig. 17.15

(a) $g/2$ (b) $g/3$
(c) $g/4$ (d) $g/5$.

17.3 A rod AB of mass m and length L is released from rest where angle $\theta = 60^\circ$. What is angular acceleration of the rod about an axis through end A (**Fig. 17.16**)?

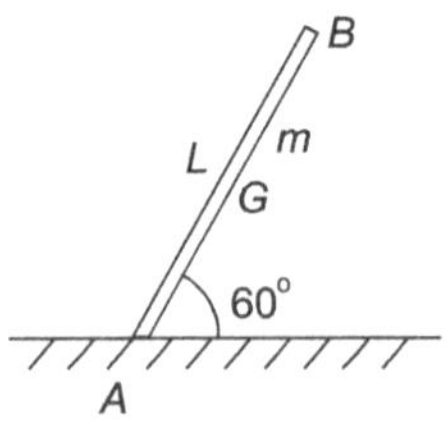

Fig. 17.16

(a) $0.433\frac{g}{L}$ (b) $0.433\frac{g}{L}$

(c) $0.75\frac{g}{L}$ (d) None of these

17.4 A sphere of mass m and radius R is pulled along a rough horizontal plane by a horizontal force P applied at the end of a bar attached to centre of sphere as shown in **Fig. 17.17**. What is acceleration of mass centre?

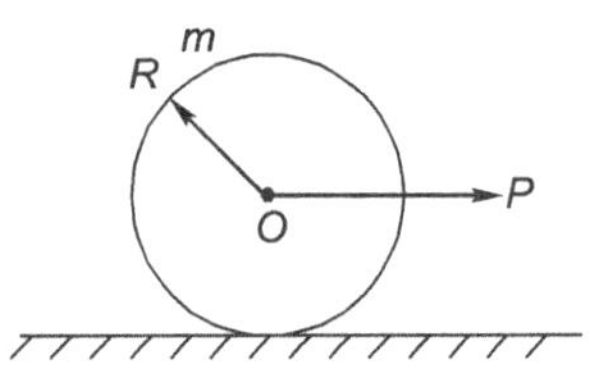

Fig. 17.17

(a) P/m (b) $P/1.4m$
(c) $1.4P/m$ (d) $P/2m$

17.5 A uniform bar of mass m, length L is supported by a string at end A and a hinge at B as shown in **Fig. 17.18**. Suddenly the string breaks at end A what is angular acceleration of the bar?

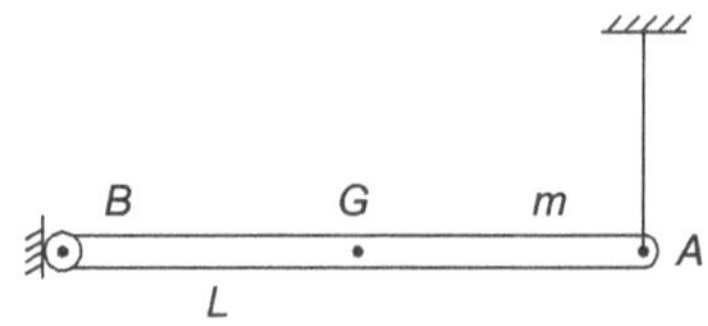

Fig. 17.18

(a) $\frac{3g}{L}$ (b) $\frac{1.5g}{L}$

(c) $\frac{g}{L}$ (d) $\frac{0.5g}{L}$

17.6 A thin hoop of mass m rolls horizontally under action of a horizontal force P applied at top as shown in **Fig. 17.19**. What is $\bar{a}$, acceleration of centre of hoop?

(a) P/m (b) $P/2m$
(c) $2P/m$ (d) Zero.

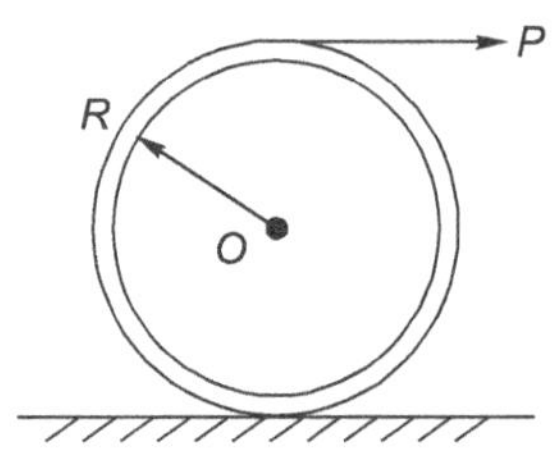

Fig. 17.19

17.7 A wheel of mass m and radius r, is in accelerated rolling motion slip under a steady axle torque T. If the coefficient of friction is μ, the friction force from the ground in the wheel is

(a) μ mg (b) $\frac{T}{r}$

(c) zero (d) None of these

[GATE 1996 : 1 Mark]

Answers

17.1 (a)	17.2 (c)	17.3 (c)	17.4 (b)	17.5 (b)
17.6 (a)	17.7 (b)			

EXPLANATIONS

17.1 (a)

For rolling without slipping

$$\frac{2}{3}g\sin\theta < 2\mu g\cos\theta$$

or
$$\frac{\sin\theta}{3} < \mu\cos\theta$$

$$\tan\theta < 3\mu$$
$$\tan\theta < 0.36$$
$$\theta < 19.8^\circ$$

17.2 (c)

Hoop rolling down inclined plane without slipping

$$a = \frac{g\sin\theta}{2} = \frac{g\times 0.5}{2} = \frac{g}{4}.$$

17.3 (c)

$$mg\frac{L}{2}\cos\theta = \frac{mg^2}{3}\times\alpha$$

$$g\times\frac{1}{2}\times 0.5 = \frac{L}{3}\times\alpha$$

$$\frac{3}{4}g\times\frac{L}{L} = \alpha, \quad \alpha = \frac{3}{4}\frac{g}{L}.$$

17.4 (b)

$$P = ma + \frac{I.\alpha}{R} = ma + \frac{0.4mR^2}{R}\times\alpha$$

$$\alpha = \text{Angular acceleration} = \frac{a}{R}$$

$$P = ma + 0.4ma = 1.4ma$$

$$a = \frac{P}{1.4m}.$$

17.5 (b)

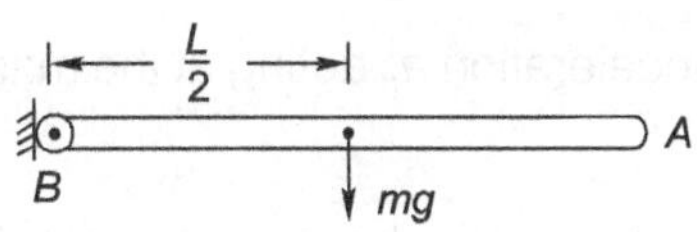

Fig. 17.20

$$I_B = \frac{mL^2}{3} \text{ (Fig. 17.20)}$$

$$T = mg\cdot\frac{L}{2} = \frac{mL^2}{3}\times\alpha$$

$$\alpha = mg\frac{L}{2}, \quad \alpha = 1.5\frac{g}{L}\text{ rad/s}^2.$$

17.6 (a)

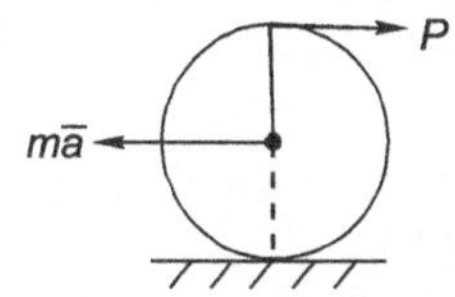

Fig. 17.21

When α = Angular acceleration (Fig. 17.21)

$$= \frac{\bar{a}}{R}$$

$$2PR - m\bar{a}R = mR^2\cdot\frac{\bar{a}}{R} = m\bar{a}.R$$

$$2P = 2m\bar{a}$$

$$\bar{a} = \frac{P}{m}.$$

17.7 (b)

If force of friction

$$\mu mg < \frac{T}{r}\frac{\text{Torque}}{\text{Radius}}$$

Rolling will occurs with sure slip

So, $F = \dfrac{T}{r}$

■■■■

18

CHAPTER

Rotation about Fixed Axis

18.1 Introduction

In motion of rotation there are normal acceleration, a_n and tangential acceleration a_t, acting at the mass centre of the rigid body rotating about an axis other than centroidal axis.

These acceleration components are $a_n = \omega^2 \bar{r}$ and $a_t = \alpha . \bar{r}$ where ω is the angular velocity, α is the angular acceleration and $\bar{r}$ *is the distance of mass centre from the axis of rotation*. But if the body rotates about an axis passing through the mass centre of the body, then both these components of acceleration *i.e.*, a_n and a_t become zero because $\bar{r} = 0$.

In this chapter we will study many more cases of rigid bodies rotating about non-centroidal axis and a few cases of motion of rotation of rigid bodies about their centroidal axis as flywheel, discs, pulleys, gears etc.

Consider a rigid body of mass *m* subjected to a number of forces F_1, F_2, F_3, F_4, ..., F_6 etc. as shown in the **Fig. 18.1 (*a*)**. *G* is the mass centre of the rigid body and *O* is point, and body is rotating about an axis passing through point *O*. The distance between *O* and *G i.e.*, between axis of rotation and mass centre is $\bar{r}$. The direction from *G to O extended is the n-direction* and direction perpendicular to *n* direction is *t* direction as shown.

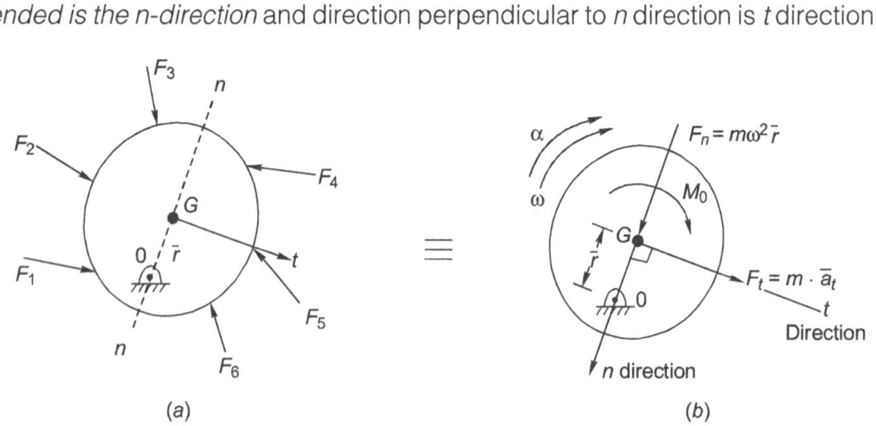

Fig. 18.1

Resultant of all external forces acting on the rigid body have following *three* components:

F_n = algebraic summation of all external forces having components in *n* direction

F_t = algebraic summation of all external forces having components in *t* direction, positive sense of F_t is in the direction of angular acceleration

M_0 = algebraic summation of the moments of the external forces about the axis of rotation through O. Positive sense of M_0 is in the direction of angular acceleration α

Under the action of M_0, the body is rotating with angular acceleration α and say at a particular instant angular velocity is ω.

As we have stated earlier,

Normal force, $F_n = m\bar{a}_n = m\omega^2\bar{r}$, where $\bar{r}$ is distance from O to G.

Tangential force, $F_t = m\bar{a}_t = m\alpha\bar{r}$

Moment, $M_0 = I_0 . \alpha$

where bar on a_n, a_t, $\bar{r}$ represent the acceleration and distance OG for the *mass centre G of the body.*

Moment of inertia, $I_0 = \bar{I} + m\bar{r}^2$

where $\bar{I} = I_G$, moment of inertia of the body about an axis through mass centre.

Example 18.1 A thin uniform bar of mass m and length l is suspended from two vertical inextensible strings as shown in **Fig. 18.2**. If the string CD is cut, find the tension in the string AO.

[GATE : 2013]

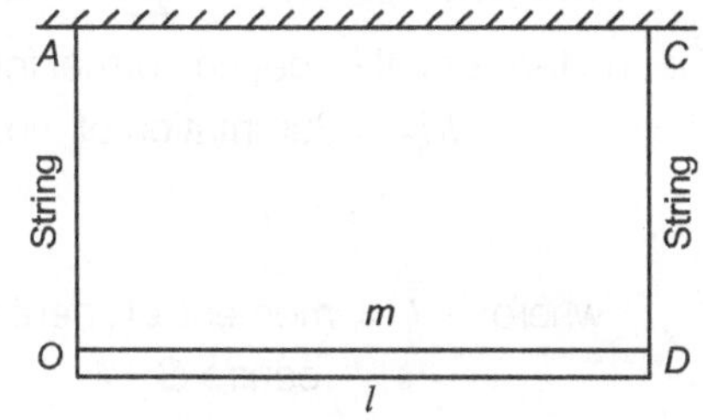

Fig. 18.2

Solution Fig. 18.3 shows the bar OD when string CD is cut-off.

Moments about O, axis of rotation

$$M_0 = mg \cdot \frac{l}{2}$$

where $\bar{r} = \frac{l}{2}$, distance of mass centre from axis of rotation

$M_0 = I_0 . \alpha$, where α is the angular acceleration

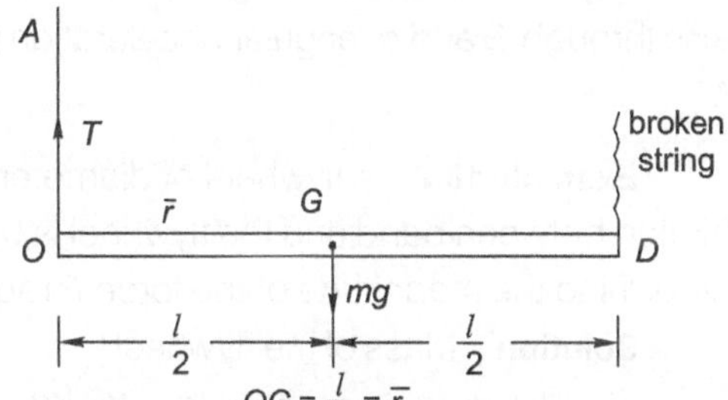

Fig. 18.3

Moment of inertia,

$$I_0 = \bar{I} + m\bar{r}^2 = \frac{ml^2}{12} + m\left(\frac{l}{2}\right)^2 = \frac{ml^2}{3}$$

So angular acceleration, $$\alpha = \frac{M_0}{I_0} = \frac{mgl}{2} \times \frac{3}{ml^2} = \frac{3}{2}\frac{g}{l} \quad ...(1)$$

acceleration of mass centre,

$$\bar{a} = \alpha . \bar{r} = \alpha \cdot \frac{l}{2} \quad ...(2)$$

Moreover $$mg - T = m\bar{a} \text{ or } T = mg - m\bar{a} \quad ...(3)$$

$$T = mg - m\alpha \cdot \frac{l}{2}, \text{ putting the value of } \alpha$$

$$T = mg - m \cdot \frac{3}{2} \times \frac{g}{l} \times \frac{l}{2}$$

$$= mg - \frac{3}{4}mg = \frac{mg}{4}.$$

Exercise 18.1 Three bars each of mass 2 kg and length 0.6 m are pinned together to form an equilateral triangle as shown in **Fig. 18.4**. These bars rotate in a horizontal plane about a vertical axis through *A*. What torque is required to produce a constant angular acceleration of 10 rad/sec^2?

[**Hint:** Moment of inertia of 3 bars about axis of rotation *A*

$$I_A = \frac{ml^2}{3} + \frac{ml^2}{3} + \frac{ml^2}{12} + m\,(0.866l)^2].$$

[**Ans:** Torque = 10.8 Nm].

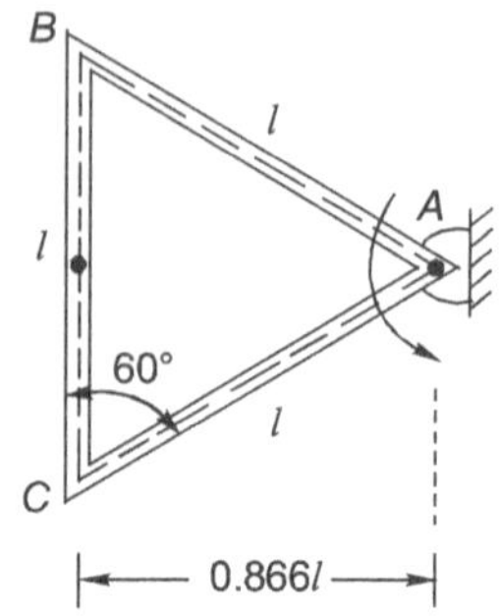

Fig. 18.4

18.2 Rotation about Centroidal Axis

If a body rotates about an axis through its mass centre *G*, then distance $\bar{r}$ becomes zero and forces:

Normal force, $\sum F_n = 0$ and

Tangential force, $\sum F_t = 0$

i.e., no tangential force, no normal force. Only

M_G = Summation of moments of external forces about mass centre *G*

$= \bar{I}.\alpha$

where $\bar{I}$ = moment of inertia of body about an axis through mass centre *G*

α = angular acceleration

Figure 18.5 shows a rigid body with mass centre *G*. Moment, M_G about axis through *G* and α, angular acceleration in the direction of applied moment, M_G.

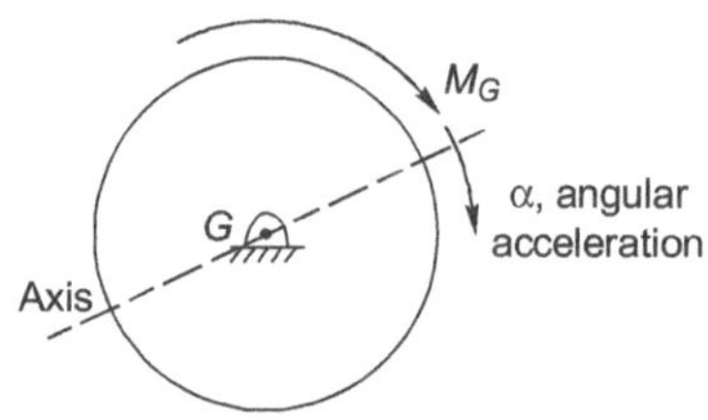

Fig. 18.5

Example 18.2 A flywheel of diameter 0.7 m has a mass of 80 kg as shown in **Fig. 18.6**. The coefficient of friction between band and the flywheel is 0.40. If the initial angular velocity of the flywheel is 400 rpm clockwise, determine the magnitude of the force *P* required to stop the flywheel with retardation of 6 rad/s^2.

Solution Mass of the flywheel,

$$m = 80 \text{ kg}$$

Radius of flywheel disc, $R = 0.35$ m

Moment of inertia of flywheel,

$$\bar{I} = \frac{mR^2}{2} = \frac{80 \times 0.35^2}{2} = 4.9 \text{ kg m}^2$$

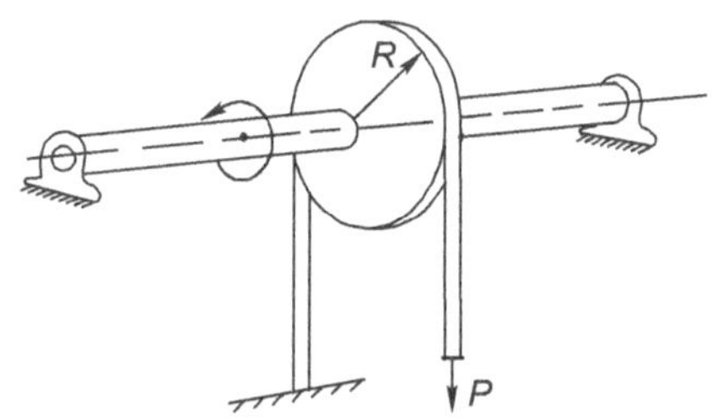

Fig. 18.6

Angle of contact between band and flywheel,

$$\theta = \pi \text{ radian}$$

Coefficient of friction between band and flywheel,

$$\mu = 0.40$$

Say tension on two sides of band are *P* and P' as shown in **Fig. 18.7**.

Ratio of belt tensions, $\frac{P}{P'} = e^{\mu\theta} = e^{0.4\pi} = 3.513$

$$P = 3.513P'$$

or $P' = 0.285P$ ...(2)

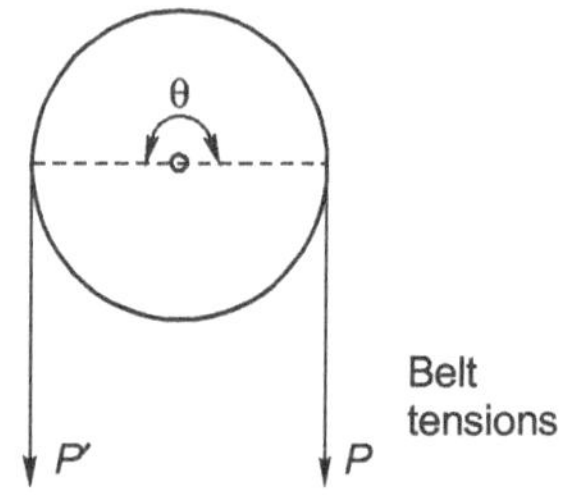

Fig. 18.7

Frictional torque, $T = (P - P')R = (P - 0.285P) \times 0.35$

$= 0.250P$ Nm if P is in N

angular retardation $\alpha = 6 \text{ rad/s}^2$

Frictional torque on flywheel, $T = \bar{I}\alpha = 4.9 \times 6 = 29.4 \text{ Nm} = 0.250P$

$$\text{Force } P \text{ on band} = \frac{29.4}{0.25} = 117.6 \text{ N.}$$

Exercise 18.2 A pulley and its rotating accessories has a mass of 500 kg and a radius of gyration of 0.15 m. If the kinetic coefficient of friction between pulley and belt is 0.25, how much force P must be applied to reduce the speed of the pulley from 1500 rpm to 500 rpm in 60 seconds (**Fig. 18.8**)?

[**Hint:** $\frac{T_2}{T_1} = e^{\mu\theta}$, $\theta = \pi$ $(T_2 - T_1)R = \text{torque} = I\alpha\,(ccw)$].

[**Ans:** $P = 72.18$ Newton].

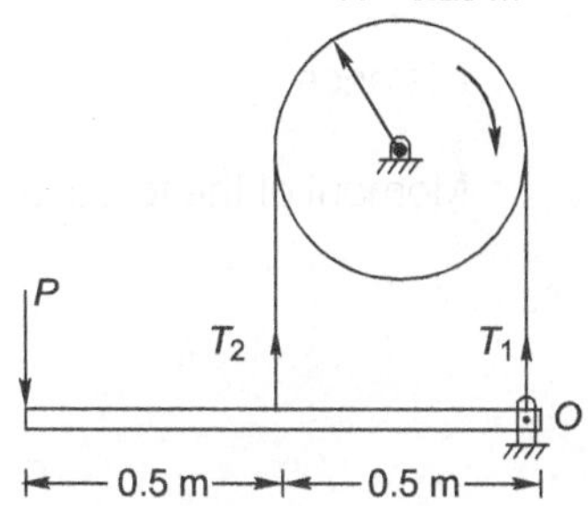

Fig. 18.8

18.3 Inertia Force Method

For a rigid body rotating about a non-centroidal axis, angular velocity ω, angular acceleration α etc. are determined by taking the components of forces in n and t directions and moment about axis of rotation (as discussed in Article 18.1).

$$\sum F_n = m\omega^2\bar{r}, \text{ effective force in } n \text{ direction}$$

$$\sum F_t = m\bar{a}, \text{ effective force in } t \text{ direction}$$

$$\sum M_o = I_o.\alpha$$

But to determine support or bearing reactions, the rigid body is brought under dynamic equilibrium by applying the reversed forces $-m\omega^2\bar{r}$ and $-m\bar{a}_t$ as shown in **Fig. 18.9** and the body is brought to rest only for the purpose of study and determining bearing reactions

The reversed force $-m\omega^2\bar{r}$ is applied away from the centre of rotation O and reversed force. $-F_t = -m\bar{a}_t$ is applied at a point P at a distance of b from axis of rotation O, and parallel to t-axis as shown and in opposite sense to the sense of α, angular acceleration, such that

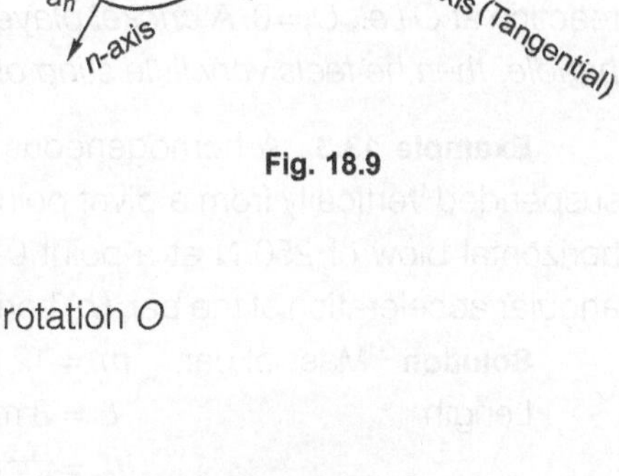

Fig. 18.9

$$b = \frac{k_0^2}{\bar{r}}$$

where k_0 = radius of gyration of the body about axis of rotation O

$\bar{r}$ = distance OG

Now $$k_0^2 = \frac{I_0}{m}$$

Distance $$b = \frac{I_0}{m\bar{r}}.$$

18.4 Centre of Percussion

Centre of percussion is that point on the body along the *n*-axis (normal axis), as shown in **Fig. 18.10**, through which the resultant of the effective forces F_n and F_t act. It is at a distance of $b = \frac{k_0^2}{\bar{r}}$ from the centre of rotation.

Consider a bar *OB* of length *l* and mass *m* suspended at *O*.

A force *F* is applied at point *P* such that,

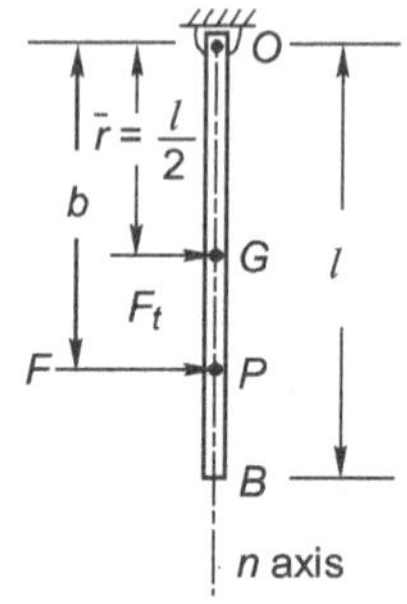

Fig. 18.10

Distance, $$b = \frac{k_0^2}{\bar{r}}$$

Moment of the force, about *O*

$$M = F.b = F \times \frac{k_0^2}{\bar{r}}$$

$$F \cdot \frac{k_0^2}{\bar{r}} = I_0 . \alpha \text{ where } \alpha \text{ is angular acceleration}$$

Initial velocity $\omega = 0$, so force $F_n = 0$

$$F \cdot \frac{k_0^2}{\bar{r}} = m(k^2 + \bar{r}^2)\,\alpha$$

but $$k^2 = \frac{l^2}{12} \quad \text{and } \bar{r}^2 = \frac{l^2}{4}$$

Distance $$b = \frac{k_0^2}{\bar{r}} = \frac{l^2}{3} \times \frac{2}{l} = \frac{2l}{3}$$

$$F \times \frac{2l}{3} = m \cdot \frac{l^2}{3} \alpha = I_0 . \alpha$$

Angular acceleration, $\alpha = 2F/ml$

Tangential force, $$F_t = m\bar{a}_t = m\frac{l}{2} \cdot \alpha = \frac{ml}{2} \times \frac{2F}{ml} = F$$

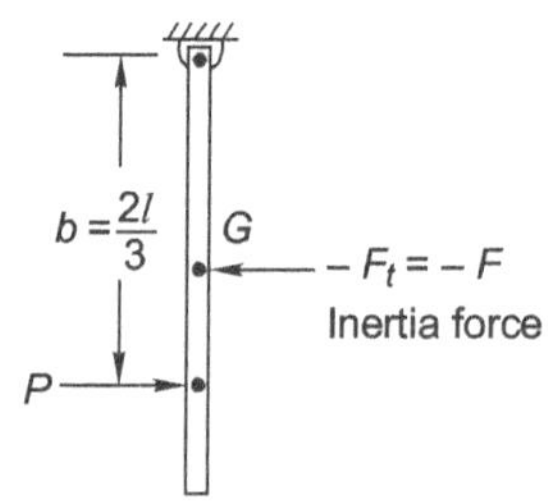

Fig. 18.11

Inertia force at *G*, $= -F$ (**Fig. 18.11**)

Reaction at *O*, $O_t = 0$

This shows that if a force is applied at *P* (called the *centre of percussion*) then tangential component of reaction at *O i.e.*, $O_t = 0$. A *cricket player knows that if a ball hits at a distance of 2/3rd of the length of the bat from handle, then he feels very little sting of the force of the ball.*

Example 18.3 A homogeneous bar 3 m long has a mass of 12 kg. It is suspended vertically from a pivot point located at one end. The bar is struck by a horizontal blow of 250 N at a point 0.75 m below the pivot point. Determine (*a*) angular acceleration of the bar, (*b*) horizontal reaction at pivot (**Fig. 18.12**).

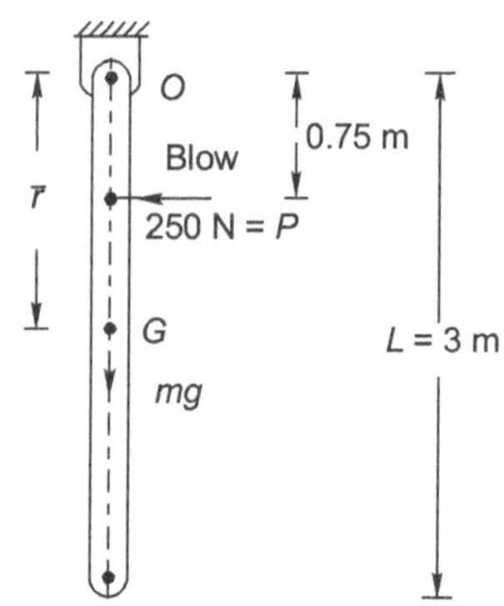

Fig. 18.12

Solution Mass of bar, $m = 12$ kg

Length $L = 3$ m

I_0 = moment inertia about pivot point, at one end

$$= \frac{mL^2}{3} = \frac{12 \times 3^2}{3} = 36 \text{ kg-m}^2$$

Blow, $P = 250$ N

M_0 = moment about pivot O

$= P \times 0.75 = 0.75 \times 250$

$= 187.5$ Nm

$= I_0 . \alpha = 36 \times \alpha$

α = angular acceleration

$$\alpha = \frac{187.5}{36} = \mathbf{5.2083\ rad/s^2}, \textbf{ angular acceleration}$$

$\bar{a}$ = linear acceleration at mass centre

$$= \frac{L}{2} \times \alpha = 1.5 \times 5.2083 = 7.8125 \text{ m/s}^2$$

$m\bar{a} = 12 \times 7.8125 = 93.75$ N

Inertia force at G = 93.75 N in opposite direction

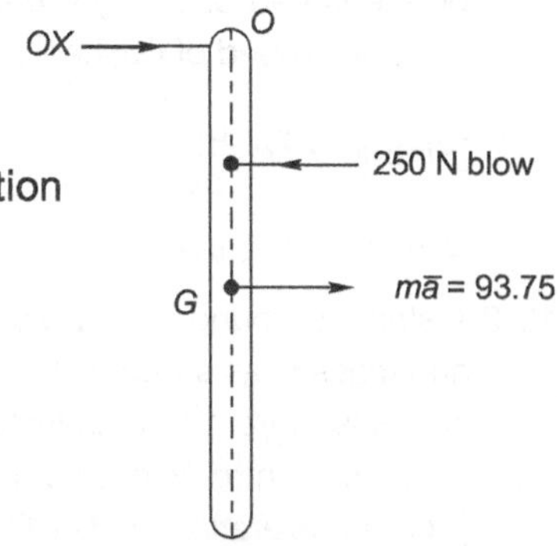

Fig. 18.13

Reaction at O (Fig. 18.13)

$$O_x = 250 - 93.75 = \overrightarrow{156.25} \text{ N.}$$

Remember

- For a body rotating about non-centroidal fixed axis.

  Normal force, $F_n = m\omega^2 \bar{r}$, directed towards the axis of rotation

  Tangential force, $F_t = m.\alpha\bar{r}$, directed in the direction of angular acceleration, α

  $\bar{r}$ = distance of mass centre from axis of rotation

  ω = angular velocity at a particular instant

  M_0 = moment about axis of rotation, O

  $= I_0 . \alpha$, where I_0 is the mass moment of inertia of the body about the axis of rotation

- For rotation about centroidal axis for a rigid body

  $\bar{r} = 0$

  $F_n = 0$ and $F_t = 0$, as mass is located at axis.

  Moment about centroidal axis,

  $M_a = \bar{I} . \alpha$

  where $\bar{I}$ = moment of inertia about centroidal axis.

- To determine reaction components at bearings, the rigid body is put under dynamic equilibrium by applying inertia forces, $F_n = m\omega^2 \bar{r}$, $F_t = m\alpha . \bar{r}$ in the reversed directions.
- Centre of percussion is that point on the body along n-axis, through which the resultant of the effective forces acts. If a force is applied at centre of percussion of the rigid body, then reaction at centre of rotation in the t-direction (tangential direction) is zero.

PRACTICE PROBLEMS

18.1 A flywheel of mass 60 kg and radius of gyration 0.5 m is subjected to a torque $3(1-e^{-0.1\theta})$ Nm. What will be the speed of the flywheel after 5 revolutions? At the start, speed of flywheel is zero.

[**Hint:** $M = \bar{I}\omega \cdot \frac{d\omega}{d\theta}$].

[**Ans:** 2.956 rad/s].

18.2 A string is wound over two uniform discs of radius R and mass m as shown in the **Fig. 18.16**. The system is released from rest, determine the acceleration of the mass centre G_2 of disc II.

[**Hint:** Acceleration of I is αR Acceleration of II is $G_2 = 2\alpha R$

$$TR = \frac{mR^2}{2}\cdot\alpha,\ T = \frac{mR}{2}\alpha,\quad mg - T = m(2\alpha R)$$

[**Ans:** Acceleration of $G_2 = 0.8g$].

18.3 A circular rotor of mass m and radius R which can rotate about its geometric axis is braked by the arrangement shown in the **Fig. 18.14**. The rotor is rotating with initial angular velocity ω, when the brake is applied. If the coefficient of kinetic friction between rotor and shoe brake is μ, in how many revolutions will the rotor come to rest?

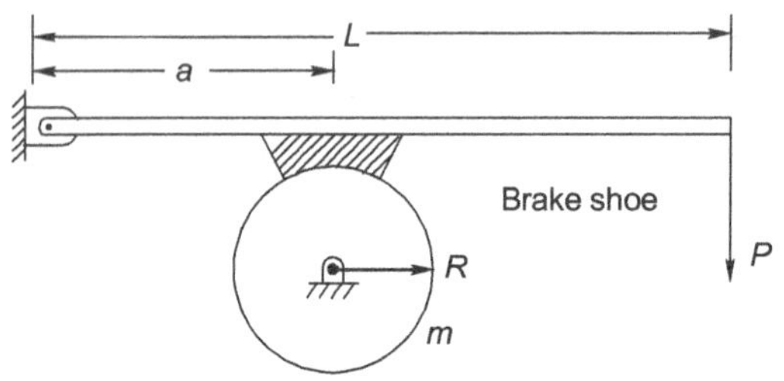

Fig. 18.14

[**Hint:** Normal reaction at brake shoe, $N = PL/a$, Frictional torque $= \mu N.R = I.\alpha$].

[**Ans:** $\frac{ma\omega^2 R}{8\pi\mu PL}$].

18.4 A flywheel of mass 160 kg is shown in the **Fig. 18.15**. A rope is wrapped around the hub of the flywheel. Hub radius is 0.15 m. A stationery weight of mass 60 kg descends 1 m in 5 seconds. What is the radius of gyration of the flywheel?

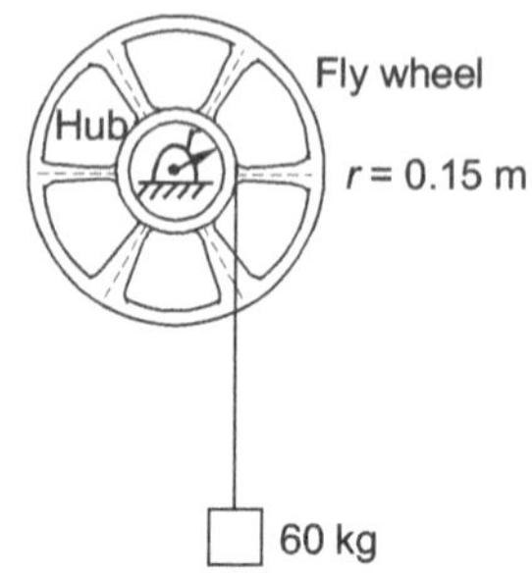

Fig. 18.15

[**Hint:** $60g - T = 60a$, $T \times r = I.\alpha$, $T = Ia/r$].

[**Ans:** $k = 1.013$ m].

18.5 The mass of a homogeneous sphere of radius 0.16 m is 30 kg. The coefficient of friction between sphere and inclined plane is 0.10. Sphere is on inclined plane as shown in **Fig. 18.16**. Determine the angular acceleration of the sphere and linear acceleration of its mass (I_g of sphere = $0.4mR^2$).

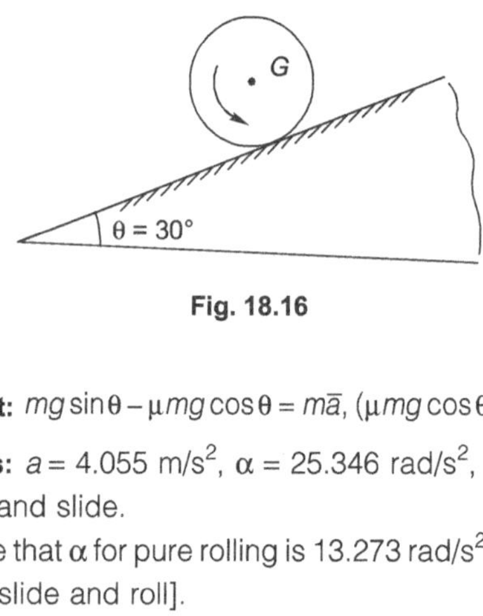

Fig. 18.16

[**Hint:** $mg\sin\theta - \mu mg\cos\theta = m\bar{a}$, $(\mu mg\cos\theta)\,r = I_g.\alpha$].

[**Ans:** $a = 4.055$ m/s^2, $\alpha = 25.346$ rad/s^2, sphere will roll and slide.

Note that α for pure rolling is 13.273 rad/s^2, so sphere will slide and roll].

18.6 Two blocks m_1 and m_2 are connected by a light string which passes over a pulley of radius r and moment of inertia I. If the string does not slip on the pulley, find the difference in tensions on two sides of the pulley, when the system is released from rest.

[**Ans:** $\frac{(m_1 - m_2)\,gI}{I + (m_1 + m_2)\,r^2}$].

MULTIPLE CHOICE QUESTIONS

18.1 A slender prismatic bar bent in at 90° is of length $2L$ and mass 2 m as shown in **Fig. 18.17**. Bar is hinged at point B. It is released from rest, what is angular acceleration of bar?

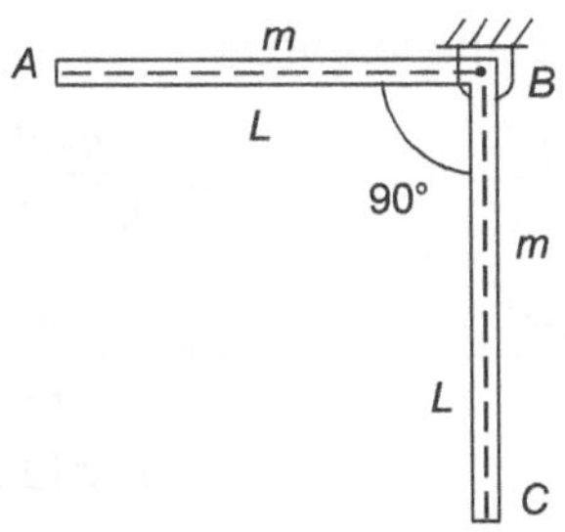

Fig. 18.17

(a) $\frac{3}{2} \times \frac{g}{L}$ (b) $\frac{3}{4}\frac{g}{L}$

(c) $\frac{g}{4L}$ (d) None of these

18.2 A flywheel of mass 400 kg and radius of gyration of 0.5 m is subjected to a torque of 100 Nm. What is speed of flywheel after 5 seconds from start?

(a) 50 rad/s (b) 10 rad/s

(c) 5 rad/s (d) None of these

18.3 A uniform disc of radius R and mass m is released from rest when O and G are on a horizontal line as shown in **Fig. 18.18**. What is angular acceleration of disc?

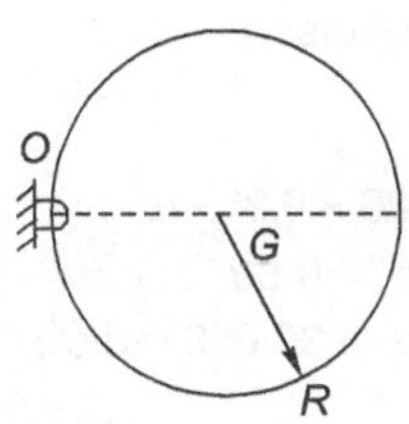

Fig. 18.18

(a) $\frac{2g}{R}$ (b) $\frac{2g}{3R}$

(c) $\frac{g}{R}$ (d) None of these.

18.4 A string is wound over two uniform discs of radius R and mass m as shown in **Fig. 18.19**. System is released from rest. If angular acceleration of disc I is α, what is angular acceleration of disc II?

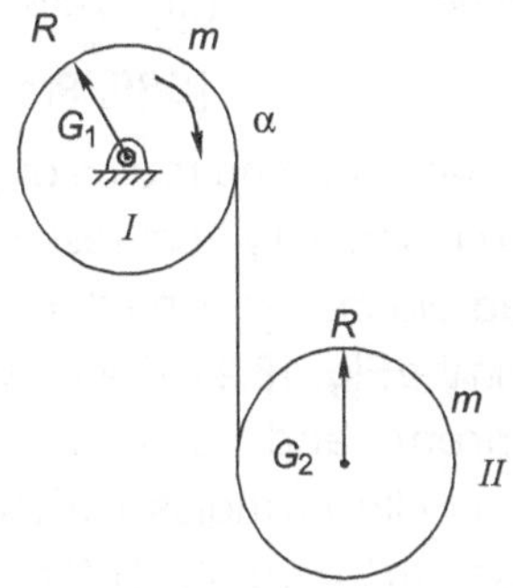

Fig. 18.19

(a) 0.5α (b) α

(c) 2α (d) 2.5α.

18.5 A pulley of mass 20 kg and radius 0.25 m start from rest. A force of $10g$ N is applied at the end of the string. If k = 0.2m for pulley, what is angular acceleration of the pulley (**Fig. 18.20**)?

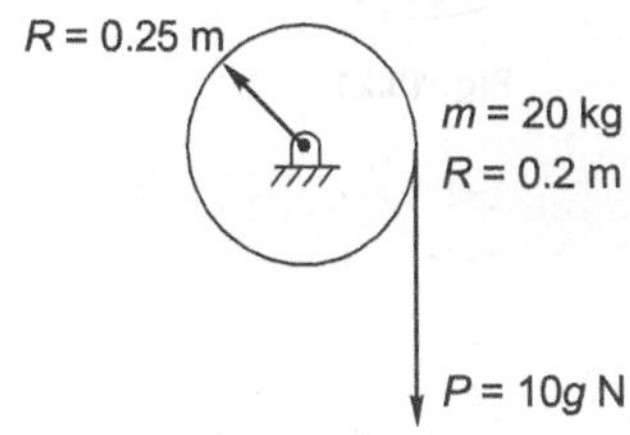

Fig. 18.20

(a) 30.7 rad/s^2 (b) 19.62 rad/s^2

(c) 15.4 rad/s^2 (d) None of these

18.6 A turbine generator unit is shut off when it was rotating at 3000 rpm. The turbine unit came to rest in 10 minutes. If mass of rotor is 2000 kg and radius of gyration is 0.25 m, what is average friction couple?

(a) 130.9 Nm (b) 92.56 Nm

(c) 65.44 Nm (d) None of these

18.7 The coefficient of friction between the horizontal road and a trackman's shoe is 0.6. What is the minimum radius of curvature of circular path which he can travel at a constant speed of 5 m/s without slipping?

(a) 5.886 m (b) 5.06 m

(c) 4.247 m (d) None of these

18.8 The rotation of rigid body is governed by

$\phi = 2.5\sin\left(\frac{\pi t}{4}\right)$ what is angular velocity at t = 2s

(a) 18 rad/s (b) $\frac{\pi}{4}$ rad/s

(c) $\frac{\pi}{8}$ rad/s (d) zero

[CSE, Prelim, CE : 2007]

18.9 A reel of mass "m" and radius of gyration "k" is rolling down smoothly from rest with one end of the thread wound on it held in the ceiling as depicted in the **Fig. 18.21**. Consider the thickness of the thread and its mass negligible in comparison with the radius 'r' of the hub and the reel mass "m". Symbol "g" represents the acceleration due to gravity.

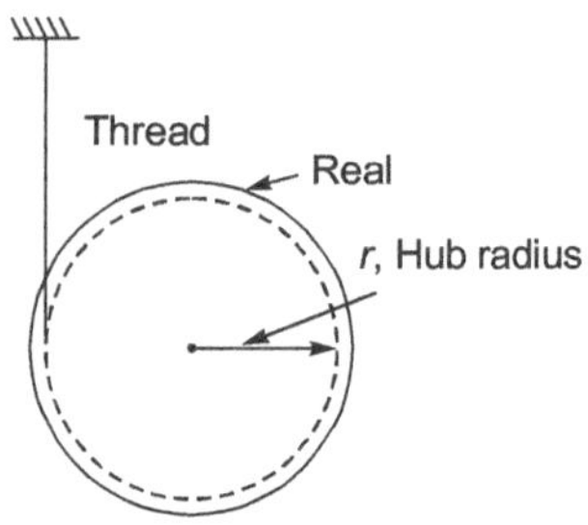

Fig. 18.21

The linear acceleration of the reel is

(a) $\frac{gr^2}{(r^2+k^2)}$ (b) $\frac{gk^2}{(r^2+k^2)}$

(c) $\frac{grk}{(r^2+k^2)}$ (d) $\frac{mgr^2}{(r^2+k^2)}$

[GATE 2003 : 1 Mark]

18.10 Tension in the thread is

(a) $\frac{mgr^2}{(r^2+k^2)}$ (b) $\frac{mgrk}{(r^2+k^2)}$

(c) $\frac{mgk^2}{(r^2+k^2)}$ (d) $\frac{mg}{(r^2+k^2)}$

[GATE 2003 : 1 Mark]

Answers

18.1 (b)	18.2 (c)	18.3 (b)	18.4 (c)	18.5 (a)
18.6 (c)	18.7 (c)	18.8 (d)	18.9 (a)	18.10 (c)

EXPLANATIONS

18.1 (b)

$$I_B = \frac{mL^2}{3} + \frac{mL^2}{3} = \frac{2mL^2}{3}$$

$$T = mg\frac{L}{2} = I_B.\alpha = \frac{2}{3}mL^2$$

$$\alpha = \frac{3}{4}\frac{g}{L}.$$

18.2 (c)

$$I = mk^2 = 400 \times 0.5^2 = 100 \text{ kg-m}^2$$

$$T = I\alpha$$

$$100 \text{ Nm} = 100 \times \alpha$$

$$\alpha = 1 \text{ rad/s}^2$$

$$\omega_5 = 0 + 5 \times 1 = 5 \text{ rad/s}.$$

18.3 (b)

$$I_0 = mR^2 + \frac{mR^2}{2} = 1.5mR^2$$

$$M = mg.R = I_0\alpha = 1.5mR^2.\alpha$$

$$\alpha = \frac{g}{1.5R} = \frac{2g}{3R}.$$

18.4 (c)

Angular acceleration of disc II = 2 × angular acceleration of disc I.

18.5 (a)

$$98.1 \times 0.25 = 20 \times 0.2^2 \times \alpha$$

$$24.525 = 0.8\alpha$$

$$\alpha = 30.656 \text{ rad/s}^2.$$

18.6 (c)

$$I = 2000 \times 0.25^2 = 125 \text{ kg-m}^2$$

$$\alpha = \frac{2\pi \times 3000}{60 \times 600} = 0.5236 \text{ rad/s}$$

$$I\alpha = 65.44 \text{ Nm}.$$

18.7 (c)

$$a_n = 0.6 \times g = 5.886 \text{ m/s}^2 = \frac{v^2}{R} = \frac{5^2}{R}$$

$$R = 4.247 \text{ m}.$$

18.8 (d)

$$\frac{dQ}{dt} = 2.5\times\frac{\pi}{4}\cos\left(\frac{\pi t}{4}\right)$$

$$t = 2\text{s}$$

$$\cos\left(\frac{\pi t}{4}\right) = \cos\left(\frac{\pi}{2}\right) = 0$$

Velocity $= 0$

18.9 (a)

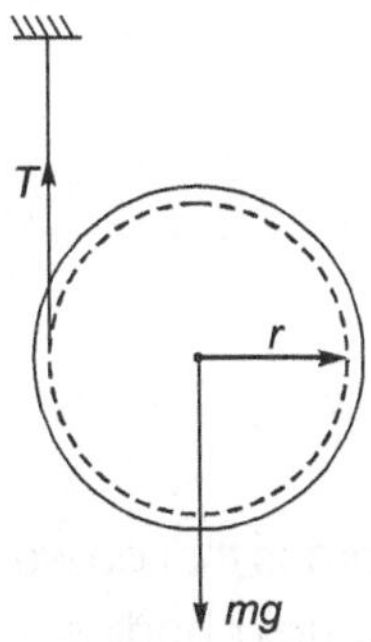

Fig. 18.22

$$mg - T = Ia$$

$$a = \text{linear acceleration}$$

Position, $\quad Tr = I\alpha = mk^2\frac{a}{r}$

$$T = \frac{mk^2}{r^2}\times a$$

$$mg - \frac{mk^2}{r^2}\times a = ma$$

$$a = \frac{gr^2}{k^2+r^2}$$

18.10 (c)

$$T = \frac{mk^2}{r^2}\times a = \frac{mk^2}{k^2}\times\frac{gr^2}{t^2+r^2} = \frac{mgk^2}{k^2+r^2}$$

■■■■

19
CHAPTER

Work and Energy

19.1 Introduction

The principle of work and energy provides a direct relationship between the forces that do work during motion of rigid body or bodies and the corresponding changes in the parameters of rigid bodies in motion. For finite displacements, work-energy principle eliminates the necessity of determining acceleration and integrating it over time interval to get velocity changes. There are various examples of work done by force applied through a particular distance as of a man climbing up a hill, or a moment applied through a specific angular displacement, as function of a flywheel in punching or pressing operation. Consider a rigid body of mass m with its mass centre at G. The body is subjected to a number of external forces and couples and resultant of all external forces and couples is a force F and a moment M as shown in the **Fig. 19.1**. Under the action of resultant force F and resultant moment M, the mass centre of the body is having a velocity V and body is rotating about its mass centre with angular velocity ω as shown. The body has moved by a distance ds along the direction of velocity V and body has rotated by an angle $d\theta$ during an interval of time dt and the body has moved and rotated from position 1 to position 2 (**Fig. 19.2**).

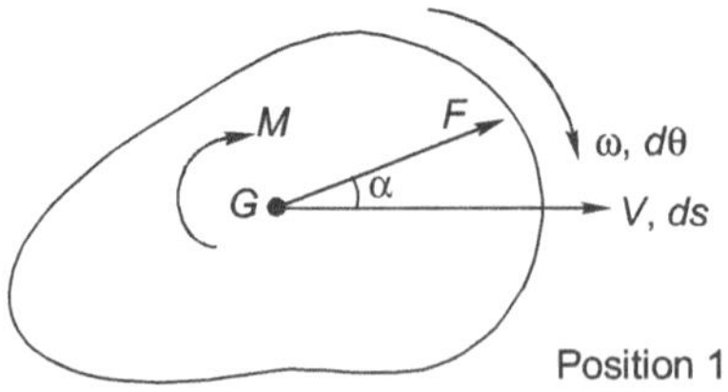

Fig. 19.1

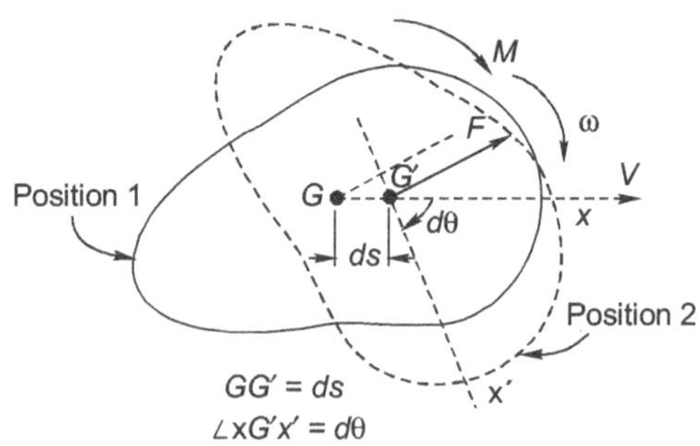

Fig. 19.2

Work done by the force and couple from position 1 to position 2 will be

$$U_{1-2} = \int_{S_1}^{S_2} F.ds + \int_{\theta_1}^{\theta_2} M.d\theta$$

where $\qquad S_2 - S_1 = ds$ and $\theta_2 - \theta_1 = d\theta$

Work is a scalar quantity therefore $F.dS$ and $M.d\theta$ are dot products of F and dS, and M and $d\theta$ respectively.

Work done $U_{1-2} = \frac{1}{2} m (V_2^2 - V_1^2) + \frac{1}{2} I (\omega_2^2 - \omega_1^2)$

During the interval, work done on the body has changed its linear speed from V_1 to V_2 and its angular speed from ω_1 to ω_2 or in other words there is change in the kinetic energy of the body *i.e.*,

$$U_{1-2} = \text{change in translational kinetic energy} + \text{change in rotational kinetic energy}$$

$$= \frac{1}{2} m (V_2^2 - V_1^2), \text{ change in kinetic energy due to translation}$$

$$+ \frac{1}{2} I (\omega_2^2 - \omega_1^2), \text{ change in kinetic energy of the body due to rotation}$$

or
$$U_{1-2} = \left(\frac{1}{2} mV_2^2 + \frac{1}{2} I\omega_2^2\right) - \left(\frac{1}{2} mV_1^2 + \frac{1}{2} I\omega_1^2\right)$$

$$= \text{Final kinetic energy} - \text{initial kinetic energy of the body}$$

$$= KE_2 - KE_1$$

In the case of conservative forces, the work U_{1-2} depends only on the initial and final positions of the body and does not depend upon the path followed by the force.

Figures 19.3 (*a*) and **(*b*)** show the work performed by a conservative force *mg* on behalf of mass *m*. In **Fig. 19.3 (*a*)** the ball travels the vertical distance *h* and finally the ball possesses horizontal velocity $V = \sqrt{2gh}$ and **Fig. 19.3 (*b*)** shows the ball falling straight under gravity, force *mg* acts through height *h* and ball possesses the vertical velocity, $V = \sqrt{2gh}$. In both these cases work performed is the same and independent of the path followed.

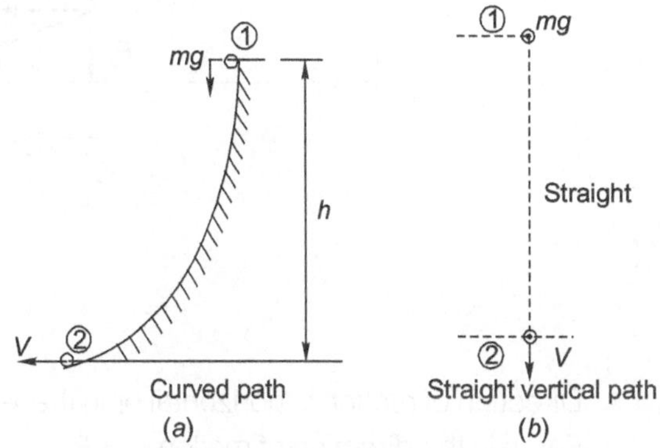

Fig. 19.3

Work done by a conservative force can be expressed in terms of change in potential energy

$$U_{1-2} = -(PE_2 - PE_1) \quad \text{...(1)}$$

$$= KE_2 - KE_1 \text{ from Equation (1)}$$

or
$$PE_1 + KE_1 = PE_2 + KE_2 \text{ constant} \quad \text{...(2)}$$

This equation expresses the conservation of energy. *Principle of conservation of energy states that if a system of rigid bodies moves under the action of conservative forces then sum of kinetic energy and potential energy of the system remains constant.*

If datum is chosen at position 2, then

Potential energy at position (2) is zero.

Potential energy at position (1) is $+mgh$.

Kinetic energy at position (1) is zero.

Kinetic energy at position (2) is $\frac{mV^2}{2}$

or
$$mgh + 0 = 0 + \frac{mV^2}{2}$$

or velocity at position (2) is $V = \sqrt{2gh}$. *In other words loss in potential energy of the system results in the gain in kinetic energy.*

Exercise 19.1 A car travels down a smooth inclined plane and moves inside a loop of radius 12 m as shown in **Fig. 19.4**. Determine the minimum value of h so that the car will remain in contact with the track.

[**Hint:** $V_A = \sqrt{2gh} = V_A^2 - 2g \times 24$, $\dfrac{mV_B^2}{R} > mg$].

[**Ans:** $h = 30$ m].

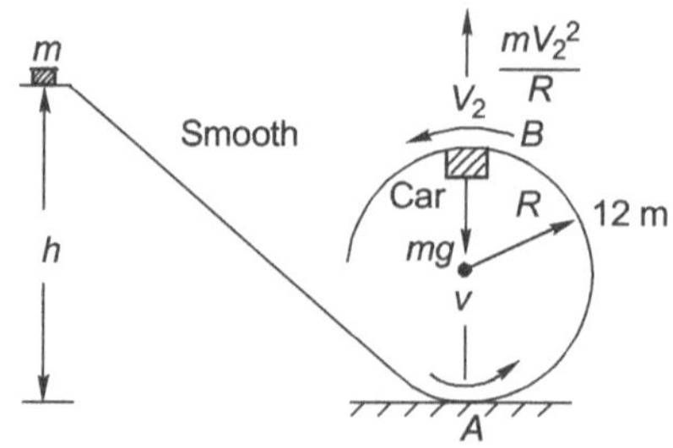

Fig. 19.4

19.2 Work done by a Force during Sliding Motion

Consider a block of weight W on a horizontal rough surface. A force P is inclined on the block. N is the normal reaction on the block and $F = \mu_k \cdot N$, is the force of friction on the block, where μ_k is the coefficient of kinetic friction. The body has moved from position (1) to position (2). At position (2) the block has gained velocity V, (**Fig. 19.5**).

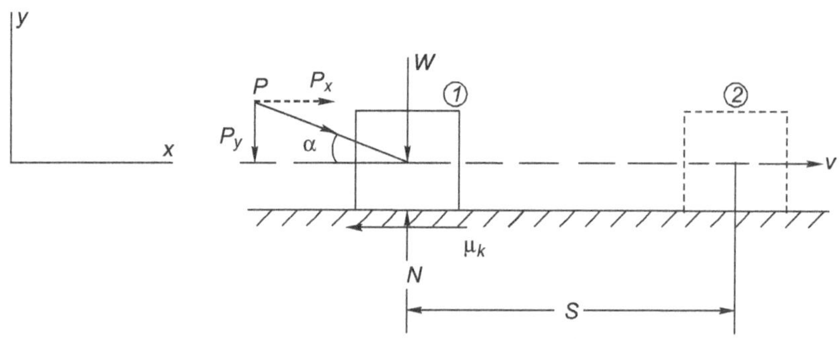

Fig. 19.5

Direction of motion is horizontal or in the x-direction.

Force in the direction of motion $= P_x$

Positive work done by force, $= P_x \cdot S$ ...(1)

Force of friction, $F = \mu_k . N = \mu_k (W + P_y)$

where, P_y is component of P in vertical direction.

Negative work done by the force of friction

$$= -\mu_k (W + P_y).S$$

Net work done on block, $U_{1-2} = P_x.S - \mu_k (W + P_y) S$ ...(1)

No work will be done by normal reaction N, because there is no movement of the block in the direction of normal reaction N.

so $$[P_x - \mu_k (W + P_y)].S = \frac{1}{2} \times \frac{W}{g} V^2 \quad ...(2)$$

The velocity of the block after travelling a distance S can be worked out from the above equation.

Example 19.1 A block of mass 50 kg is placed on a rough horizontal floor as shown in the **Fig. 19.6**. A force P inclined at an angle of 30° is applied on the block as shown. If the block reaches a velocity of 8 m/s in covering a distance of 10 m, what is the magnitude of the force P? Take coefficient of kinetic friction between block and floor equal to 0.25.

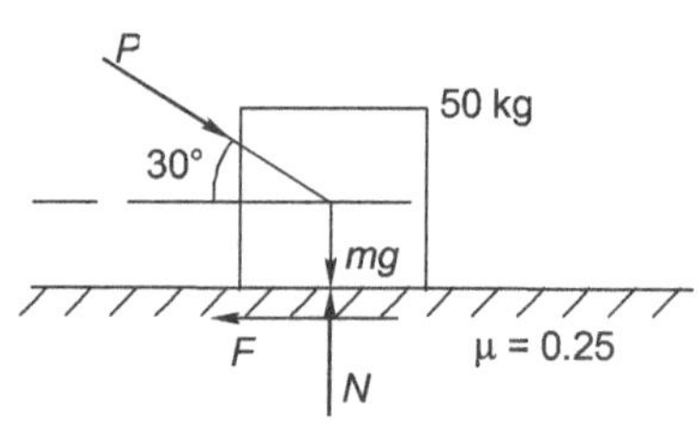

Fig. 19.6

Solution Say the force applied is P Newton.

Horizontal component of P,

$$P_x = P\cos 30° = \overrightarrow{0.866P} \text{ N}$$

Vertical component of P, $P_y = P\sin 30° = 0.5P$ N ↓

Weight of the block, $W = mg = 50 \times 9.81 = 490.5$ N

Normal reaction of the plane on the body

$$N = W + P\sin 30° = 490.5 + 0.5P \text{ N}$$

Force of friction, $F = \mu_k \times N$

$$= 0.25\,(490.5 + 0.5P) = 122.625 + 0.125P \quad ...(1)$$

Net force applied on block $= P_x - F$

$$= 0.866P - 122.625 - 0.125P = 0.741P - 122.625$$

Work done by the force on moving the block by 10 m

$$U = (0.741P - 122.625) \times 10 = 7.41P - 1226.25 \text{ Nm}$$

Kinetic energy of the block,

$$KE = \frac{1}{2}mV^2 = \frac{1}{2} \times 50 \times 8^2 = 1600 \text{ Nm}$$

Using the work-energy principle

$$7.41P - 1226.25 = 1600 \text{ Nm}$$

Force, $P = 381.4$ N.

Exercise 19.2 A block of mass 10 kg is pushed upwards an inclined plane by a horizontal force P equal to 150 N. After covering a distance S, block acquires a velocity of 5 m/s. If the coefficient of kinetic friction between block and plane is 0.2, what is the distance S covered? (Refer to **Fig. 19.7**).

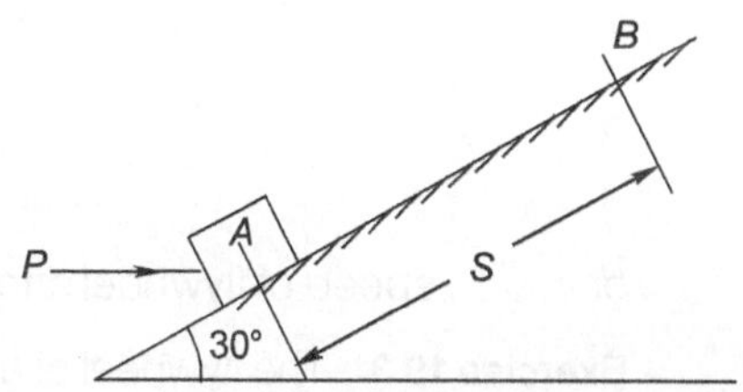

Fig. 19.7

[**Hint:** $P\cos\theta - mg\sin\theta - \mu R = ma$

$R = mg\cos\theta + P\sin\theta,\ V^2 = 2aS$].

[**Ans:** 2.56 m].

19.3 Work done by a Moment during Motion of Rotation

Consider a body of mass m suspended about an axis in a bearing. A moment M is applied on the body and the body rotates with some angular acceleration about the axis of rotation passing through the mass centre G. During rotation a frictional moment, M_f acts on the body as shown in **Fig. 19.8**. Say the body rotates by an angle θ and during this interval its speed has increased from ω_1 to ω_2 rad/sec.

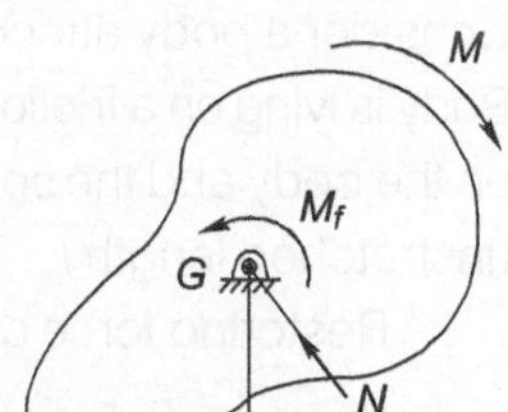

Fig. 19.8

Positive work done by M on body $= M.\theta$

Negative work done by M_f on body $= -M_f.\theta$

Net work done $U_{1\text{-}2} = (M - M_f)\,\theta \quad ...(1)$

No work will be done by normal reaction on the body at the axis, as this reaction passing through axis will not produce any moment.

Gain in kinetic energy, $KE = \frac{I}{2}(\omega_2^2 - \omega_1^2)$ ...(2)

where $I = mk^2$, mass moment of inertia

k = centroidal radius of gyration

m = mass of the body

So $(M - M_f)\,\theta = \frac{I}{2}(\omega_2^2 - \omega_1^2)$. ...(3)

Example 19.2 A flywheel connected to a punching machine has a mass of 300 kg and radius of gyration of 0.8 m. Each punching operation requires 2100 Joules of work. Before punching the speed of flywheel is 220 rpm, what will be its speed after punching operation.

Solution Mass of the flywheel, $m = 300$ kg

Radius of gyration of flywheel, $k = 0.8$ m

Moment of inertia of flywheel, $I = mk^2 = 300 \times 0.8^2 = 192 \text{ kg . m}^2$

Work done during one punching operation

$$U = 2100 \text{ Joules} = 2100 \text{ Nm/s}$$

Original speed of flywheel, $\omega_0 = \frac{2\pi \times 220}{60} = 23.04$ rad/sec

Say final speed of flywheel after punching $= \omega_f$ rad/sec

then
$$-U = I\left(\frac{\omega_f^2}{2} - \frac{\omega_0^2}{2}\right)$$

$$2U = I(\omega_0^2 - \omega_f^2)$$

$$2 \times 2100 = 192\,(23.04^2 - \omega_f^2)$$

or
$$\omega_f^2 = 23.04^2 - 21.875 = 530.841 - 21.875 = 508.96$$

$$\omega_f = 22.56 \text{ rad/sec}$$

or speed of flywheel after punching = 215.44 rpm

Exercise 19.3 The flywheel of a small punching machine operates at 250 rpm. For each punching operation 600 Joules of work is required. It is desired that during punching operation the speed of the flywheel should be reduced only to 230 rpm. Determine required moment of inertia of flywheel.

[**Hint:** $U = \frac{I}{2}(\omega_f^2 - \omega_0^2) = 600$ Nm].

[**Ans:** 11.42 kg-m^2].

19.4 Change of Potential Energy in a Spring

Consider a body attached to a spring of stiffness k as shown in **Fig. 19.9**. Body is lying on a frictionless horizontal plane. An external force F is applied on the body and the spring is stretched by a distance x as shown. From its unstretched length l.

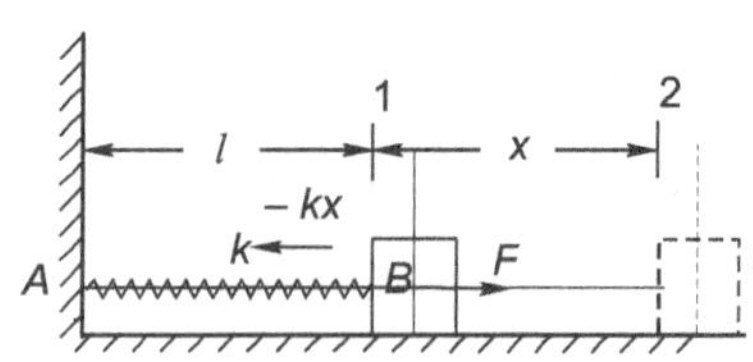

Frictionless horizontal plane

Fig. 19.9

Restoring force developed by spring,

$$F = -kx$$

For equilibrium

$$F + kx = 0$$

As x increases, restoring force ($-kx$) also increases as shown in the **Fig. 19.10** as the force on spring is proportional to the deformation in spring. Work done on the spring in extending it by an amount x

$$U_{1-2} = +\frac{1}{2}kx(x) = \frac{1}{2}kx^2 \text{ (area under the curve shown)}$$

= Potential energy because of pull of the spring from unstretched length

This is because of change in position of the end B of the spring.

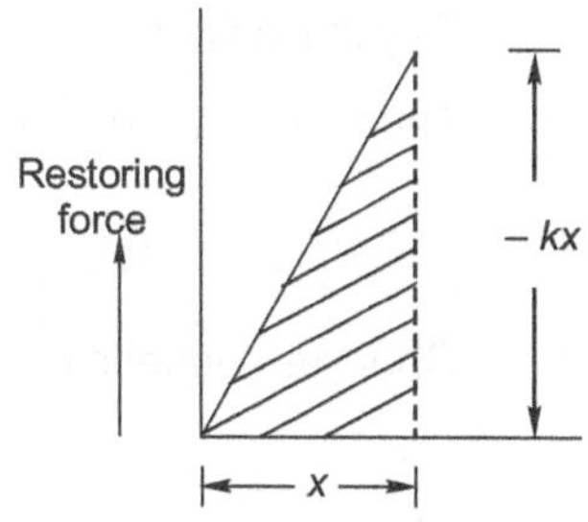

Fig. 19.10

Example 19.3 A spring of initial unstressed length of 300 mm is compressed to a length of 200 mm making a net initial compression of 100 mm. What additional work is done in compressing the spring to 150 mm length? Spring constant is $k = 10$ N/m.

Solution Initial compression $x_1 = 100$ mm

Final compression $x_2 = 300 - 150 = 150$ mm

Spring constant $k = 10$ N/mm

Additional work done in compressing the spring by further 50 mm

$$U_{1-2} = \int_{100}^{150} F\,dx = \int_{100}^{150} kx\,dx = \left|\frac{kx^2}{2}\right|_{100}^{150},$$

putting the value of k

$$= \frac{10}{2}[150^2 - 100^2]$$

$$= 62500 \text{ Nmm} = 62.5 \text{ Nm}$$

Exercise 19.4 A block of mass 6 kg freely falls from a height of 2 m on the top of spring whose stiffness is 12 N/mm. The spring is compressed by 100 mm as shown in **Fig. 19.11**. What is the velocity of the block when the spring is deformed by 100 mm?

[**Hint:** Loss of $PE = W(h + \delta)$, $= \frac{1}{2}mV^2$].

[**Ans:** $V = 4.6$ m/s]

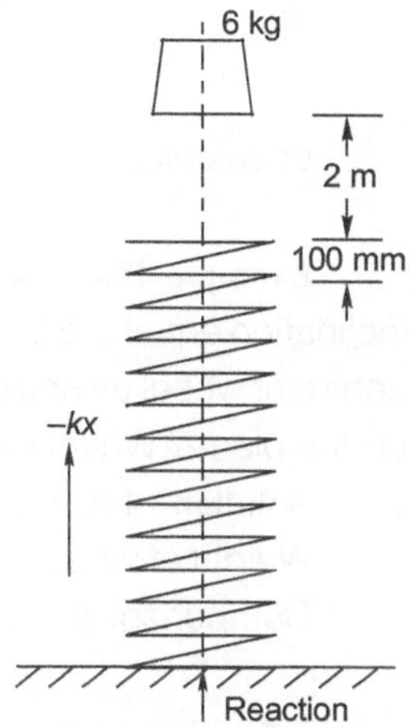

Fig. 19.11

19.5 Rolling Motion under Gravity

Consider a cylinder of radius r and mass m released from the position of rest at position 1 and covering a distance S along the inclined plane and coming to position 2 (**Fig. 19.12**). Weight of the cylinder is mg and force on the cylinder along the plane is $mg\sin\theta$. Force of friction F will act upwards and provides a moment $F \times r$ to rotate the cylinder in the anticlockwise direction.

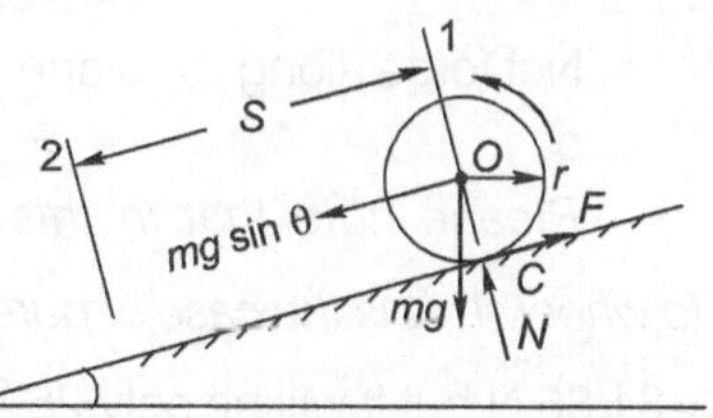

Fig. 19.12

Normal reaction of plane on cylinder

$$= N = mg\cos\theta$$

Normal reaction N does no work as there is no motion of the cylinder in the direction perpendicular to the plane.

Motion is rolling without slipping, i.e., there is no relative motion between the point of contact on the wheel and the surface. Instantaneously the point of contact C is at rest as it becomes the instantaneous centre of rotation. *Therefore frictional force acting at contact point C does no work.*

Initial velocity of centre O of cylinder $= 0$

Initial angular velocity of cylinder $= 0$

Say in position (2) final velocity of mass centre of cylinder is V and final angular velocity of cylinder is ω.

Then work done by force $mg\sin\theta$ along the plane is

$$U_{1-2} = mg.\sin\theta.S$$

Change in kinetic energy $= \frac{1}{2}mV^2 + \frac{1}{2}I\omega^2$

where I = mass moment of inertia of cylinder about axis of rotation $= \frac{mr^2}{2}$

Using the work-energy principle

$$mg\sin\theta.S = \frac{1}{2}mV^2 + \frac{I}{2}\omega^2, \text{ where } I = \frac{mr^2}{2}$$

Therefore $mg.\sin\theta.S = \frac{1}{2}mV^2 + \frac{1}{2}\times\frac{mr^2}{2}\times\omega^2$

but $V = \omega r$

So $S.mg\sin\theta = \frac{1}{2}mV^2 + \frac{1}{4}mV^2 = \frac{3}{4}mV^2$

or velocity, $V = \sqrt{\frac{4S}{3}g\sin\theta}.$

Example 19.4 Find the work done in moving a 15 kg wheel by 2 m up an incline plane with angle of inclination equal to 30°. Coefficient of friction between wheel and plane is 0.25. A force of 100 N is applied at the centre of wheel as shown in **Fig. 19.13**. What will be angular velocity of wheel after the wheel has travelled 3 m up the plane? Wheel radius is 0.1 m.

Solution Mass of wheel $m = 15$ kg

Weight of wheel, $mg = 15 \times 9.81 = 147.15$ N

Component along the plane,

$$mg \sin\theta = 147.15 \times \sin 30^\circ = 73.57 \text{ N}$$

Normal reaction, $N = mg\cos\theta = 147.15 \times 0.866 = 127.43$ N

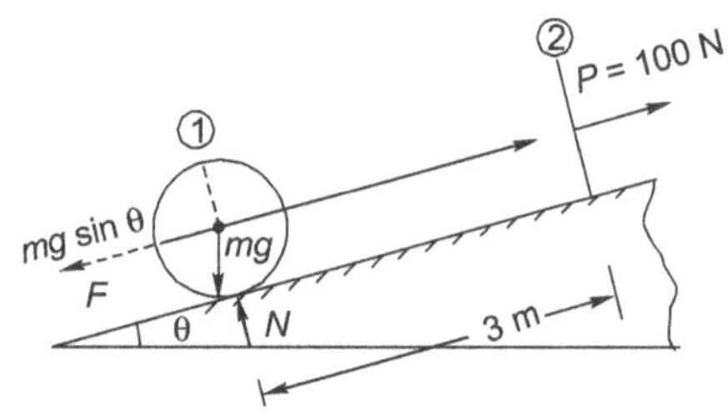

Fig. 19.13

Net force along the plane $= 100 - 73.57$

or $F_{net} = 26.43$ N ...(1)

Please note that in this case force of friction will be equal to F_{net} and provides turning moment to wheel. This is the case of pure rolling without slipping. Force of friction will not be equal to, $\mu N = 0.25 \times 127.43 = 31.86$ N but it will be only 26.43 N.

Work done in rolling the wheel up the plane by 3 m

$$U_{1-2} = (100 - mg\sin\theta)\,S = 26.43 \times 3 = 79.29 \text{ Nm}$$

Say the angular velocity is ω and linear velocity of wheel centre is V at the position 2.

$$U_{1-2} = \frac{1}{2}mV^2 + \frac{I}{2}\omega^2 \quad ...(2)$$

where $V = \omega r = 0.1 \times \omega$ m/s

$$I = \frac{mr^2}{2} = \frac{15 \times 0.1^2}{2} = 0.075 \text{ kg-m}^2$$

Substituting these values in Equation (2)

$$79.29 = \frac{1}{2}\times 15\times(0.1\omega)^2 + \frac{0.075}{2}\times\omega^2$$

$$= 0.075\omega^2 + 0.0375\omega^2 = 0.1125\omega^2$$

Angular velocity, $\omega = \sqrt{\frac{79.29}{0.1125}} = \sqrt{704.8} = 26.54$ rad/s.

Exercise 19.5 A roller of mass 30 kg and radius 0.2 m is rolled up a plane without slipping by a horizontal force *P* as shown in the **Fig. 19.14**. The roller gains a velocity of 3 m/s in a distance of 6 m along the plane. Determine magnitude of *P*.

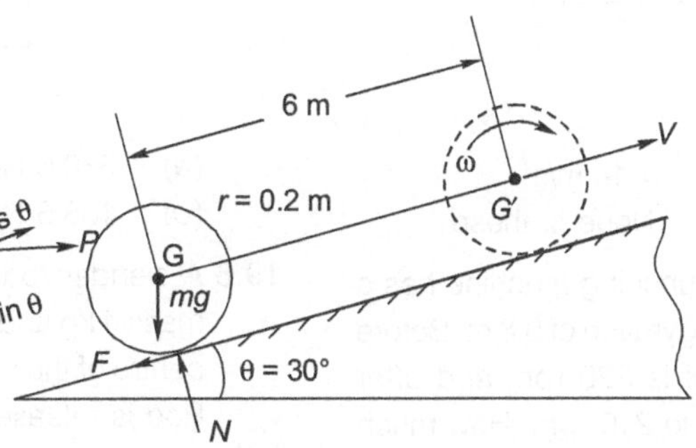

Fig. 19.14

[**Hint:** $U = [P\cos\theta - mg\sin\theta]\times 6$, $U = \frac{1}{2}mV^2 + \frac{I}{2}\omega^2$, $\omega = V/r$].

[**Ans:** $P = 208.9$ N].

PRACTICE PROBLEMS

19.1 A flywheel of centroidal radius of gyration 0.5 m is rigidly attached to a shaft of radius 25 mm which may roll along parallel rails inclined at an angle of 20° as shown in **Fig. 19.15**. The system is released from rest. What is the velocity of the centre of the shaft after it has moved 2 m along the rails?

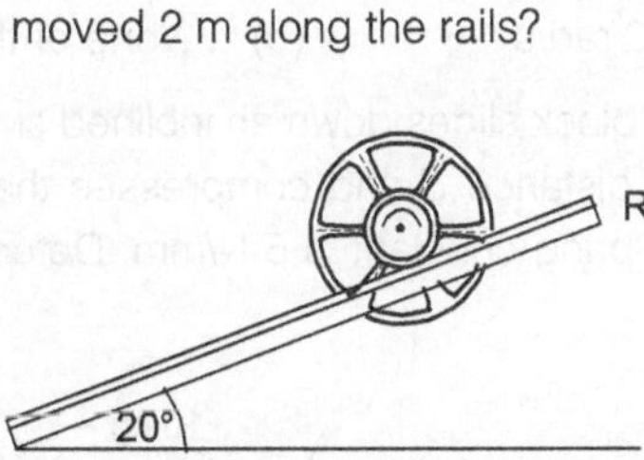

Fig. 19.15

[**Hint:** $V = \omega r = 0.025\omega$ m/s]

$\frac{1}{2}mk^2\omega^2 + \frac{1}{2}mV^2 = mg\sin\theta\times 2$ m].

[**Ans:** $V = 0.183$ m/s].

19.2 A string is wound over a cylinder of mass *m* and radius *R*. At a particular instant the cylinder is released and string unwinds over the cylinder, allowing the cylinder to fall through a height *h* as shown in **Fig. 19.16**. What is the angular velocity of the cylinder? What is the linear velocity of the mass centre *G* of the cylinder?

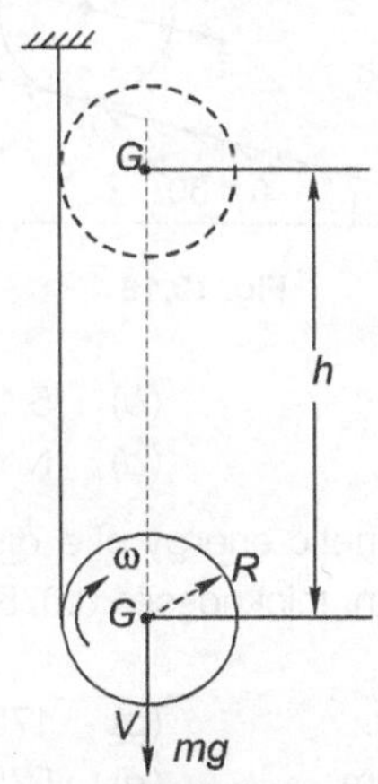

Fig. 19.16

[**Hint:** $PE = KE$, $mgh = \frac{1}{2}mV^2 + \frac{1}{2}I_G\omega^2$, and $\omega = \frac{V}{R}$].

[**Ans:** $V = \sqrt{\frac{4gh}{3}}$, $\omega = \frac{V}{R} = \sqrt{\frac{4gh}{3R^2}}$].

MULTIPLE CHOICE QUESTIONS

19.1 A block of mass 10 kg is placed on a rough horizontal floor as shown in **Fig. 19.17**. A force of 50 N is applied on block and block covers a distance of 10 m. μ between block and floor is 0.25, what is the velocity gained by block?

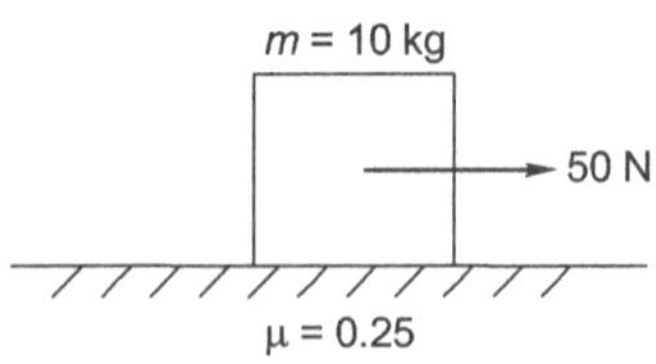

Fig. 19.17

(*a*) 10 m/s (*b*) 7.14 m/s
(*c*) 5 m/s (*d*) None of these

19.2 A flywheel connected to a punching machine has a mass of 300 kg and radius of gyration of 0.8 m. Before punching speed of flywheel is 220 rpm and after punching speed is reduced to 210 rpm. How much energy is consumed in punching operation?

(*a*) 4526 Nm (*b*) 6402 Nm
(*c*) 9052 Nm (*d*) None of these

19.3 A disc of radius 0.1 m, mass 10 kg rolls down an inclined plane with $\theta = 30^\circ$, without slipping and gains a velocity of 5 m/s. How much distance along the plane is covered? $g = 9.81$ m/s (**Fig. 19.18**)

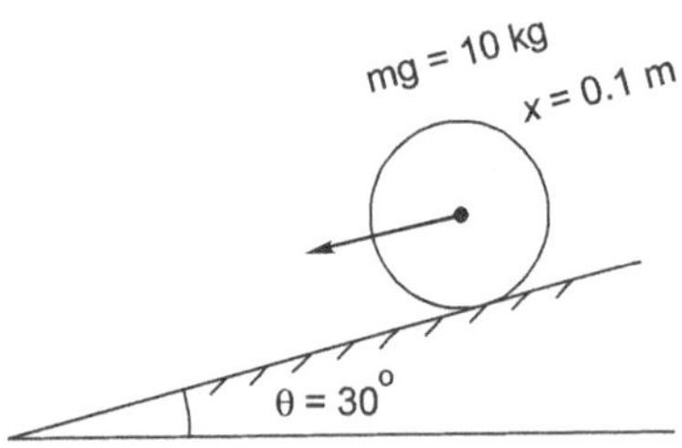

Fig. 19.18

(*a*) 10 m (*b*) 5.1 m/s
(*c*) 3.82 m (*d*) None of these

19.4 What is the kinetic energy of a disc of mass 70 kg, diameter 0.6 m, thicknesses 0.075 m and rotating at 100 rpm?

(*a*) 86.4 Nm (*b*) 172.7 Nm
(*c*) 345.4 Nm (*d*) 575.5 Nm

19.5 Drum shown in **Fig. 19.19** rotates in frictionless bearings *A* mass of 30 kg attached to the string wound as shown falls through a distance of 2 m. A constant torque of M = 80 Nm is applied on drum of radius 0.4 m. What work is done on system?

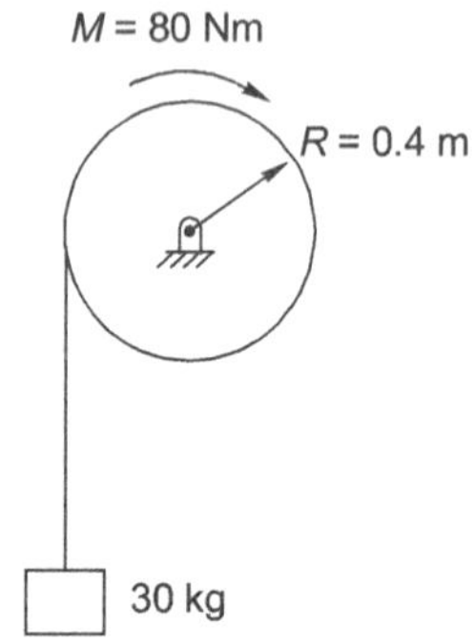

Fig. 19.19

(*a*) 588.6 Nm (*b*) 428.6 Nm
(*c*) 188.6 Nm (*d*) None of these

19.6 A slender rod *OA*, hinged at *O*, of length 1 m is of mass 4 kg and a coller of mass 2 kg is attached to the centre of the rod. Horizontal position shown is at rest. Rod is released and reaches vertical position, what is angular speed of rod in vertical position (**Fig. 19.20**)?

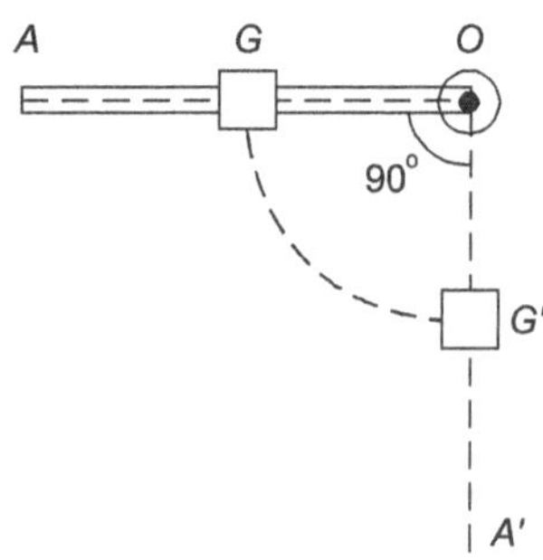

Fig. 19.20

(*a*) 4.7 rad/s (*b*) 5.667 rad/s
(*c*) 4.2 rad/s (*d*) None of these

19.7 A 10 kg block slides down an inclined smooth plane through distance *S* and compresses the spring by 0.1 m. Spring constant is 5 N/mm. Determine value of *S*

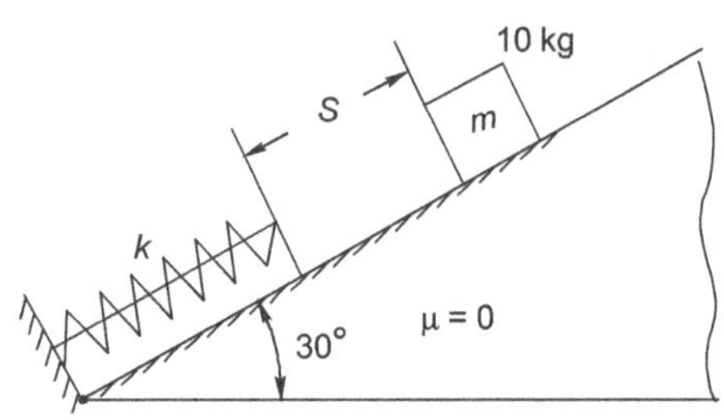

Fig. 19.21

(*a*) 0.51 m (*b*) 0.41 m
(*c*) 0.9 m (*d*) None of these

19.8 A mass m is moving horizontally with velocity V_0 strikes ideal pendulum of mass m. If the two masses stick together after collision, what is the maximum height attainable by the pendulum.

(a) $\frac{V_0^2}{8g}$ (b) $\frac{V_0^2}{4g}$

(c) $\sqrt{V_0 g}$ (d) $\sqrt{V_0 \frac{9}{2}}$

[CSE, Prelim, CE : 2009]

19.9 A thin uniform hoop string of radius R is rolling without slipping such that mass center moves at a speed of V, if the hoop weight W, what is the corresponding kinetic energy of the hoop relative to the ground?

(a) $\frac{WV^2}{4g}$ (b) $\frac{WV^2}{2g}$

(c) $\frac{WV^2}{g}$ (d) $\frac{2WV^2}{g}$

[CSE, Prelim, CE : 2009]

19.10 An ejector mechanism consists of a helical compression spring having a spring constant of $K = 981 \times 10^3$ N/m. It is pre-compressed by 100 mm from its free state. If it is used to eject a mass of 100 kg held on it, the mass will move up through a distance of (**Fig. 19.22**)

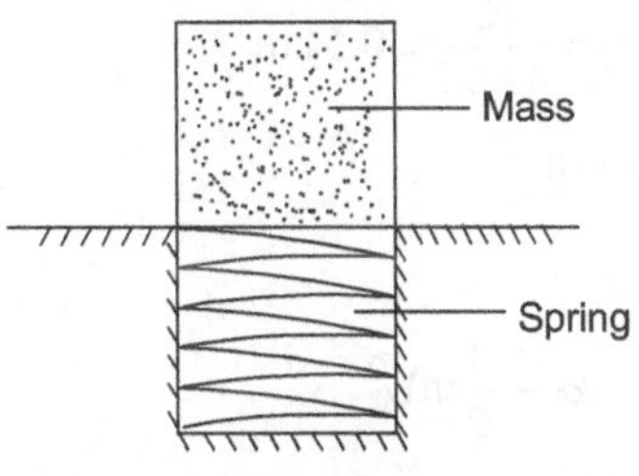

Fig. 19.22

(a) 100 mm (b) 500 mm
(c) 581 mm (d) 1000 mm

[GATE 2004 : 2 Marks]

12.11 A 1 kg mass of clay, moving with a velocity of 10 m/s, strikes a stationary wheel and sticks to it. The solid wheel has a mass of 20 kg and a radius of 1m. Assuming that the wheel is set into pure rolling motion, the angular velocity of the wheel immediately after the impact is approximately (**Fig. 19.23**).

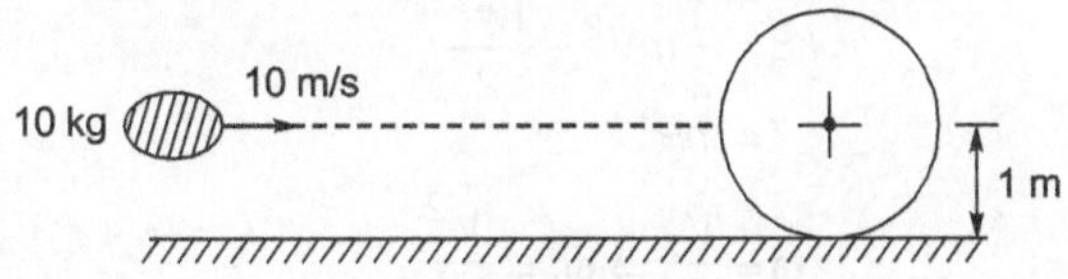

Fig. 19.23

(a) Zero (b) $\frac{1}{3}$ rad/s

(c) $\sqrt{\frac{10}{3}}$ rad/s (d) $\frac{10}{3}$ rad/s

[GATE 2005: 2 Marks]

12.12 A circular solid disc of uniform thickness 20 mm, radius 200 mm and mass 20 kg, is used as a flywheel. If it rotates at 600 rpm, the kinetic energy of the flywheel, in Joules is

(a) 395 (b) 790
(c) 1580 (d) 3160

[GATE 2012 : 1 Mark]

12.13 An annular disc has a mass m, inner radius R and outer radius 2R. The disc rolls on a flat surface without slipping. If the velocity of the center of mass is v, the kinetic energy of the disc is

(a) $\frac{9}{16}mv^2$ (b) $\frac{11}{16}mv^2$

(c) $\frac{13}{16}mv^2$ (d) $\frac{15}{16}mv^2$

[GATE 2014 : 2 Marks, Set-3]

12.14 When 100J of work is done upon a flywheel, its angular speed increase from 60 rpm to 180 rpm. What is the moment of inertia of the flywheel

(a) 0.21 kg m^2 (b) 0.63 kg m^2
(c) 0.84 kg m^2 (d) 0.15 kg m^2

[CSE, Prelim, CE : 2010]

12.15 A car accelerates from rest to a speed of 10 m/s and the energy spent in doing there is E. If the car is further accelerated from 10 m/s to 20 m/s, what is the amount of energy to be spent further?

(a) E (b) 2E
(c) 3E (d) 4E

[CSE, Prelim, CE : 2007]

12.16 An object of mass 5.0 kg falls from rest through a vertical distance of 20 m and acquires a velocity of 10 m/s. What is the work done by the air resistance as the object? (Take g = 10 m/s^2)

(a) 75 J (b) 150 J
(c) 750 J (d) 850 J

[CSE, Prelim, CE : 2006]

Answers

19.1 (b)	19.2 (a)	19.3 (c)	19.4 (b)	19.5 (c)
19.6 (b)	19.7 (b)	19.8 (b)	19.9 (c)	19.10 (a)
19.11 (b)	19.12 (b)	19.13 (c)	19.14 (b)	19.15 (c)
19.16 (c)				

EXPLANATIONS

19.1 (b)

$$(50 - \mu\, mg) \times 10 = \frac{1}{2} mV^2$$

$$(50 - 0.25 \times 10 \times 9.81) \times 10 = \frac{1}{2} \times 10 \times V^2 = 5V^2$$

$$254.75 = 5V^2$$

$$V = 7.138 \text{ m/s.}$$

19.2 (a)

$$I = 300 \times 0.8^2 = 192 \text{ kg-m}^2$$

$$\omega_1 = \frac{220 \times 2\pi}{60} = 23.038, \quad \omega_1^2 = 530.76$$

$$\omega_2 = 210 \times \frac{2\pi}{60} = 21.99, \quad \omega_2^2 = 483.61$$

U = energy consumed

$$= \frac{1}{2} \times I(\omega_1^2 - \omega_2^2) = \frac{192}{2} \times 47.15$$

= 4526 Nm = 4526 Joule.

19.3 (c)

$$I = \frac{mr^2}{2} = \frac{10 \times .1^2}{2} = .05 \text{ kg m}^2$$

V = 5 m/s

$\omega = V/r = 5/0.1 = 50$ rad/s

$$KE = \frac{1}{2} mV^2 + \frac{1}{2} I\omega^2$$

$$= \frac{1}{2} \times 10 \times 5^2 + \frac{1}{2} \times 0.05 \times 50^2 = 125 + 62.5 = 187.5$$

$187.5 = mg \,.\, S \sin\theta$

$= 10 \times 9.81 \times S \times 0.5$

$$S = \frac{187.5}{5 \times 9.81} = 3.82 \text{ m.}$$

19.4 (b)

$$I = \frac{mr^2}{2} = \frac{70 \times 0.3^2}{2} = 3.15 \text{ kg-m}^2$$

$$\omega = \frac{2\pi \times 100}{60} = 10.47 \text{ rad/s}$$

$$U = \frac{I}{2} \times \omega^2 = \frac{3.15}{2} \times 10.47^2 = 172.7 \text{ Nm.}$$

19.5 (c)

$$U = mgh - M\theta, \qquad \theta = \frac{h}{R} = \frac{h}{0.4} = 2.5h$$

$= mgh - M \times 2.5h$

$= 30 \times g \times 2 - 80 \times 2.5 \times 2$

$= 60 \times 9.81 - 400 = 588.6 - 400 = 188.6$ Nm.

19.6 (b)

PE loss $= (2 + 4)\, g \times 0.5 = 3g = 3 \times 9.81$

= 29.43 Nm

$$I_0 = \frac{4\,(1^2)}{3} + 2\left(\frac{1}{2}\right)^2 = \frac{4}{3} + 0.5 = 1.833 \text{ kg-m}^2$$

$29.43 = 1.833 \times \omega^2/2$

$\omega = 5.667$ rad/s.

19.7 (b)

$$(S + 0.1)\, mg \sin\theta = \frac{1}{2} \times 5000 \times 0.1^2,$$

note k = 5000 N/m

$(S + 0.1) \times 10 \times 9.81 \times 0.5 = 25$

$$(S + 0.1) = \frac{25}{49.05} = 0.51 \text{ m}$$

S = 0.41 m

19.8 (b)

$$kE = \frac{1}{2} mV_0^2$$

$$pE = 2\, mgh$$

$$mV_0^2 = 4\, mgh$$

$$V_0^2 = 4\, gh, \; h = \frac{V_0^2}{4g}$$

19.9 (c)

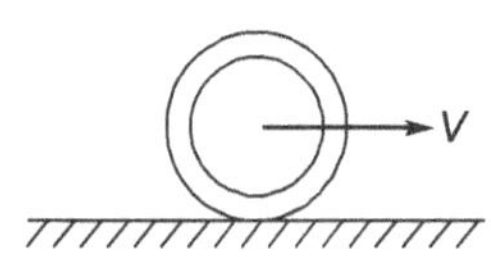

$$kE = \frac{1}{2} mV^2 + \frac{Iw^2}{2}$$

$$I = mR^2$$

$$\omega = \frac{V}{R} \Rightarrow \omega^2 = \frac{V^2}{R^2}$$

$$kE = \frac{1}{2}mV^2 + \frac{mR^2}{2} \times \frac{V^2}{R^2}$$

$$mV^2 = \frac{W}{g}V^2$$

19.10 (a)

PE of spring

$$\frac{1}{2}k\delta^2 = \frac{1}{2} \times 981 \times 10^3 \times (0.1)^2 = 4905 \text{ Nm}$$

Say spring recods by x m

$$4905 = \frac{1}{2} \times 981 \times 10^3 x^2 + 100 \times 9.81 \times (x + 0.1)$$

$$I = 100x^2 + 0.2(x + 0.1)$$

$$I = 100x^2 + 0.2 + 0.2$$

$$100x^2 + 0.2x - 0.98 = 0$$

$$x = \frac{-0.2 + \sqrt{0.04 + 392}}{200}$$

$$= \frac{-0.2 + 19.8}{200} = 0.098 \text{ m}$$

$$\simeq 98 \text{ mm or } 100 \text{ mm}$$

19.11 (b)

$$mU = (m + M)V + I\omega$$

$$1 \times 10 = (1 + 20)V + \frac{20 \times 1^2}{2} \times \omega$$

$$10 = 21V + 10\omega = 21V + 10 \times \frac{V}{1}$$

$$= 31V = 31 \times \omega \times R = 3100$$

$$\omega = \frac{10}{35} \approx \frac{1}{3} \text{ rad/sec}$$

19.12 (b)

$$I = \frac{20 \times 0.2^2}{2} = 0.4 \text{ kg m}^2$$

$$\omega = \frac{2 \times \pi \times 600}{60} = 6283 \text{ rad/sec}$$

$$kE = \frac{I}{2}\omega^2 = \frac{0.4}{2}(62.83)^2 = 790 \text{ Nm}$$

19.13 (c)

Disc, mass = m

Inner value = R, center value $2R$

$$I = \frac{m(4R^2 + R^2)}{2} = 2.5\, mR^2$$

$$kE = \frac{1}{2}mV^2 + \frac{1}{2}I\omega^2$$

$$\omega = \frac{V}{2R}, \text{ an outer value is 2R}$$

$$\omega^2 = \frac{V^2}{4R^2}$$

$$kE = \frac{1}{2}mV^2 + \frac{1}{2} \times 2.5mR^2 \times \frac{V^2}{4R^2}$$

$$= \frac{1}{2}mV^2 + \frac{5}{16}mV^2 = \frac{13}{16}mV^2$$

19.14 (b)

$$100 = \frac{I}{2}(\omega_2^2 - \omega_1^2)$$

$$\omega_2 = 18.85 \text{ ra/dsec}$$

$$\omega_1 = 6.28 \text{ rad/sec}$$

$$100 = \frac{I}{2}[355.305 - 39.48] = 157.9\, I$$

$$I = 0.633 \text{ kg m}^2$$

19.15 (c)

$$E = \frac{1}{2}m(10)^2 = 50 \text{ m Nm}$$

$$\frac{1}{2}m(20)^2 - \frac{1}{2}m(10)^2 = 150 \text{ m NM}$$

$$= 3E, \text{ additional}$$

19.16 (c)

$$H = 20\text{m}$$

$$v = 10 \text{ m/sec}$$

$$H' = \frac{V^2}{2g} = \frac{10^2}{2 \times 10} = 5 \text{ m}$$

Resistance = $mg(H - H') = 5 \times 10 \times 15$

$= 750 \text{ Nm} = 750 \text{ J}$

■■■■

20

CHAPTER

Impulse and Momentum

20.1 Introduction

Newton's second law of motion states that the externally applied force is responsible for the rate of change of momentum in a body. In other words this law gives relationship between force, mass and acceleration *i.e.*, rate of change of velocity. The principle of impulse and momentum is derived again from the Newton's second law of motion. This principle is generally used in problems of rigid body dynamics where large forces act for a short interval of time producing change of momentum of the body, such as the forces acting during hammer blow action in smithy work, press work, driving of piles into the ground, firing of shell from a gun etc. The most important examples of industrial use is (*a*) Impulse turbine, in which principle of impulse-momentum is used for power generation. (*b*) Hammer blows—used in forging operation and making connecting rods, crank shafts and many other engineering components.

As per Newton's second law, force = mass × acceleration

or $$F = m\frac{dV}{dt}$$

or $$F\,dt = m\,dV$$

Integrating both the sides $$\int_{t_1}^{t_2} F\,dt = m\int_{V_1}^{V_2} dV$$

or $$F(t_2 - t_1) = m(V_2 - V_1)$$

Linear impulse = Linear momentum, a large force applied momentarily

When no external force is acting on the system, the principle of impulse-momentum is reduced to principle of conservation of momentum.

Two masses m_1 and m_2 collide with each other with velocities U_1 and U_2 respectively and after impact they separate with velocity V_1 and V_2 respectively as shown in **Fig. 20.1**. Then as per principle of conservation of momentum

$$m_1U_1 + m_2U_2 = m_1V_1 + m_2V_2$$

Initial momentum before impact = Final momentum after impact.

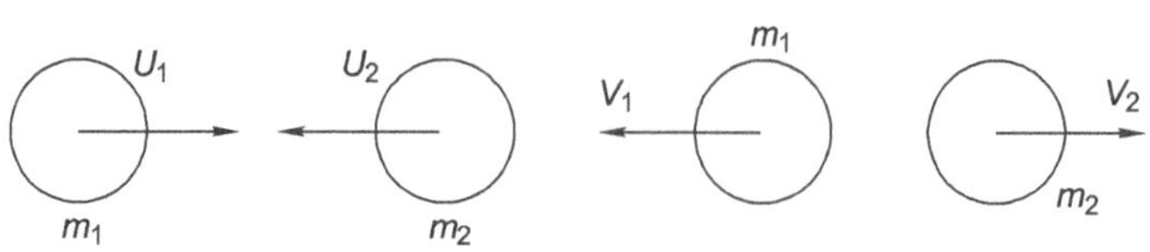

Fig. 20.1

Example 20.1 A particle A of mass 2 kg with velocity $U = 10i - 3j + 4k$ m/s strikes another particle B and its velocity is changed to $V = 18i + 3j - 2k$ m/s. The impact between the two particles takes place for 15 millisecond. Determine (*a*) average force exerted on particle A; (*b*) change in linear momentum of particle B.

Solution *Particle A:*

Initial velocity, $U = 10i - 3j + 4k$ m/s

Final velocity, $V = 18i + 3j - 2k$ m/s

Mass of particle A, $m = 2$ kg

Change of linear momentum of particle A

$$I_{mA} = m_A[V - U]$$

$$= 2[18i + 3j - 2k - 10i + 3j - 4k] \text{ N sec}$$

$$= 2[8i + 6j - 6k] = 16i + 12j - 12k \text{ N sec}$$

Say the average force on $A, = F_A$

Time of impact, $t = 0.015$ s

Using the linear impulse momentum equation

$$I_{mA} = \text{Impulse on } A = F_A \cdot t = m_A(V - U)$$

$$F_A \times 0.015 = 16i + 12j - 12k \text{ N}$$

$$F_A = \frac{16i + 12j - 12k}{0.015} = 1066.66i + 800j - 800k \text{ Newton}$$

On the principle of equal and opposite reaction,

Impulse on particle A = –Impulse on particle B

So impulse on particle B $= -I_{MA} = -16i - 12j + 12k$ N sec.

Example 20.2 A gun of mass 4000 kg is designed to fire a 20 kg shell with an initial velocity of 560 m/s. Determine the average force required to hold the gun motionless if the shell leaves the gun 20 ms (milli second) after being fired.

[**Hint:** $t = 0.02$ s, change in momentum = 20 × 560 kg m/s].

[**Ans:** $F = 560$ kN].

Example 20.3 A shell of mass 40 kg is fired horizontally from a gun of mass 2800 kg, at a velocity of 250 m/s. With what velocity gun will recoil? If the gun is brought to rest in a distance of 0.7 metres, what is the average force acting on the gun?

Solution Figure 20.2 shows the sketch of a gun and shell. When the gun powder explodes it imparts velocity v to the shell and the shell moves out with the momentum mv, where m is the mass of the shell. On the system of shell and gun, no external force is applied, *the principle of impulse momentum is reduced to the principle of conservation of momentum*, therefore the momentum of the system is conserved.

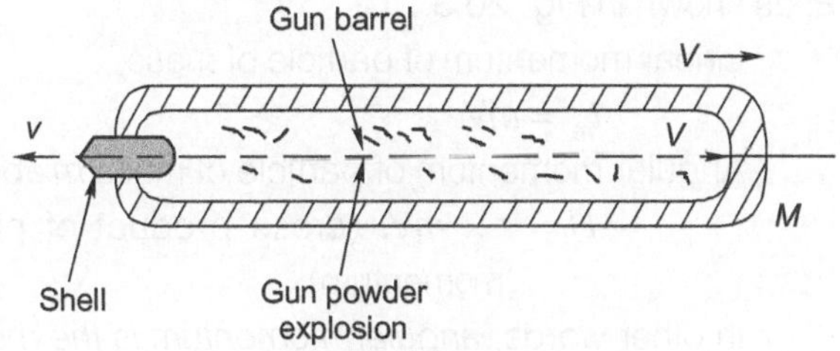

Fig. 20.2

Say M = Mass of the gun

V = Velocity of the gun at the time shell is fired

Then using the principle of conservation of momentum

$$mv + MV = 0$$

or $$MV = -mv \quad \text{...(1)}$$

Velocity of recoil, $V = -\dfrac{mv}{M}$

Now $m = 40$ kg, $M = 2800$ kg

and $v = 250$ m/s, velocity of shell

Velocity of gun recoil, $V = -\dfrac{40 \times 250}{2800} = -3.57$ m/s

Gun is to be stopped in a distance of 0.7 m, during this distance say average force acting on the gun is R Newton, then using the principle of work-energy

$$RS = \frac{M}{2}(V^2 - 0) \text{ where final velocity of gun is zero.}$$

S = Distance in which gun is stopped

Substituting the values $R \times 0.7 = \dfrac{2800}{2}(3.57^2 - 0)$

Average force on gun, $R = \dfrac{17842.86}{0.7} = 25.5$ kN

Deceleration of the gun, $a = -\dfrac{25.5 \times 1000}{2800} = -9.10$ m/s^2

Time in which gun is brought to rest,

$$t = \frac{V}{a} = \frac{-3.57}{-9.10} = 0.39 \text{ sec.}$$

Exercise 20.1 A particle of mass 2 kg is at rest. A force varying with time acts on the particle $F = t^2\, i + (5t - 6)\, j + 1.2t^3\, k$ N where t is in seconds. What is the velocity of the particle after 8 seconds?

[**Hint:** $\int_0^8 F\,dt = I_m = m(V - u)$] $V = 85.33i + 56j + 614.4k$ m/s].

[**Ans:** $|V| = 622.82$ m/s].

20.2 Angular Impulse and Angular Momentum Principle

Let us first define what is angular momentum of a particle. Consider a particle of mass m traversing a curved path ab and at a particular instant its position vector is $OP = r$ and velocity V is tangential to the curved path at point P, as shown in **Fig. 20.3**.

Linear momentum of particle of mass,

$$I_m = mV$$

Angular momentum of particle of mass m about the axis of rotation OZ

$H = r \times mV$, (cross product of position vector and linear momentum)

In other words, *angular momentum is the moment of linear momentum* about the axis of rotation. As per Euler's equation

momentum, $M = I_G . \alpha$ = mass moment of inertia about mass centre of body × angular acceleration

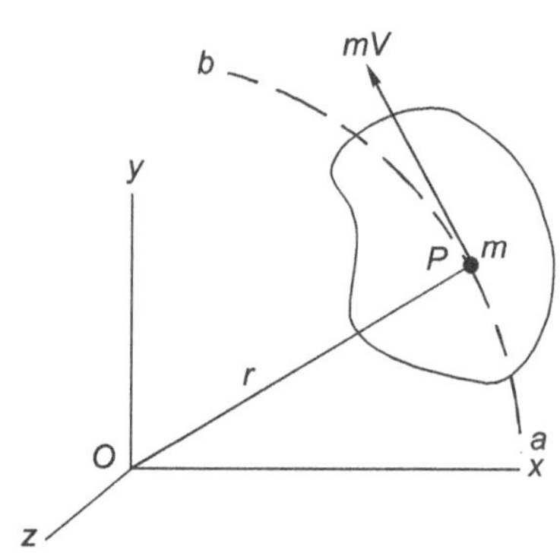

Fig. 20.3

Moment, $M = I_G \cdot \dfrac{d\omega}{dt}$

$$\text{or} \qquad \int_{t_1}^{t_2} M\,dt = \int_{\omega_1}^{\omega_2} I_G . d\omega$$

$$\text{or} \qquad M(t_2 - t_1) = I_G(\omega_2 - \omega_1)$$

Angular impulse, $\qquad H = I_G(\omega_2 - \omega_1)$ = angular momentum

If the external moment on a body is zero, then principle of angular impulse momentum is reduced to the principle of conservation of angular momentum.

Example 20.4 At a certain instant the linear momentum of a particle is given by

$$I_m = -4i - 3j + 4k \text{ kg m/s}$$

and its position vector is $r = 4i - 6j + 2k$ m. Determine the magnitude of angular momentum H_0 of the particle about the origin of co-ordinates.

Solution Linear momentum, $I_m = -4i - 3j + 4k$ kg m/s

Position vector, $r = 4i - 6j + 2k$ m

Angular momentum, $H_0 = r \times I_m$

$$= (4i - 6j + 2k) \times (-4i - 3j + 4k) \text{ kg m}^2\text{/s}$$

$$= \begin{vmatrix} i & j & k \\ 4 & -6 & +2 \\ -4 & -3 & +4 \end{vmatrix} = (-24 + 6)\,i + (-8 - 16)\,j + (-12 - 24)\,k$$

$$= -18i - 24j - 36k \text{ kg m}^2\text{/s.}$$

Exercise 20.2 A mass of 6 kg is attached at the end of a bar of length 0.75 m but of negligible mass. The mass is rotating at 80 rpm as shown in Fig. 20.4. If the angular speed of the bar is increased to 160 rpm from 80 rpm in 3 revolutions, what is the torque applied on the bar?

[**Hint:** Torque of bar $M \qquad M.t = I(\omega_f - \omega_0)$

$I = ml^2$].

[**Ans:** M = 18.85 Nm].

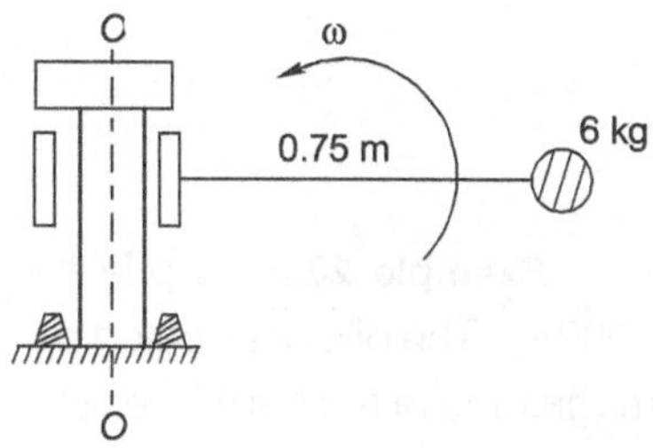

Fig. 20.4

20.3 Pile and Pile Hammer

During the construction of bridges, high rising buildings etc. earth or the soil is strengthened by steel girders. These steel girders or piles are driven into the ground with the help of hammer blows. The hammer is of definite shape and is allowed to fall freely under gravity and strikes the pile (or the girder) and the pile is pushed into the ground by a small distance. Each time the pile is given hammer blow, each time the pile is pushed into the ground depending upon the mass of the hammer, height through which hammer is allowed to fall and the resistance offered by the soil.

Let us consider a pile of mass m and pile hammer of mass M. The hammer is allowed to fall through a height h under gravity (**Fig. 20.5**). Velocity of the hammer when it strikes the top edge ab of the pile

$$v = \sqrt{2gh}$$

Momentum of hammer $= Mv$

Now the pile and hammer move together with velocity V, using the principle of conservation of momentum

$$Mv = (M + m)\,V$$

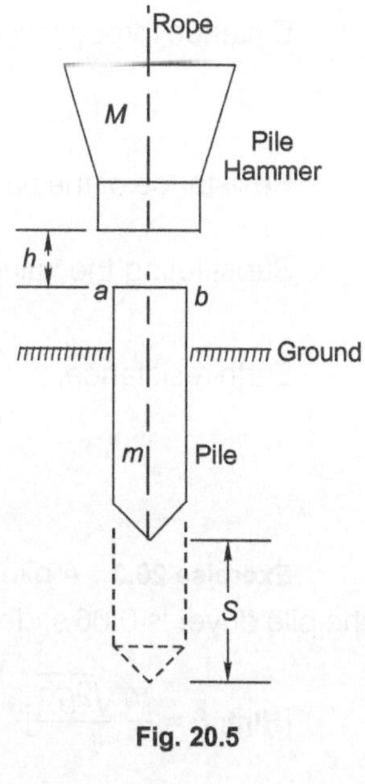

Fig. 20.5

or Velocity, $V = \frac{Mv}{M+m}$ ...(1)

After the impact, pile and hammer move together with velocity V, the pile being pushed into the ground for the purpose of reinforcing the soil resistance. The ground offers resistance say R to the motion of the pile and hammer.

The pile comes to rest by the time it has been pushed into the ground by a depth S.

Work done by ground resistance R, $U = R \, . \, S$

Loss of potential energy of hammer and pile $= (M+m)\, g \, . \, S$

Loss of kinetic energy of hammer and pile $= \frac{1}{2}(M+m)\, V^2$

Using the principle of work and energy, $R \, . \, S = (M+m)\, g . S + \frac{(M+m)}{2} V^2$

Resistance offered by the ground $R = (M+m)\, g + \frac{(M+m)\, V^2}{2S}$ ...(2)

where $V = \frac{Mv}{(M+m)} = \frac{M}{(M+m)} \times \sqrt{2gh}$

Substituting the value of V^2 in equation (2)

$$R = (M+m)\, g + \frac{M+m}{2S} \times \frac{M^2}{(M+m)^2} \times 2gh$$

$$= (M+m)\, g + \frac{M^2}{(M+m)} \times \frac{gh}{S}. \quad ...(3)$$

Example 20.5 A pile hammer of mass 2000 kg drops from a height of 0.7 m on a pile of mass 1000 kg. The pile penetrated by 120 mm per blow. Assuming that the resistance of earth is uniform, determine the magnitude of the earth's resistance.

Solution Mass of pile hammer, $M = 2000$ kg

Height through which hammer drops, $h = 0.7$ m

Mass of the pile, $m = 1000$ kg

Distance through which pile is penetrated into the ground,

$S = 120$ mm $= 0.12$ m

Resistance of the earth, $R = (M+m)\, g + \frac{M^2}{(M+m)} \times \frac{gh}{S}$

Substituting the values of M, m, g, h and S

Earth resistance, $R = (2000+1000)\, g + \frac{2000^2}{(2000+1000)} \times g \times \frac{0.7}{0.120}$

$= 3000g + 7777.77g = 10777.77g$

$= 10777.77 \times 9.81 = 105.73$ kN.

Exercise 20.3 A pile hammer of mass 500 kg is dropped from a height of 5 m. If the time required to stop the pile driver is 0.06 s, determine average force acting, during the time.

[**Hint:** $F = \frac{m\sqrt{2gh}}{t}$]. [**Ans:** 82.5 kN].

PROBLEMS

Problem 20.1 A trolley of mass 500 kg can move along a horizontal frictionless track. A man of mass 60 kg sits on the trolley. Initially the trolley with a man on it is moving towards right at speed of 8 m/s. What will be the velocity of the trolley if the man walks on the trolley at a speed of 2.5 m/s towards left as shown in **Fig. 20.6**?

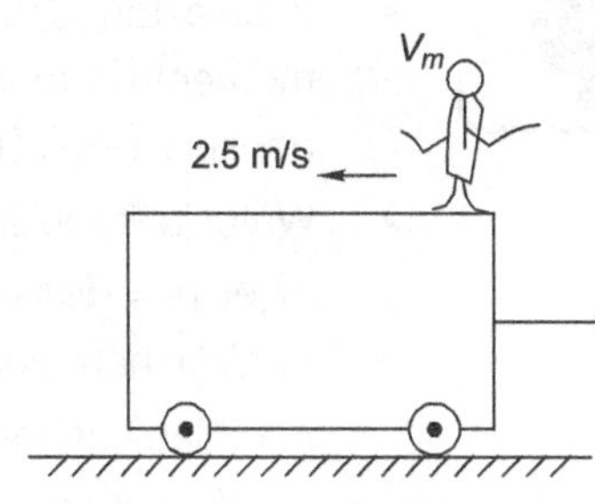

Fig. 20.6

Solution M, mass of trolley= 500 kg

m, mass of man = 60 kg

Initial velocity, V = 8 m/s

Initial momentum, = (500 + 60) 8 Ns = 4480 Ns

Now the man starts moving in opposite direction with velocity V_m = 2.5 m/s.

Total momentum remain the same, and velocity of trolley is increased to V' m/s, so

$$4480 = (M+m)V' - mV_m = 560V' - 60\times 2.5$$

$$V' = \frac{4480}{560} + \frac{150}{560} = 8 + 0.27 \text{ m/s} = 8.27 \text{ m/s}$$

Velocity of trolley is increased by 0.27 m/s.

Problem 20.2 A hammer of mass 0.5 kg drives a nail of mass 20 gm with a velocity of 5 m/s, horizontally into a fixed wooden block. If the nail penetrates by 8 mm per blow, calculate the resistance of the block, assuming it to be uniform.

Solution **Fig. 20.7** shows a nail being driven hori-zontally into a wooden block with the help of hammer blow.

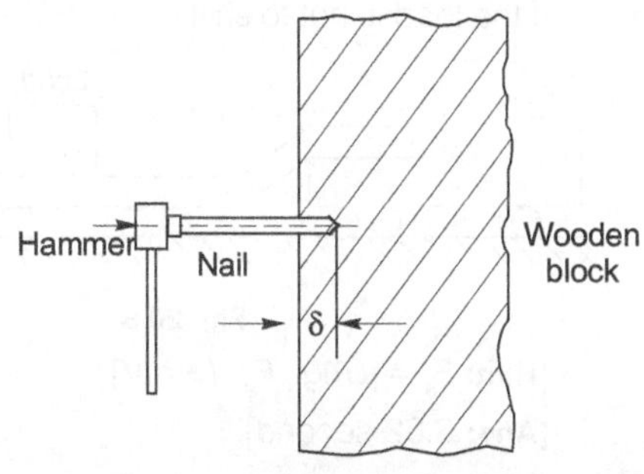

Fig. 20.7

Mass of hammer, M = 0.5 kg

Mass of nail, m = 0.020 kg

Initial velocity of hammer,

V_0 = 5 m/s

Final velocity of hammer and nail,

$$V = \frac{MV_0}{M+m} = \frac{0.5\times 5}{0.5+0.020} = 4.808 \text{ m/s}$$

Since the nail is driven horizontally into the wooden block, there is no loss of potential energy, which remains constant.

$$\text{Loss of kinetic energy, } KE_l = \frac{1}{2}(M+m)V^2 = \frac{1}{2}\times 0.52\times 4.808^2 = 6.0096 \text{ Nm}$$

Work done by resistance of the block = RS

where R = Resistance of the block

S = Distance of penetration of nail

= 8 mm = 0.008 m

$$\text{Resistance of wooden block, } R = \frac{KE_l}{S} = \frac{6.0096}{0.008} = 751.2 \text{ N.}$$

Remember

- Linear impulse, $F \cdot t = m(V_2 - V_1)$, change of linear momentum.
- Angular impulse, $M \cdot t = I(\omega_2 - \omega_1)$, where I—mass moment of inertia.
- If the external force or external moment on a body is zero, then principle of linear impulse-momentum or principle of angular impulse-momentum is reduced to the principle of conservation of linear/angular momentum respectively.
- Water jet impulsive force $F_x = A\rho(V-U)^2(1-\cos\theta)$, $F_y = A\rho(V-U)^2\sin\theta$ where A = area of jet, ρ = density of liquid, V = velocity of jet, U = velocity of moving vane, θ = angle of jet.
- Pile/pile hammer $MV' = (M+m)V$
  M = Mass of hammer, m = mass of pile
  V = Velocity of hammer, V' = velocity of pile and hammer.
- Resistance from ground
  $$R = (M+m)g + \frac{M^2}{(M+m)} \times \frac{gh}{S}, \text{ in pile driving}$$
  h = Height through which hammer falls
  S = Depth through which pile is pushed into the ground.

PRACTICE PROBLEMS

20.1 The coefficient of friction between the load and flat bed trailer shown in **Fig. 20.8**, are $\mu_s = 0.40$ and $\mu_k = 0.30$. Speed of the rig is 15 m/s, determine the shortest time in which the rig can be brought to a stop if the load is not to shift.

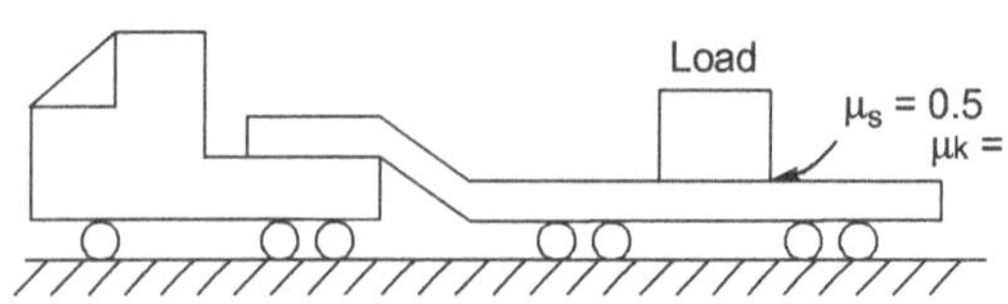

Fig. 20.8

[**Hint:** $F_s = \mu mg$, $F_s \cdot t = mV$].

[**Ans:** 3.82 second].

20.2 After scaling a wall, a man lets himself drop by 3.2 m to the ground. If his body comes to rest to a complete stop in 0.12 second, after his feet first touch the ground, determine the vertical average impulse force exerted by ground on his feet. The mass of man is 65 kg.

[**Hint:** $V = \sqrt{2gh}$].

[**Ans:** 4292 Newton].

20.3 A man of mass 70 kg stands in an aluminium canoe of 40 kg. He fires a bullet of 25 gm horizontally over the bow of the canoe to hit a wooden block of mass 2 kg resting on a smooth horizontal surface. If the wooden block and bullet move together with a velocity of 5 m/s, find the velocity of the canoe.

[**Hint:** (40 + 70) V = (2 + 0.025) × 5), use principle of conservation of momentum].

[**Ans:** 0.092 m/s].

20.4 A cylinder of radius R, mass moment and moment of inertia I about centre C rolls down the inclined plane as shown in **Fig. 20.9**. Cylinder starts from rest, what will be its velocity after t seconds if $\theta = 30^o$?

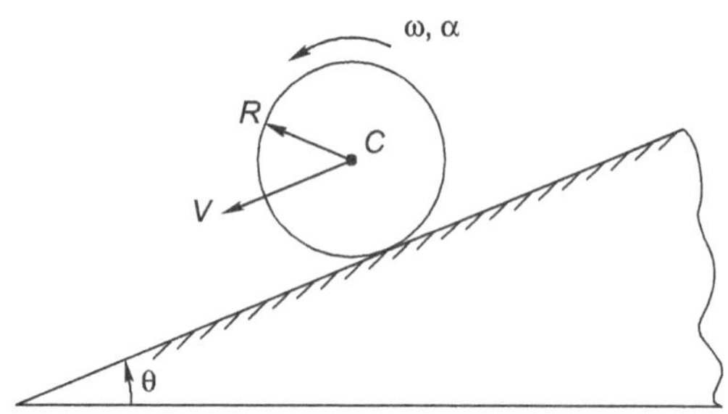

Fig. 20.9

[**Hint:** Friction force F at contact point,

$t(mg\sin\theta - F) = m(V-U)$

$(F \times R)\,t = I(\omega - 0)$ $\omega = \frac{V}{R}$].

[**Ans:** $V = \frac{g}{3}t$].

20.5 A particle of mass m rests on a rough horizontal table. It is given an initial velocity of V_0 along a circular path of radius R. The coefficient of kinetic friction between particle and table is μ_k. In how much time particle will come to rest?

[**Hint:** $M_f = \mu_k \cdot mg \cdot R$]. [**Ans:** $t = \frac{V_0}{\mu_k \cdot g}$].

MULTIPLE CHOICE QUESTIONS

20.1 A trolley of 600 kg can move along a horizontal frictionless track, as shown in **Fig. 20.10**. Initially the trolley with a man on it of mass 70 kg is moving towards right at a speed of 5 m/s. If the man starts walking on the trolley with a speed of 2 m/s towards left, what is velocity of travel?

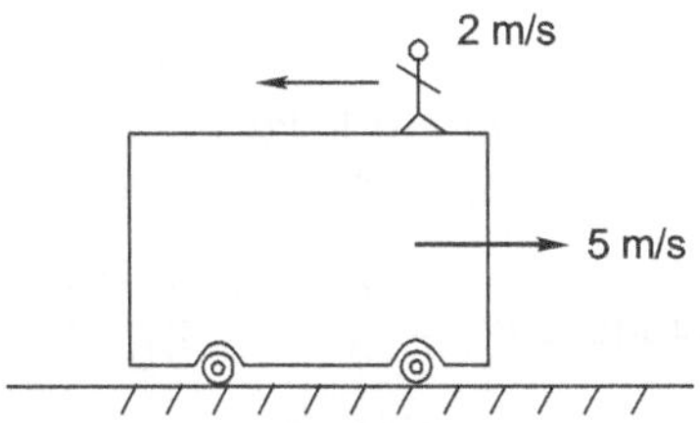

Fig. 20.10

(*a*) 4.79 m/s (*b*) 5.21 m/s
(*c*) 5.81 m/s (*d*) None of these.

20.2 A 32 gram bullet is fired with a velocity of 500 m/s into a wooden block which rests against a rigid vertical wall. The bullet is brought to rest in 0.1 s. What is average impulsive force exerted by the bullet on the block?

(*a*) 160 N (*b*) 1600 N
(*c*) 16 kN (*d*) None of these

20.3 A pile hammer of weight 12 kN strikes on pile of weight 4 kN with a velocity of 3 m/s. How much does a single blow of hammer drive the pile, if the resistance of ground is 200 kN, assume ground resistance to much the uniform

(*a*) 0.12 (*b*) 0.15 m
(*c*) 0.18 m (*d*) 0.36 m

20.4 During service, a tennis player hits the ball when it is on the top of its trajectory (when thrown by hand). The speed of the ball often being hit is 150 km/hour. If the contact of the ball with the racket is for 0.02 s, and weight of ball is 50 gm, what is average force applied by the player on the ball?

(*a*) 69.4 N (*b*) 104.16 N
(*c*) 10.41 N (*d*) None of these

20.5 A 5 kg homogeneous rotor with radius of gyration 0.4 m comes to rest in 60 seconds from a speed of 200 rpm, what was the frictional torque that stopped the rotor?

(*a*) 0.28 Nm (*b*) 0.56 Nm
(*c*) 0.7 Nm (*d*) None of these

20.6 A single degree of freedom system having mass 1 kg and stiffness 10 kN/m initially at rest is subjected to an impulse force of magnitude 5 kN for 10^{-4} seconds. The amplitude in mm of the resulting free vibration is

(*a*) 0.5 (*b*) 1.0
(*c*) 5.0 (*d*) 10

20.7 A player catcher a cricket ball of mass 0.1 kg moving with a speed of 20 m/s If the ball is in constant with his hand for 0.1 s, what is the impulse (approximate) exerted by the ball on the hand of the player?

(*a*) 2 Ns (*b*) 5.2 Ns
(*c*) 10.8 Ns (*d*) 12.5 Ns

[CSE, Prelim, CE : 2007]

20.8 A cricket ball of mass 150 gram moving with a velocity of 12 m/s is hit by a bat so that the ball is turned back with a velocity of 20 m/s. The force of blow acts for 0.01 s on the ball. What is the average force exerted by the bat on the ball?

(*a*) 480 N (*b*) 48 N
(*c*) 248 N (*d*) 48 N

[CSE, Prelim, CE : 2009]

20.9 Which one of the following is stated by the moment of momentum principle of a rotating system

(*a*) angular momentum is conserved
(*b*) vector sum of all external forces acting on a control volume in a fluid flow equals rate of change of linear momentum
(*c*) the resultant force exerted on a body is equal to the rate of change of angular momentum
(*d*) the torque due to resultant force is equal to the rate of change of angular momentum

[CSE, Prelim, CE : 2002]

20.10 A 100 kg flywheel having a radius of gyration of 1 m is rotating at 100 rpm. What is its angular momentum (approximate) of the flywheel about its axes of rotation is kgm^2/s

(*a*) 10000 (*b*) 10470
(*c*) 1000 (*d*) 12000

[CSE, Prelim, CE : 2007]

20.11 A bullet of mass 'm' travels at a very high velocity v (as shown in the figure) and gets embedded inside the block of mass "M" initially at rest on a rough horizontal floor. The block with the bullet is seen to move a distance "s" along the floor. Assuming μ to be the coefficient of kinetic friction between the block and the floor and "g" the acceleration due to gravity what is the velocity v of the bullet?

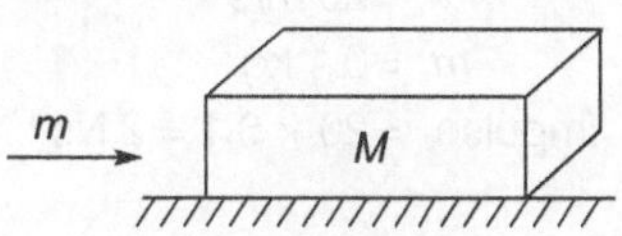

Fig. 20.11

(a) $\frac{M+m}{m}\sqrt{2\mu gs}$ (b) $\frac{M-m}{m}\sqrt{2\mu gs}$

(c) $\frac{\mu(M+m)}{mm}\sqrt{2\mu gs}$ (d) $\frac{M}{m}\sqrt{2\mu gs}$

[GATE 2003 : 1 Mark]

Answers

20.1 (b)	20.2 (a)	20.3 (c)	20.4 (b)	20.5 (a)
20.6 (c)	20.7 (a)	20.8 (a)	20.9 (d)	20.10 (b)
20.11 (a)				

EXPLANATIONS

20.1 (b)

$$(600 + 70)\,5 = (600 + 70 - 70 \times 2)\,V'$$

$$V' = \frac{3490}{670} = 5.21 \text{ m/s.}$$

20.2 (a)

$$F = \frac{mV}{dt} = \frac{0.032 \times 500}{0.1} = 160 \text{ N.}$$

20.3 (c)

$$V = \frac{MV_0}{M+m} = \frac{12 \times 3}{12+6} = 2 \text{ m/s}$$

$$\frac{1}{2}(M+m)\,V^2 = R.S$$

$$\frac{1}{2}(18) \times 2^2 = 200 \times S$$

$$36 = 200\text{ s}$$

$$S = 0.18 \text{ m.}$$

20.4 (b)

$$F = \frac{mV}{t} = 0.050 \times \frac{150{,}000}{3600} \times \frac{1}{0.02} = 104.16 \text{ N.}$$

20.5 (a)

$$\text{Torque} = I\alpha = 5 \times 0.4^2 \times \left(\frac{2\pi \times 200}{60 \times 60}\right) = 0.28 \text{ Nm.}$$

20.6 (c)

Impulse= $5000 \times 10^{-4} = 0.5$ Ns $= m.A.\omega$,

where $\omega = \sqrt{k/m}$, = 100 rad/s^2

A = amplitude, m = mass

0.5= $1 \times 100 \times A$, A = .005 m = 5 mm

20.7 (a)

$$v = 20 \text{ m/s}$$

$$m = 0.1 \text{ kg}$$

$$\text{Impulse} = 20 \times 0.1 = 2 \text{ Ns}$$

20.8 (a)

$$m = 0.15 \text{ kg}$$

$$v - u = 12 - (-20) = 32 \text{ m/s}$$

$$\text{Average force} = \frac{mV}{dt} = \frac{0.15 \times 32}{0.01} \text{ Ns}$$

$$= 480 \text{ Ns}$$

20.9 (d)

Angular momentum $= I\,\omega$

$$\text{Torque, } T = \frac{A}{dt}(I\omega)$$

$$= I - \frac{d\omega}{dt}$$

$$= I\alpha$$

= rate of change of angular momentum

20.10 (b)

$$I = 100 \times 1^2 = 100 \text{ kg m}^2$$

$$\omega = \frac{2\pi \times 1000}{60} = 104.72 \text{ rad/sec}$$

$$I_\omega = I\omega = 100 \times 104.72$$

$$= 10472 \text{ kgm}^2\text{/s}$$

20.11 (a)

$$mV = (m + M)V'$$

$$V'^2 = \left(\frac{mV}{m+M}\right)^2$$

$$0 = V'^2 = 2\mu gs = \left(\frac{mV}{m+M}\right)^2 - 2\mu gs$$

$$\left(\frac{mV}{m+M}\right)^2 = 2\mu gs$$

$$V = \frac{M+m}{m}\sqrt{2\mu gs}$$

■■■■

21

CHAPTER

Impact : Collision of Elastic Bodies

21.1 Introduction

Impact or collision is a *short duration phenomenon* which occurs when two bodies travelling with different velocities U_1 and U_2 collide with each other with a large impact force at the point of contact as shown in **Fig. 21.1 (*a*)** at contact point *P*.

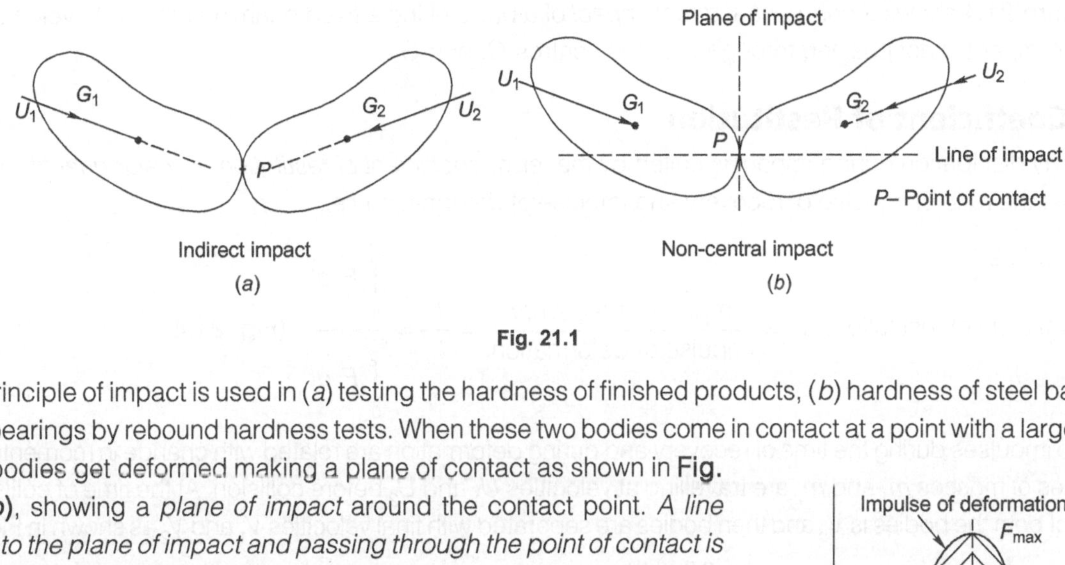

Fig. 21.1

Principle of impact is used in (*a*) testing the hardness of finished products, (*b*) hardness of steel balls used in ball bearings by rebound hardness tests. When these two bodies come in contact at a point with a large impact force, bodies get deformed making a plane of contact as shown in **Fig. 21.1 (*b*)**, showing a *plane of impact* around the contact point. *A line normal to the plane of impact and passing through the point of contact is known as* ***line of impact***. This deformation is partly or fully recovered and bodies get separated travelling with final velocities V_1 and V_2, other than initial velocities U_1 and U_2 as shown. Total time of impact dt is composed of time of deformation dt_1 and time of recovery dt_2 *i.e.*,

$$dt = dt_1 + dt_2 \quad ...(1)$$

Total time of impact = time of deformation + time of recovery

If the impact is elastic, then all the deformation developed during impact is fully recovered, but if the impact is fully plastic, then no deformation is recovered. In the case of elasto plastic impact, a part of deformation is recovered as shown in **Fig. 21.2.** High impulsive forces are generated during the short duration of deformation and recovery. These

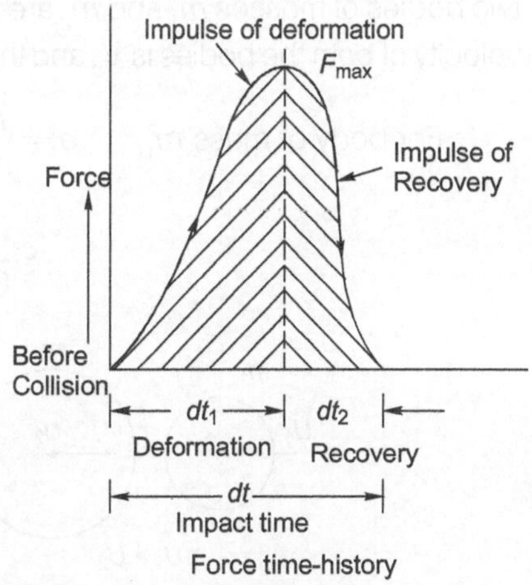

Fig. 21.2

impulses of deformation and recovery are extremely small but finite. There is a step change in the velocities of two bodies.

During collision, the bodies experience different dissipative phenomena such as wave propagation, vibrations, partial plastic deformation (as explained above), slipping under friction, radiation of sound, heat or light. *During the period of deformation the contact force develops from zero to maximum value when collision begins and this force diminishes to zero value during the period of recovery when collision ends.*

Figure 21.1 (*a*) shows the bodies colliding, while travelling with velocities U_1 and U_2. This type of impact is known as **indirect** or **oblique impact**. If both the velocities are collinear as shown in **Fig. 21.3**, then the impact is *direct*. Moreover *the line of impact passes through the centres of both the bodies*, then it is known as **central impact**. But if the line of impact does not pass through centres of both the bodies the impact is known as **non-central impact** as shown in **Fig. 21.3 (*b*)**.

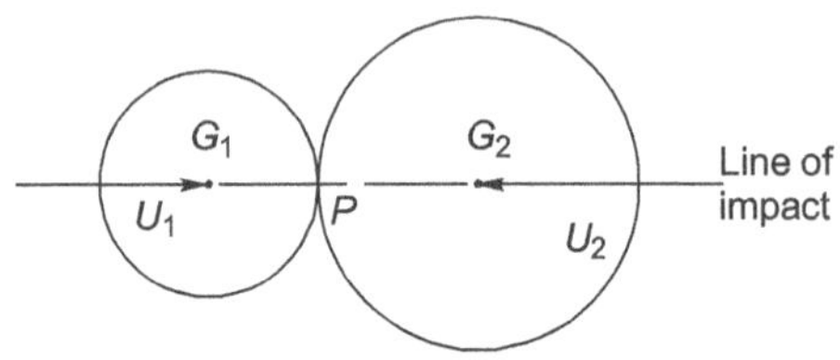

Direct impact and central impact

Fig. 21.3

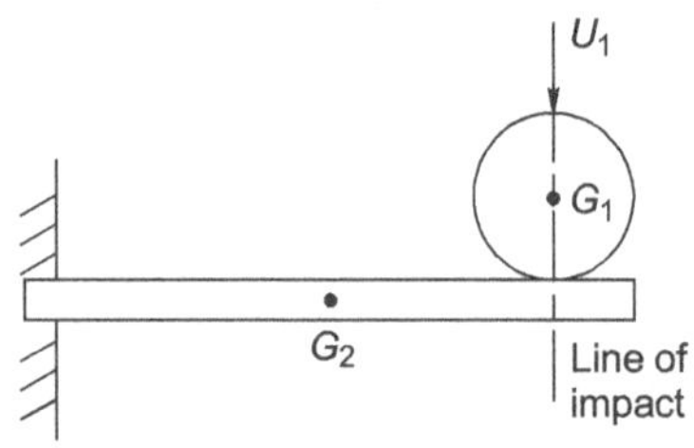

Non-central impact

Fig. 21.4

Figure 21.4 shows a *non-central direct impact* of a ball striking a fixed beam member with velocity U_1, as the line of impact is not passing through both the centres G_1 and G_2.

21.2 Coefficient of Restitution

The energy dissipation during impact is called by the term, *coefficient of restitution, a scalar quantity*, which is defined as the ratio of impulse of recovery and impulse of deformation *i.e.*,

$$\text{Coefficient of restitution, } e = \frac{\text{Impulse of recovery}}{\text{Impulse of deformation}} = \frac{I_r}{I_d} = \frac{\int_{dt_1}^{dt} F.dt}{\int_{0}^{dt_1} F.dt}, \text{ (Fig. 21.4)}$$

The impulses during the time of recovery and during deformation are related with change in momentum. Say two bodies of masses m_1 and m_2 are travelling at velocities U_1 and U_2 before collision. At the time of collision say velocity of both the bodies is V_c and then bodies are separated with final velocities V_1 and V_2 as shown in **Fig. 21.5.**

$$\text{For body of mass } m_1, \quad e = \frac{m_1(V_1 - V_c)}{m_1(V_c - U_1)}, \text{ along the line of impact}$$

$$= \frac{\text{Change of momentum during recovery}}{\text{Change of momentum during deformation}}$$

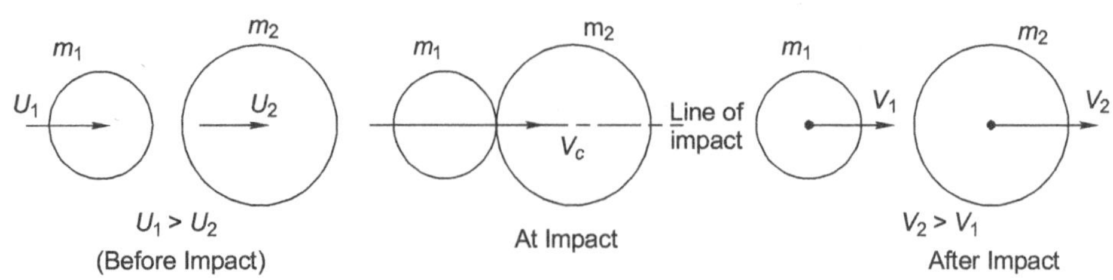

Fig. 21.5

For body of mass m_2, $e = \dfrac{m_2(V_2 - V_c)}{m_2(V_c - U_2)}$, along the line of impact

or $e = \dfrac{V_1 - V_c}{V_c - U_1} = \dfrac{V_2 - V_c}{V_c - U_2}$, eliminating, V_c

$$= \frac{V_1 - V_c - V_2 + V_c}{V_c - U_1 - V_c + U_2} = \frac{V_1 - V_2}{U_2 - U_1} \quad ...(2)$$

$$= \frac{V_2 - V_1}{U_1 - U_2}$$

$$= \frac{\text{Velocity of separation along the line of impact}}{\text{Velocity of approach along the line of impact}}$$

Using the principle of conservation of momentum.

Momentum of bodies before collision

= Momentum of bodies after impact

i.e., $m_1U_1 + m_2U_2 = m_1V_1 + m_2V_2$ ...(3)

For a direct impact, final velocities of the bodies after impact can be determined from Equations (2) and (3).

Example 21.1 Two steel balls of masses 2 kg and 4 kg respectively collide with each other with initial velocities of 3 m/s and 2 m/s as shown in **Fig. 21.6**. If the impact is perfectly elastic, find the velocities of the balls after impact. Determine final velocities of two masess.

Solution Masses $m_1 = 2$ kg

$m_2 = 4$ kg

Initial velocities $U_1 = 3$ m/s

$U_2 = 2$ m/s

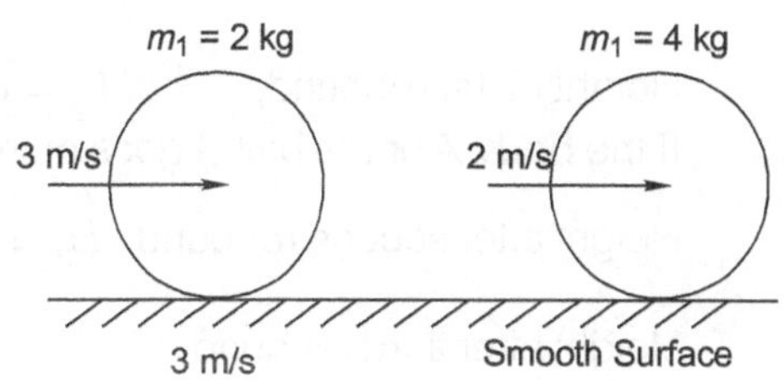

Fig. 21.6

Impact is elastic, therefore coefficient of restitution, $e = 1$.

Say the final velocities of the bodies are V_1 and V_2 m/s, then using the principle of conservation of momentum

$$m_1U_1 + m_2U_2 = m_1V_1 + m_2V_2$$

putting the values of masses

or $2 \times 3 + 4 \times 2 = 2V_1 + 4V_2$

or $2V_1 + 4V_2 = 14$ ...(1)

Coefficient of restitution, $e = \dfrac{V_2 - V_1}{U_1 - U_2} = \dfrac{\text{Velocity of separation}}{\text{Velocity of approach}}$, because impact is perfectly elastic

$$1 = \frac{V_2 - V_1}{3 - 2}$$

or $V_2 - V_1 = 1$ ...(2)

From Equations (1) and (2)

$V_2 = 2.667$ m/s; $V_1 = 1.667$ m/s

Exercise 21.1 Two identical steel balls of same mass collide with velocities 3 m/s and –3 m/s as shown in the **Fig. 21.7**. If coefficient of restitution is 0.7. What is the final velocity of each ball?

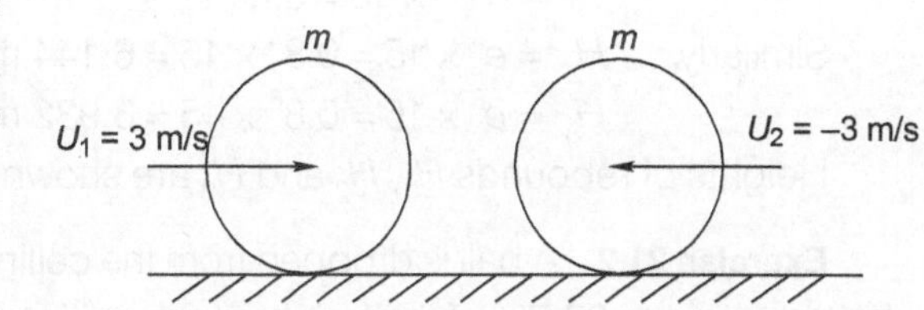

Fig. 21.7

[**Hint:** $e = 0.7 = \dfrac{V_2 - V_1}{U_1 - U_2}$].

[**Ans:** $V_2 = 2.1$ m/s, $V_1 = -2.1$ m/s].

21.3 Impact of a Body on another Stationery Body

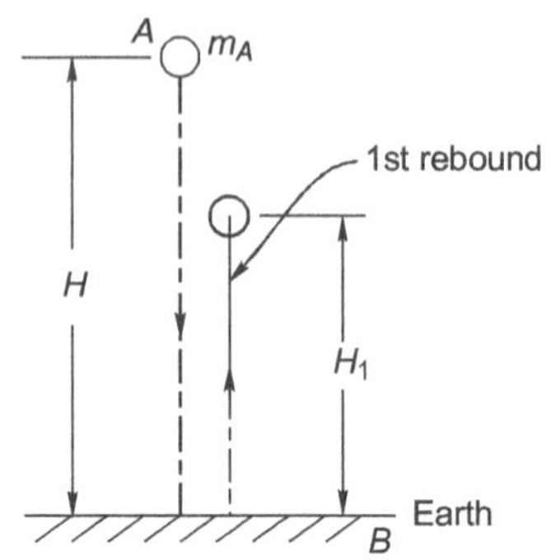

Fig. 21.8

Let us consider a body A of mass m_A, falling under gravity through height H and colliding with the ground *i.e.*, earth surface whichis stationery (**Fig. 21.8**). Initial velocity of body A is

$$U_A = \sqrt{2gH} \quad \text{(taking downward direction positive)}$$

Initial velocity of earth

$$U_B = 0$$

Say V_A is the final velocity of the body A after impact and V_B is the final velocity of the earth which is zero.

Coefficient of restitution, $$e = \frac{V_B - V_A}{U_A - U_B} = \frac{-V_A}{U_A}$$

as stationery ground $V_B = U_B = 0$

If $e = 1$, elastic impact.

Final velocity of body A after impact

$$V_A = -U_A \quad \text{...(1)}$$

Otherwise final velocity of body A after impact

$$V_A = -e \,.\, U_A \quad \text{...(2)}$$

Height of rebound $$H_1 = \frac{V_A^2}{2g} = \frac{e^2 U_A^2}{2g}$$

Height of 1st rebound, $H_1 = e^2 H$

If the body A or the ball A goes on rebounding again and again.

Height after second rebound, $H_2 = e^2 H_1 = e^4 H$

Height after third rebound, $H_3 = e^2 H_2 = e^6 H$

or height after n rebounds, $H_n = e^{2n} \,.\, H.$

Example 21.2 A golf ball is dropped from a height of 15 m on to a fixed steel plate. The coefficient of restitution is 0.8. Find the height to which the ball rebounds as the first, second and third bounces.

Solution Height from which the ball is allowed to fall

$$H = 15 \text{ m}$$

Coefficient of restitution of ball and stationery steel plate

$$e = 0.80$$

Height of the ball after n rebounds

$$H_n = e^{2n} \,.\, H$$

Therefore height of the ball after 1st rebound

$$H_1 = e^2 \times 15$$
$$= 0.8^2 \times 15 = 9.6 \text{ m}$$

Similarly $$H_2 = e^4 \times 15 = 0.8^4 \times 15 = 6.144 \text{ m}$$
$$H_3 = e^6 \times 15 = 0.8^6 \times 15 = 3.932 \text{ m}$$

Heights of rebounds H_1, H_2 and H_3 are shown in **Fig. 21.10.**

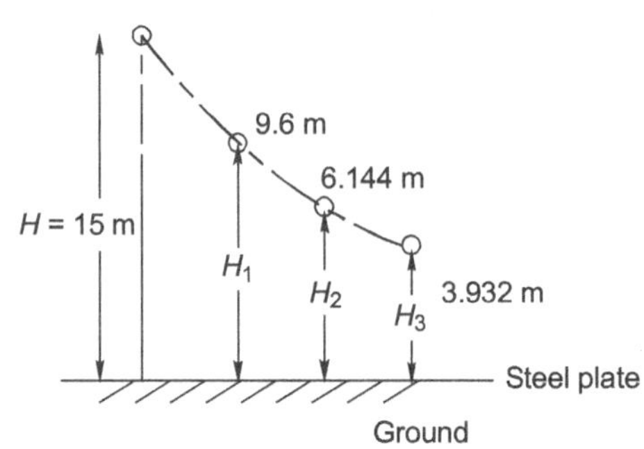

Fig. 21.9

Exercise 21.2 A ball is dropped from the ceiling of a room. After rebounding thrice from the floor it reaches a height equal to 37.7% of the height of the ceiling from the floor. What is the coefficient of restitution?
[**Ans:** $e = 0.85$].

PROBLEMS

Problem 21.1 A bullet of mass 20 gram strikes and gets embedded in a stationery block of mass 5 kg which moves horizontally after impact. If the bullet is travelling horizontally at 500 m/s, what is the velocity of the block after impact? What percentage of kinetic energy is lost in impact?

Solution Mass of bullet, $m = 0.020$ kg

Velocity of bullet, $V = 500$ m/s

Mass of block, $M = 5$ kg

Velocity of block and bullet. Say V'

$$(m+M)\,V' = mV = 0.020 \times 500$$

$$V' = \frac{10}{5.020} = 1.992 \text{ m/s}$$

Total mass, $m + M = 0.02 + 5 = 5.02$ kg

Velocity, $V' = 1.992$ m/s

Final KE $= \frac{1}{2}(M+m)\,V'^2 = \frac{1}{2} \times 5.02 \times 1.992^2 = 9.96$ Nm

Initial KE $= \frac{1}{2}mV^2 = \frac{1}{2} \times 0.02 \times 500^2 = 2500$ Nm

% of energy lost $= \frac{2500 - 9.96}{2500} = 0.996 = 99.6\%.$

Problem 21.2 A sphere mass 1 kg moving with velocity of 3 m/s overtakes another sphere of mass 5 kg moving in the same line at 0.6 m/s. Find the loss of kinetic energy during inspact and show that the direction of motion of the first sphere is reversed. Take coefficient of restittion as 0.75.

(UPSC Civil Services 1985)

Solution $m_1 = 1$ kg $u_1 = 3$ m/s $e = 0.75$

$m_2 = 5$ kg $u_2 = 0.6$ m/s

$$m_1 v_1 + m_2 v_2 = m_1 V_1 + m_2 V_2$$

$$1 \times 3 + 5 \times 0.6 = v_1 + 5v_2$$

$$v_1 + 5v_2 = 6 \qquad ...(1)$$

$$e = \frac{V_2 - V_1}{u_1 - u_2} = 0.75$$

$$\frac{V_2 - V_1}{3 - 0.6} = 0.75$$

$$V_2 - V_1 = 1.8 \qquad ...(2)$$

$$\mathbf{5V_2 + V_1 = 6}$$

$$V_2 = 1.3 \text{ m/s}$$

$$V_1 = -0.5 \text{ m/s}$$

Loss of KE $= \frac{1}{2}[1 \times 3^2 + 5 \times 0.6^2] - \frac{1}{2}[1 \times (-0.5)^2 + 5\,(1.3)^2]$

$= 5.4 - 4.35 = 1.05$ Nm.

Problem 21.3 A 8 kg block *A* moves on a rough horizontal floor and hits a sphere *B* of mass 16 kg, which is suspended vertically by a string of length 2 m as shown in **Fig. 21.10.** The coefficient of friction between block *A* and floor is 0.25. The coefficient of restitution *e* for impact between block and sphere is 0.75. If the duration of impact is 10 ms. Determine impulsive force.

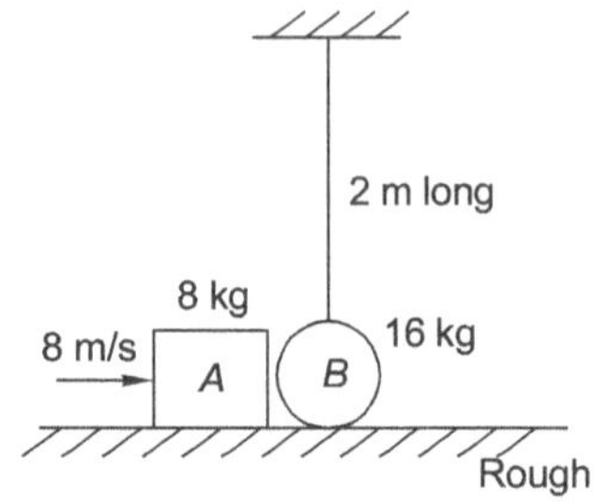

Fig. 21.10

Solution Say initial and final velocities of block *A* are U_A and V_A

Initial and final velocities of sphere *B* are U_B and V_B

Now velocity, $U_B = 0$

$U_A = 8$ m/s

$m_A = 8$ kg, $m_B = 16$ kg

Using the principle of conservation of momentum

$$m_A U_A + m_B U_B = m_A V_A + m_B V_B$$

$$8 \times 8 + 16 \times 0 = 8V_A + 16V_B$$

$$V_A + 2V_B = 8 \quad \text{...(1)}$$

Coefficient of restitution, $$e = -\left(\frac{V_B - V_A}{U_B - U_A}\right)$$

$$0.75 = -\left(\frac{V_B - V_A}{0 - 8}\right)$$

$$V_B - V_A = 6 \quad \text{...(2)}$$

$$2V_B + V_A = 8 \quad \text{...(1)}$$

Velocities after impact $$V_B = \frac{14}{3} \text{ m/s}$$

$$V_A = 8 - \frac{28}{3} = -\frac{4}{3} \text{ m/s}$$

(*a*) Impulsive force on sphere,

$$P = m(V_B - U_B), \text{ but } u_B = 0$$

$$= 16 \times \frac{14}{3} = 74.66 \text{ Ns.}$$

Problem 21.4 A ball is dropped from a height $h_0 = 2$ m on a smooth floor as shown in **Fig. 21.11**. For the first bounce $h_1 = 1.5$ m, determine coefficient of restitution

Solution Ball is allowed to fall from height,

$h_0 = 2.0$ m

Vertical component of velocity before impact,

$$U_1 = \sqrt{2gh} = \sqrt{2 \times 9.81 \times 2.0}$$

$$= 6.264 \text{ m/s}$$

Let the vertical component of velocity after first bounce is $V_1 y$

then $$V_{1y} = \sqrt{2gh_1} = \sqrt{2 \times 9.81 \times 1.5}$$

$= 5.425$ m/s (putting the value of h_1)

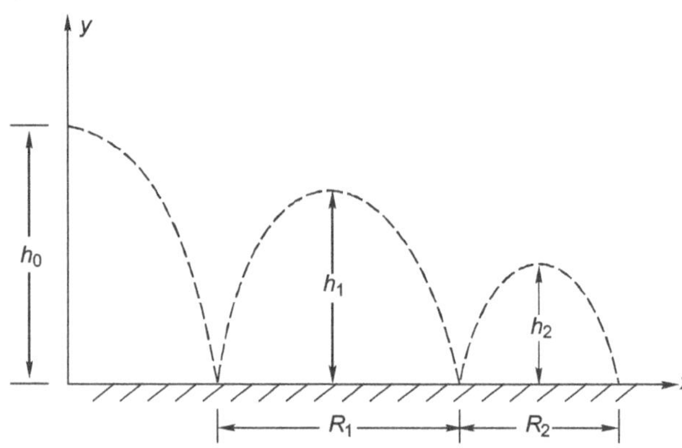

Fig. 21.11

Coefficient of restitution,

$$e = \frac{\text{Velocity of separation}}{\text{Velocity of approach}} \text{ (floor remains stationary)}$$

$$e = \frac{5.425}{6.264} = 0.866$$

Remember

- Impact or collision is a short duration phenomenon which occurs when two bodies travelling with different velocities collide with each other with a large impact force at the point of contact.
- During the period of deformation the contact force develops from zero to maximum value when collision starts and this force diminishes to zero value during the period of recovery when collision ends.
- Coefficient of restitution, $e = \dfrac{\text{Impulse of recovery}}{\text{Impulse of deformation}}$

$$= \frac{\text{Velocity of separation}}{\text{Velocity of approach}}$$

$$= \frac{V_1 - V_2}{U_2 - U_1} = \frac{V_2 - V_1}{U_1 - U_2}$$

where U_1, V_1 and U_2, U_2 are initial and final velocities of two bodies.

PRACTICE PROBLEMS

21.1 From what height H, a ball A of mass 1 kg should be released from rest sliding down along the smooth surface of a bowl, so that after impact ball B of mass 0.35 kg resting at the bottom of the bowl just leaves the bowl covering a height of 0.25 m as shown in **Fig. 21.12**. Take coefficient of restitution is 0.75.

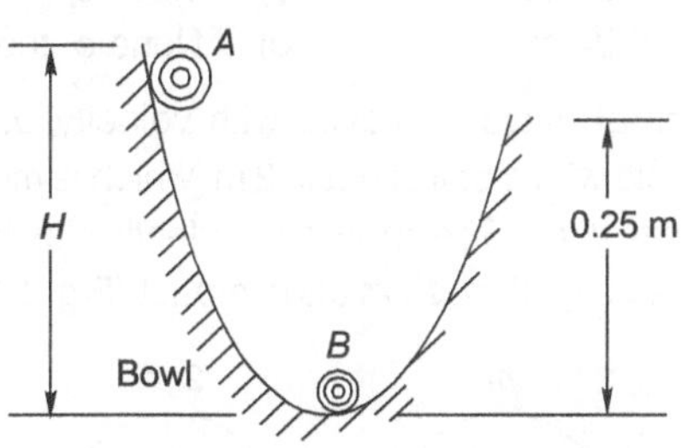

Fig. 21.12

[**Hint:** $U_A = \sqrt{2gH}$, $V_B = \sqrt{2g \times 0.25}$, $U_B - 0$]

[**Ans:** $H = 0.1487$ m].

21.2 The masses of two balls are in the ratio of 2:1 and their respective velocities are in the ratio of 1:2 but in opposite direction before impact. If the coefficient of restitution for these two balls is 0.5, show that after the impact, each ball will move back with half of its original velocity.

21.3 A 8 kg block A moves on a rough horizontal floor and hits a sphere of mass 10 kg and suspended by a string 1.6 m long as shown in the **Fig. 21.13**. The coefficient of friction between block A and floor is 0.20. The coefficient of restitution e for impact between block and sphere is 0.70. If the duration of impact is 0.02 second, determine impulsive force.

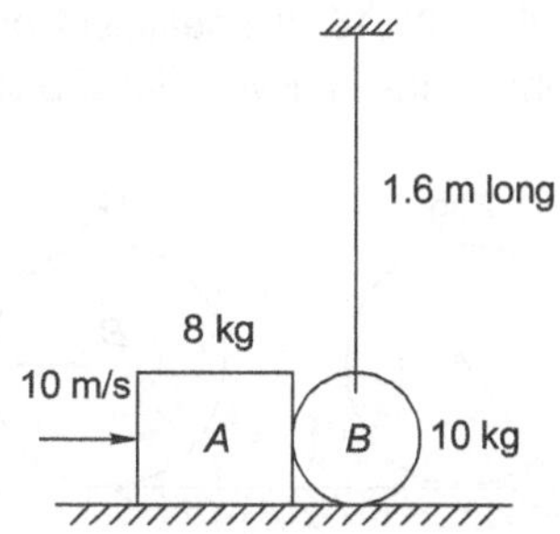

Fig. 21.13

[**Ans:** $P = 3777.5$ N].

21.4 A ball falls from a height H on a fixed horizontal floor. If e is the coefficient of restitution, show that the total distance travelled by the ball after it has finished rebounding is $\left(\dfrac{1+e^2}{1-e^2}\right)H$.

21.5 A ball impinges directly on a similar ball at rest. The first ball is reduced to rest by the impact. Find the coefficient of restitution, if half of the initial kinetic energy is lost by impact (**Fig. 21.14**).

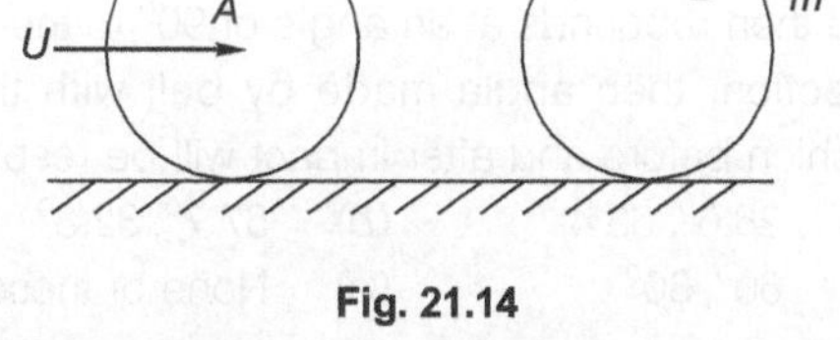

Fig. 21.14

[**Ans:** Note $U_B = 0$, you may find out that $V_A = 0$].

[**Ans:** $e = 0.707$].

MULTIPLE CHOICE QUESTIONS

21.1 Two steel balls of masses 2 kg each collide with each other with initial velocities of 3 m/s and 2 m/s as shown. If impact is perfectly elastic, what is the velocity of ball *A* after impact (**Fig. 21.15**)?

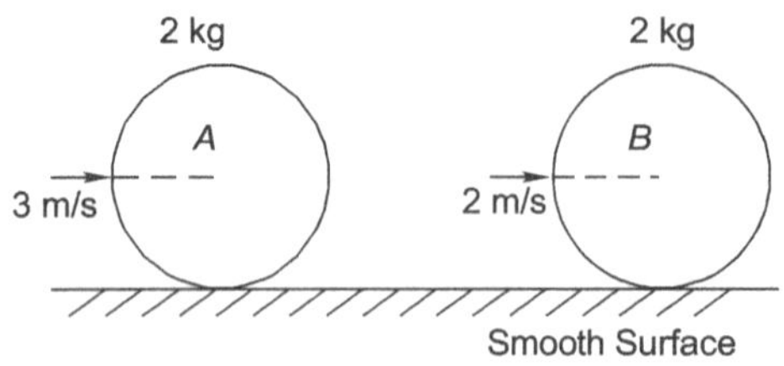

Fig. 21.15

(*a*) 2.5 m/s (*b*) 2 m/s
(*c*) –2 m/s (*d*) None of these.

21.2 Two steel balls of mass 2 kg each collide with each other with initial velocities of 3 m/s and –2 m/s as shown in **Fig. 21.16**. If coefficient of restitution is $e = 0.6$, what is the final velocity of ball *B*?

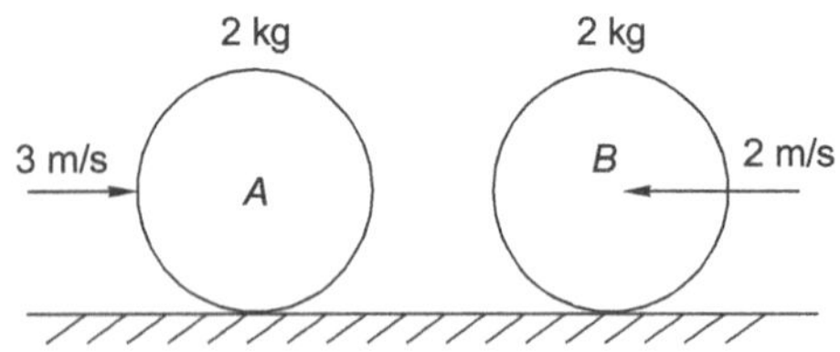

Fig. 21.16

(*a*) 2 m/s (*b*) 2.5 m/s
(*c*) 3 m/s (*d*) None of these

21.3 For a perfectly elastic collision between two bodies, what is the value of coefficient of restitution?
(*a*) 0 (*b*) 0.5
(*c*) 1 (*d*) None of these

21.4 A ball is dropped from a height of 3 m on a smooth floor which attains a bounce of 1.3 m. What is the coefficient of restitution between ball and floor?
(*a*) 0.43 (*b*) 0.66
(*c*) 0.81 (*d*) None of these

21.5 The coefficient of restitution between a snooker ball and side cushion is 0.4. If the ball hits the cushion and then rebounds at an angle of 90° to the original direction, then angle made by ball with the side cushion before and after impact will be respectively
(*a*) 26.6°, 63.4° (*b*) 57.7°, 32.3°
(*c*) 60°, 30° (*d*) None of these

21.6 After perfectly inelastic collision between two balls of same mass and same speed moving in different directions, the speed of the balls becomes one-third of the initial speed. What is the angle between the two balls before collision?
(*a*) 90° (*b*) 120°
(*c*) 141° (*d*) 150°

21.7 A ball is moving towards a thick masonry wall at 3 m/s, while the wall is moving towards the ball at the speed of 1 m/s. Assuming the collision between ball and wall to be elastic, what is the velocity of the ball immediately after impact?
(*a*) +2 m/s (*b*) –3 m/s
(*c*) –4 m/s (*d*) None of these

21.8 A golf ball is dropped on a cement side pavement from a height *H*. After impact, it rebounds by 0.9*H*, what is coefficient of restitution?
(*a*) 0.9 (*b*) 0.925
(*c*) 0.95 (*d*) None of these

21.9 A man weighing 600 N runs and jumps from a pier into a boat of weight 1000 N. Man jumps into boat with a horizontal velocity of 2.5 m/s. Assuming the impact to be perfectly plastic, man and boat will move away from the pier with what velocity?
(*a*) 1.5 m/s (*b*) 1.2 m/s
(*c*) 0.94 m/s (*d*) None of these

21.10 A ball of mass *m* moves with velocity u_1 = 4 m/s impacts with a ball of mass 2 m, which is moving with velocity 2 m/s. The impact is perfectly elastic. What is the velocity of mass *m* after impact (**Fig. 21.17**)

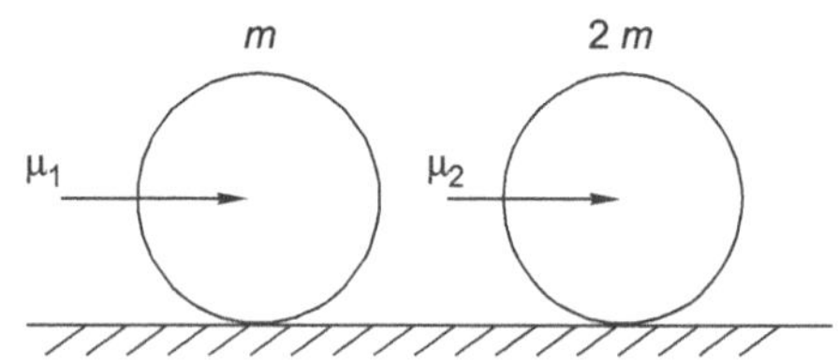

Fig. 21.17

(*a*) 1.33 m/s (*b*) 2.33 m/s
(*c*) 3.33 m/s (*d*) None of these

21.11 Time taken by bodies to regain original shape after compression due to impact is known as
(*a*) time of deformation
(*b*) time of restitution
(*c*) time of collision
(*d*) None of these.

21.12 If the coefficient of restitution is zero during the impact of two bodies, the two bodies are
(*a*) inelastic (*b*) elastic
(*c*) elastic plastic (*d*) None of these.

21.13 A toy ball is shot to hit the wall of a room as shown in the **Fig. 21.18**. The ball then returns back to the point of projection. If the time taken by the ball is returning is twice the time taken in reaching the wall, what is coefficient of restitution?

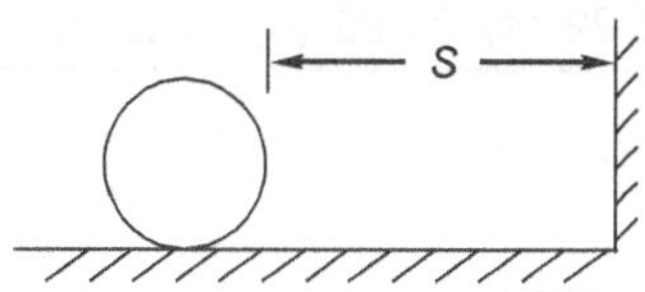

Fig. 21.18

(*a*) 1.0 (*b*) 0.707
(*c*) 0.50 (*d*) 0.25

21.14 A bullet of mass 20 gram strikes and gets embedded in a stationary block of 5 kg which moves horizontally after impact. If the bullet is travelling horizontally at 500 m/s, what is the velocity of the block after impact? What is the percentage of energy loss?

(*a*) 1.942 m/s, 91% (*b*) 1.992 m/s, 99.6%
(*c*) 2 m/s, 96% (*d*) None of these.

21.15 Ball *A* of mass 2 kg moving with a velocity of 2 m/s impinges directly on a ball *B* of mass 4 kg at rest as shown in **Fig. 21.19**. If coefficient of restitution is 0.5, what is the velocity of two balls after impact?

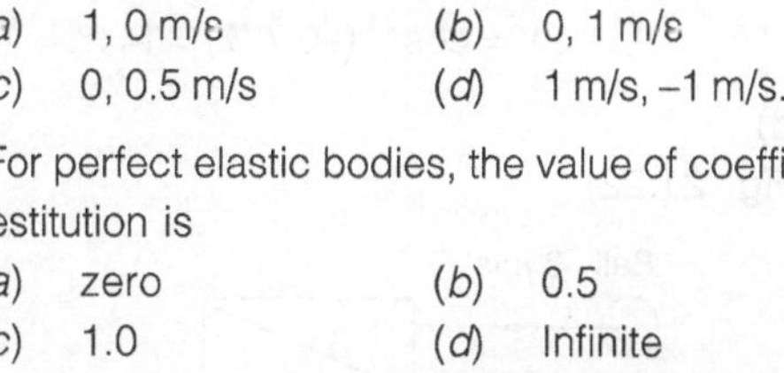

Fig. 21.19

(*a*) 1, 0 m/s (*b*) 0, 1 m/s
(*c*) 0, 0.5 m/s (*d*) 1 m/s, –1 m/s.

21.16 For perfect elastic bodies, the value of coefficient of restitution is

(*a*) zero (*b*) 0.5
(*c*) 1.0 (*d*) Infinite

[CSE, Prelim, CE : 2002]

21.17 A bomb of 10 kg explodes into two pieces of makes 3 kg and 7 kg with negligible shock effect. If the velocity of the first part is 210 m/s, the velocity of the other part is

(*a*) 210 m/s (*b*) 105 m/s
(*c*) 90 m/s (*d*) zero

[CSE, Prelim, CE : 2003]

21.18 A ball of mass 1 kg moving with a velocity of 2 m/s collides directly with a stationary ball of mass 2 kg and comes to rest after impact. The velocity of the second ball after impact will be

(*a*) 0 (*b*) 1 m/s
(*c*) 2 m/s (*d*) 4 m/s

[CSE, Prelim, CE : 2003]

21.19 A 80 N body moving to the right at a speed of 3 m/s strikes a 10 N body moving to the left at a speed of 10 m/s. The final velocity of 10 N body is 4 m/s to the right. Then the final velocity of 80 N body will be

(*a*) 1.25 m/s (*b*) 3 m/s
(*c*) 4 m/s (*d*) 10 m/s

[CSE, Prelim, CE : 2006]

21.20 A ball of mass 1 kg moving with a velocity of 2 m/s collides directly with a stationary ball of mass 2 kg. If the first ball comes to rest after impact, what is the velocity of the second ball after impact?

(*a*) 1.0 m/s (*b*) zero
(*c*) 2 m/s (*d*) 0.5 m/s

[CSE, Prelim, CE : 2005]

21.21 A ball is allowed to fall freely from a height 'h' on fixed state. What is the total distance traversed by the ball before coming to rest if e = 0.5?

(*a*) $\frac{h}{4}$ (*b*) $\frac{5h}{3}$
(*c*) $\frac{h}{2}$ (*d*) $\frac{3h}{4}$

[CSE, Prelim, CE : 2010]

21.22 The coefficient of restitution of a perfectly plastic impact is

(a) 0 (b) 1
(c) 2 (d) ∞

[GATE 2011 : 1 Mark]

21.23 A mass m_1 of 100 kg travelling with a uniform velocity of 5 m/s along a line collides with a stationary mass m_2 of 1000 kg. After the collision, both the masses travel together with the same velocity. The coefficient of restitution is

(a) 0.6 (b) 0.1
(c) 0.01 (d) 0

[GATE 2014 : 1 Mark (set-3)]

21.24 A ball A of mass m falls under gravity from a height h and strikes another ball B of mass m which is supported at rest on a spring of stiffness k. Assume perfect elastic impact, immediately after the impact

(a) the velocity of ball A is $\frac{1}{2}\sqrt{2gh}$

(b) the velocity of ball A is zero

(c) the velocity of both balls is $\frac{1}{2}\sqrt{2gh}$

(d) None of the above

[GATE 1996 : 1 Mark]

Answers

21.1 (b)	21.2 (a)	21.3 (c)	21.4 (b)	21.5 (b)
21.6 (c)	21.7 (d)	21.8 (c)	21.9 (c)	21.10 (a)
21.11 (b)	21.12 (a)	21.13 (c)	21.14 (b)	21.15 (b)
21.16 (c)	21.17 (c)	21.18 (b)	21.19 (a)	21.20 (a)
21.21 (b)	21.22 (a)	21.23 (d)	21.24 (b)	

EXPLANATIONS

21.1 (b)

$$m_1 = 2\text{ kg} \quad m_2 = 2\text{ kg}$$
$$u_1 = 3\text{ m/s} \quad u_2 = 2\text{ m/s}$$
$$m_1 u_1 + m_2 u_2 = 2\times 3 + 2\times 2 = 10$$
$$e = \frac{V_2 - V_1}{u_1 - u_2} = \frac{V_2 - V_1}{3-2}$$
$$1 = V_2 - V_1 \quad V_2 = V_1 + 1$$

Now, $m_1 V_1 + m_2 V_2 = m_1 u_1 + m_2 u_2 = 10$

$$2V_1 + 2V_2 = 10$$
$$V_1 + V_2 = 5$$
$$V_1 + V_1 + 1 = 5$$
$$V_1 = 2\text{ m/s}$$
$$V_2 = 3\text{ m/s.}$$

21.2 (a)

$$m_1 u_1 + m_2 u_2 = m_1 V_1 + m_2 V_2$$
$$2\times 3 - 2\times 2 = 2V_1 + 2V_2$$
$$2 = 2V_1 + 2V_2$$
$$V_1 + V_2 = 1 \quad (1)$$
$$e = 0.6 = \frac{V_2 - V_1}{u_1 - u_2} = \frac{V_2 - V_1}{3-(-2)}$$
$$V_2 - V_1 = 3$$
$$V_2 + V_1 = 1$$
$$V_2 = 2\text{ m/s}$$
$$V_1 = -1\text{ m/s.}$$

21.3 (c)

For perfectly elastic collison, $e = 1$.

21.4 (b)

$$1.3 = e^2 (3)$$
$$e = \sqrt{\frac{1.3}{3}} = 0.66.$$

21.5 (b)

(Fig. 21.20)

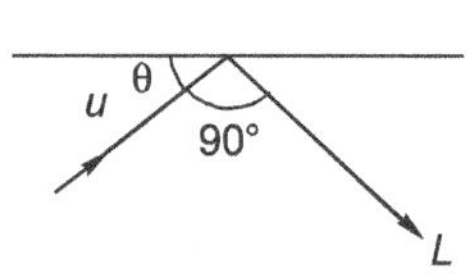

Fig. 21.20

$$\frac{u\cos\theta}{V} = \tan\theta$$
$$V = 0.4u\sin\theta$$
$$\tan\theta = \frac{u\cos\theta}{0.4u\sin\theta} = \frac{1}{0.4\tan\theta}$$
$$\tan^2\theta = 2.5,\ \tan\theta = 1.5811,\ \theta = 57.7^\circ.$$

21.6 (c)

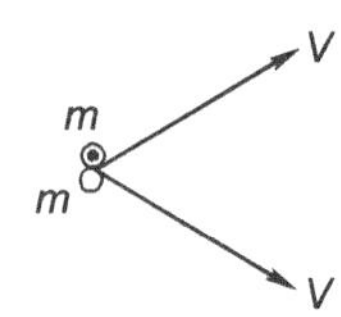

Fig. 21.21

$$\left[2m\left(\frac{V}{3}\right)\right]^2 = (mv)^2 + (mv)^2 + 2(mv)\cos\theta$$
$$\frac{4}{9}m^2v^2 = m^2v^2 + m^2v^2 + 2m^2v^2\cos\theta$$

or $4/9 = 1 + 1 + 2\cos\theta$

$$2\cos\theta = -2 + \frac{4}{9} = -\frac{14}{9}$$
$$\cos\theta = -\frac{7}{9} = -0.777$$
$$\theta = \cos^{-1}(-0.777) = 141^\circ.$$

21.7 (d)

(Fig. 21.22)

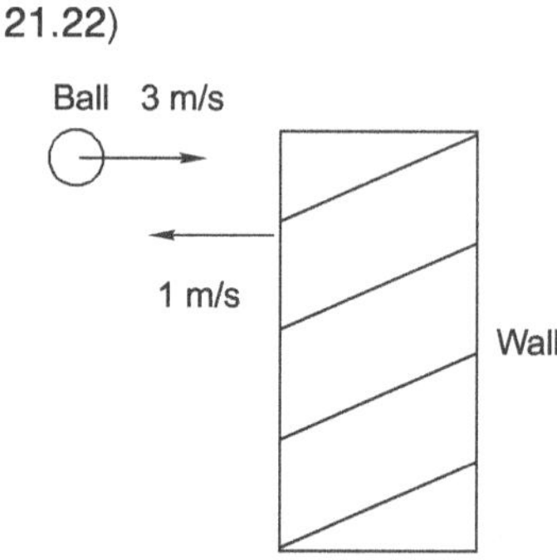

Fig. 21.22

$$e = \frac{3+1}{-1-v_1} = 1$$

$$4 = -1 - V_1$$
$$5 = -V_1$$
$$V_1 = -5 \text{ m/s.}$$

21.8 (c)

$$e = \sqrt{\frac{0.95H}{H}} = 0.95.$$

21.9 (c)

$$600 \times 2.5 = (600 + 1000)\ V$$
$$V = 0.94 \text{ m/s.}$$

21.10 (a)

$$e = 1 = \frac{u_1 - u_2}{V_2 - V_1} = \frac{4-2}{V_2 - V_1} = 1 \text{ or } V_2 - V_1 = 2 \qquad (1)$$

$$4m + 4m = mV_1 + 2mV_2$$
$$8 = V_1 + 2V_2$$
$$2 = V_2 - V_1$$

or $V_2 = 10/3$ m/s. $V_1 = V_2 - 2 = 4/3$ m/s.

21.11 (b)

Time of restitution to regain original shape.

21.12 (a)

$e = 0$, inelastic.

21.13 (c)

$$e = \frac{V_2 - V_1}{u_1 - u_2} = \frac{-V_1}{u_1} \text{ as } V_2 = u_2 = 0$$

$$V_1 = -0.5u_1$$
$$e = 0.5.$$

21.14 (b)

$$V' = \frac{102 \times 500}{50 + 0.02} = 1.992 \text{ m/s}$$

$$KE_e = \frac{1}{2} \times 0.02 \times 500^2 - \frac{1}{2} \times 5.02 \times 1.992^2$$
$$= 2500 - 9.96 = 2490.04$$

$$\% \text{ of energy loss} = \frac{2490.04}{2500} = 99.6\%$$

21.15 (b)

$$2 \times 2 = 2V_1 + 4V_2 \qquad ...(1)$$
$$V_1 + 2V_2 = 2 \qquad ...(1)$$

$$e = 0.5 = \frac{V_2 - V_1}{u_1 - u_2} = \frac{V_2 - V_1}{2 - 0}$$

$$V_2 - V_1 = 1$$
$$V_1 + 2V_2 = 2$$

So $V_2 = 1$ m/s, $V_1 = 0$.

21.16 (c)

Perfectly elastic bodies, coefficient of restitution is 1.

21.17 (c)

$$m_1V_1 = m_2V_2$$
$$3 \times 210 = 7 \times V_2$$
$$V_2 = 90 \text{ m/s}$$

21.18 (b)

$$m_1V_1 + m_2V_2 = m_1V_1 + m_2V_2$$
$$u_1 = 2 \text{ m/s}, m_1 = 1 \text{ kg}, u_2 = 0, V_1 = 0$$
$$1 \times 2 = 2 \times V_2$$
$$V_2 = 1 \text{ m/s}$$

$u_1 = 2$ m/s

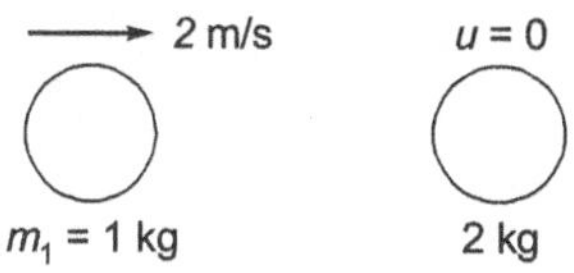

21.19 (a)

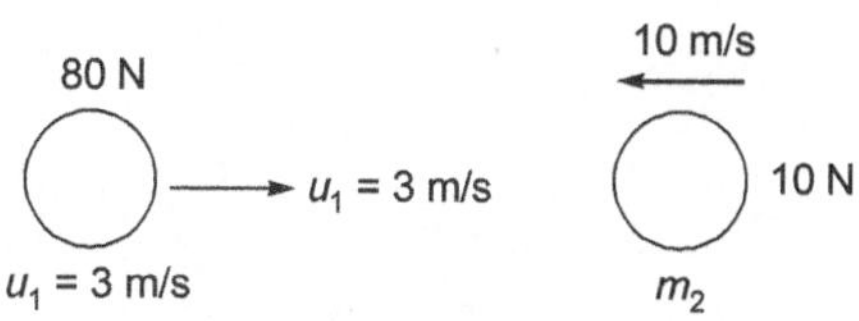

$$80 \times 3 - 10 \times 10 = 240 = m_1V_2 + m_2V_2$$
$$V_2 = \overline{4 \text{ m/s}}$$
$$140 = m_1V_1 + 10 \times 4 = 40 + m_1 V_1$$
$$m_1V_1 = 100$$
$$V_1 = \frac{100}{80} = 1.25 \text{ m/s}$$

21.20 (a)

$$m_1 = 1 \text{ kg}, u_1 = 2 \text{ m/s}$$
$$m_2 = 2\ V_1, u_2 = 0$$
$$2 \times 1 = m_2 \times V_2 = 2 \times V_2$$
$$V_2 = 1 \text{ m/s}$$

21.21 (b)

$e = 0.5$ (Total distance covered

$$= \left(\frac{1+e^2}{1-e^2}\right)h = \frac{1.25}{0.75} = \frac{5}{3}h$$

21.22 (a)

Perfectly plastic impact, $e = 0$.

21.23 (d)

$$u_1 = 5 \text{ m/s}$$
$$u_2 = 0$$
$$m_1 u_1 + m_2 u_2 = 5 \times 100$$
$$e = \frac{V_2 - V_1}{u_1 - u_2} = \frac{V_2 - V_1}{5 - 0}$$

But $V_2 = V_1$

$$e = 0$$

21.24 (b)

Perfectly elastic collision velocity of ball A is zero after impact.

■■■■